Energy Systems

Series Editor:

Panos M. Pardalos, University of Florida, USA

For further volumes:
http://www.springer.com/series/8368

Alexey Sorokin • Steffen Rebennack • Panos M. Pardalos • Niko A. Iliadis • Mario V.F. Pereira
Editors

Handbook of Networks in Power Systems I

Springer

Editors
Alexey Sorokin
University of Florida
Industrial and Systems
Engineering
Weil Hall 303
32611 Gainesville Florida
USA
sorokin@ufl.edu

Steffen Rebennack
Colorado School of Mines
Division of Economics and Business
Engineering Hall
15th Street 816
80401 Golden Colorado
USA
srebennack@mines.edu

Panos M. Pardalos
University of Florida
Dept. Industrial & Systems
Engineering
Weil Hall 303
32611-6595 Gainesville Florida
USA
pardalos@ufl.edu

Niko A. Iliadis
EnerCoRD - Energy Consulting
Research & Development
Plastira Street 4
171 21 Athens
Nea Smyrni
Greece
niko.iliadis@enercord.com

Mario V.F. Pereira
Centro Empresarial
Rio Praia de Botafogo
-A-Botafogo 2281701
22250-040 Rio de Janeiro Rio de
Janeiro
Brazil
mario@psr-inc.com

ISSN 1867-8998 e-ISSN 1867-9005
ISBN 978-3-642-23192-6 e-ISBN 978-3-642-23193-3
DOI 10.1007/978-3-642-23193-3
Springer Heidelberg Dordrecht London New York

Library of Congress Control Number: 2012930379

Printed on acid-free paper

Springer is part of Springer Science+Business Media (www.springer.com)

Handbook of Networks in Power Systems: Optimization, Modeling, Simulation and Economic Aspects

This handbook is a continuation of our efforts to gather state-of-the-art research on power systems topics in Operations Research. Specifically, this handbook focuses on aspects of power system networks optimization and is, as such, a specialization of the broader "Handbook of Power Systems I & II," published by Springer in 2010.

For decades, power systems have been playing an important role in humanity. Industrialization has made energy consumption an inevitable part of daily life. Due to our dependence on fuel sources and our large demand for energy, power systems have become interdependent networks rather than remaining independent energy producers.

Such dependence has revealed many potential economic and operational challenges with energy usage and the need for scientific research in this area. In addition to fundamental difficulties arising in power systems operation, the industry has experienced significant economic changes; specifically, the power industry has transformed from being controlled by government monopolies to becoming deregulated in many countries. Such substantial changes have brought new challenges in that many market participants maximize their own profit.

The challenges mentioned above are categorized in this book according to network type: Electricity Network, Gas Network, and Network Interactions.

Electricity Networks constitute the largest and most varied section of the handbook. Electricity has become an inevitable component of human life. An overwhelming human dependence on electricity presents the challenge of determining a reliable and secure energy supply. The deregulation of the electricity sector in many countries introduces financial aspects such as forecasting electricity prices, determining future investments and increasing the efficiency of the current power grid through network expansion and transmission switching.

The Gas Networks section of the book addresses the problem of modeling gas flow, based on the type of gas, through a pipeline network. The section describes the problem of long-term network expansion as well as the optimal location of network supplies. Deregulation of the gas sector is becoming common in many countries. The deregulation presents new decisions to the gas industry including determining optimal market dispatch and nodal prices.

Network Interactions are common in power systems. This section of the book addresses the interaction between gas and electricity networks. The development of natural gas fired power plants has significantly increased interdependence between these two types of networks.

This handbook is divided into two volumes. The first volume focuses solely on electricity networks, while the second volume covers gas networks, and network interactions.

We thank all contributors and anonymous referees for their expertise in providing constructive comments, which helped to improve the quality of this volume. Furthermore, we thank the publisher for helping to produce this handbook.

Alexey Sorokin
Steffen Rebennack
Panos M. Pardalos
Niko A. Iliadis
Mario V.F. Pereira

Contents

Part I
Electricity Network

Models of Strategic Bidding in Electricity Markets Under Network Constraints

Ettore Bompard and Yuchao Ma

Abstract Starting from the nineties of the last century, competition has been introduced in the electricity industry around the world, as a tool to increase market efficiency and decrease prices.

Electricity is a commodity that needs to be traded over a physical network with strict physical and operational constraints that cannot be found in other commodity markets.

Present electricity markets may be better described in terms of oligopoly than of perfect competition from which they may be rather far. In an oligopoly market, the producer is a market player that shows strategic behavior, submitting offers higher than the marginal costs, as they under perfect competition, with the aim to maximize its individual surpluses.

The market clearing price, quantities and the market efficiency depending on the strategic interactions among producers must be taken into account in modeling competitive electricity markets.

The network constraints provide very specific opportunities of exercising strategic behaviors to the market participants. Game theory provides a conceptual framework and analytical tool to model such a context.

The modeling of electricity markets will be presented by discussing the traditional Game Theory models, such as bertrand, cournot, conjecture supply function, supply function equilibrium, adapted to be able to capture, in determining the Nash equilibrium, the network structure of the system in which the market is

E. Bompard (✉)
Department of Electrical Engineering, Polytechnic di Torino, Torino, Italy

CERIS-CNR (Institute for Economic Research on Firms and Growth of the National Research Council), Moncalieri (TO), Italy
e-mail: ettore.bompard@polito.it

Y. Ma
Department of Electronic & Electrical Engineering, University of Strathclyde, Glasgow, UK
e-mail: yuchao.ma@eee.strath.ac.uk

A. Sorokin et al. (eds.), *Handbook of Networks in Power Systems I*,
Energy Systems, DOI 10.1007/978-3-642-23193-3_1,

implemented. A formalized representation and a comparison of some of the most common game theory models will be provided with some conceptual examples. In addition, some newly proposed approaches for strategic bidding modeling based on the complex systems techniques such as Multi Agent systems and Complex Networks will be mentioned and some related references provided.

Keywords Electricity markets • Game theory • Network constraints • Strategic bidding

1 Introduction

The electric power industry has over the years been dominated by large state-owned monopolies that had an overall authority over all the activities in generation, transmission and distribution of power within their jurisdiction. Chile is often considered as the first Country to introduce liberalization in the electricity sector in 1982. Regulatory reforms of the industry in the United States started in 1978 with the passage of the Public Utility Regulatory Policies Act; regulatory reform was accelerated over the latter half of the 1990s with the advent of the open access transmission regime in 1996, the subsequent formation of several large regional spot markets, later, regional transmission organizations e.g. PJM, 1997, CAISO, 1998, Midwest 2002, etc. [1]. The first initial steps of liberalization of electricity markets in Scandinavia started in Norway in 1990. Through the subsequent steps of development through market expansion, the Nordic market became the world's first multi-national market that was quite well-functioning [2]. Since 1996 the generation, distribution and supply of electricity in eastern and southern Australian states has been amalgamated under the National Electricity Market and in 2009 the Australian Energy Market Operator (AEMO) has been established. Australia took the forefront of energy industry reform worldwide, one of the first countries to establish highly competitive and transparent electricity markets underpinned by strong governance structures [3]. In UK, The Electricity Pool of England and Wales was created in 1990 to balance electricity supply and demand, acting as a clearing house between generation and wholesale. In March 2001 the electricity pool is replaced by the New Electricity Trading Arrangements (NETA). On 1 April 2005, British Electricity Trading and Transmission Arrangements (BETTA) is introduced to replace the NETA in England and Wales, and the separate arrangements that existed in Scotland and the British Grid System Agreement, to create fully-competitive, British-wide wholesale market for the trading of electricity generation [4].

The justification for introducing competition in the electricity sectors is that in a monopoly it is not possible to achieve, no matter which market rules are designed, two important objectives at the same time: to hold down prices to marginal costs and to maximize efficiency [5]. On the contrary, in a competitive market those two objectives may be reached by a proper market design and the specification of a proper set of rules. In this respect, the goal of electric industry restructuring is to achieve a better, more efficient allocation of resources by increasing the role of market forces, and simultaneously decreasing the role of regulation. The main objectives of the reforms are achieved through a clear separation between production and sale of electricity and the operation of electric power grid. Based on the market reform, many market based roles have been penetrating into the electricity industry in progress with a slight relaxation of the obligation to serve the loads that has been segregated and assigned to various entities. Among the new roles *energy producers* and *retailers*, *brokers*, *independent system operator (ISO)* or *transmission system operator* (TSO) are the most popular entities existing in real electricity markets around the world.

There are two reference paradigms proposed for electricity markets: *pool* and *bilateral*. *Pool electricity markets* coordinate the selling and buying activities through a centralized market place administrated by third entity that may coincide with the ISO, whilst the *bilateral markets* the transactions to be contracted between the seller and buyer directly on a private basis. In the pool paradigm the market optimum is reached a central decision making run after collecting the offers from the producers and the bids from the customers while in the bilateral paradigm the decision making process is distributed among various sellers and buyer that meet in the marketplace.

Electricity markets are pretty different from other commodity markets mainly due to the physical constraints related to the network structure that may impact the market performance. The network constraints and the special features of the electricity provide the market players an opportunity to behave strategically, gaming the market, which is very specific this context and cannot be found in other commodity markets. Strategic bidding behaviours of electricity producers are widely studied in open literature [6, 7].

Game theory is popularly used in investigating the strategic bidding interactions between the electricity producers. Models based on various games such as *Bertrand* [8, 9], *Cournot* [10–12], *Stackelberg* [13, 14], *Supply Function Equilibrium* [15–17], as well as *conjectural supply function* [18, 19] have been proposed. By taking into account the network constraints, the strategic bidding behaviors analysis based on game theory models usually involves a bi-level optimization problem modeled as a mathematical program with equilibrium constraints (MPEC) [20, 21]. The lower level problem of market clearing with the consideration of the network constraints, i.e. the equilibrium constraints, is inserted in the upper level problem of the maximization of the producer surplus [22–26]. There has been an intensive research for efficient solution methods for the MPEC problem, with proposed solution schemes ranging from specific analytic algorithms to heuristics procedures [27–32].

However, the strategic interactions among participants in today's electricity markets can be very complicated, due to various aspects such as supply and demand uncertainties, unit commitment arrangement, multi-rounds auctions in both energy and ancillary service markets, which are not conveniently modeled by game theory techniques. An alternative efficient approach for analyzing the strategic bidding and decision support of the market participants is provided by the multi agent system environment [33]. Based on the computational economics approach, several autonomous adaptive agent models have been proposed, including those created by Anthony J. Bagnall and George D. Smith,[34], Athina C. Tellidou and Anastasios G. Bakirtzis [35], Sun J. and Tesfatsion L. [36], as well as Isabel Praca and Carlos Ramos [37]. Recently, the application of evolutionary complex network has been as well proposed by Ettore B. and Ma Y.C. [38] for modeling the bilateral electricity markets. Stable network structures that can be used to anticipate possible bilateral transactions in the real market place are developed by the improving path rule of evolutionary complex network principles. Different stable network structures can produce the same maximum value of global utility, reflecting the complex and disordered individual behavior has self-organization properties that produce the highest market efficiency in terms of the social welfare without the need for a centralized decision-making authority.

In this chapter, we first outline the basic features of electricity as a commodity and of electricity markets recalling the basic metrics used for assessing the market performance. Then we propose a formalized representation of some of the most common game theory models by taking into account network constraints. A comparison study on those game theory models through a unified conceptual example is investigated, which has not been discussed in open literature so far. Such comparison study provides a quantitative assessment on the electricity market performance affected by the different strategic gaming behaviors of the electricity producers.

Notations

N:	Number of network buses
N_l:	Number of network lines
N_G:	Number of generators
N_D:	Number of demand consumers
n:	Index of the network bus set $\mathcal{N}$ $\mathcal{N} = [1\ 2,,\ldots, n-1, n, n+1, \ldots, N]$
l:	Index of the network line set $\mathcal{L}$ $\mathcal{L} = [1\ 2,,\ldots, l-1, l, l+1, \ldots, N_l]$
g:	Index of the electricity producer set $\mathcal{G}$ $\mathcal{G} = [1, 2, \ldots, g-1, g, g+1, \ldots, N_G]$
d:	Index of the electricity consumer set $\mathcal{D}$ $\mathcal{D} = [1, 2, \ldots, d-1, d, d+1, \ldots, N_D]$
$\boldsymbol{a}/\boldsymbol{b}$	Intercept ($/MW) and slope ($/MW/MW) parameters of the marginal cost curves of the electricity producers, dim ($\boldsymbol{a}$) = dim ($\boldsymbol{b}$) = $N_G \times 1$ ($\boldsymbol{a}^m$, $\boldsymbol{b}^m$ refer to marginal cost curve parameters)
$\boldsymbol{e}/\boldsymbol{h}$	Intercept ($/MW) and slope ($/MW/MW) parameters of the demand curve of the electricity consumers, dim ($\boldsymbol{e}$) = dim ($\boldsymbol{h}$) = $N_D \times 1$
$\boldsymbol{p}/\boldsymbol{q}$	Power production and demand vector, MW, dim($\boldsymbol{p}$) = $N_G \times 1$, dim ($\boldsymbol{q}$) = $N_D \times 1$
$\boldsymbol{I}_G/\boldsymbol{I}_D$	All-one-element vector for producers/consumers, dim ($\boldsymbol{I}_G$) = $N_G \times 1$, dim ($\boldsymbol{I}_D$) = $N_D \times 1$
$\boldsymbol{P}^+/\boldsymbol{P}^-$	Upper and lower production limits of the producers, MW, dim ($\boldsymbol{P}^+$) = dim ($\boldsymbol{P}^-$) = $N_G \times 1$
$\boldsymbol{H}$	Diagonal matrix formed by the vector $\boldsymbol{h}$, dim ($\boldsymbol{H}$) = $N_D \times N_D$
$\boldsymbol{B}^m$	Diagonal matrix formed by the vector $\boldsymbol{b}^m$, dim ($\boldsymbol{B}^m$) = $N_G \times N_G$

(continued)

$\boldsymbol{F}$	Flow limits of the transmission lines, MW, $\dim(\boldsymbol{F}) = N_l \times 1$
$\boldsymbol{\mu}^+/\boldsymbol{\mu}^-$	Lagrange multipliers corresponding to the inequality expressions of the line flow, \$/MW, $\dim(\boldsymbol{\mu}^+) = \dim(\boldsymbol{\mu}^-) = \dim(\boldsymbol{F})$
$\boldsymbol{J}$	Matrix of power transfer distribution factors, $N_l \times N - 1$
$\boldsymbol{J_G}^T, \boldsymbol{J_D}^T$	Generator and load buses rows of the transpose of $\boldsymbol{J}$ matrix, respectively, $\dim(\boldsymbol{J_G}^T) = N_G \times N_l$, $\dim(\boldsymbol{J_D}^T) = N_D \times N_l$
$\boldsymbol{\lambda_G}, \boldsymbol{\lambda_D}$	Nodal prices at the generator and load buses, \$/MW, $\dim(\boldsymbol{\lambda_G}) = N_G \times 1$, $\dim(\boldsymbol{\lambda_D}) = N_D \times 1$
λ_N	Price at the reference bus N, \$/MW
λ	Average price weighted by the quantity (\$/MW), $\lambda = (\boldsymbol{\lambda_G}^T \boldsymbol{p} + \boldsymbol{\lambda_D}^T \boldsymbol{q})/(\boldsymbol{I_G}^T\boldsymbol{p} + \boldsymbol{I_D}^T\boldsymbol{q})$
S_g^G	Surplus of producer g, $\forall g \in \mathcal{G}$, (\$)
S^G	Total surplus of producers (\$)
S_d^D	Surplus of consumer d, $\forall d \in \mathcal{D}$, (\$)
S^D	Total surplus of consumers, (\$)
S^M	Merchandise surplus (\$)
S^S	Social surplus, $S^S = S^G + S^D + S^M$, (\$)

Special operators

$\boldsymbol{m}_1\ \boldsymbol{m}_2$	Matrix multiplication of $\boldsymbol{m}_1$ and $\boldsymbol{m}_2$
$\boldsymbol{v}_1 \cdot \boldsymbol{v}_2$	Element by element multiplication of vector $\boldsymbol{v}_1$ and vector $\boldsymbol{v}_2$
$(\boldsymbol{v}_1) \pm (\boldsymbol{v}_2)$	Element by element addition or subtraction of vector $\boldsymbol{v}_1$ and vector $\boldsymbol{v}_2$
$(\boldsymbol{v}_1) \leq (\boldsymbol{v}_2)$:	Element by element inequality between vector $\boldsymbol{v}_1$ and vector $\boldsymbol{v}_2$
$\boldsymbol{v}_1 = \boldsymbol{v}_2$:	Element by element equality between vector $\boldsymbol{v}_1$ and vector $\boldsymbol{v}_2$
$\boldsymbol{v}^T/\boldsymbol{m}^T$:	Transpose of vector $\boldsymbol{v}$/matrix $\boldsymbol{m}$
$\boldsymbol{m}^{-1}$:	Inverse of the matrix $\boldsymbol{m}$

2 Electricity as a Commodity

Liberalization has been introduced in many economic sectors such as air transportation, telecommunication with the goal of achieving efficiency though competition. Electricity is one of the last sectors in which liberalization and competition has been introduced. Those very specific features of electricity as a commodity, from one side, and its criticality to the society, from the other, need to be considered when switching from regulated monopoly to competition. The market modeling and simulation need to capture those specificities related to the physical constraints and network structure of the power systems. In this section, we first review the specific electric power system operation. Then, paradigms of the electricity markets are elaborated with two reference models, pool and bilateral markets whilst the poor model is used in this chapter to study the strategic bidding models in an electricity market.

2.1 Power Systems Operation

Power systems are composed a transmission system with buses, to which are connected the generators and the loads, interconnected by lines and transformers over the meshed network where the power injected by the generators is delivered to load centers.

The operation of the system should satisfy a set of boundary conditions which can be addressed by a set of equality and inequality constraints.The equality constraints assure power balance between the electricity generation and load demand while inequality constraints define a feasible operation region represented by the line transmission capacity limits, system frequency and bus voltage ranges. If all the equality and inequality constraints are satisfied the system is operated in its *normal state*. The inequality constraints are satisfied with certain security margins in terms of generation spinning reserve and/or transmission capacity reserve, etc. If the reserve margin, due to some disturbances, is reduced, the system enters into the *alert state* in which the constraints are still satisfied. *Preventive control* then takes place to secure again proper security margins. If the preventive control fails or the disturbance was severe enough the system goes into the *emergency state* in which the inequality constraints are not; *emergency control* actions such as cutting of faults, rerouting of generation, excitation control, fast-valving, generation tripping, generation run-back, HVDC (high voltage dc) modulation, and load shedding are undertaken. If emergency control fails the system will go into the *extremis state* with cascading outages and system islanding. From this state the system operator to push the system to *restorative state*, in which the system, matching again generation and load, is driven back to the normal state [39].

Mathematically, the steady-state power system operation can be expressed as:

$$\boldsymbol{s} = \boldsymbol{f}(\boldsymbol{x}) = 0 \quad \forall n \in \mathcal{N} \tag{1}$$

$$\boldsymbol{g}(\boldsymbol{x}) \leq \boldsymbol{0} \tag{2}$$

where $\boldsymbol{s} = \boldsymbol{s}_P + j\boldsymbol{s}_Q$ is the vector of complex power (real and reactive) injection at each bus n, $n \in N$, $\boldsymbol{f}(\cdot)$ is the vector of functions that express the complex power transferred over the lines connected to each bus and $\boldsymbol{x}$ is the vector of unknown phasor voltages (magnitudes and angles) at each bus; the group of inequities $\boldsymbol{g}(\cdot)$ represent the network constraints.

The group of Eq. 1 represents the so called AC power flow model where real and reactive power determined by both the voltage magnitude and voltage angle at each bus. The real power balance enforced at each bus is to keep the electricity frequency at the expected value. Any unbalance of the real power between the electricity generated and consumed causes the frequency drifting to a new value. The reactive power balance at each bus is to govern the voltage magnitude, i.e. to generate capacitive reactive power to restore the voltage magnitude from low value to high value or to absorb the inductive reactive power to reverse the process.

The group inequalities (2) include the various limits with respect to the problem size to be considered. Generally, the voltage magnitude limits and real power flow limits of the network branches are the two typical conditions imposed to the power system for operating in a secure normal state. In addition, real and reactive power generation limits are incorporated to capture the production characteristics of the electricity generators.

The model can be simplified with some assumptions that lead to a linearized model that will be introduced in Sect. 3.

2.2 Specific Features of Electricity

Apart from its economic features such as the *no-direct-storability*, *lack of good substitutes* and *inelastic demand*, the technical peculiarities of the electricity as a commodity the electricity market a very specific one. Electricity can only be delivered by wires over a transmission and distribution networks to customers at the same time when the electricity is generated. Actual power flows of the network wires are governed by the Kirchhoff laws, which makes transmission of power different from the transportation of an ordinary commodity in a spatial market. In addition, network constraints need to be enforced on the trading activities and affect the market clearing results. The specific features of the electricity are summarized by using the example shown in Fig. 1 where all the parameters are in per unit of system base value.

- Need for an instantaneous balance between power production and power consumption plus losses. In the example, the total power generated, 0.6024 + j0.1842 p.u. is equal to the total power demanded, 0.6 + j 0.1721 p.u., plus total power loss, 0.0024 + j0.0121 p.u.
- Power *flow path depending on the system physical* parameters. The paths followed by the power flows do not coincide with the contract paths of the economic transactions; almost all the lines, other than that connecting the generator and the load in the transaction, are involved. If the values of the line impedances change the power flows over the network change as well.
- *Transmission losses.* The total real power generated exceeds the real power demanded by the loads of 0.0024 p.u. and that correspond to the losses on the transmission system. Gen 1, in this case, need to produce additional 0.0024 p.u. to balance the real power loss of the network.
- *Reactive power support.* To allow for the transactions scheduled in terms of real power, a reactive power support, to balance the reactive power demanded by the loads and used by the transmission systems, needs to be provided by the generators. The total reactive power demanded by the loads is 0.1721 p.u. while those totally generated by the two generators is 0.1842 p.u.; the difference is the reactive power needed by the system to feasibly allow for the transaction (0.0121 p.u.).

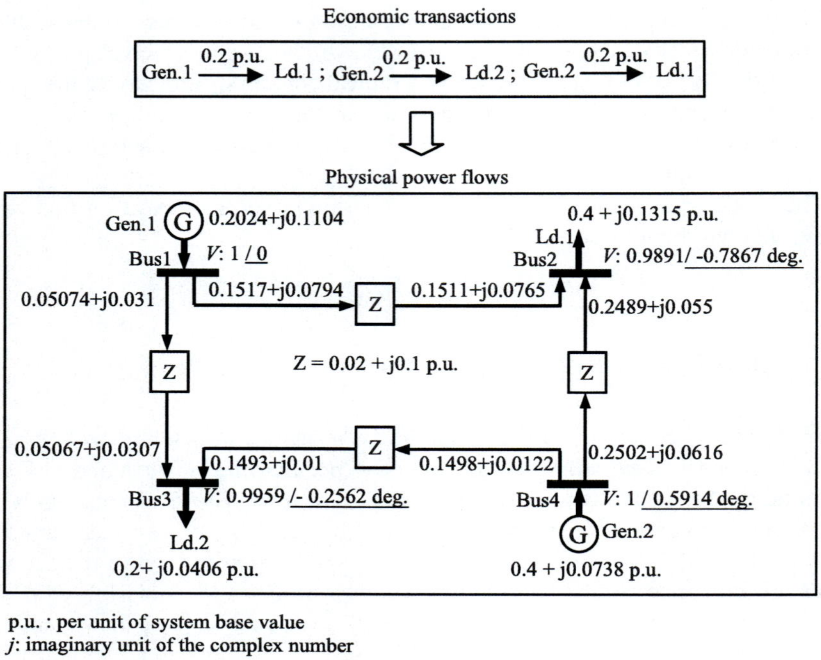

Fig. 1 Network impacts on electric power transactions

- *Line flow limits*. Lines have some limitation on the maximum power they can transmit due to thermal, voltage drop and stability limits. If the power flow of the branch from bus 1 to bus 2 reaches its limit, the system should re-route the power through other lines to be kept feasible.

2.3 *Electricity Markets*

The energy market in most countries is organized as a day-ahead market (DA) where the electricity energy transactions are cleared for each hour of the next day. In day-ahead market, demand is forecasted for each trading interval, e.g. 1 h period, 24 h ahead and offers and bids are received from the market participants. The market clearing is conducted by an independent body, which may coincide with the Independent System Operator (ISO), to match demand and supply. In the pool models the dispatching of injected and withdrawn power quantities are assigned considering the transmission limits and providing, as a by-product, the locational marginal prices (LMPs), such as in PJM Interconnection, in order to capture the network impacts on the market clearing [40].

Several related markets concur to make the electricity transactions possible, including *ancillary service market* and *transmission right market*. *Ancillary market* is organized to acquire on the market all the services needed for the operation of the power system as scheduling, system control and dispatch, reactive and voltage support, regulation and frequency response, energy imbalance, spinning reserves and supplemental reserves. *Transmission right market* is designed for auctioning the right to assure the availability of transfer capability over the network or edge against the risk of change in transmission cost due to the scarcity of transfer capability. There are two types of transmission rights: physical transmission rights (PTRs) and financial transmission rights (FTRs). However, PTRs are gradually replaced by FTRs, defined upon the locational marginal prices, due to its superiorities over PTRs by opening the network access to all the market participants. Several of the restructured US electricity markets have already experimented with styles of the transmission rights markets in the last decade [41].

Strategic bidding behaviors are extensively investigated in the day-ahead energy market whilst, due to their functional complexity, ancillary markets attract more technical concerns than the economic behaviors of the market participants. Therefore, in this chapter we study the strategic bidding behaviors of the electricity producers in the day-ahead market clearing process.

2.4 Reference Paradigms of the Electricity Markets

A major objective of electricity deregulation is to achieve a workably competitive wholesale market. Wholesale electricity markets are organized with several generation companies that compete to sell their electricity in a centralized pool and/or through bilateral contracts with large electricity buyers [42–44]. The transactions among sellers and buyers can take place in an organized market ("power exchange") that collects all the offers from the generators and the bids from the loads and performs a centralized market clearing compatible with the network constraints, as in the "pool paradigm"; otherwise sellers and buyers arrange, on a bilateral basis, their own transactions, submitting afterward to an ISO to check their feasibility in terms of the network constraints as in the "bilateral paradigm."

A vast majority of studies on electricity markets to date either explicitly or implicitly assume a centralized auction process, administered by a pool, through which generators sell energy to consumers. A growing number of studies typically assume a decentralized trading process by which generators sell to consumers bilaterally through power exchanges or arbitragers [38, 45, 46].

2.4.1 Pool Markets

Many of the restructuring experiences (e.g., in the UK, Argentina, Chile, Australia) have been based on pool trading with centralized coordination in the Power

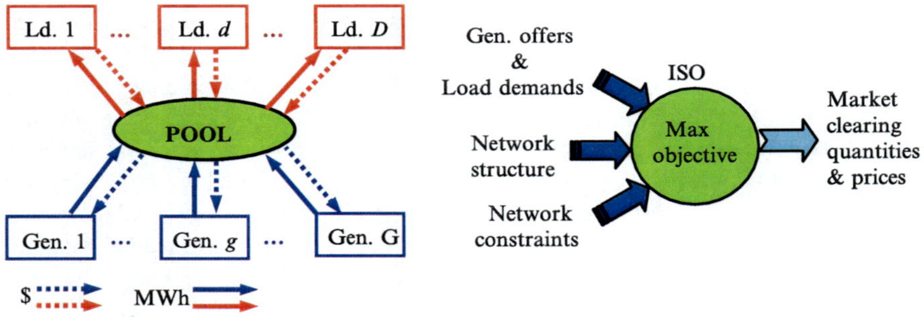

Fig. 2 Pool type market

Exchange (PX) by an Independent System Operator (ISO). Examples include Australia, Argentina, the PJM Interconnection and the New England Power Pool.

In the pool market all producers sell into a pool run by an independent entity. The ISO has also the responsibility for system reliability and transmission congestion management. Producers' physical sales of power and energy trades are all within the pool. The pool is the only buyer (for the producers) and the only seller (for the electricity consumers). The ISO holds central auction in which each generator bids different prices for different quantities (from specific plants or as a portfolio) or an offer curve for the trading period; for example, for each hour of the following day. Based on the bids and the considered demand quantities, the ISO uses a security based dispatch process to set the market price and the generation quantities [42–44].

2.4.2 Bilateral Markets

Under bilateral trading model buyers and sellers individually contract with each other for power quantities at negotiated prices, terms, and conditions. All transactions must be announced to the ISO, which analyzes all the trades in each period and determines, without discrimination, which ones are infeasible for grid security constraints. The ISO does not need to know the prices and demand side bids may co-exist with the generation bids. Generally, the bilateral types of markets are split into two markets named forward contract market (PX) and spot market (ISO). Two steps are needed to arrange the bilateral trading:

- Step 1: Dispatch without network constraints;
- Step 2: Re-dispatch with consideration of transmission constraints based on the adjustment offers (demands) from generators and loads, in case that the dispatch would lead to network constraints violations.

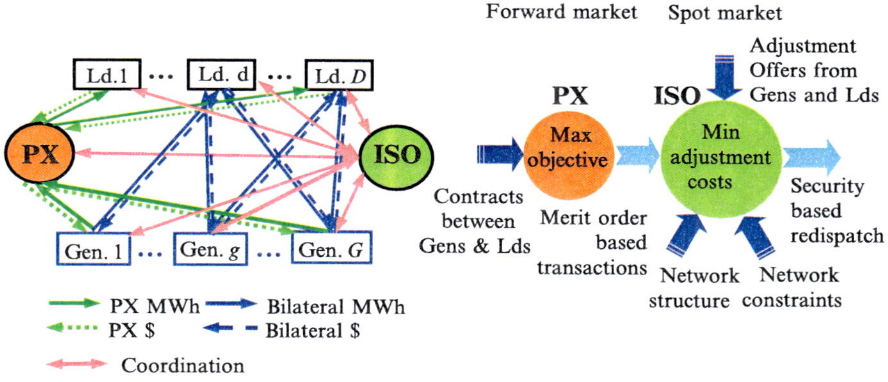

Fig. 3 Bilateral type market

3 Market Equilibrium and Market Performance

Like any commodity or, electrical energy can be bought and sold in an established market place. Bids, from buyers, and offers, from sellers, set the price in the electricity markets on the basis of principle of supply/demand intersection. The difference among electrical energy and other commodities is the delivery system and the technical features of the framework in which the market transactions are undertaken.

3.1 Format of the Electricity Supply and Demand

Although it may be different with respect to the functional rules and scopes, most day-ahead markets consist of unbundled Generation, Transmission and Retailing sectors that are corresponding to the electricity suppliers, transmission organization and the electricity customers. A neutral entity called Independent System operator is introduced to undertake the role of the market coordinator.

For the electricity suppliers, the supply curve is usually represented as an upward liner or stepwise function that expresses the quantity that all the sellers in a market are willing to sell as the function of price, as shown in Fig. 4. The increasing trend of the curve is explained by the fact that as the power offered raises more expensive unit need to be committed. It is important to note that the supply curve is not necessarily equal to the marginal cost curve, the additional cost incurred in producing one extra unit of output, from which the strategic bidding behavior of the electricity suppliers is originated. The demand curve shows the relationship between the quantity demanded and the price of a commodity. All other factors held constant, almost all commodities obey the law of downward-sloping demand,

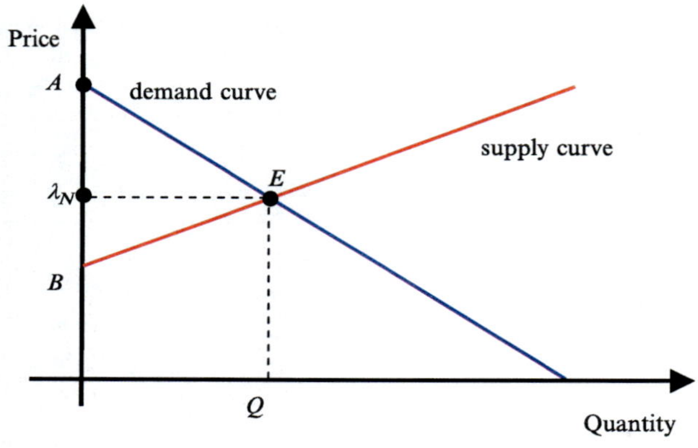

Fig. 4 Electricity supply and demand curves

which states that quantity demanded falls as a price rises, Fig. 7. The degree of such price responsiveness is called demand elasticity. Under regulation, electricity demand was considered inelastic (fixed amount with no price responsiveness) and new capacity was built to cover the projected demand to minimize investment plus operation costs. Under deregulation, the consumers' demands for electricity are encouraged to be price responsive to enhance the wholesale market efficiency and system reliability [47, 48]. Strategic behaviors from the supply side are more evident when the demand elasticity of the electricity consumers is low. Improving demand elasticity plays a positive contribution in mitigating the strategic bidding behaviors, pushing the uncompetitive electricity market performance towards to a high level competitive one [49].

3.2 *Market Equilibrium*

A market is a real or virtual environment in which buyers and sellers interact to exchange goods, services or commodities; the outcome of a market, from a macroscopic viewpoint is the quantity and the price of the good traded.

Given the supply and demand curves, the electricity market is cleared at the equilibrium at which the market clearing price is established such that the amount of goods or services sought by buyers is equal to the amount of goods or services produced by sellers. Graphically, the market equilibrium is represented by the intersection of the supply curve and the demand curve, point E in Fig. 4. The equilibrium is established in a price-quantity adjustment process where incremental quantity of electricity is supplied at the price that demand customers willing to buy, i.e. the supplier offer price is lower than the demand bid price. Equivalently, the market clearing can be expressed mathematically as

$$Max\ {}^1\!/\!_2\, \boldsymbol{q}^T\boldsymbol{H}\boldsymbol{q} + \boldsymbol{q}^T\mathbf{e} - \left({}^1\!/\!_2\boldsymbol{p}^T\boldsymbol{B}\boldsymbol{p} + \boldsymbol{p}^T\boldsymbol{a}\right) \tag{3}$$

$$s.t.\ \boldsymbol{I_D}^T\boldsymbol{q} - \boldsymbol{I_G}^T\boldsymbol{p} = 0 \tag{4}$$

The market clearing price, λ_N, in Fig. 4. is the Lagrange multiplier value of the optimization problem (19)–(20) and can be expressed as

$$\lambda_N = \frac{\boldsymbol{I}_G^T\boldsymbol{B}^{-1}\boldsymbol{a} - \boldsymbol{I}_D^T\boldsymbol{H}^{-1}\boldsymbol{e}}{\boldsymbol{I}_D^T\boldsymbol{H}^{-1}\boldsymbol{I}_D - \boldsymbol{I}_G^T\boldsymbol{B}^{-1}\boldsymbol{I}_G} \tag{5}$$

The market clearing quantities are

$$p_g = (\lambda_N - a_g)/b_g \quad \forall g \in \mathcal{G} \tag{6}$$

$$q_d = (\lambda_N - e_d)/h_d \quad \forall d \in \mathcal{D} \tag{7}$$

In Eqs. 3–7, the vector of $\boldsymbol{a}$ and the diagonal matrix $\boldsymbol{B}$ indicate the parameters of the supply curves of the electricity producers. Those parameters are not necessarily equal to the parameters of the marginal cost curves, i.e. $\boldsymbol{a}_m$ and $\boldsymbol{B}_m$. The strategic bidding behaviour of the electricity producers can be represented by choosing the parameters of $\boldsymbol{a}$ and/or $\boldsymbol{B}$ to achieve the economic goals in the market clearing.

3.3 Levels of Competition in the Markets

According to the competition level, market is classified as perfect one or imperfect. In a perfect competitive market, all sellers and buyers are "price-takers" who assume that their own production and purchase decisions do not affect the market price [50]. According to the classic economic theory, a price-taking firm that wishes to maximize its profits would bid the products at its own marginal cost [51]. Each supplier submits the marginal cost as supply function and the social surplus, S^S, is maximized in the market clearing, as mentioned in the last section. Price-taking behaviors will lead to the most efficient market operation characterized by the least cost of the production while meeting the demand of the consumers. In reality, perfect competition is difficult to be implemented in whichever real market due to the strict conditions need to be satisfied, such as a large number of price-taking producers with a very small market share produce homogeneous and perfectly substitutable products. Nevertheless it can serve as a reference case to identify the market power behaviors in a real implemented market [22, 52, 53].

The opposite situation to perfect competition is monopoly in which just one producer faces all the market demand. Monopoly market is thought to have no competition. Perfect competition and monopoly represent the two extreme cases of

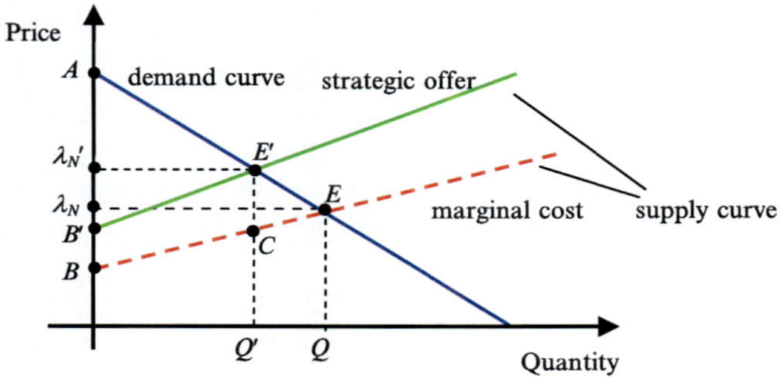

Fig. 5 Market clearing under strategic bidding of the supply side

market structures. A more common case is the oligopoly in which the market is dominated by a small numbers of the sellers and the market equilibrium is in between the two preceding cases. Current electricity markets are oligopoly in generation competition where the electricity suppliers will adopt the strategic behaviors, different from the marginal cost curves shown in Fig. 5, striving to get the maximum profit, the area $E'\lambda_N'BC$ in Fig. 5. The main goal of the restructuring the electricity industry for the market regulators is to force the market toward perfect competition, from point E' to point E, while monitoring continuously the distance from such a condition. Under imperfect competition, the market clearing produces the maximization not of the social surplus, area ABE in Fig. 5, but of the similar quantity that we denote as the system surplus, area $AB'E'$ in Fig. 5.

The key part of studying the imperfect competition of the electricity markets is to derive the oligopoly equilibrium at which each individual electricity supplier's optimal strategy is established. Although to obtain the market clearing results at oligopoly equilibrium is straightforward by inserting the optimal strategy of each electricity producer into the market clearing model (3) and (4), to derive the equilibrium of optimal strategies of the electricity suppliers is a complex problem where each producer will compete against other's strategies. Such problem can be addressed by game theory models which will be discussed in detail in the next sections.

3.4 The Impacts of the Scarcity in the Transfer Capability

Due to the capacity limits of the network lines, electricity transactions have to be settled according to the power flow constraints. The impacts of the scarcity of the transfer capability of the network lines on the electricity market performance can be briefly introduced by a simple two bus system, Fig. 6. One supplier located at bus 1 is to deliver the electricity to the consumer at bus 2. Without the network constraint; the transaction will be clearing at the point E, the left hand side of Fig. 6. The market clearing price is unique at which electricity supplied is equal to demanded.

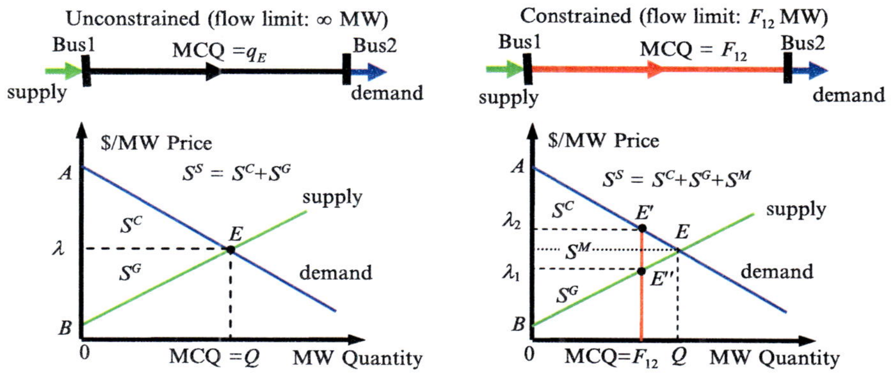

Fig. 6 Market clearing without (*left hand side*) and with (*right hand side*) network constraints

Table 1 Network congestion impacts on the market clearing

	Unconstrained network	Constrained network
Market clearing price, \$/MW	λ ($\lambda = \lambda_1 = \lambda_2$)	Bus 1: λ_1 Bus 2: λ_2
Market clearing quantity, MW	Q	F_{12}
Social surplus, S^S, \$	area ABE	area $AB\,E''\,E'$
Producer surplus, S^G, \$	area $E\lambda B$	area $E''\lambda_1\,B$
Consumer surplus, S^C, \$	area $E\lambda A$	area $E'\lambda_2\,A$
Merchandise surplus, S^M, \$	0	area $E'\lambda_2\lambda_1\,E''$

If the real power flow limit, F_{12}, is imposed, the market clearing process splits the unique market clearing price into supply price and demand price. The market clearing quantity of the electricity is limited at the real power flow limit, F_{12}.

Based on Fig. 6, the market performance can be summarized in Table 1.

3.5 Metrics for Assessing the Equilibrium and the Level of Competition

Given the market clearing results of the price, λ_g and λ_d, and quantity, p_g and q_d of the electricity transactions, (6) and (7), the social surplus, S^S, producer surplus of producer g, ${S_d}^G$, consumer surplus demander d, ${S_d}^D$, and merchandise surplus, S^M, can be determined as

$$S^S = {}^1\!/\!_2\,\boldsymbol{q}^T\boldsymbol{H}\boldsymbol{q} + \boldsymbol{q}^T\boldsymbol{e} - {}^1\!/\!_2\,\boldsymbol{p}^T\boldsymbol{B_m}\boldsymbol{p} - \boldsymbol{p}^T\boldsymbol{a_m} \tag{8}$$

$${S_g}^G = \lambda_g p_g - {}^1\!/\!_2\, b_{g,m}{p_g}^2 - a_{g,m}p_g \quad \forall g \in \mathcal{G} \tag{9}$$

$$S_d{}^D = \lambda_d q_d - {}^1\!/\!_2 h_d q_d{}^2 - e_d q_d \quad \forall d \in \mathcal{D} \tag{10}$$

$$S^M = \boldsymbol{\lambda}_D{}^T \boldsymbol{q} - \boldsymbol{\lambda}_G{}^T \boldsymbol{p} \tag{11}$$

According to (8)–(11) and Fig. 6, social surplus is defined by the difference between the total benefit to consumers minus the total cost of production. Producer surplus is the difference between the producer sales revenue and the producer variable cost while the consumer surplus is the difference between the amounts that a consumer would be willing to pay for a commodity and the amount actually paid. The merchandise surplus is non zero when the price charged for the buyers is not equal to the selling price of the suppliers, in which case the electricity market is cleared under congested network, as shown in the right part of the Fig. 6.

With the market clearing results derived under the competition, comparison indices are introduced to assess the imperfect market performance against the reference case of the perfect competition market. We will use the superscript * for the values associated with perfect competition.

3.5.1 Efficiency and Allocation Indices

The efficiency and allocation indices are expressed in per unit and have the general form as

$$K = (S - S^*)/S^* \tag{12}$$

This index can be used for the social surplus, the total producer surplus S^G, and the total consumer surplus S^D

$$K_S = (S^S - S^{S*})/S^{S*} \tag{13}$$

$$K_G = (S^G - S^{G*})/S^{G*} \tag{14}$$

$$K_D = (S^D - S^{D*})/S^{D*} \tag{15}$$

3.5.2 Price and Quantity Indices

To have a reference value for the prices under congestion, we define the weighted average price as

$$\underline{\lambda} = (\boldsymbol{\lambda}_G{}^T \boldsymbol{p} + \boldsymbol{\lambda}_D{}^T \boldsymbol{q})/(\boldsymbol{I}_G{}^T \boldsymbol{p} + \boldsymbol{I}_D{}^T \boldsymbol{q}) \tag{16}$$

The price distortion may be then measured by

$$K_\lambda = (\underline{\lambda} - \lambda_N{}^*)/\lambda_N{}^* \tag{17}$$

For the index for quantity, the general form (28) is adapted as

$$K_P = (\boldsymbol{I}_G{}^T \boldsymbol{p} - \boldsymbol{I}_G{}^T \boldsymbol{p}^*)/\boldsymbol{I}_G{}^T \boldsymbol{p}^* \tag{18}$$

4 Modeling Strategic Bidding Under Network Constraints in Pool Model

The analysis of strategic bidding behaviors seeks to answer basic questions; including how a firm exercises strategic biddings and to what extent the strategic bidding behavior affects equilibrium quantities, prices, and market efficiency. Such assessment, at least implicitly, requires a comparison of observations of real world market prices and quantities to the comparable values of the variables that the perfectly competitive model predicts.

Game theory provides an efficient tool to model the strategic bidding behaviors of the market participants. The solution of the game models is to derive the market clearing results at Nash Equilibrium (NE) by which the distortion of the perfect competition can be predicted and measured for assessing the strategic bidding impacts. The key point in formulating the game models to represent the strategic bidding behavior is a bi-level programming problem where the market clearing is inserted, as subject conditions, into the individual Producer Surplus Maximization (PSM) problem [11, 24–26, 49]. The market clearing is to maximize the system surplus taking into account the physical constraints of electricity networks represented by the power flow model. According to the Karush-Kuhn-Tucker (KKT) conditions, the market clearing is transformed into a group of equality and inequality constraints of the individual PSM problem. The optimal variable derived from the individual PSM problem is the optimal strategies that the producer will submit to the market coordinator.

With the individual PSM problems, the Nash Equilibrium is derived at the point no player can be benefited from changing his/her strategy when his/her competitors do not. Very often, the individual PSM problem can be quite complicated with the large size of the electricity network to be considered. The derivation of the NE and the unique/existence of the NE are general concerns in many related references [22, 23, 49, 54]. The unique/existence NE may be guaranteed in an analytical way under simple systems while for the complex systems, numerical approaches is usually employed and ex-post check of the NE is needed.

4.1 Electricity Network Model

In the pool operated electricity markets, the Independent System Operator (ISO) is responsible the aggregate offers from the supply side and the aggregate demands from the demand side for a specified time interval, usually 1 h.

Due to the peculiarities of the electricity transmission, the transactions must be settled according to the physical conditions of the electricity network and different nodal prices may arise when the flow limits are binding. In addressing the network features of the electricity transactions, the DC power flow model is popularly employed in studying the strategic interactions in the competitive electricity markets [22, 26, 49]. DC power flow provides a fairly good approximation of the AC power flow in terms of the real power flow due to the fact of the large ratio of X/R and invariant of the voltage magnitude of the power transmission network where wholesale electricity transaction is accommodated. The characteristics of the DC power flow is summarized as

- Reactive power balance at each bus $n, f_n{}^Q$, is disregarded

$$f_n{}^Q = 0 \quad \forall n \in \mathcal{N} \tag{19}$$

- Lossless network

$$P_l{}^{loss} = 0 \quad \forall l \in \mathcal{L} \tag{20}$$

- A group of linear expressions of the real power flow of the branches, $f_l{}^P$, in terms of the bus voltage angles

$$f_l^P = \boldsymbol{Y}_l\boldsymbol{\theta} \quad \forall l \in \mathcal{L} \tag{21}$$

Where $\boldsymbol{Y}_l$ is the row vector of the branch matrix $\boldsymbol{Y}$ ($N_l \times N_{b-1}$) with the positive admittance value at the element of row index l (line l) and column index of from-bus f and negative admittance value of to-bus t; $\boldsymbol{\theta}$ is a column vector of the bus voltage angles. Both of the $\boldsymbol{Y}$ and $\boldsymbol{\theta}$ matrix are formulated without the elements related to the reference bus.

In terms of the net injection of the active power at each bus, the real power flow of the branches can be expressed as

$$\boldsymbol{f}_l{}^P = \boldsymbol{Y}\theta = \boldsymbol{Y}\boldsymbol{B}^{-1}(\boldsymbol{p} - \boldsymbol{q}) = \boldsymbol{J}(\boldsymbol{p} - \boldsymbol{q}) \tag{22}$$

Where, $\boldsymbol{J} = \boldsymbol{Y}\boldsymbol{B}^{-1}$, is called power transfer distribution factor (PTDF) matrix, $\boldsymbol{J}$, $N_l \times (N-1)$; matrix $\boldsymbol{B}$ is the admittance matrix, $(N-1) \times (N-1)$, and elements of the column and row related to the reference bus are not included; $\boldsymbol{p} - \boldsymbol{q}$ is vector of the net injection of the active power at each bus.

4.2 Market Clearing Model Under Network Constraints

The market clearing can be modelled with an optimization problem subject to the electricity network constraints represented by the DC power flow, as

$$max\, S^S = 1/2\, \boldsymbol{q}^T \boldsymbol{H} \boldsymbol{q} + \boldsymbol{q}^T \boldsymbol{e} - 1/2\, \boldsymbol{p}^T \boldsymbol{B}_m \boldsymbol{p} - \boldsymbol{p}^T \boldsymbol{a}_m \tag{23}$$

s.t.

$$\boldsymbol{I}_G{}^T \boldsymbol{p} - \boldsymbol{I}_D{}^T \boldsymbol{q} = 0 \tag{24}$$

$$-\boldsymbol{F} \leq \boldsymbol{J}(\boldsymbol{p} - \boldsymbol{q}) \leq \boldsymbol{F} \tag{25}$$

The equality constraint (24) indicates the total power production and consumption balanced at the reference bus. The inequality constraints (25) represent the line flow limits. Note that the inequality and equality symbols in (24) and (25) represent the inequality and equality between two vectors.

The solution of the above optimization problem provides the nodal price at bus n, as

$$\lambda_n = \lambda_N - \boldsymbol{J}_n{}^T(\boldsymbol{\mu}^+ - \boldsymbol{\mu}^-) \quad \forall n \in \mathcal{N} \text{ and } n \neq N \tag{26}$$

Where, λ_N, is the price at the reference bus N, i.e., the Lagrange multiplier of the equality constraint (24); $\boldsymbol{J}_n{}^T$ is the nth row vector of the transposed matrix $\boldsymbol{J}$; $\boldsymbol{\mu}^+$ and $\boldsymbol{\mu}^-$ are Lagrange multipliers of the power flow inequities (25).

The price at the reference bus N_b, λ_N is equal to

$$\lambda_N = \frac{\boldsymbol{I}_G^T(\boldsymbol{B}^m)^{-1}[\boldsymbol{J}_G^T(\boldsymbol{\mu}^+ - \boldsymbol{\mu}^-) + \boldsymbol{a}_m] - \boldsymbol{I}_D^T \boldsymbol{H}^{-1}[\boldsymbol{J}_D^T(\boldsymbol{\mu}^+ - \boldsymbol{\mu}^-) + \boldsymbol{e}]}{\boldsymbol{I}_G^T(\boldsymbol{B}_m)^{-1}\boldsymbol{I}_G - \boldsymbol{I}_D^T \boldsymbol{H}^{-1}\boldsymbol{I}_D} \tag{27}$$

The nodal price differs from the price, λ_N, by the values of $\boldsymbol{\mu}^+/\boldsymbol{\mu}^-$.When network constraints are not considered, which means $\boldsymbol{\mu}^+ = \boldsymbol{\mu}^- = 0$, the prices at all buses are equal to λ_N that is the market clearing price derived in the (5). Equivalently, the problem (23)~(25) is reduced to the problem (3)~(4).

4.3 The Point of View of Each Producer

The model of PSM is a bi-level mathematical programming problem in which the lower level of market clearing is taken into account to get the price and quantity values to compute the objective function for the upper level of producer surplus optimization problem. The mathematical model is generally expressed as

$$Max\ S_g{}^G \quad \forall g \in \mathcal{G} \tag{28}$$

$$s.t.\ max\, S^S \tag{29}$$

$$s.t.\ \boldsymbol{I_G}^T\boldsymbol{p} - \boldsymbol{I_D}^T\boldsymbol{q} = 0 \tag{30}$$

$$-\boldsymbol{F} \leq \boldsymbol{J}(\boldsymbol{p}-\boldsymbol{q}) \leq \boldsymbol{F} \tag{31}$$

Due to the convexity property of the market clearing problem (quadratic programming with the DC flow model), *KKT* conditions outline the optimal solution and can be used as the constraint functions of the maximum producer surplus problem to get the optimal strategy.

By using the KKT conditions of the market clearing, the above problem can be transformed into

$$Max\, S_g{}^G = (\lambda_N - \boldsymbol{J}_g{}^T(\boldsymbol{\mu}^+ - \boldsymbol{\mu}^-))\, p_g - (a_m + b_m p_g)\ \forall g \in \mathcal{G} \tag{32}$$

s.t.

$$\boldsymbol{a}_m + \boldsymbol{B}_m\boldsymbol{p} = \lambda_N \boldsymbol{I}_G - \boldsymbol{J}_G{}^T(\boldsymbol{\mu}^+ - \boldsymbol{\mu}^-) \tag{33}$$

$$\boldsymbol{e} + \boldsymbol{H}\boldsymbol{q} = \lambda_N \boldsymbol{I}_G - \boldsymbol{J}_G{}^T(\boldsymbol{\mu}^+ - \boldsymbol{\mu}^-) \tag{34}$$

$$\boldsymbol{I_G}^T\boldsymbol{p} - \boldsymbol{I_D}^T\boldsymbol{q} = 0 \tag{35}$$

$$-\boldsymbol{F} \leq \boldsymbol{J}(\boldsymbol{p}-\boldsymbol{q}) \leq \boldsymbol{F} \tag{36}$$

$$\boldsymbol{\mu}^+ \cdot [\boldsymbol{J}(\boldsymbol{p}-\boldsymbol{q}) - \boldsymbol{F}] = \boldsymbol{0} \tag{37}$$

$$\boldsymbol{\mu}^- \cdot [\boldsymbol{J}(\boldsymbol{p}-\boldsymbol{q}) + \boldsymbol{F}] = \boldsymbol{0} \tag{38}$$

$$\boldsymbol{\mu}^+ \leq \boldsymbol{0}, \boldsymbol{\mu}^- \leq \boldsymbol{0}, \lambda_N > 0 \tag{39}$$

$$\boldsymbol{P}^- \leq \boldsymbol{p} \leq \boldsymbol{P}^+ \tag{40}$$

Where λ_N is the nodal price at the reference bus *N*.

4.4 Strategic Interaction Among Producers: Game-Theory Model for Oligopoly

In the last 50 years, game theory provides an efficient tool to model the strategic interactions among individuals aware that the behaviour of their competitors can affect their results in the market. Market power analysis in terms of the strategic bidding behaviours can be gained from the use of game-theoretic models through simulating the competition between a given set of competitors in a well-specified market environment, taking into account the network constraints that may provide additional possibilities of market power arising that are very specific of this in electricity markets [52, 55].

According to the classification of the strategic variables, there are three types of game models which are *bidding in price*, *bidding in quantity* and *supply function bidding* models.

The price bidding models include Bertrand and Forchheimer models. However, taking into account network constraints for the analysis of hourly electricity markets, so far there no literature using price bidding game models as an efficient tool. Another reason for the Bertrand model has not been the focus in the literature would be that Bertrand model might correspond to perfect competition case [56].

The quantity bidding game models include Cournot, Stackelberg and Conjectural Supply Function (CSF) models. An essential assumption of the former two models is that the individual player's own output decision will not have an effect on the decisions of its competitors, i.e. the optimal offer quantities are assumed fixed in formulating the optimal strategy of the considered player. As for the CSF model, the basic assumption is that the output of the other competitors can be estimated to change in an expected way with respect to the output decision of the considered player. Since the strategic variable is the quantity of the electricity transacted, those game models do not give meaningful equilibrium when price elasticity of the demand curve is low (the demand quantity is fixed with the zero value of the price elasticity).

The supply function bidding models choose a strategic supply function from the marginal cost curve with the aim of maximizing individual producer surplus. Different the quantity bidding game models, for the optimal strategy formulation of one producer, the given strategies are the supply functions of other competitors but the production quantities. The dispatched quantities of the competitors are determined by the supply function of the considered producer through current decision making process.

Taking into account the strategic bidding behaviors of the electricity suppliers, the quantity, $\boldsymbol{p}$ and the price at the reference bus, λ_N, become the variables with respect to the strategic variables, $\boldsymbol{\rho}$ ($N_G \times 1$), that are $\boldsymbol{p}(\boldsymbol{\rho})$ and $\lambda_N(\boldsymbol{\rho})$ in problem (32)$\sim$(40). According to the strategic variable of the different game modes, the of the $\lambda_N(\boldsymbol{\rho})$ and $\boldsymbol{p}(\boldsymbol{\rho})$ are summarized in Table 2.

The information given in Table constitutes the core part of the game theory models in a formalized way. Such formulations will be used in the conceptual examples to obtain the market results under game theory models.

Table 2 Price at the reference bus and production of the players according to different game models

Model	Strategy variable ρ_g	Formulation
Cournot	$p_g \ \forall\, g \in G$	$\lambda_N = \dfrac{\boldsymbol{I}_G^T \boldsymbol{p}' + p_g + \boldsymbol{I}_D^T \boldsymbol{H}^{-1}[\boldsymbol{J}_D^T(\boldsymbol{\mu}^+ - \boldsymbol{\mu}^-) + \boldsymbol{e}]}{\boldsymbol{I}_D^T \boldsymbol{H}^{-1} \boldsymbol{I}_D}$ where, $\boldsymbol{p}' = [p_1', \ldots, p_{g-1}', 0, p_{g+1}', \ldots, p_{NG}]^T$, p_i' ($i \in G$, $i \neq g$) is the optimal offered quantity derived from the last move of the producer i, p_g is the decision variable in the optimal problem
Stackelberg	$p_g^{\,l} \ \forall\, g \in G_l \subset G$	$\lambda_N = \dfrac{\boldsymbol{I}_F^T \boldsymbol{p}^F + \boldsymbol{I}_L^T \boldsymbol{P}'^L + p_g^L + \boldsymbol{I}_D^T \boldsymbol{H}^{-1}[\boldsymbol{J}_D^T(\boldsymbol{\mu}^+ - \boldsymbol{\mu}^-) + \boldsymbol{e}]}{\boldsymbol{I}_D^T \boldsymbol{H}^{-1} \boldsymbol{I}_D}$ where $\boldsymbol{P}'^{\,L} = [P_1', \ldots, P_{i-1}', 0, P_{i+1}', \ldots, P_{NGl}']^T$, $P_i'^L$ ($i \in G_l \subset G$, $i \neq g \in G_l$, G_l is the set of the leader producers and N_{Gl} is the number of the leaders) is the optimal offered quantity of the leader i, $p_g^{\,L}$ is the decision variable in the optimal problem; $\boldsymbol{p}^F$ is the vector of the optimal production quantities of the followers If the slope parameter of the marginal cost curve of the follower producers are identical, $b_1^{\,m} = b_2^{\,m} =, \ldots, b_i^{\,m} =, \ldots, = b_{NGf}^{\,m} = b^m$, ($i \in G_f \subset G$, G_f is set of the follower producers and $G_f \cap G_l = G$, N_{Gf} is the number of the followers), the optimal production quantities of the followers can be derived from the first order rule of the producer surplus maximization as $\boldsymbol{I}_F^T \boldsymbol{p}_F = \dfrac{(\boldsymbol{I}_D^T \boldsymbol{H}^{-1} \boldsymbol{I}_D)(\boldsymbol{I}_F^T \boldsymbol{a}_F^m) - N_{Gf}(p_l + \boldsymbol{I}_L^T \boldsymbol{P}'_L + \boldsymbol{I}_D^T \boldsymbol{H}^{-1} \boldsymbol{e})}{N_{Gf} + 1 - b^m \boldsymbol{I}_D^T \boldsymbol{H}^{-1} \boldsymbol{I}_D}$
Conjecture supply function (CSF)	$p_g, \ \forall\, g \in G$	$\lambda_N = \dfrac{p_g + \boldsymbol{I}_D^T \boldsymbol{H}^{-1}[\boldsymbol{J}_D^T(\boldsymbol{\mu}^+ - \boldsymbol{\mu}^-) + \boldsymbol{e}] - \boldsymbol{r}_g^T \boldsymbol{J}_G^T(\boldsymbol{\mu}^+ - \boldsymbol{\mu}^-) + \boldsymbol{I}_G^T \boldsymbol{p}^{g-1} - \boldsymbol{r}_g^T \boldsymbol{\Lambda}_g^{g-1}}{\boldsymbol{I}_D^T \boldsymbol{H}^{-1} \boldsymbol{I}_D - \boldsymbol{I}_G^T \boldsymbol{r}_g}$ where $\boldsymbol{r}_g = [r_{1,g}, r_{2,g}, \ldots, r_{g-1,g}, 0, r_{g+1,\,g}, r_{NG,\,g}]^T$ $\boldsymbol{p}^{\,g-1} = [p_1^{\,g-1}, p_2^{\,g-1}, \ldots, p_{g-1}^{\,g-1}, 0, p_{g+1}^{\,g-1}, \ldots, p_{NG}^{\,g-1}]^T$ $\boldsymbol{\Lambda}_g^{\,g-1} = [\Lambda_1^{\,g-1}, \ldots, \Lambda_{g-1}^{\,g-1}, 0, \Lambda_{g+1}^{\,g-1}, \ldots, \Lambda_{NG}^{\,g-1}]^T$ $p_i^{\,g-1}$ and Λ_i^{g-1} are the dispatched quantity and the nodal price of producer i derived from the last move of the producer $g-1$, respectively. $r_{g,i}$, $\forall g$, $i \in G$, $i \neq g$, represents the assumed rate of change in competitor supply per unit price. The *CSF* function is $p_i = p_i^{\,g-1} + r_{g,\,i}(\lambda_i - \Lambda_i^{\,g-1}) \ \forall i \in G, i \neq g$
SFE Intercept	$a_g \ \forall\, g \in G$	$\lambda_N = \dfrac{\boldsymbol{I}_G^T(\boldsymbol{B}^m)^{-1}[\boldsymbol{J}_G^T(\boldsymbol{\mu}^+ - \boldsymbol{\mu}^-) + \boldsymbol{a}'] - \boldsymbol{I}_D^T \boldsymbol{H}^{-1}[\boldsymbol{J}_D^T(\boldsymbol{\mu}^+ - \boldsymbol{\mu}^-) + \boldsymbol{e}]}{\boldsymbol{I}_G^T(\boldsymbol{B}^m)^{-1} \boldsymbol{I}_G - \boldsymbol{I}_D^T \boldsymbol{H}^{-1} \boldsymbol{I}_D}$
The supply function is $a_g + b_g^{\,m} p_g$		$p_g = \dfrac{\lambda_N - \boldsymbol{J}_g^T(\boldsymbol{\mu}^+ - \boldsymbol{\mu}^-) - a_g}{b_g^m}$ where $\boldsymbol{a}' = [a_1', a_2', \ldots, a_{g-1}', a_g, a_{g+1}', \ldots, a_{NG}']^T$; a_i', $i \neq g$, $\forall\, i, g \in G$, is the obtained value derived from the last move of producer i
SFE slope	$b_g \ \forall\, g \in G$	$\lambda_N = \dfrac{\boldsymbol{I}_G^T(\boldsymbol{B}')^{-1}[\boldsymbol{J}_G^T(\boldsymbol{\mu}^+ - \boldsymbol{\mu}^-) + \boldsymbol{a}^m] - \boldsymbol{I}_D^T \boldsymbol{H}^{-1}[\boldsymbol{J}_D^T(\boldsymbol{\mu}^+ - \boldsymbol{\mu}^-) + \boldsymbol{e}]}{\boldsymbol{I}_G^T(\boldsymbol{B}')^{-1} \boldsymbol{I}_G - \boldsymbol{I}_D^T \boldsymbol{H}^{-1} \boldsymbol{I}_D}$
The supply function is $a_g^{\,m} + b_g\, p_g$		$p_g = \dfrac{\lambda_N - \boldsymbol{J}_g^T(\boldsymbol{\mu}^+ - \boldsymbol{\mu}^-) - a_g^m}{b_g}$ where $\boldsymbol{B}'$ is the diagonal matrix formulated with the vector of $[b_1', \ldots, b_{g-1}', b_g, b_{g+1}', \ldots, b_{NG}']^T$; b_i', $i \neq g$, $\forall\, i, g \in G$, is the obtained value derived from the last move of producer i

(continued)

Table 2 (continued)

Model	Strategy variable ρ_g	Formulation
SFE k multiplier The supply function is $k_g(a_g{}^m + b_g{}^m p_g)$	$k_g\ \forall\ g\in G$	$\lambda_N = \frac{\boldsymbol{I}_G^T\boldsymbol{K}[\boldsymbol{J}_G^T(\boldsymbol{\mu}^+ - \boldsymbol{\mu}^-)] + \boldsymbol{I}_G^T(\boldsymbol{B}^m)^{-1}\boldsymbol{a}^m + \boldsymbol{I}_D^T\boldsymbol{H}^{-1}[\boldsymbol{J}_D^T(\boldsymbol{\mu}^+ - \boldsymbol{\mu}^-) + \boldsymbol{e}]}{\boldsymbol{I}_G^T\boldsymbol{K}\boldsymbol{i}_G - \boldsymbol{I}_D^T\boldsymbol{H}^{-1}\boldsymbol{i}_D}$ $p_g = \frac{\lambda_N - \boldsymbol{J}_g^T(\boldsymbol{\mu}^+ - \boldsymbol{\mu}^-) - k_g a_g^m}{k_g b_g^m}$ where $\boldsymbol{K}$ is a diagonal matrix formulated by the vector $[1/(k_1{}'\, b_1{}^m), \ldots, 1/(k_{g-1}{}'\, b_{g-1}{}^m), 1/(k_g\, b_g{}^m), 1/(k_{g+1}{}'\, b_{g+1}{}^m), \ldots, 1/(k_{NG}{}'\, b_{NG}{}^m)]^T$; $k_i{}', i \neq g, \forall\ i, g\in G$, is the obtained value derived from the last move of producer i

4.5 *Nash Equilibrium and Search Methods*

The scope of the game is to get the optimal strategy of each market player derived at the Nash Equilibrium (NE). In terms of solving techniques for the NE, there are two different to be considered: best response functions and iterative search algorithm. The former is an analytic approach that can be used to analyze simple duopoly games without network consideration while the latter is a numerical approach that is suitable for studying muli-player games with network consideration.

4.5.1 Best Response Functions

Best response function is an analytic approach to obtain the oligopoly market equilibrium. The idea behind the best response function approach is that all the players get the maximum surplus concurrently. Mathematically, define the best response function of producer g, in terms of strategies of all the producers, $\boldsymbol{\rho}$, as the first derivative of his/her producer surplus with respect to his/her own strategic variable, ρ_g, as

$$B_g(\boldsymbol{\rho}) := \partial S_g{}^G(\boldsymbol{\rho})/\partial \rho_g = 0\,,\ \forall g \in \mathcal{G} \tag{41}$$

Based on, (34) we have

$$\begin{aligned} B_g(\rho) = {} & (\lambda_N(\boldsymbol{\rho}) - \boldsymbol{J}_g{}^T(\boldsymbol{\mu}^+ - \boldsymbol{\mu}^-))\partial p_g(\boldsymbol{\rho})/\partial \rho_g + (\partial \lambda_N(\boldsymbol{\rho})/\partial \rho_g)p_g(\boldsymbol{\rho}) \\ & - a_g{}^m - b_g{}^m p_g(\boldsymbol{\rho}),\ \forall g \in \mathcal{G} \end{aligned} \tag{42}$$

The Nash equilibrium is derived by solving the group of the best response functions simultaneously, i.e. the intersection point of the best response functions.

This approach is suitable for simple problems especially in duopoly game models (two players) and no power system operation considered. In case of network constraints to be considered, a specific branch power flow state can be integrated into the (41). In this respect, the state of the line (not congested or congested in the two possible directions) can be determined by comparison of the line flow's amplitude and direction, derived in the case of no network constraints, with the line flow limit.

In duopoly markets where the power system operational constraints are represented by one line power flow constraints, the NE of ρ_1* and ρ_2* can be derived from the following three steps.

Step 1, Game solution without the network constraints

$$\begin{cases} B_1(\rho_1^*, \rho_1^*) = 0 \\ B_2(\rho_1^*, \rho_1^*) = 0 \\ \mu_l^+ = \mu_l^- = 0 \end{cases} \tag{43}$$

Step 2, If: $\boldsymbol{J}_l\,(\boldsymbol{p}(\rho_1^*, \rho_2^*) - \boldsymbol{q}(\rho_1^*, \rho_2^*)) > F_l$
Then

$$\begin{cases} B_1(\rho_1^*, \rho_2^*) = 0 \\ B_2(\rho_1^*, \rho_2^*) = 0 \\ J_l(\boldsymbol{p}(\rho_1^*, \rho_2^*) - \boldsymbol{q}(\rho_1^*, \rho_2^*)) = F_l \\ \mu_l^+ < 0; \mu_l^- = 0 \end{cases} \tag{44}$$

Else if : $\boldsymbol{J}_l\,(\boldsymbol{p}(\rho_1^*, \rho_2^*) - \boldsymbol{q}(\rho_1^*, \rho_2^*)) < -F_l$
Then

$$\begin{cases} B_1(\rho_1^*, \rho_2^*) = 0 \\ B_2(\rho_1^*, \rho_2^*) = 0 \\ J_l(\boldsymbol{p}(\rho_1^*, \rho_2^*) - \boldsymbol{q}(\rho_1^*, \rho_2^*)) = -F_l \\ \mu_l^+ = 0; \mu_l^- < 0 \end{cases} \tag{45}$$

Where J_l is the row vector of the matrix $\boldsymbol{J}$, corresponding to the power transfer distribution factors for the selected branch l.

4.5.2 Iterative Search Algorithm

For the iterative search algorithm, while keeping the competitor strategies as given the players solve the PSM problem one by one. The gaming behaviours of the market players converge to the equilibrium point through a process characterised by a set of moves. Each move is a PSM problem where the optimal strategies of the individual player is derived based on the updated optimal strategies of the competitors. NE is found when the changing process, in terms of optimal strategies of the market players, is stabilized, i.e. no player can improve his/her payoff by changing his/her strategy if his/her competitors do not. The best response function approach, the iterative search algorithm is suitable for the multi player games with consideration of the power system network constraints. Fig. 7 the flow chart of NE derivation based on the iteration search algorithm.

As shown the innermost iteration of the flow chart, Fig. 7, complementary terms of the power flows of the considered lines are managed in a way of traversing all the states that may be *uncongested (0), congested in the positive direction (1),*

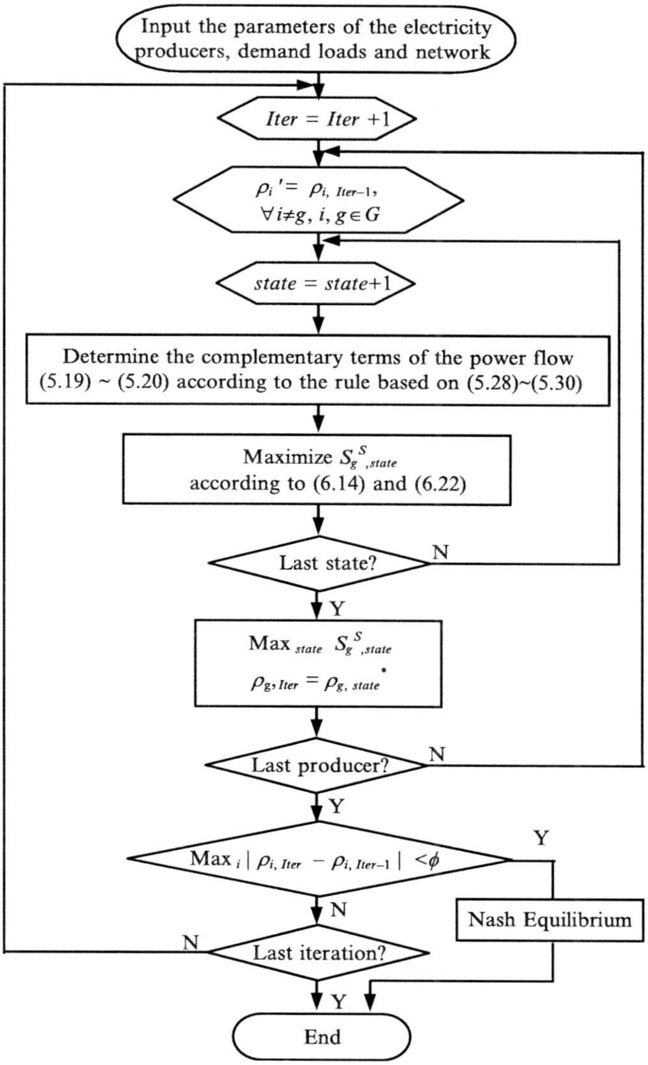

Fig. 7 Flow chart of the iterative search of the Nash equilibrium

or congested in the negative direction (−1). Therefore, each player, in choosing a move, should consider 3^z alternatives if there are z lines to be considered in the power system network. *In each state, the complementary relationships can be determined. For instance, given the state of the three lines of interest as 1, −1, 0, (37)~(39) are transformed as*

$$-F_3 < \boldsymbol{J}_3(\boldsymbol{p}-\boldsymbol{q}) < \boldsymbol{F}_3 \tag{46}$$

$$\boldsymbol{J}_1(\boldsymbol{p}-\boldsymbol{q}) = F_1 \tag{47}$$

$$-\boldsymbol{J}_2(\boldsymbol{p}-\boldsymbol{q}) = F_2 \tag{48}$$

$$\mu_1{}^+ \leq 0, \mu_2{}^+ \leq 0, \ \mu_3{}^+ = 0$$

Where the subscript of 1, 2, 3 in (46)–(47) are the indices of the three considered lines.

The PSM problem is formulated by model of (32)–(40) and the price at the reference bus λ_N and the production of the players, p_g, listed in Table 2. The iteration process is terminated by either the Nash Equilibrium derived by checking the maximal derivation from the last move strategy, smaller than a gate value of ϕ, or by exceeding the predefined iteration steps. The latter case the Nash Equilibrium has not been found.

An important note is that the unique/existence of the NE is a general concern in many related references [12, 17, 22, 23, 25, 26]. So far the unique NE can be guaranteed under simple test system by using the best response functions in analytical way [16, 49]. As for the complicate game problems where iterative search algorithm is used, the existence of equilibrium can not be guaranteed analytically and ex-post check is needed [22].

4.6 Conceptual Examples

In this section, two examples are designed for comparison study of different game theory models in assessing the strategic bidding behaviors of the electricity producers. The first example is a duopoly game in an electricity market modeled by a simple three-bus power system whilst the is a relative complicated one with multi-players in the electricity market represented by a large size power system.

4.6.1 Test Case1: A Conceptual Application for Duopoly Game

The three-bus test system is shown in Fig. 8. The system is characterized by two generators and one load respectively connected at the bus 1, 2 and 3 (reference bus),

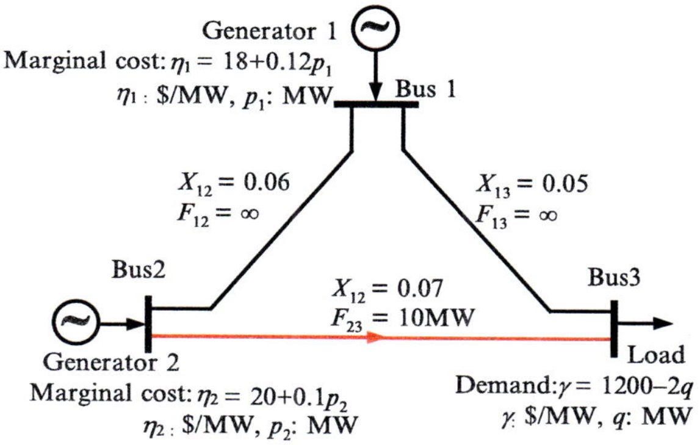

Fig. 8 Three buses test system

Table 3 Reference case: market clearing under perfect competition without network constraints

p_1	p_2	q	λ_1	λ_2	λ_3	S_1^G	S_2^G	S^C	S^M	S^S
31.4	17.7	49.1	21.8	21.8	21.8	60.2	16.2	2410.8	0	2487.2

Table 4 Market clearings under different models without network constraints

Models \ Variables	p_1	p_2	q	λ_1	λ_2	λ_3	S_1^G	S_2^G	S^C	S^M	S^S
Cournot	16.9	16.1	33	53.9	53.9	53.9	590	533	1092	0	2214.7
Stackelberg	24.5	12.4	36.9	46.1	46.1	46.1	652.4	652.4	1365	0	2333.7
CSF ($r_{12} = r_{21} = 1$)	22.16	20.1	42.3	35.4	35.4	35.4	356.1	289.3	1789.3	0	2436.2
SFE-intercept	28.5	19.4	47.9	24.1	24.1	24.1	125.1	60.7	2299.2	0	2485
SFE-slope	23.9	20.7	44.6	30.8	30.8	30.8	271.3	201.3	1990.5	0	2463.1
SFE-k parameter	28.4	19.3	47.8	24.4	24.4	24.4	134	66.7	2284.2	0	2484.8

respectively. The line from bus 2 to bus 3 has line flow limit of 10 MW, shown in red line in Fig. 8, while other lines have no line flow limits. Since no demand loads are located at bus 1 and 2, the line 3 can only be possibly congested the positive direction, from bus 2 to bus 3. The market clearing under perfect competitive without network constraints is summarized in Table 3.

The market clearings under different game models, without and with the network constraints, are reported in Tables 4 and 5, respectively. Without network consideration, the nodal prices λ_1, λ_2 and λ_3 are identical, as shown in the Tables 3 and 4. The merchandise surplus aroused from the different nodal prices is equal to zero. With the introduction of the network constraints, the NEs of the game models can only be obtained with the power flow of line binding at its limit value. The reason is that the power flow at the market equilibriums without consideration of the network constraints is beyond the line flow limit, 10 MW. Due to the network congestion, the nodal prices, λ_1, λ_2 and λ_3, are different with each other, as shown in Table 5.

Due to the strategic bidding behavior from the supply side, the market performance is deviated from the perfect competition case (Table 3) with the increased

Table 5 Market clearings under different models with network constraints

Models \ Variables	p_1	p_2	q	λ_1	λ_2	λ_3	λ	$S_1{}^G$	$S_2{}^G$	S^C	S^M	S^S
Cournot	17.2	8.5	25.7	54.6	37.9	68.5	58.8	612.8	148.5	663.1	499.2	1922.5
Stackelberg	25	5	30	46.6	30.5	60	52	677.5	51.3	900	482.5	2111.3
CSF ($r_{12} = r_{21} = 1$)	28.8	3.26	32.1	40.7	22.5	55.8	47.4	604	7.6	1030.4	545.7	2187.7
SFE-intercept	31.5	2	33.5	41.2	27.2	52.9	46.6	671.3	14.2	1125.6	420	2231
SFE-k parameter	31.6	1.98	33.6	41.5	28.1	52.8	46.7	684.7	15.8	1130.6	403.7	2235

prices, by comparison of the values in column 5 and 7 in Tables 5 to the reference values in column 4 and 6 in Table 3, at the decreased transacted amount of the electricity, by comparison of the values in the column 4 in Tables 4 and 5 to the reference values in the column 3 in Table 3. The total producer surplus is changed from the reference value, the addition of the values in column 7 and 8 of Table 3, with a significant increment as shown Tables 4 and 5. However, the consumer surplus of demand side is decreased from the reference case; by comparison the values in column 10 in Table 3 to the values in column 10 in the Table 4 and column 11 in the Table 5.

Another important observation is that the values in Table 4 and the values in Table 5, network congestion pushes the market performance toward lower level of efficiency. The market players can utilize the network constraints, very specific to the electricity market, to obtain an even higher producer surplus at higher nodal prices.

4.6.2 Test Case 2: IEEE57 Bus System for Multiple player's Game

In this test case, the IEEE57 bus system is used to model the network constraints. The IEEE57 bus system is composed of 7 generators, 50 demand loads and 80 transmission branches, as shown in the appendix Fig. 10. The seven generators are assumed as individual players to have strategic biddings. The parameters of the marginal cost curves of the generators and the demand curves of the loads are reported in the appendix Tables 9 and 10. The network structure parameters are listed in the appendix Table 11. The flow limits of the selected lines are shown as the red lines in the Fig. 10.

Cournot model and SFE intercept model are run based on the iterative search method in this test case. To get the comparison indices, we consider the following four different cases:

- C0 unconstrained market under perfect competition
- C1 constrained market under perfect competition
- C2 unconstrained market under strategic bidding
- C3 constrained market under strategic bidding

The optimal strategies of the electricity producers under *Cournot* and *SFE_Intercept* game modes of test case C2 and test case C3 are shown in Table 6.

Table 6 Optimal strategies of the electricity producers at NE

Test case	Producers model	G1	G2	G3	G4	G5	G6	G7
C2	Cournot (optimal quantity)	96.11	108.05	95.56	95.72	131.04	101.89	107.18
	SFE_intercept (optimal intercept parameter of the supply curve)	22.643	25.144	26.269	26.681	24.111	28.976	21.031
C3	Cournot (optimal quantity)	91.95	95.55	71.26	25.94	139.28	123.81	119.18
	SFE_intercept (optimal intercept parameter of the supply curve)	23.029	45.511	40.180	44.713	20	26	18

Table 7 Market clearing under different test cases

Test case	Game model	Market clearing price $/MW	Market clearing production MW	Producer surplus $	Consumer surplus $	Merchandise surplus $	Social surplus $
C0	–	48.5	815.1	10821.0	32707.8	0	43529
C1	–	50.4	781.3	10793.5	30255.2	1560.7	42609
C2	Cournot	56.1	735.6	16306.5	26794.7	0	43101
	SFE_intercept	50.7	792.5	12518.7	30967.2	0	43486
C3	Cournot	62.3	667	18870.6	22187.9	467.3	41526
	SFE_intercept	55	747.2	14118.9	27621.5	0	41740

For *Cournot* model, the optimal strategies are derived by the flow chart of Fig. where the equilibrium is attainted when the variation of the optimal quantities of each player is smaller than 1 MW. For *SFE_intercept* model, the equilibrium optimal strategies is attained when the variation of the optimal intercept of supply curve of each player is smaller than 0.001 $/MW.

The market clearing results under the four test cases, C0 ~ C3, are reported in the Table 7. The market performance indices, introduced in the Sect. 3.5, are computed based on the market clearing results and are shown in Table 8. Under strategic bidding test cases, the market is cleared at the higher price with lower transacted quantities, causing the producer surplus increased at the expense of the decrement of the consumer surplus. Due to the surplus share transferred from the demand side to the producer side, the total social surplus is changed from the perfect case of C0 in a minor value. From Tables 7 and 8, one can find that the competition level of *Cournot* game model is lower than the *SFE_intercept* model, by comparison of the index values in row 3 to row 6 of Table 8. Such uncompetitive market level is more intensive when the network constraints taken into account, the market is cleared at even higher price and lower quantities in test case C3 that in test case C2, the *Cournot* model and *SFE_intercept* model. The network constraints can provide opportunities for the market players, very specific to the electricity market, to obtain even higher individual producer surplus.

Table 8 Market performance indices (in percentage) under different test cases

Test case	Game models	K_G	K_D	K_S	K_P	K_λ
C1	–	−0.25	−7.50	−2.11	−4.15	3.82
C2	Cournot	50.69	−18.08	−0.98	−9.76	15.72
	SFE_intercept	15.69	−5.32	−0.10	−2.77	4.46
C3	Cournot	74.39	−32.16	−4.60	−18.18	28.49
	SFE_intercept	30.48	−15.55	−4.11	−8.33	13.42

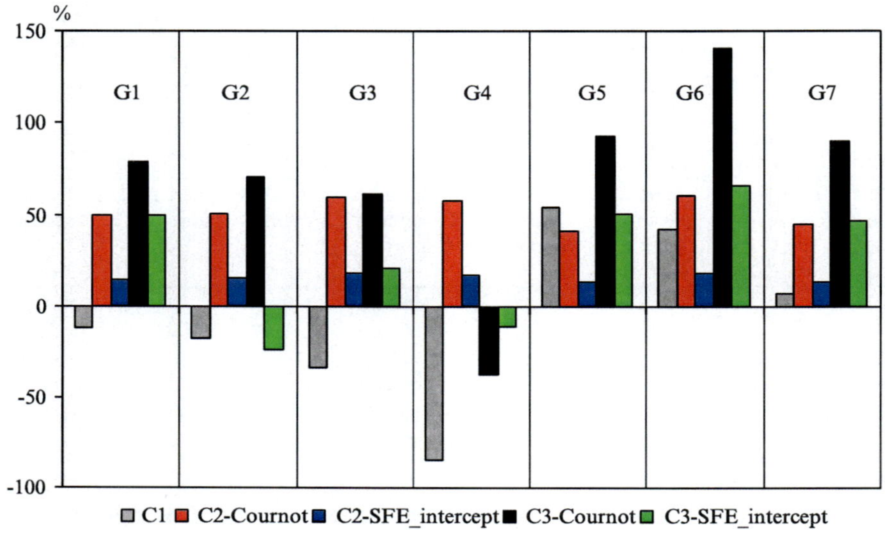

Fig. 9 Individual producer surplus deviation from the reference case C0

For the individual producer surplus, Fig. 9 the deviation percentage computed against the reference case of C0. When the network constraints are not considered, test case C2, all the producers can gain increased individual surplus both under *Cournot* and *SFE_intercept* model, the red and black bars shown in Fig. 9. However, such case of increased surpluses of all the electricity producers due to strategic bidding behaviors is not applied to the case where the network constraints are considered, the blue and green bars shown in Fig. 9 for test case C3. There is producer G4, at bus 6, obtains decreased individual surplus under *Cournot* model while producer G2, at bus 3, gets decreased value under *SFE_intercept* model. The network constraints cause the electricity market to have geographic property that lend its self to benefit some producers at specific locations at the expense of other producers at other locations, the grey bars shown in Fig. 9. However, from the whole point of view of supply side, the strategic bidding behavior with the network constraints result in getting high producer surplus from the demand side, causing the market cleared at a uncompetitive level that against the deregulation goal of the electricity industry. Positive approaches, such as transmission network reinforcement, introduction of more electricity producers and encouragement of demand side bidding, may be adopted to push the electricity markets toward a high competitive level one.

5 Conclusion

For the specific features of the electricity industry, the present electricity markets may be better described in terms of oligopoly than of perfect competition from which they may be rather far. In an oligopoly, the producer as a market player adopts strategic behavior, offerring submitted different from the marginal cost, to obtain the maximization of the producer surplus. The producer surplus maximization problem is outlined by a bi-level mathematic programming problem where the optimal DC power flow model, the lower level optimization problem, is inserted as the subject conditions of the upper level programming problem.

With the surplus maximization problem of each producer, the oligopoly competition in the electricity markets can be addressed by the game theory models. The solution of the game model is to derive the Nash Equilibrium at which point no individual player can improve his/her payoff by changing his/her strategy if his/her competitors do not. The approaches to obtain the Nash Equilibrium are best response functions and iterative search algorithm. The former can be used to simple games while the latter is suitable for the complicated games. One shortcoming of the game theory models is the assumption of the rational market players, with the aim of maximization producer surplus, which is not always applicable to the real markets. The other disadvantage is the unique/existence of the Nash Equilibrium, especially when electricity network constraints are taken into account, can not be guaranteed for different markets.

However, game theory models can provide some instructive results on the oligopoly electricity market performance if the Nash Equilibrium can be obtained by ex-post check. At the oligopoly equilibrium, social surplus is transferred from the consumer side to the supply side. The main effects of the strategic behavior from the supply side are remarkably presented in obtaining extra surplus from the consumer side with higher market clearing prices and production withdrawn. Due to the network constraints, the transmission network plays a major role in determining the oligopoly equilibrium. With the consideration of network constraints, the market clearing price is higher and the cleared demand is lower than the corresponding values under unconstrained network. As for the producer surpluses, the network constraints provide some producers with opportunities to obtain higher values at the expense of consumer surplus, leading to a higher level of market inefficiency compared with the unconstrained electricity market.

Appendix

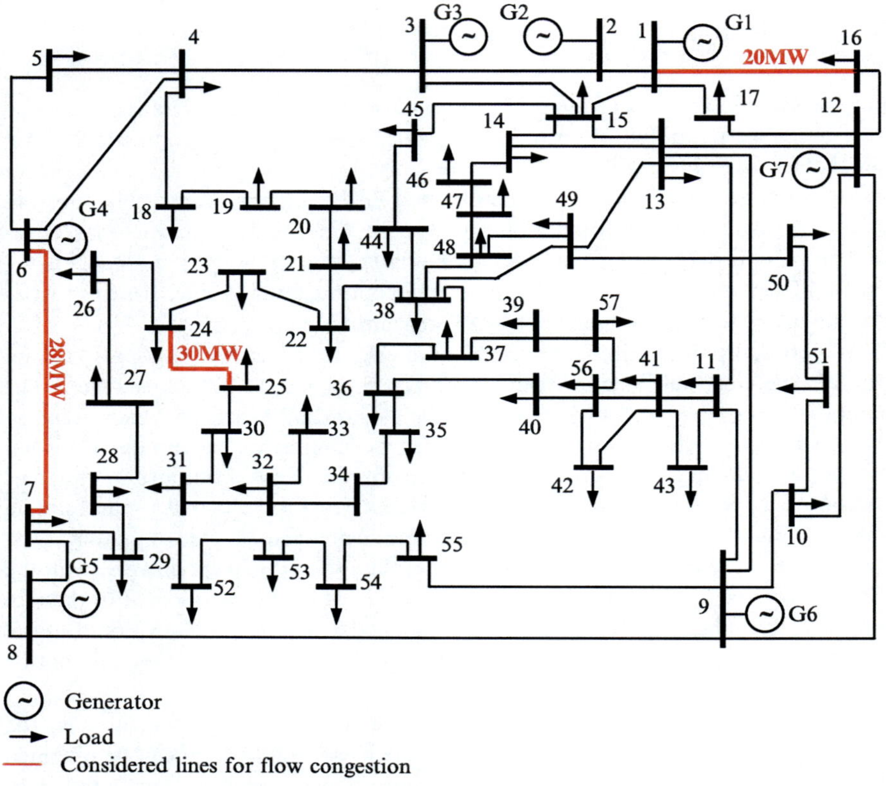

Fig. 10 IEEE 57 bus system for test case 2

Table 9 Parameters of the electricity producers

Bus Number	Producer	Marginal cost curve parameters		P^+ MW	P^- MW
		Intercept \$/MW	Slope \$/MW^2		
1	G1	20	0.28	300	0
2	G2	22	0.22	150	0
3	G3	25	0.23	150	0
6	G4	24	0.24	100	0
8	G5	20	0.18	250	0
9	G6	26	0.2	250	0
12	G7	18	0.26	300	0

Table 10 Parameters of the electricity demand loads

Bus	Parameters of demand curve		Bus	Parameters of demand curve		Bus	Parameters of demand curve		Bus	Parameters of demand curve	
	Intercept \$/MW	Slope \$/MW^2		Intercept \$/MW	Slope \$/MW^2		Intercept (\$/MW)	Slope \$/MW^2		Intercept \$/MW	Slope \$/MW^2
4	120	−5	22	120	−5.5	36	135	−6	50	125	−4.5
5	130	−4.5	23	125	−6.5	37	150	−5.5	51	120	−4
7	135	−6	24	130	−5.5	38	125	−4.5	52	135	−4.5
10	150	−5.5	25	135	−4	39	130	−5	53	130	−5
11	135	−6	26	150	−5.5	40	110	−3.5	54	150	−5.5
13	120	−5	27	120	−4	41	120	−4.5	55	125	−4.5
14	135	−6	28	110	−4	42	120	−4	56	115	−4.5
15	150	−5.5	29	135	−6	43	150	−5.5	57	110	−3.5
16	110	−4	30	120	−5.5	44	125	−4.5	–	–	–
17	135	−6.5	31	110	−3.5	45	135	−6	–	–	–
18	135	−6	32	135	−4.5	46	110	−3.5	–	–	–
19	150	−5.5	33	150	−5.5	47	125	−4.5	–	–	–
20	110	−4	34	115	−4.5	48	135	−4.5	–	–	–
21	115	−4.5	35	135	−6.5	49	110	−4.5	–	–	–

Table 11 Parameters of the IEEE 57 bus network

Line number	From bus	To bus	x (p.u.)	Line number	From bus	To bus	x (p.u.)	Line number	From bus	To bus	x (p.u.)	Line number	From bus	To bus	x (p.u.)
1	1	2	0.028	21	5	6	0.0641	41	7	29	0.0648	61	47	48	0.0233
2	2	3	0.085	22	7	8	0.0712	42	25	30	0.202	62	48	49	0.129
3	3	4	0.0366	23	10	12	0.1262	43	30	31	0.497	63	49	50	0.128
4	4	5	0.132	24	11	13	0.0732	44	31	32	0.755	64	50	51	0.22
5	4	6	0.148	25	12	13	0.058	45	32	33	0.036	65	10	51	0.0712
6	6	7	0.102	26	12	16	0.0813	46	34	32	0.953	66	13	49	0.191
7	6	8	0.173	27	12	17	0.179	47	34	35	0.078	67	29	52	0.187
8	8	9	0.0505	28	14	15	0.0547	48	35	36	0.0537	68	52	53	0.0984
9	9	10	0.1679	29	18	19	0.685	49	36	37	0.0366	69	53	54	0.232
10	9	11	0.0848	30	19	20	0.434	50	37	38	0.1009	70	54	55	0.2265
11	9	12	0.295	31	21	20	0.7767	51	37	39	0.0379	71	11	43	0.153
12	9	13	0.158	32	21	22	0.117	52	36	40	0.0466	72	44	45	0.1242
13	13	14	0.0434	33	22	23	0.0152	53	22	38	0.0295	73	40	56	1.195
14	13	15	0.0869	34	23	24	0.256	54	11	41	0.749	74	56	41	0.549
15	1	15	0.091	35	24	25	1.182	55	41	42	0.352	75	56	42	0.354
16	1	16	0.206	36	24	25	1.23	56	41	43	0.412	76	39	57	1.355
17	1	17	0.108	37	24	26	0.0473	57	38	44	0.0585	77	57	56	0.26
18	3	15	0.053	38	26	27	0.254	58	15	45	0.1042	78	38	49	0.177
19	4	18	0.555	39	27	28	0.0954	59	14	46	0.0735	79	38	48	0.0482
20	4	18	0.43	40	28	29	0.0587	60	46	47	0.068	80	9	55	0.1205

References

1. Federal Energy Regulatory Commission 2011. Electric power markets: national overview. http://www.ferc.gov/market-oversight/mkt-electric/overview.asp
2. Nordic Energy Regulators. The nordic electricity market 2011. https://www.nordicenergyregulators.org/The-Development-on-the-Nordic-Electricity-Market/ December 2011
3. Overview of Australian Energy Market Operator- AEMO, Dec. 2011 http://www.aemo.com.au/corporate/aboutaemo.html
4. National Grid. GB seven year statement 2009, chapter 10, market overview. http://www.nationalgrid.com/uk/Electricity/SYS/current/
5. Stoft S (2002) Power system economics. IEEE Press/Wiley, New york
6. Ventosa M, Baillo A, Ramos A, Rivier M (2005) Electricity market modeling trends. Energ Policy 33(7):897–913
7. David AK, Wen F (2002) Strategic bidding in competitive electricity markets: a literature survey. In: Power engineering society summer meeting, 2000. IEEE Seattle, WA , USA, vol 4, pp 2168–2173
8. Bompard E, Carpaneto E, Ciwei G, Napoli R, Benini M, Gallanti M, Migliavacca G (2005) A game theory simulator for assessing the performances of competitive electricity markets. In: Power tech, 27–30 June 2005 IEEE Russia, St. Petersburg pp 1–9
9. Lee K-H, Baldick R (2003) Solving three-player games by the matrix approach with application to an electric power market. IEEE Trans Power Syst 18(4):1573–1580
10. Neuhoff K, Barquin J, Boots MG (2005) Network-constrained cournot models of liberalized electricity markets: the devil is in the details. Energ Econ 27(3):495–525
11. Hobbs BE (2001) Linear complementarity models of nash-cournot competition in bilateral and poolco power markets. IEEE Trans Power Syst 16(2):194–202
12. Contreras J, Klusch M, Krawczyk JB (2004) Numerical solutions to nash-cournot equilibria in coupled constraint electricity markets. IEEE Trans Power Syst 19(1):195–206
13. Yu Z, Sparrow FT, Morin TL, Nderitu G (2000) A stackelberg price leadership model with application to deregulated electricity markets. In:Power engineering society winter meeting, 23–27 Jan 2000. IEEE, vol 3, pp 1814–1819
14. de Lujan Latorre M, Granville S (2003) The stackelberg equilibrium applied to AC power systems-a noninterior point algorithm. IEEE Trans Power Syst 18(2):611–618
15. Baldick R, Grant R, Kahn E (2004) Theory and application of linear supply function equilibrium in electricity markets. J Regul Econ 25(2):143–167
16. Baldick R (2002) Electricity market equilibrium models: the effect of parameterization, power engineering review. IEEE Trans Power Syst 17(4):1170–1176
17. Petoussis SG, Zhang XP, Godfrey KR (2007) Electricity market equilibrium analysis based on nonlinear interior point algorithm with complementarity constraints. IET Gener Transm Distrib 1(4):603–612
18. Hobbs BF, Rijkers FAM (2004) Strategic generation with conjectured transmission price responses in a mixed transmission pricing system-part I: formulation. IEEE Trans Power Syst 19(2):707–717
19. Day CJ, Hobbs BF, Pang J-S (2002) Oligopolistic competition in power networks: a conjectured supply function approach. IEEE Trans Power Syst 17(3):597–607
20. Ferris MC, Pang JS (1997) Engineering and economic applications of complementarity problems, SIAM Rev 39(4):669–713
21. Luo ZQ, Pang JS, Ralph D (1996) Mathematical programming with equilibrium constraints. Cambridge University Press, Cambridge
22. Bompard E, Wene Lu, Napoli R (2006) Network constraint impacts on the competitive electricity markets under supply-side strategic bidding. IEEE Trans Power Syst 21(1):160–170
23. Liu YF, Wu FF (2007) Impacts of network constraints on electricity market equilibrium. IEEE Trans Power Syst 22(1):126–135

24. Fampa M, Barroso LA, Candal D, Simonetti L (2008) Bilevel optimization applied to strategic pricing in competitive electricity markets. Comput Optim Appl 39(2):121–142
25. Hobbs BF, Metzler CB, Pang J-S (2000) Strategic gaming analysis for electric power systems: an MPEC approach. IEEE Trans Power Syst 15(2):638–645
26. Li T, Shahidehpour M (2005) Strategic bidding of transmission-constrained GENCOs with incomplete information. IEEE Trans Power Syst 20(1):437–447
27. Yao J, Adler I, Shmuel SO (2008) Modeling and computing Two-settlement oligopolistic equilibrium in a congested electricity network. Oper Res 56(1):34–47
28. Hu X, Ralph M (2004) Convergence of a penalty method for mathematical programming with complementarity constraints. J Optim Theory Appl 123(2):365–390
29. Ralph D, Wright SJ (2004) Some properties of regularization and penalization schemes for MPECs. Optim Methods Softw 19(5):527–556
30. Scheel H, Scholtes S (2000) Mathematical programs with equilibrium constraints: stationarity, optimality, and sensitivity. Math Oper Res 25(1):1–22
31. Jean-Pierre D, Patrice M, Sebastien R, Gilles S (2006) A smoothing heuristic for a bilevel pricing problem. Eur J Oper Res 174(3):1396–1413
32. Ma YC, Jiang CW, Hou ZJ, Wang CM (2006) The formulation of the optimal strategies for the electricity producers based on the particle swarm optimization algorithm. IEEE Trans Power Syst 21(4):1663–1671
33. Yu Nm, Liu CC (2008) Multi-agent systems and electricity markets: state-of-the-art and the future. In: Power and energy society general meeting - Conversion and delivery of electrical energy in the 21st century, 20–24 July 2008. IEEE, pp 1–2
34. Bagnall AJ, Smith GD (2005) A multiagent model of the UK market in electricity generation. IEEE Trans Evol Comput 9(5):522–536
35. Tellidou AC, Bakirtzis AG (2007) Agent-based analysis of capacity withholding and tacit collusion in electricity markets. IEEE Trans Power Syst 22(4):1735–1742
36. Koritarov VS (2004) Real-world market representation with agents. IEEE Power Energy Mag 2(4):39–46
37. Praca I, Ramos C, Vale Z, Cordeiro M (2003) MASCEM: a multiagent system that simulates competitive electricity markets. IEEE Intell Syst 18(6):54–60
38. Bompard E, Ma YC (2008) Modeling bilateral electricity markets: a complex network approach. IEEE Trans Power Syst 23(4):1590–1660
39. Elgerd OI (1983) Electric energy systems theory. Tata McGraw-Hill, New Delhi
40. Conejo AJ, Castillo E, Minguez R, Milano F (2005) Locational marginal price sensitivities. IEEE Trans Power Syst 20(4):2026–2033
41. Ma XW, Sun DI, Ott A (2002) Implementation of the PJM financial transmission rights auction market system. In: Power engineering society summer meeting, 25–25 July 2002, vol 3. IEEE, pp 1360–1365
42. Shahidehpour M, Yamin H, Li Z (2002) Market operation in electric power systems, forecasting, scheduling and risk management. Wiley, New york
43. Rothwell G, Gomez T (2003) Electricity economics: regulation and deregulation. Wiley, Hoboken
44. Lai LL (2001) Power system restructuring and deregulation. Wiley, Chichester
45. Song HL, Liu CC, Lawarree J (2002) Nash equilibrium bidding strategies in a bilateral electricity market. IEEE Trans Power Syst 17(1):73–79
46. Vaisakh K, Rao GVS (2007) Determination of optimal bilateral power contracts with line flow constraints. In: 39th North American power symposium NAPS '07, 30 Sept– 2 Oct 2007, pp 448–455
47. Borenstein S, Bushnell J, Knittel CR (1999) Market power in electricity markets: beyond concentration measures. PWP-059r, Feb. 1999, University of California Energy Institute, Berkeley, California
48. Strbac G, Farmer ED, Cory BJ (1996) Framework for the incorporation of demand-side in a competitive electricity market. IET Gener Transm Distrib 143(3):232–237

49. Bompard E, Ma Y, Napoli R, Abrate G (2007) The demand elasticity impacts on the strategic bidding behavior of the electricity producers. IEEE Trans Power Syst 22(1):188–197
50. U.S. Department of Energy (2000)Horizontal market power in restructured electricity markets. Office of Economic, Electricity and Natural Gas Analysis. http://www.emlf.org/Archives/Resources/press/hmp-0308.pdf. Washington, DC 20585
51. Borenstein S (2000) Understanding competitive pricing and market power in wholesale electricity markets. Electr J 13(6):49–57
52. Berry C, Hobbs B (1999) Understanding how market power can arise in network competition: a game theoretic approach. Util. Pol. 8(3):139–158
53. Nicolaisen J, Petrov V, Tesfatsion L (2001) Market power and efficiency in a computational electricity market with discriminatory double-auction pricing. IEEE Trans Evol Comput 5 (5):504–523
54. Klemperer P, Meyer M (1989) Supply function equilibria in oligopoly under uncertainty. Econom., Eco. Soc. 57(6):1243–1277
55. Singh H (1999) Introduction to game theory and its applications in electric power markets. IEEE Comput Appl Power 12(2):18–20
56. Valenzuela J, Mazumdar M (2005) A probability model for the electricity price duration curve under an oligopoly market. IEEE Trans Power Syst 20(3):1250–1256

Optimization-Based Bidding in Day-Ahead Electricity Auction Markets: A Review of Models for Power Producers

Roy H. Kwon and Daniel Frances

Abstract We review some mathematical programming models that capture the optimal bidding problem that power producers face in day-ahead electricity auction markets. The models consider both price-taking and non-price taking assumptions. The models include linear and non-linear integer programming models, mathematical programs with equilibrium constraints, and stochastic programming models with recourse. Models are emphasized where the producer must self-schedule units and therefore must integrate optimal bidding with unit commitment decisions. We classify models according to whether competition from competing producers is directly incorporated in the model.

Keywords Auctions • Bidding • Day-ahead electricity markets • Day-ahead markets • Mathematical programming • Unit commitment

1 Introduction

The transformation from regulation to competition in power industries around the world have led to the development of markets for power. Day-ahead electricity markets are emerging as an important medium through which power is allocated in many de-regulated environments. A day-ahead electricity market is a short term hedge market that operates a day in advance of the actual physical delivery of power. In these environments, the generation decisions for the next day are in most cases the result of a double (two-sided) auction where producing (selling) and consuming (buying) agents submit a set of price-quantity curves (bids). The bids must be submitted by a deadline on the day before actual delivery of power.

R.H. Kwon (✉) • D. Frances
Department of Mechanical and Industrial Engineering, University of Toronto, Toronto, ON, Canada
e-mail: rkwon@mie.utoronto.ca; frances@mie.utoronto.ca

A. Sorokin et al. (eds.), *Handbook of Networks in Power Systems I*,
Energy Systems, DOI 10.1007/978-3-642-23193-3_2,

A clearing price based on the submitted bids is determined by the ISO (Independent System Operator) or market making agent and all subsequent trades are settled at this price.

A significant amount of power is allocated through day-ahead markets. For example in the Nord Pool day-ahead market in Scandinavia, Elspot, the volume of power traded in 2007 was more than 290 TWh which amounts to more than 65% of all consumption in the Nordic countries for that year [1]. Irastorza and Fraser [2] find that in the United States most electricity is traded through day-ahead markets or through bilateral forward contracts. Day-ahead markets are beneficial since they act as a short term forward market for power that in conjunction with a real-time market offers significant benefits to both producers and consumers of power through price transparency, the reduction of price uncertainty, reduction of strategic gaming, unit commitment certainty, and facilitation of demand-side (consumer) participation.

In a day-ahead market, a producer of power must decide their offer curve and a consumer must decide their bid curve. In addition, the ISO or market maker must clear the market by finding the equilibrium clearing price of the auction based on the submitted bids. Given the importance of the role of day-ahead markets in the generation and allocation of power normative models for the agents have been emerging in the literature over the last decade on optimal bid construction, unit commitment, and payment/pricing and other decisions in the context of day-ahead markets. In this paper, we give a review of the literature on the various types of optimization modeling of the power producers i.e. those agents that generate and supply power into the market.

The purpose of this chapter is to classify and characterize the emerging literature of optimization models for producers participating in day-ahead markets. Producer models considered in this paper involve a diverse set of mathematical programming approaches including non-linear integer programs, stochastic programming with recourse, and mathematical programs with equilibrium constraints.

We focus on models where producers and not the ISO have the responsibility of unit commitment decisions and so must integrate these decisions with the offer or bidding strategy. There are advantages to not having an ISO perform a centralized unit commitment as cost information of competing producers must be revealed to the ISO. Also, a centralized unit commitment in a decentralized market setting can be problematic for individual units. In particular, small cost changes in the centralized unit commitment can result in large differences to individual units/ generators [3]. Transmission and congestion issues when included are dealt within models and we do not consider separate for models for a congestion or transmission entity. We do not intend to give an exhaustive coverage of the literature, but select models that represent the current state of the art or represent the major issues in modeling of optimal producer bidding strategy in the context of day-ahead markets.

Modeling approaches for producers in the day-ahead market environment reflect the decentralized nature of the market. In this setting separate models for entities (agents) of the market are developed e.g. models for optimal producer bidding. This is in marked contrast to models for energy under the older regulated environments where for example a public utility would schedule generation and decide unit commitment without the presence of competitive bids. Most of these models are centralized cost minimization models. For a good review of this class models see Hobbs [4]. For an excellent review of general market modeling trends see Ventosa et al. [5]. In addition, see Wallace and Fleten [6] for stochastic programming models in regulated environments. Wallace and Fleten [6] discuss models in de-regulated environments e.g. day-ahead models as well but in the current paper we address and focus on day-ahead markets in a more comprehensive and detailed manner and consider other modeling approaches in addition to stochastic programming models.

2 Producer Models

The main focus of models of producers in day-ahead markets is on constructing optimal offer curves to submit to the market coordinator e.g. ISO or market maker. The producers of power in the market have a multitude of important considerations in developing offers for power in the day-ahead market. A bidding model for a producer will depend on the particular market structure e.g. auction rules and protocols. In addition, producers face uncertainty in demand for power and in many cases uncertainty in the behavior of competing power producers. In addition, in some environments the unit commitment decisions may be the responsibility of the producer and thus any bids for power must consider the cost of operating generation units as well as inter-temporal operating constraints. The integration will depend on whether an ISO or other agent decides unit commitment.

It is typical in day-ahead markets to require producers to submit supply functions (offer curves) for each of the 24 h of the day-ahead schedule. The supply functions give for each hour the price per unit of power associated with a volume of power that the producer is willing to sell at. Offer curves are typically non-decreasing. We classify the models we consider in this chapter broadly into two classes. The first class of models directly incorporates bidding behavior of competing producers. The second class does not incorporate competition directly into the model. We call the first class "Producer models with strategy" and the second class "Producer models without strategy".

3 Producer Models with Strategy

This class of models considers a producer's optimal bidding problem where important model attributes include self-scheduling, integration with unit-commitment, and incorporation of demand load and competing producer bidding behavior. We emphasize the structural aspects of all models in general but note that the incorporation of competing producers will involve important estimation techniques (these techniques are complementary to the structural development/description of models and will not be pursued in detail in this chapter.) We give detailed examples for two representative models. They differ dramatically in how competition is incorporated. The first example is a non-equilibrium approach (i.e. does not use mathematical programming formulations with equilibrium constraints, MPEC) and the second example is an MPEC model i.e. an explicit equilibrium programming approach.

3.1 Non-equilibrium Example

The first model we present is by Wen and David [7]. The modeling framework was developed for day-ahead markets in California that pre-dates the California energy crisis of 2000–2001. In the pre-crisis environment, there was an ISO that managed the grid, and a separate market maker called PX (Power Exchange), that coordinated the day-ahead market. The structure of the newly emerging post-crisis energy markets in California will include a day-ahead market, however, the ISO now will act as the market coordinator. The framework of Wen and David [7] is nevertheless instructive for environments where ISOs are not the market coordinators and producers consider the strategies of other producers and self-unit commitment in the construction of offer curves. It is assumed that the producers are thermal producers of electricity.

In this framework, producer i submits linear non-decreasing offer curve (bid price) of the form $B^{(t)}{}_{(i)}(P_{(i)}{}^{(t)}) = \alpha_i{}^{(t)} + \beta_i{}^{(t)} \times P_{(i)}{}^{(t)}$ for each hour of in the day ahead market (there are 24 h) where $\alpha_i{}^{(t)}$ and $\beta_i{}^{(t)}$ are bidding coefficients for producer i and $P_{(i)}{}^{(t)}$ is the generation output.

3.1.1 Market Maker Model (i.e. PX Model)

The market maker, PX, after receiving bids from the producers computes the market clearing price R^t and solves the following problem in a manner similar to solving a classical economic dispatch problem to compute the generation output for each producer.

$$R^t = \alpha_i{}^{(t)} + \beta_i{}^{(t)} \times P_{(i)}{}^{(t)} \quad t = 1, 2, 3, ..., 24 \tag{1}$$

$$\sum_{j=1}^{n} P_j{}^{(t)} = Q_t \quad t = 1, 2, 3, ..., 24 \tag{2}$$

$$P_{j\min}{}^{(t)} \leq P_j{}^{(t)} \leq P_{j\max}{}^{(t)} \quad j = 1, 2, , ..., n \quad t = 1, 2, 3, ..., 24 \tag{3}$$

In (1), generation is assigned to each producer so that bid prices coincide with the market clearing price for each hour. Equation 2 ensures that generation is assigned so that the load for each hour Q_t is met by total of generation from all producers. Equation 3 ensure that all generation is assigned this is between the lower and upper bounds for each producer where ${P_j}_{\min}^{(t)}$ and ${P_j}_{\max}^{(t)}$ are the respective bounds for each producer.

3.1.2 Producer Strategies

The strategy of a producer in this framework is to determine offer curves for each hour with the aim of "maximizing hourly benefit" or providing "minimum stable output". The two strategies ensure that enough generation is dispatched so that offer curves are profitable of at least enough generation output dispatched for the generator to remain in continuous operation. The producer considers an optimization model for each of these problems and then formulates a unit commitment model that that incorporates the strategies from the previous two models.

3.1.3 The Producer Hourly Benefit Model

This model seeks to find generation offers that would maximize hourly benefit given data about estimated loads and estimated bidding behavior of other bidders (power producers). The model is as follows:

$$\text{maximize } \psi^{(t)}({\alpha_i}^{(t)}, {\beta_i}^{(t)}) = R_t P_i^{(t)} - C_i({P_i}^{(t)}) \tag{4}$$
$$\text{subject to (1) to (3)}$$

where $C_i({P_i}^{(t)})$ is the cost of generation for producer i which is a function of the generation. $\psi^{(t)}({\alpha_i}^{(t)}, {\beta_i}^{(t)})$ is the hourly benefit objective function which is a measure of hourly benefit or profit. It should be noted that constraint (1) in the context of the producer model requires that the bidding coefficients of other producers to be estimated since a producer would not have access to this information directly see Wen and David [7] for details. In this approach, they use a joint probability distribution to estimate all other producers and then the hourly benefit model becomes a stochastic optimization problem.

3.1.4 Minimum Stable Output Bidding Strategy

This model aims to ensure that average output from a producer achieves near the minimum generation level. The model is as follows:

$$\text{Minimize } \varsigma^{(t)}(\alpha, \beta) = |\bar{P}_i^{(t)} - {P_{\min}}^{(t)}| + \gamma(\bar{P}_i^{(t)} - P_{\min}^{(t)})^2 \tag{5}$$

$$\text{subject to} \, P_{i\min}{}^{(t)} \leq \bar{P}_i^{(t)} \leq P_{i\max}{}^{(t)} \tag{6}$$

where γ is a positive penalty parameter.

3.1.5 Overall Producer Bidding Model

This model integrates the strategies from the hourly benefit and the minimum stable output models to determine which generators are to be on for each time period.

$$\text{Maximize} : \Omega(\mu_t) = M + \sum_{t=1}^{24} [\mu_t \psi^{(t)}(\alpha_i^{(t)}, \beta_i^{(t)}) - S(\tau)\mu_t(1 - \mu_{t-1})] \tag{7}$$

$$\text{subject to} \sum_{t=1}^{24} (\mu_t - \mu_{t-1})^2 \leq N \tag{8}$$

where the objective function in (7) is constructed to be positive (M is a sufficiently large number) in all cases since the formulation will be solved be a genetic algorithm representing the fitness of bidding strategies. μ_t is a binary variable that is 1 if a unit is up for hour t, 0 otherwise and $S(\tau)$ is the start up cost of the generator. Constraint (8) ensures that a unit has a maximum number of start ups and shutdowns in a day.

Recent papers similar to Wen and David [7] that make estimates of load forecast and competitor behavior have emerged that also incorporate reserve markets as well as the spot (day-ahead) markets. Attaviriyanupap et al. [8] consider thermal producer optimization models that incorporate self-scheduling and unit commitment based on estimates of competing producers in both spot and reserve markets. The resulting producer models are non-convex and non-differentiable and evolutionary programming heuristics are used to solve the models. An alternative approach by Swinder [9] considers the spot market to be a price-taking market but assumes that bidders behave strategically in the reserve market and that the behavior is captured in a joint probability distribution. A simultaneous bidding model is developed that is a stochastic non-linear profit maximization model.

Other models that incorporate competitor behavior and load estimates include Zhang et al. [10] in which they consider a Lagrangian relaxation-based approach after obtaining a closed form solution for the ISO problem. A Lagrangian relaxation approach is also adopted in Gross and Finlay [11] where the load forecast and competitor behavior is incorporated. An ordinal optimization approach is presented in Guan et al. [12] where the idea is to generate good enough solutions. Anderson and Philpott [13] consider the more general problem of a producer that makes offers into a wholesale power market for which the prices are determined by a sealed-bid auction (this case would be applicable to day-ahead markets). Demand and behavior of competing producers are represented in a probability distribution for which models are then defined and necessary optimality conditions are derived for these models.

3.2 Equilibrium Approaches for Producer Models

Next, we consider an MPEC (mathematical program with equilibrium constraints) model by Bakirtzis et al. [14] for a generator's offering strategy with step-wise offers. An MPEC model is a non-linear programming problem in which there are constraints defined by a parametric variational inequality or complementarity system [15]. The MPEC framework is useful for modeling strategic interaction that follows a Stackleberg game [16]. In this model, a "leader" producer can affect the market price and estimates the demand declarations as well as the supply functions of competing producers to be used in the leader's optimal offering strategy problem. The competing producers are the "followers" in the sense that the "leader" producer faces a residual demand function of the aggregate of the competing producers. An MPEC generator model will have an outer problem of maximizing profit given that the ISO will solve an economic dispatch problem (this latter problem is the inner problem). The ISO problem (given supply functions from all generators) is to minimize revealed cost while determining the dispatched quantity for each generator. In the model of Bakirtzis et al. [14], a producer j constructs optimal energy offers in the form of non-decreasing step-wise bids i.e. selects a numberB^j of steps and a set of quantity-price pairs (Q_{jb}, π_{jb}) for each step $b \in \{1, ..., B^j\}$ where Q_{jb} is the quantity offered by producer j for block b and π_{jb} is the corresponding offer price.

The deterministic ISO problem is of the following form:

$$Minimize \sum_j \sum_b \pi_{jb} \cdot q_{jb} \tag{9}$$

subject to (10) to (12)

The objective (9) is to minimize the cost of dispatched energy with respect to the revealed price of energy where q_{jb} is the dispatched power for unit j step b. The constraints are defined as follows:

$$\sum_j \sum_b q_{jb} = d \tag{10}$$

$$q_{jb} \leq Q_{jb} \text{ for all } j \in A, b \in B^j \tag{11}$$

$$q_{jb} \geq 0 \text{ for all } j \in A, b \in B^j \tag{12}$$

Constraint (10) ensures that demand d is met through all dispatched power and (11) ensures that accepted power does not exceed the offered volume (i.e. a producer is never dispatched more power then bid for).

The producer model assumes a 1 h horizon and pertains to a producer with a set of thermal units$A$ and is defined as follows:

$$Maximize \sum_{s\in S}\left\{p^s \sum_{j\in A}\left[\sum_{b\in B^j}\lambda^s q^s_{jb} - c_j\left(\sum_{b\in B} q^s_{jb}\right)\right]\right\} \tag{13}$$

subject to (14) to (28)

The objective (13) is to maximize the expected profit where the system marginal price for power in scenario s is λ^s and q^s_{jb} is the quantity of step b of unit j offer accepted by the ISO in scenario s. $c_j(\bullet)$ is the hourly non-linear cost of unit j as a function of generation level.

3.2.1 Producer Constraints

$$0 \le Q_{jb} \le Q^{\max}_j \text{ for all } j \in A, b \in B^j \tag{14}$$

$$\sum_b Q_{jb} = Q^{\max}_j \text{ for all } j \in A \tag{15}$$

$$Q_{jq} \le \delta \cdot Q^{\max}_j \text{ for all } j \in A \tag{16}$$

Constraints (14) ensure that any offer is not greater than the capacity of a unit j and (15) ensures that all of the capacity of a unit j is offered. Constraint (16) ensures that the first step of an offer is a fraction of the available capacity of a unit j (as required by Greek power markets).

$$0 \le \pi_{jb} \le \pi^{\max} \text{ for all } j \in A, b \in B^j \tag{17}$$

$$\pi_{jb} \le \pi_{j(b+1)} \text{ for all } j \in A, b = 1, ..., B^j - 1 \tag{18}$$

$$\pi_{jb} \ge MVC_j \text{ for all } j \in A, b = 1, ..., B^j \tag{19}$$

$$\sum_{b\in B^j} Q_{jb} \cdot \pi_{jb} \ge MVC_j \cdot \sum_{b\in B^j} Q_{jb} \text{ for all } j \in A \tag{20}$$

Constraint (17) places a price cap of all price offers and (18) ensures the offer prices are non-decreasing. Constraints (19) and (20) ensure that offer prices are at least the minimum variable cost, MVC_j, for a unit j.

3.2.2 ISO Market Clearing Problem

$$\sum_{j,b} q^s_{jb} = d^s \text{ for all } s \in S \tag{21}$$

$$q^s_{jb} \leq Q_{jb} \text{ for all } j \in A, b \in B^j, s \in S \tag{22}$$

$$q^s_{jb} \leq \bar{Q}^s_{jb} \text{ for all } j \in \bar{A}, b \in B^j, s \in S \tag{23}$$

$$q^s_{jb} \geq 0 \text{ for all } j \in J, b \in B^j, s \in S \tag{24}$$

$$\mu^s_{jb} \geq 0 \text{ for all } j \in J, b \in B^j, s \in S \tag{25}$$

$$\pi_{jb} + \mu^s_{jb} \geq \lambda^s \text{ for all } j \in A, b \in B^j, s \in S \tag{26}$$

$$\bar{\pi}_{jb} + \mu^s_{jb} \geq \lambda^s \text{ for all } j \in \bar{A}, b \in B^j, s \in S \tag{27}$$

$$\sum_{j\in A, b\in B^j} \pi_{jb} q^s_{jb} + \sum_{j\in A, b\in B^j} \mu^s_{jb} Q_{jb} + \sum_{j\in \bar{A}, b\in B^j} \bar{\pi}^s_{jb} q^s_{jb} + \sum_{j\in \bar{A}, b\in B^j} \mu^s_{jb} \bar{Q}^s_{jb} = \lambda^s d^s \text{ for all } s \in S \tag{28}$$

Constraints (21) to (28) represent the expected value (scenario-based) ISO problem via the Karush-Kuhn-Tucker conditions whereλ^s is the dual multiplier associated with demand balance constraint (10) for scenario s and μ^s_{jb} is the dual multiplier for the constraint (11) for scenarios s.

It should be noted that the producer MPEC formulation is converted to a mixed integer linear program by a binary expansion of the offer prices and quantities using the techniques in Pereira et al. [17]. Other bi-level level mathematical programming-based approaches which involve an outer and an inner problem corresponding to producer and ISO, respectively, include Gountis et al. (2004) [29] where each producer submits a linear supply curve and estimates the competing producers and demand behavior using Monte Carlo approaches. In addition, risk aversion is modeled and is seen to have an impact on optimal offer strategies. Earlier bi-level model is given in Weber and Overbye [18].

4 Producer Models Without Strategy

In this section, we consider producer models where there is no explicit incorporation of competing producers in the models. We detail two models. The first is a deterministic mixed integer linear programming modeling framework and the second a stochastic integer programming model. Both models are price-taking in the sense that the model assumes that the producer can not impact the market clearing price.

4.1 Mixed Integer Linear Programming Model

The first modeling framework is by Conejo et al. [19] for constructing offer curves for price-taking thermal producers (producers). In this approach, hourly prices are forecast and a self-schedule is obtained by an optimization model. The framework is for general pool type electricity markets. Generators submit a bid for each hour in the day-ahead time frame and consists a set of blocks of power along with corresponding unit prices. Then, hourly offer curves are constructed based on the optimal self-schedule by the use of a simple bidding strategy.

The producer (price-taking) optimization model is

$$\text{Maximize}\, E_{\lambda_1,\ldots,\lambda_T}\left\{\sum_{t=1}^{T}\lambda_t p_t\right\} - \sum_{t=1}^{T} c_t \tag{29}$$

$$\text{subject to}\, p_t \in \Pi \tag{30}$$

where λ_t are the random prices of power for hour t (the distributions are approximately lognormal), c_t is the (non-linear) cost of generating for hour t (see Conejo et al. [19] for details), p_t is the power generation for hour t, and Π is the set of feasible generation outputs that satisfy operational constraints with respect to minimum up and down times, ramping, and power output limits. Thus, the model seeks to maximize optimal expected profit subject to operating constraints. This model is seen to be equivalent to a mixed integer program (MIP), where Π is defined by the following constraints:

$$p_t \geq \underline{P}\, v(t) \text{ for all } t = 1, \ldots, T \tag{31}$$

$$p_t \leq \bar{P}\,[v(t) - z(t+1)] + z(t+1)SD \text{ for all } t = 1, \ldots, T \tag{32}$$

these constraints ensure the lower and upper limits on power production $(z(t)(v(t))$is a binary variable which is equal to 1 is the generator is shut down (on-line) at the start of hour t, SD is the shut down ramp rate limit in MW/h.)

$$p_t \leq p_{t-1} + RUv(t-1) + SUy(t) \text{ for all } t = 1, \ldots, T \tag{33}$$

$$p_{t-1} - p_t + RDv(t) + SDz(t) \text{ for all } t = 1, \ldots, T \tag{34}$$

these constraints ensure that ramp rates are obeyed (where RU (RD) is the ramp-up (ramp-down) rate limit and SD is the shut down ramp rate limit)

$$\sum_{t=1}^{G}[1 - v(t)] = 0 \tag{35}$$

$$\sum_{j=t}^{t+UT-1} v(j) \geq UTy(t) \quad t = G+1, \ldots, T - UT + 1 \tag{36}$$

$$\sum_{j=t}^{T}[v(j) - y(t)] \geq 0 \quad t = T - UT + 2, \ldots, T \tag{37}$$

$$\sum_{t=1}^{F} v(t) = 0 \tag{38}$$

$$\sum_{j=t}^{t+DT-1} [1 - v(j)] \geq DTz(t) \quad t = F + 1, \ldots, T - DT + 1 \tag{39}$$

$$\sum_{j=t}^{T}[1 - v(j) - z(t)] \geq 0 \quad t = T - DT + 2, \ldots, T \tag{40}$$

Constraints (35) to (40) ensure that the minimum up and down times of a generator are satisfied where DT (UT) is the minimum downtime (uptime) and F (G) is the required number of time intervals that a generating unit must be off-line (on-line) because of downtime (uptime) constraints.

$$y(t) - z(t) = v(t) - v(t-1) \quad t = 1, \ldots, T \tag{41}$$

$$y(t) + z(t) \leq 1 \quad t = 1, \ldots, T \tag{42}$$

$$z(t) \in \{0, 1\} \quad t = 1, \ldots, T \tag{43}$$

Constraints (41) to (43) ensure the correct logical relationship between a generator' shut-down, start-up, and running states.

The offer curve for a generator is obtained by solving the MIP model to obtain the optimal production schedulep_t^* for each of the 24 h in the day-ahead time frame. Then the bidding strategy to realizep_t^* for hour t is summarized as follows:

1. If $p_t^* = 0$ then offer curve will consist of a single block of power equivalent to maximum capacity of a thermal generator, $\bar{P}$, at unit price of $\lambda_t^{est} + a_t\sigma_t^{est}$
2. If $0 < p_t^* < \bar{P}$, then the offer curve will consist of two blocks of power p_t^* and $\bar{P} - p_t^*$ at prices $\lambda_t^{est} - b_t\sigma_t^{est}$ and $\lambda_t^{est} + a_t\sigma_t^{est}$, respectively.
3. If $p_t^* = \bar{P}$, then the offer curve will be a block of power $p_t^* = \bar{P}$ at a price per unit of $\lambda_t^{est} - b_t\sigma_t^{est}$.

where λ_t^{est} is the expected value of λ_t and σ_t^{est} the estimate of the standard deviation of λ_t. The prices (i.e. coefficients a_t and b_t) are chosen so that the offer curves guarantee with a level of 99% confidence that the power accepted by the market maker is the specified power amount in the offer curve.

Other price–taking producer models without strategy include Ladurantaye et al. [20] where a bidding model for a price-taking hydro-producer is formulated. Gonzalez et al. [21] consider a profit-based hydro-producer model for day-ahead markets. Risk aversion is incorporated into the model via the conditional Value at Risk (CVaR) measure [22]. Scenarios are generated for market prices via hidden markov models. Also considered are minimum profit-based models. The resulting formulations are mixed integer linear programs. Conejo et al. [23], Yamin et al. (2004) [31] and Dicorato et al. [24] also incorporate risk aversion into producer bidding models where the first two consider as a measure of risk the variance of market clearing price and the latter the CVaR measure.

4.2 Stochastic Programming Model

Stochastic programming has been emerging as an important modeling framework for problems arising in the energy sector such as hydro and thermal scheduling [6], unit commitment (Takriti et al. 1996) [30] and structuring energy forward contracts [25] among many other classes of problems. Uncertainty in deregulated energy markets often takes the form of uncertainty in spot prices and uncertainty in weather which relates to uncertainty in demand (load). For an excellent survey of the use of stochastic programming in both regulated and deregulated energy markets see Wallace and Fleten [6].

We detail the model of Fleten and Kristoffersen [26] where a two-stage stochastic programming model is developed for optimizing the offer strategy for a Nordic hydropower producer for the day-ahead market in the Nord Pool, the Elspot. The hydropower producer is assumed to be a price-taker and the model incorporates price uncertainty. Another stochastic programming model in the same spirit of Fleten and Kristoffersen [26] is given in Ladurantaye et al. [20]. The rationale for stochastic programming with recourse stems from the fact that the clearing price cannot be known before dispatch making it a challenge to commit generators before the actual trading price is known. In particular, a decision (offer curve) must be made now before uncertainty of price is resolved. Ideally, the "here and now" decision should reflect the possible *recourse* (corrective actions) required after price uncertainty has resolved e.g. the actual production decisions occurring in the future based on realized prices at that time. Furthermore, the "here and now" decision should be constructed as to minimize the costs of recourse actions over all possible random outcomes. Stochastic programming with recourse is a natural framework for this situation. The stochastic programming model incorporates a finite set of scenarios that capture different price possibilities to generate offers that are robust to price uncertainty. There are two major decision stages where the first stage involves decisions related to day-ahead bid construction i.e. offer volume and the second stage "*recourse*" decisions reflect production of power e.g. hydro power production and dispatch subject to operating and balance constraints. The model assumes a simple hydro-plant with two reservoirs in a cascade, one larger upper reservoir and one smaller reservoir downstream.

4.2.1 Bid Structure

A hydro-producer can submit hourly bids, block bids, of flexible hourly bids. An hourly bid has the form (x_{it}, p_i) where the first component is the amount of power offered (i.e. bid volume) by a producer for hour t and the second component is the associated unit price of power (the i indexes a finite set of predetermined prices). These bids are seen to be points on a bidding curve constructed by making an linear interpolation between the points. For example, the unit price ρ_t for hour t associated with a volume y_t is given by

$$\rho_t = p_{i-1} + \frac{p_2 - p_1}{x_{2t} - x_{1t}}(y_t - x_{1t}) \text{ for } x_{i-1,t} \leq y_t \leq x_{it}. \tag{44}$$

4.2.2 Two-stage Stochastic Programming Model

First Stage decisions: x_{it} (x_{ib}) represents offer volume for hour t (block b)

Second Stage decisions: (Scenario dependent decisions) $y_t^s(y_b^s)$ volume dispatched for hourly (block) bids, $z_t^{+,s}(z_t^{-,s})$ is the positive (negative) imbalance between volume dispatched and volume produced, l_{jt}^s is storage level for reservoir j for time period t, μ_{jt}^s is the on or off state of a generator, and v_{jt}^s is the volume of discharge from generator j for time period t. (Scenario independent decisions) w_{jt} is the generation level for generator j for time period t, and l_{jt} is the storage level for reservoir j for time period t.

Parameters: π^s is the probability of occurrence of scenario s, ρ_t^s is the unit price of power for hour t in scenario s, $\overline{\rho_b^s}$ is the average unit price of power in block b in scenario s, $\mu_t^+(\mu_t^-)$ are the penalty (reward) for power imbalances.

$$\begin{aligned} Maximize \sum_{s \in S} \pi^s \Bigg(& \sum_{t \in S} \rho_t^s y_t^s + \sum_{b \in B} \overline{\rho_b^s} y_b^s - \sum_{t \in T} (\mu_t^+ z_t^{+,s} - \mu_t^- z_t^{-,s}) \\ & - \sum_{j \in T} (V_j(l_{j0}^s) - V_j(l_{jT}^s)) - \sum_{t \in T} \sum_{j \in J} S_j(\mu_{jt-1}^s, \mu_{jt}^s) \Bigg) \end{aligned} \tag{45}$$

subject to (46) to (59)

The objective (45) is to maximize the expected profit from offers and power production where the function $V_j(\bullet)$ is such that $\sum_{j \in T} (V_j(l_{j0}^s) - V_j(l_{jT}^s))$ gives the opportunity costs of storing water for generator j and $S_j(\mu^s_{jt-1}, \mu^s_{jt})$ is the direct cost function of starting up a generator (the functions are defined to so that the model will correspond to a linear MIP).

$$y^s_t = \frac{\rho^s_t - \rho_{i(t,s)}}{\rho_{i(t,s)+1} - \rho_{i(t,s)}} x_{i(t,s)+1t} + \frac{\rho_{i(t,s)+1} - \rho^s_t}{\rho_{i(t,s)+1} - \rho_{i(t,s)}} x_{i(t,s)t} \text{ for all } t \in T, s \in S \quad (46)$$

Constraint (46) gives the representation of the actual dispatch in hour t under scenario s implied by the piecewise linear offer curve (44) where $i(t,s) = \max\{i \in I : p_i \le \rho^s_t\}$.

$$x_{it} \le x_{i+1t} \text{ for } i \in \mathrm{I} \backslash I,\ b \in B \quad (47)$$

$$y^s_b = \sum_{j \le i(b,s)} x_{jb} \text{ for } b \in B,\ s \in S \quad (48)$$

Constraint (47) ensures that bid (offer) curve is non-decreasing, and (48) relates the actual power dispatched for block bids to offer volumes under all scenarios.

$$w^s_{1t} = \mu^s_{1t} w^{\max}_1 \text{ for } t \in \mathrm{T} \text{ and } s \in S \quad (49)$$

$$\mu^s_{2t} w^{\min}_2 \le w^s_{2t} \le \mu^s_{2t} w^{\max}_2 \text{ for } t \in \mathrm{T}, s \in S \quad (50)$$

$$v^{\min}_j \le v^s_{jt} \le v^{\max}_j \text{ for } j \in \mathrm{J}, t \in \mathrm{T}, s \in S \quad (51)$$

$$l^{\min}_j \le l^s_{jt} \le l^{\max}_j \text{ for } j \in \mathrm{J}, t \in \mathrm{T}, s \in S \quad (52)$$

Constraints (49) to (52) impose water discharge bounds e.g. (49) enforces that the maximum amount of water is discharged or no water is discharged whereas (50) allows the second reservoir to discharge any amount between specified upper and lower bounds.

$$l^s_{1t} - l^s_{1t-1} + v^s_{1t} + r^s_{1t} = v^s_{1t} \text{ for } t \in \mathrm{T}, s \in S \quad (53)$$

$$l^s_{2t} - l^s_{2t-1} + v^s_{2t} + r^s_{2t} = v^s_{1t-\tau} \text{ for } t \in \mathrm{T}, s \in S \quad (54)$$

Constraints (53) to (54) are the reservoir balance equations.

$$w^s_{jt} = \gamma_j v^s_{jt} \text{ for } j \in \mathrm{J}, t \in \mathrm{T}, s \in S \quad (55)$$

Constraint (55) gives the generation efficiency of reservoirs.

$$y^s_t + \sum_{b \in B : t \in b} y^s_b - \sum_{j \in J} w^s_{jt} = z^{+,s}_t - z^{-,s}_t \text{ for } t \in \mathrm{T}, s \in S \quad (56)$$

Constraint (56) measures the imbalances between volumes produced and volumes dispatched.

$$x_{it}, x_{ib} \in R_{+} \text{ for } i \in \mathrm{I}, t \in \mathrm{T}, b \in \mathrm{B} \tag{57}$$

$$\mu_{jt}^{s} = 0 \text{ or } 1 \tag{58}$$

$$y_{t}^{s}, y_{b}^{s}, z_{t}^{+,s}, z_{t}^{-,s}, v_{jt}^{s}, w_{jt}^{s}, l_{jt}^{s} \in R_{+} \text{ for } j \in \mathrm{J}, t \in \mathrm{T},\ b \in \mathrm{B}, s \in S \tag{59}$$

The stochastic programming model above has been extended by Faria and Fleten [1] to incorporate the adjustment market called the Elbas in NordPool. The Elbas market allows adjustment in accepted bids up to 1 h before scheduled dispatch. The rationale is that accepted offers are made before prices, loads, and inflow are known so after realization of these uncertainties a rebalancing or recourse should occur. It is found, however, that the incorporation of the Elbas does not significantly change bidding. Nowak et al. [27] consider a stochastic integer program for incorporating day-ahead trading in hydro-thermal unit commitment decisions made for a week ahead for a German power utility. The main sources of uncertainty are the bids made by competitors. The stochastic model is fully linear which allows a Lagrangian-based branch and bound procedure to be applied.

5 Discussion of Model Features

Common to most models that incorporate strategy is the need for estimation of demand load and competitor bidding behavior while models without competitive behavior need estimation of market clearing prices. In addition, Table 1 lists additional features (as done in a manner analogous to Ventosa et al. [5]) for models that incorporate strategy and Table 2 lists features for those models without strategy. Besides the classification of a model as one with or without strategy and author names, the features in Tables 1 and 2 include the type of optimization model e.g. integer program, solution methodology, features particular to a specific model, and intended market.

Almost all day-ahead producer models are seen to be non-convex and non-differentiable e.g. mixed-integer programming models with the exception of Dicorato et al. [24], where the unit commitment decisions are assumed to have been made *ex ante*. Thus, the computational tractability is an issue for most day-ahead producer models as bidding and unit commitment (which alone is a difficult problem to solve) are combined and is dealt with using a variety of methods as seen in Tables 1 and 2. The most common techniques involve commercial branch and bound solvers, Lagrangian relaxation, and evolutionary heuristics such as genetic algorithms. Some models deal with the complexity of the producer optimization by decoupling and solving separately the bidding strategy and scheduling of power

Table 1 Models with strategy

Strategic models	Year	Model-type/solution technique	Features	Intended market
Wen and David	2001	Mixed-integer program/ genetic algorithms	Integrates two bidding strategies: (1) hourly benefit and (2) minimum stable output	California (pre-crisis i.e. before 2000)
Attaviriyanupap et al.	2005	Mixed-integer program/ evolutionary heuristics	Power and reserve markets are incorporated	
Swider	2007	Stochastic non-linear optimization	Power markets assumed to be price-taking; strategic behavior in reserve markets	Germany
Zhang et al.	2000	Mixed-integer programming/ Lagrangian relaxation	ISO problem is analytically solved	New England
Gross and Finlay	2000	Mixed-integer programming/ Lagrangian relaxation	Analytic solution under perfect competition	England and Wales
Guan et al.	2001	Mixed-integer program/ Lagrangian relaxation	Approximate solutions obtained via ordinal optimization theory	California
Bakirtzis et al.	2007	Mathematical program with equilibrium constraints (MPEC)/mixed integer programming	MPEC model is converted into a mixed integer program	Greece
Gountis et al.	2004	Bi-level program/genetic algorithms	Incorporates risk aversion and Monte Carlo simulation is used to compute expected profit	
Weber and Overbye	1999	Bi-level program	Transmission constraints are incorporated	

producing units with subsequent integration of these sets of decisions e.g. Wen and David [7] and Conejo et al. [19]. It is also seen that most producer models that include risk aversion are in models that have price-taking assumptions i.e. models without strategy.

6 Conclusion

Some representative models for producers (producers) for power in day-ahead markets were given in this chapter. The models have spanned across price-taking and non-price-taking assumptions. The primary focus has been on models that emphasize self-scheduling by a producer i.e. models that integrate offer decisions with unit commitment decisions. These models take of the form of non-linear

Table 2 Models without strategy

Non-strategic models	Year	Model-type/solution technique	Features	Intended market
Conejo et al.	2002	Mixed-integer program/branch and bound	Derives bidding strategy that achieves optimal self-schedule; requires estimation of day ahead hourly prices (probability distribution)	Spain/general pool type markets
Ladurantaye et al.	2007	Stochastic program/ successive linear programming	Integrates bidding with hydro-electric production	
Gonzalez et al.	2007	Mixed-integer linear program/under relaxation with branch and bound	Risk aversion is incorporated through the conditional value at risk (CVaR) measure	
Conejo et al.	2004	Mixed-integer quadratic program/branch and bound	Risk averse version of [19] where variance of market clearing price is a measure of risk	Spain/general pool type market
Yamin et al.	2004	Mixed-integer quadratic program/ Lagrangian relaxation	Risk aversion incorporated in self-scheduling where variance of market clearing price is a measure of risk	
Dicorato et al.	2009	Convex optimization model	Risk aversion is incorporated by using the CVaR measure. Hydro-electric and thermal units are considered.	
Fleten and Kristoffersen	2007	Stochastic integer program/ branch and bound	Integrates bidding with hydro-electric production	Nord Pool
Faria and Fleten	2009	Stochastic integer program/branch and bound	Similar to Fleten et al. 2007 but with incorporation of reserve markets	Nord Pool
Nowak et al.	2005	Stochastic integer program/ Lagrangian-based branch and bound	Incorporating bidding into hydro-thermal unit commitment. Main source of uncertainty is bids by competitors	Germany

integer programs, MPECs, and stochastic programming with recourse models. An important facet of many of the models is the incorporation of demand load and competitor behavior estimates. In addition, the incorporation of risk aversion into producer models is emerging and will continue to be an important development as preferences and utilities of producing agents are in general not the same. A universal assumption of the models was that the clearing mechanism by an ISO was performed as a single round auction, thus a produce model would need to be solved only once given all the relevant input parameters. An interesting future development will be day-ahead markets that have multiple round auction formats [28]. The benefits of such an auction would be in the information provided by results of a single round of the auction which could then be used to improve bidding for the producers in subsequent auctions. In such a case, a producer would have to repeatedly solve offer models based on updated results for a round of the auction.

References

1. Faria E, Fleten S-E (2009) Day-ahead market bidding for a Nordic hydropower producer: taking the Elbas market into account, Comput Manage Sci. doi:10.1007/s10287- 009-0108-5, 8:75–101
2. Irastorza V, Fraser H (2002) Are ITP-run day-ahead markets needed? *Electricity J* **15**(9):25–33
3. Oren SS, Svoboda AJ, Johnson RB (1997) Volatility of unit commitment in competitive electricity markets. In: Proceedings of the thirtieth Hawaii international conference on system sciences, Maui, vol 5, pp 594–601
4. Hobbs BF (1985) Optimization methods for electric utility resource planning. *Eur J Oper Res* **83**(1):1–20
5. Ventosa M, Baillo A, Ramos A, Rivier M (2005) Electricity market modeling trends. *Energ Policy* **33**:897–913
6. Wallace SW, Fleten S-E (2003) Stochastic programming models in energy. In: Ruszczynski A, Shapiro A (eds) Handbooks in OR&MS, vol 10. Elsevier Science, Amsterdam, pp 637–677
7. Wen FS, David AK (2001) Strategic bidding for electricity supply in a day-ahead energy market. *Electrical Power Syst Res* **59**:197–206
8. Attaviriyanupap P, Kita H, Tanaka E, Hasegawa J (2005) New bidding strategy formulation for day-ahead energy and reserve markets based on evolutionary programming. *Electrical Power Energy Syst* **27**:157–167
9. Swinder DJ (2007) Simultaneous bidding in day-ahead auctions for spot energy and power systems reserve. *Electrical Power Energy Syst* **29**:470–479
10. Zhang D, Wang Y, Luh PB (2000) Optimization based bidding strategies in the deregulated market. *IEEE Trans Power Syst* **15**(3):981–986
11. Gross G, Finlay D (2000) Generation supply bidding in perfectly competitive electricity markets. *Comput Math Organ Theory* **6**(1):83–98
12. Guan X, Ho Y-C, Lai F (2001) An ordinal optimization based bidding strategy for electric power producers in the daily energy market. *IEEE Trans Power Syst* **16** (4):788–797
13. Anderson EJ, Philpott AB (2002) Optimal offer construction in electricity markets. *Math Oper Res* **27**(1):82–100
14. Bakirtzis AG, Ziogos NP, Tellidou AC, Bakirtzis GA (2007) Electricity producer offering strategies in day-ahead energy market with step-wise offers. *IEEE Trans Power Syst* **22**(4):1804–1818
15. Lou ZQ, Pang JS, Ralph D (1996) Mathematical programming with equilibrium constraints. Cambridge University Press, New York
16. Hobbs BF, Metzler CB, Pang J-S (2000) Strategic gaming analysis for electric power systems: an MPEC approach. *IEEE Trans Power Syst* **15**(2):638–645
17. Pereira MV, Granville S, Fampa MC, Dix R, Barraso RA (2005) Strategic bidding under uncertainty. *IEEE Trans Power Syst* **20**(1):180–188
18. Weber JD, Overbye TJ (1999) A two-level optimization problem for analysis of market bidding strategies, In: Proceedings of the IEEE power engineering society summer meeting, vol 2. Edmonton, pp 1845–1849
19. Conejo AJ, Nogales FJ, Arroyo JM (2002) Price-taker bidding strategy under price uncertainty. *IEEE Trans Power Syst* **17**(4):1081–1087
20. De Ladurantaye D, Gendreau M, Potvin J-Y (2007) Strategic bidding for price-taking hydroelectricity producers. *IEEE Trans Power Syst* **22**(4):2187–2203
21. Gonzalez JG, Parrilla E, Mateo A (2007) Risk-averse profit – based optimal scheduling of a hydro-chain in the day-ahead electricity market. *Eur J Oper Res* 181:1354–1369
22. Rockafellar R, Uryasev S (2000) Optimization of conditional value-at risk. J Risk 2(3):21–41
23. Conejo AJ, Nogales FJ, Arroyo JM, Garcia-Bertrand R (2004) Risk-constrained self-scheduling of a thermal power producer. *IEEE Trans Power Syst* **19**(3):1569–1574
24. Dicorato M, Forte G, Trovato M, Caruso E (2009) Risk – constrained profit maximization in day-ahead electricity market. *IEEE Trans Power Syst* **24**(3):1107–1114

25. Kwon RH, Rogers JS, Yau S (2006) Stochastic programming models for replication of electricity forward contracts for industry. *Nav Res Logistics* **53**(7):713–726
26. Fleten S-E, Kristoffersen TK (2007) Stochastic programming for optimizing bidding strategies of a Nordic hydropower producer. *Eur J Oper Res* 181:916–928
27. Nowak MP, Schultz R, Westphalen M (2005) A stochastic integer programming model for incorporating day-ahead trading of electricity into hydro-thermal unit commitment. *Optimization Eng* 6:163–176
28. Contreras J, Candiles O, de la Fuente JI, Gomez T (2001) Auction design in day-ahead electricity markets. *IEEE Trans Power Syst* **16**(3):409–417
29. Gountis VP, Bakirtzis AG (2004) Bidding strategies for electricity producers in a competitive electricity marketplace. *IEEE Trans Power Syst* **19**(1):356–356
30. Takriti S, Birge JR (1996) A stochastic model for the unit commitment problem. *IEEE Trans Power Syst* **11**(3):1497–1506
31. Yamin HY, Shahidehpour M (2004) Risk and profit in self – scheduling for GenCos. *IEEE Trans Power Syst* **19**(4):2104–2106

Finding Joint Bidding Strategies for Day-Ahead Electricity and Related Markets

Patricio Rocha and Tapas K. Das

Abstract In restructured electricity markets, generators and other market participants submit bids to the system operator, who dispatches power while satisfying the system constraints. Dispatch establishes the market clearing price for electricity at each node of the network. In recent years, power market participants have begun competing in electricity-related markets for financial instruments such as financial transmission rights (FTRs) and CO_2 allowances. Benefits derived from these markets depend largely on the electricity dispatch. For example, payments received by FTR holders are determined by the differential market clearing price between the nodes; CO_2 allowances needed by the generators depend on the amount of fossil fuel-based electricity dispatched in the network. Hence, the participants must develop bidding strategies that maximize their joint profits from electricity and other related markets. This chapter, presents a multi-tier game theoretic framework that can be used to develop joint bidding strategies. In the electricity market, we focus on day-ahead and spot market bidding. Though there are many market participants (generators, loads, and third parties), the framework presented in this chapter caters directly to the needs of the generators. The multi-player matrix games underlying the framework are solved using an approach that incorporates a reinforcement learning algorithm. Application of the framework is exemplified via three example problems.

Keywords Bidding strategies • Cap and trade • CO_2 allowance • FTRs • Matrix games • Restructured electricity markets

P. Rocha
PJM Interconnection, 955 Jefferson Avenue, Norristown, PA 19403
e-mail: rochap@pjm.com

T.K. Das (✉)
University of South Florida, 4202 E. Fowler Avenue, Tampa, FL 33620, USA
e-mail: das@usf.edu

A. Sorokin et al. (eds.), *Handbook of Networks in Power Systems I*,
Energy Systems, DOI 10.1007/978-3-642-23193-3_3,

1 Introduction

Deregulation of the power industry began in the U.K. during the 1980s motivated by the success of deregulation in the airline and telecommunications industries. Since then, several countries and regions, such as some states in the U.S., the Nordic countries, Chile, Argentina, parts of Australia, among others, have introduced reforms to deregulate/restructure their respective electricity market industry. As to achieving the desired objectives of deregulation, which includes lowering consumer prices and increasing competition, the results are mixed. Joskow in [18] identifies *lessons learned* from the implementation of deregulation in electricity markets around the globe. Deregulation uncertainties have caused many States in the U.S. to put a hold in their restructuring processes [12] even though some of the deregulated power markets (e.g., Pennsylvania–New Jersey–Maryland market, PJM) seem to be working satisfactorily [14, 34]. The market designs are still evolving, and it is imperative for the designers to have the tools to assess the impact of the new market rules on market performance. A survey of critical research areas in deregulated electricity markets is presented in [28].

Restructuring of the power industry involves the functional unbundling of vertically integrated monopolies into horizontally integrated power generators having an open access to the transmission grid. The restructured power markets can have two broad market structures: a purely POOLCO type (i.e., a privately-owned marketplace for energy auctions, spot market sales, and derivatives trading) and a purely bilateral type [6]. In a POOLCO type structure, all the generators supply their price-quantity offers to the independent system operator (ISO). The retailers also submit their price sensitive demand bids to the ISO. The functions of the ISO include: matching supply and demand every minute, maintaining the transmission grid, managing congestion, and determining the nodal price and quantity of power using a merit order dispatch. A description of the POOLCO market structure can be found in [7, 13]. In a pure bilateral type structure, the generators enter into long-term contracts with the retailers and negotiate prices among themselves. The generators supply the contract information to the ISO. Here, the role of the ISO is limited to keeping the records of the contracts and managing the transmission grid. Most of the current restructured power markets operate as some combination of POOLCO and bilateral types.

Deregulated electricity markets, such as PJM, consists of two types of markets: day-ahead and real-time. The former is a forward market in which hourly prices are calculated for the next operating day based on generation offers, demand bids, and scheduled bilateral transactions. The latter is a spot market in which current prices are calculated at 5-min intervals based on actual grid operating conditions [34]. Power generators face the challenge of designing a trading portfolio of bilateral and spot market sales to maximize profit and minimize risk.

In addition to day-ahead and real-time electricity markets, there are other markets in which the power generators also participate.

- Reserve Market: used to procure supplemental power usually on a day-ahead basis to hedge against unanticipated conditions disrupting system reliability.
- Capacity Market: designed to create long-term price signals to attract needed generation investments in insuring system reliability. These investments can be for maintaining existing generation and also for encouraging new capacity development.
- Ancillary Markets: support the reliability of the transmission network when disruptions, such as unexpected demand hikes, occur on short notice.
- Financial Transmission Rights (FTR) Market: used by the market participants to hedge their price risk when delivering energy on the grid and by the system operator to redistribute excess revenue collected caused by network congestion. FTR revenues depend on the differential locational marginal prices observed in the power network. Thus, FTR and electricity markets are interdependent.
- Allowance Market: The introduction of environmental regulations (e.g., CO_2 cap-and-trade) in some regions of the U.S. has triggered the creation of allowance (pollution permits) markets in which electricity generators and other market entities compete. The CO_2 allowance market, though not under control of the system operator, directly impacts the operations of the power generators in the electricity market. An example of an operational market for CO_2 allowances is the Regional Greenhouse Gas Initiative (RGGI) [36] in the northeast U.S.

The electricity market participants (generators, loads, etc.) aim at maximizing the combined profit from the above markets. Due to the complexities of each market and the strategic interdependencies among them, the problem of finding the optimal joint bidding strategies is non-trivial. In this chapter, our objective is to provide a game theoretic modeling framework to find joint bidding strategies for deregulated electricity and related markets. In the electricity market, we focus on day-ahead and spot electricity markets. We do not consider long-term electricity forward contracts. Among the related markets, we consider only the FTR and CO_2 allowance markets. A game theoretic model for joint bidding strategies in electricity and FTR markets was presented in our recent paper [3]. Here we provide a summary of the game theoretic model and a modified version of the example problem presented in [3]. Modifications include the use of more realistic assumptions on how bidding alternatives of the participants are developed, and the presentation of insights into the bidding behavior drawn from the numerical results. We then accommodate the modeling framework to study a joint market for electricity and CO_2 allowances. This is a new contribution to the literature.

In the next section, we describe the bidding processes in the electricity, FTR, and allowance markets as well as the methods for settlement of each market. In Sect. 3 we present the game theoretic model. Its solution algorithm is presented in Sect. 4. The application of the model is demonstrated in Sect. 5 via two separate examples. This section also includes a brief discussion on how the framework can be used to develop capacity expansion plans by the generators in restructured electricity markets under a CO_2 cap-and-trade program (more details on this can be found in [37]).

2 Bidding Strategies in Electricity and Related Markets

In this section, we first describe how the generators bid in restructured electricity markets as well as how the system operator dispatches power in a network and obtains its locational marginal prices. Later in the section, we formalize the FTR and the CO_2 allowance markets and show how they are connected to the electricity market.

2.1 *Electricity Market*

Participants of an electricity market maximize their benefits by seeking optimal bidding strategies. We focus here on the day-ahead and real-time (spot) electricity markets only. Strategic bidding behavior of the participants results in different market outcomes (e.g., nodal electricity prices and generation quantity allocations) under different auction mechanisms used by the system operator [27]. Two forms of auctions commonly used in deregulated electricity markets are uniform price auction and discriminatory auction.

- In a uniform price auction, all selected suppliers are paid a uniform price, equal to the market clearing price. The selection process starts by distributing the requested electricity generation units to the highest bidder, then to the second highest bidder and so forth until all generation needs (in MWh) are allocated. The market clearing price corresponds to the bid offered by the last selected bidder.
- In a discriminatory auction, the suppliers are selected in a manner similar to the uniform auction, but are paid according to their own bids instead of the market clearing price.

A game theoretic model in [39] compares uniform-price and discriminatory auction. It is shown that the equilibrium revenues of the generators under uniform price auction and discriminatory auction are different, in particular, that the expected total auction revenue of a network with *market power* is higher under uniform price auction when compared to discriminatory auction (pay-as-bid type). Market power is defined in the microeconomics literature as the ability of a seller to maintain prices above competitive levels for a sustained period of time. Commonly used market power indices are Herfindahl-Hirschmann index, Lerner index, quantity modulated price index (QPMI), and revenue-based market power index (RPMI) [27].

In the process of developing bids for the day-ahead and real-time electricity markets, generators and loads consider market forecasts for the power network. The forecasts provide information regarding expected prices, transmission network conditions, demand, among other parameters. Generators and loads then submit their bids to the system operator which in turn solves an optimal power flow (OPF) problem to determine the quantity allocations and the prices in the network. The solution of the OPF, thus, serves as the basis to compute the profits that generators and loads make in the day-ahead and real-time markets. Since all market participants develop bids with the objective of maximizing their respective profits, the competition in a deregulated electricity market can be modeled as a multi-player matrix game.

Let $z_n = (z_{n_1}, z_{n_2}, \ldots, z_{n_Y})$ denote a supply bid vector submitted by generator n. Each element of the vector represents the supply bid for each of the Y power plants generator n owns in the network. Each individual supply bid z_{n_k} is defined by the pair(a_{n_k}, b_{n_k}), where the first element is the intercept and the second is the slope of the supply curve. Hence, each supply bid curve is characterized by $p = a_{n_k} + b_{n_k} q$, where p and q denote the price and quantity, respectively. Let z_{-n} denote the supply bids of the rest of the generators in the network. When the range of values of the supply bid vector elements are suitably discretized, the total number of action choices for each bidder is finite. Then the cardinality of the supply bid vector of bidder n for each power plant k is given as follows,

$$|Z_{n_k}| = |a_{n_k}| \times b_{n_k}|, \tag{1}$$

The cardinality of the total action space for generator n in the electricity market can be written as

$$|z_n| = \prod_{k=1}^{Y} |z_{n_k}|. \tag{2}$$

It is assumed that the loads submit linear demand bids (similar to the supply bid curves but with $b_{n_k} < 0$). The payoff for generator n in a single instance of the electricity day-ahead or spot market can be written as follows.

$$r_n[z_n] = \mathscr{R}_n + \sum_{k=1}^{Y} q_{n_k} p_h - \sum_{k=1}^{Y} \left(a_{n_k}^0 q_{n_k} + \frac{1}{2} b_{n_k}^0 q_{n_k}^2 \right), \tag{3}$$

where $\mathscr{R}_n$ is the net amount of payments made/received by generator n to settle forward electricity contracts [21], q_{n_k} is the quantity of electricity produced by plant k of generator n, p_h is the price at node h where plant k is located, and $\sum_{k=1}^{Y} \left(a_{n_k}^0 q_{n_k} + \frac{1}{2} b_{n_k}^0 q_{n_k}^2 \right)$ is the total cost for each generator n (thus, $a_{n_k}^0, b_{n_k}^0$ are the true marginal cost parameters). The values for q_{n_k} and p_h in (3) are obtained by solving the following Optimal Power Flow (OPF) problem as presented in [5],

$$\begin{aligned}
&\max \sum_h B_h[p_h] - \sum_h C_h[p_h],\\
&\text{subject to:}\\
&Q_h - \sum_{\alpha\in\alpha(h)} q_\alpha[p_h] = 0 \ \forall \text{ node h},\\
&D_h - \sum_{\beta\in\beta(h)} d_\beta[p_h] = 0 \ \forall \text{ node } h,\\
&Q_h - D_h - \sum_{l\in l(h)} (t_{hl} - t_{lh}) = 0 \ \forall \text{ node } h,\\
&\sum_{hl\in A(v)} R_{hl}(t_{hl} - t_{lh}) = 0 \ \forall \text{ voltage loop v},\\
&t_{hl} \le T_{hl} \ \forall \text{ arc } hl,\\
&t_{hl} \ge 0 \forall \text{ arc } hl,
\end{aligned} \tag{4}$$

where $B_h[p_h]$ is the total benefit to consumers at node h, $C_h[p_h]$ is the total cost to producers at node h, Q_h and D_h are the total supply and total demand at node h respectively, $\alpha(h)$ is the set of producers at node h, $\beta(h)$ is the set of consumers at node h, $q_\alpha[p_h]$, is the quantity supplied by producer α located at node h, d_β $[p_h]$ is the quantity demanded by consumer β at node h, $l(h)$ is the set of nodes directly connected through a transmission line with node h, t_{hl} is the power flow between nodes h and l, $A(v)$ is the set of links that define loop v, R_{hl} is the reactance of link hl, and T_{hl} is the fixed capacity of link hl. The formulation is a quadratic convex problem whose solution provides the quantities supplied, quantities demanded and the locational marginal prices (LMPs).

Several approaches have been developed to find equilibrium bidding strategies in restructured electricity markets. A mathematical program with equilibrium constraints (MPEC) approach is presented in [16]. The model assumes that, while preparing their own bid, all generators have complete information about rival players' bids. A bi-level optimization model is formulated, where a generator's profit maximization problem at the first level is subjected to the OPF constraints at the second level. As part of the MPEC procedure, the OPF constraints are then replaced with equivalent KKT conditions resulting in a linear complementarity problem framework (LCP). This two-level problem, known as MPEC, has a maximization problem in the first level and equilibrium constraints in the second level. This MPEC optimization approach can be extended to a game theoretic setting with multiple competing players to find equilibrium bidding strategies. This approach is known as equilibrium problems with equilibrium constraints (EPEC) [40]. In EPEC, each player is solving an MPEC problem subject to a set of common OPF constraints. According to [41], there are two general methods to solve the EPEC problem: (1) obtain the optimality conditions (KKTs) for all the MPEC problems and solve them together as a complementarity problem, or (2) iteratively solve each of the MPECs using standard MPEC algorithms until the equilibrium solution of the EPEC game is obtained. The EPEC solution does not guarantee finding a NE. If a solution does exist, it is called a sub-game perfect Nash equilibrium. Other approaches to find equilibrium bidding strategies in electricity markets have been developed using linear/nonlinear complementarity problems (LCP/NCP) [15, 23] and stochastic games [35].

2.2 FTR Market

Transmission congestion occurs when there is insufficient transmission capacity to simultaneously accommodate all requests for transmission service within a power network [20]. When transmission congestion is present, the prices at different nodes of the network vary (nodal prices are referred to as the locational marginal prices, LMPs). The variation in the LMPs depends on the network configuration and the extent of congestion. When LMPs differ, the total payment made by the loads to the independent system operator (ISO) is higher than the payment made by ISO to the generators. This is referred to as the congestion loss (or, congestion charge) to the generators. Financial Transmission Right (FTR) is an instrument that allows the ISO to redistribute the excess revenue resulting from congestion [17]. For market participants (generators and loads), FTRs provide an opportunity to hedge their price risk when delivering electricity on the grid [22]. FTR is a contract between a market participant and the ISO that entitles the former to receive part of the excess revenue. The contract is designated by a MW amount from a source to a sink node in the network. Henceforth, we use the term *path* as a short form for *source node to sink node*. Note that a path is not a transmission line connecting the nodes.

There are two types of FTRs: point-to-point and flowgate. An FTR contract specified between any two particular buses/nodes (source and sink), as alluded above, in the network is known as point-to-point FTR. A less commonly used FTR type is the flowgate FTR which grants the holder a capacity reservation or scheduling priority for using specific transmission links or flowgates. In this chapter, we only consider point-to-point FTRs, which can be further classified as *obligations* and *options*.

- The holder of an obligation FTR receives a payment when the difference between the LMPs at the source and sink nodes (ΔLMP) is positive and has to make a payment if the ΔLMP is negative.
- The holder of an option FTR receives a payment when the ΔLMP is positive but is not required to make a payment when the ΔLMP is negative.

The participants in the FTR market submit non-increasing bids for option and obligation FTRs corresponding to a quadratic concave benefit function [4]. The benefit function is represented, for bidder n, as $f_n(x) = \varepsilon_n x - \frac{\tau_n}{2}x^2$, where x is the quantity of FTR (either option or obligation), ε_n and $\frac{\tau_n}{2}$ are the linear and quadratic benefit function coefficients, respectively. This implies a linear decreasing bid function, with a maximum price of ε_n (intercept), at quantity $x = 0$, and a downward slope of τ_n. The bid of bidder n on FTR node-to-node path i is given as a tuple as follows,

$$\rho_n^i = (\ \varepsilon_n^{i,ob},\ \tau_n^{i,ob},\ \varepsilon_n^{i,op},\ \tau_n^{i,op}),$$

where

$\varepsilon_n^{i,ob}(\varepsilon_n^{i,op})$ nth bidder's obligation (option) FTR bid intercept on node-to-node path i and

$\tau_n^{i,ob}(\tau_n^{i,op})$ nth bidder's obligation (option) FTR bid slope on node-to-node path i.

Let $\mathcal{I} = \{1,2,\ldots,I\}$ denote the set of possible node-to-node paths made available by the ISO for FTR bidding. A bidder $n \in N$ is considered to bid on $i \in \mathcal{I}_n$ with the above bid/action vector ρ_n^i, where $\mathcal{I}_n \subset \mathcal{I}$. If the range of values of the bid vector elements are suitably discretized, the total number of action choices for a bidder is finite. If $|\rho_n^i|$ denotes the number of FTR bids considered by bidder n for node-to-node path i, then the cardinality $|\mathcal{F}_n|$ of the total action space for bidder n in the FTR market can be written as

$$|\mathcal{F}_n| = \prod_{i \in \mathcal{I}_n} |\rho_n^i|. \tag{5}$$

Upon receipt of the FTR bids, the ISO develops an allocation using the following optimization model [3],

$$max \sum_{n=1}^{N} \sum_{i \in \mathcal{I}_n} [\varepsilon_n^{i,ob} FTR_n^{i,ob} - \tau_n^{i,ob} (FTR_n^{i,ob})^2 + \varepsilon_n^{i,op}\, FTR_n^{i,op} - \tau_n^{i,op}\, (FTR_n^{i,op})^2] \tag{6}$$

s.t.

$$\begin{aligned}&\sum_{n=1}^{N} \sum_{i \in \mathcal{I}_n} [PTDF_{n,1}^{i,c} FTR_n^{i,ob} + max(0, PTDF_{n,1}^{i,c})\\ &FTR_n^{i,op}] \leq T_l^c \ \forall l,\ c\end{aligned} \tag{7}$$

$$\begin{aligned}&\sum_{n=1}^{N} \sum_{i \in \mathcal{I}_n} [-PTDF_{n,l}^{i,c} FTR_n^{i,ob} + max(0, -PTDF_{n,l}^{i,c})\\ &FTR_n^{i,op}] \leq T_l^c \quad \forall\, l,\ c\end{aligned} \tag{8}$$

$$FTR_n^{i,ob} \geq 0;\ FTR_n^{i,op} \geq 0 \ \ \forall n,\ i \tag{9}$$

where

$FTR_n^{i,ob}$ quantity of obligation FTR allocated to nth bidder on node-to-node path i (decision variable),

$FTR_n^{i,op}$ quantity of option FTR allocated to nth bidder on node-to-node path i (decision variable),

$PTDF_{n,l}^{i,c}$ power transfer distribution factor (PTDF) of the nth bidder's ith FTR node-to-node path on line l under contingency c,

T_l^c capacity limit of line l under contingency c.

Note that the above formulation maximizes the revenue collected from the FTR auction. The set of constraints are equivalent to the set of constraints presented in the OPF formulation (4). Since FTRs are contracts designated in MW amounts, FTRs are modeled as actual power flowing in the network. The formulation above, which employs power transfer distribution factors (PTDFs), clearly illustrates the difference between options and obligations, in that counter price flows are ignored in the case of options. PTDFs can be computed as presented in [1]. The solution of the formulation determines the net number of FTRs that the network is able to support given a set of contingency scenarios during the FTR bidding period.

Under a uniform market clearing price auction, the FTR cost for bidder n is computed as,

$$FC_n = \sum_{i \in \mathcal{I}_n} [MCP^{i,ob}(FTR_n^{i,ob}) + MCP^{i,op}(FTR_n^{i,op})], \tag{10}$$

where $MCP^{i,ob}$ and $MCP^{i,op}$ denote market clearing prices for obligation and option type FTRs, respectively, on node-to-node path i. MCPs are obtained from the shadow prices of the constraints (7) and (8). Note that, though the FTR costs are calculated at the end of an FTR auction, the FTR revenues collected by the bidders are determined only after the ISO clears the different instances of the electricity spot market and the LMPs at the buses, in each of the instances, are known. The FTR revenue for a single instance of the electricity spot market is computed as follows,

$$FP_n = \sum_{i \in \mathcal{I}_n} [\Delta LMP_n^i \ FTR_n^{i,ob} + max(\Delta LMP_n^i, 0) FTR_n^{i,op}] \tag{11}$$

Thus, the FTR profit for a single instance of the electricity spot market is obtained as the difference between Eqs. 11 and 10. In Sect. 3, we present how we use these FTR profits to obtain the joint bidding strategies for electricity day-ahead and FTR markets.

In the literature, one of the first papers that examines the impact of FTR allocation on the energy market is by Joskow and Tirole [19]. They show, by analyzing sample networks, that a monopoly generator can enhance its market power by increasing FTR holdings. Regarding obtaining equilibrium FTR bidding strategies, a bi-level optimization method is presented in [24] where each bidder considers multiple bidding strategies and models the bidding behavior of its opponents in the upper level problem. The lower level problem finds the FTR market clearing price and the respective FTR allocation. The solution of the bilevel problem is obtained by iteratively updating the bidding strategies of each bidder, one at a time, while maintaining the opponents' bidding strategies fixed. This procedure ends when the bidding strategies of all participants cease to change. An alternative approach to find FTR bidding strategies using matrix games can be found in [11]. A model presented in [31] considers the FTR and electricity markets together by examining an auction-based process that jointly includes FTR bids and forward electricity contracts.

2.3 Allowance Market

Cap-and-trade is the most discussed CO_2 emissions control scheme in the U.S. Several CO_2 cap-and-trade bills have been proposed in the U.S. Congress. A CO_2 cap-and-trade program establishes a **cap** on the total quantity of CO_2 emissions allowed in a geographic region. A certain number (consistent with the cap) of tradable **allowances** is issued. Though a cap-and-trade program could involve all economic sectors linked with CO_2 emissions, we focus our attention on the electricity generation sector. The fossil fuel generators strive to procure a sufficient number of allowances to compensate for the emissions caused by electricity generation and thus avoid costly penalties. Excess allowances procured by the generators are traded in secondary markets. This general framework is common to all CO_2 cap-and-trade programs that are either already implemented or are being considered around the world.

Allowances are distributed among generators and other entities either via auction, or for free based on historic emissions (grandfathering), or by using a hybrid approach. Per the auction or the hybrid approaches, generators submit strategic allowance bids. For example, in the Regional Greenhouse Gas Initiative (RGGI), a cap-and-trade program implemented in the Northeast U.S., generators submit bids indicating price and quantity of allowances required. The auction is cleared using a uniform price scheme [36].

Generators bid in the allowance market considering their strategies in the electricity market. Generators determine the number of allowances to procure based on the fossil fuel-driven capacity that they offer to the market. The price that the generators pay for the allowances impact their supply bids in the electricity market. When generators are not able to surrender allowances commensurate with the emissions resulting from their electricity production, they are subjected to hefty fines.

In this chapter, it is assumed that allowances are distributed via auction to which generators submit bids indicating price (ω) and quantity (ζ). Similar to the situations described earlier for the FTR and energy markets, generators can develop multiple bids by varying discretized parameters for price and quantity. The production cost curve (the integral of the supply curve) of plant k owned by generator n is given by $c = a_{n_k}q + \frac{b_{n_k}}{2}q^2$. We assume that the allowance cost is incorporated into the cost curve by increasing the linear term a_{n_k} as

$$\widehat{a}_{n_k} = a_{n_k} + \psi P, \tag{12}$$

where ψ is the emissions factor (which indicates the amount of CO_2 (in tons) generated per MWh of electricity production, depending on the technology) and P is the allowance price obtained from the allowance auction settlement. Let P' be the price at which a generator trades unused allowances in the secondary market. Assuming that there is sufficient supply and demand in the secondary market, the allowance revenue/loss X_n for generator n during an entire allowance bid period can be obtained as

$$X_n = P'(m_n^a - m_n^c), \tag{13}$$

where m_n^a and m_n^c are the number of allowances allocated and consumed, respectively. It may be noted that m_n^a is a function of the allowance bid for the bidding period, whereas m_n^c is a function of the supply function bid submitted for each instance of the day-ahead (or spot) electricity market during the allowance bidding period.

The impact of an allowance market along with a green certificate market on generation expansion decisions in the electricity market can be found in [25]. The model assumes Cournot competition and is formulated as a linear complementarity problem (LCP). Other non-game theoretic models that examine implications of CO_2 cap-and-trade programs are [32, 38]. In [32], an economy-wide analysis of different cap-and-trade proposals that were considered by the U.S. Congress in spring 2007 is presented. This analysis was performed using the MIT Emissions Prediction and Policy Analysis (EPPA) model, and the results include allowance prices and expected emissions reductions for each cap-and-trade proposal. In [38], the energy and economic implications of joining RGGI are assessed for the State of Maryland. This assessment is conducted using: a simulation model for interregional trade among regional electricity markets, a market equilibrium model that incorporates market power in regional electricity markets, and a software system to assess economic impacts by industrial sector. Other studies analyzing the implications of cap-and-trade programs can be found in [8, 9].

In the following section, we first present the modeling framework to obtain joint bidding strategies for electricity and related markets.

3 A Two-Tier Game Theoretic Model for Joint Bidding Strategies

Figure 1 presents a schematic of a two-tier game for a hypothetical scenario with three players in each game. We assume that Market 1 and Market 2 settle with frequencies f_1 and f_2, respectively, where $f_2 > f_1$ (e.g., monthly for Market 1 and every 5 min for Market 2). Upper tier depicts the competition of the three players in Market 1. Each box in the upper tier (such as the ones marked in black) represents a combination of actions (i.e., bid strategies) of the players. The outcome of the

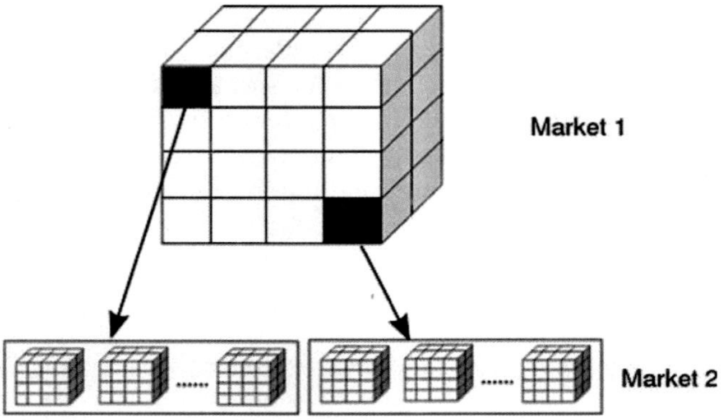

Fig. 1 Schematic of a two-tier game

actions in Market 1 impacts the actions of the players in Market 2. The smaller cubes in the figure represent multiple instances of Market 2. Thus, each action combination in Market 1 yields a different competitive scenario for the players in Market 2. The profit from Market 2 is required to construct the payoff matrices of Market 1. Consequently, the solution methodology for the two-tier model, described later, must begin by finding the equilibrium actions in Market 2 (for each instance) followed by the equilibrium actions in Market 1. Note that the inherently chronological sequence for selecting actions in Markets 1 and 2 (which will be evident from the examples discussed later) supports the tiered modeling and solution approach, where the solutions of Market 2 games are computed first, followed by solution of the Market 1 game. Also note that, all of the games in Market 2 can be solved separately and simultaneously using dedicated core processors, thus significantly reducing the computation time. The tiered solution approach eliminates the need to solve a very large comprehensive game involving both markets. Sample applications of this model can be found in [3, 30].

3.1 Joint Bidding Strategies

Since the FTR and allowance markets must settle before the electricity day-ahead and spot markets, we consider the FTR or allowance settlement in Market 1 (in the upper tier) and the day-ahead (or spot) electricity settlement in Market 2 (in the lower tier). We first provide a general description of the individual matrix game formulations followed by some details on two specific scenarios involving (1) FTR and electricity and (2) allowance and electricity.

3.1.1 Market 1 Game

Market 1 Game can be defined by a tuple $< N, A^1, \ldots, A^N, R^1, \ldots, R^N >$. The elements of the tuple are as follows.

- N denotes the number of bidders/players,
- A^n denotes the set of actions in Market 1 available to bidder n,
- $R^n: A^1 \times \ldots \times A^N \rightarrow \mathscr{R}$ is the payoff function of bidder n defined for each combination of actions of the N bidders.

Let $|\mathscr{F}_n|$ denote the cardinality of the action space for player n in Market 1 (given by (5) for the FTR market). The payoffs of the players can be modeled via N payoff matrices of size $|\mathscr{F}_1| \times |\mathscr{F}_2| \times \ldots \times |\mathscr{F}_N|$. For example, the upper tier game in Fig. 1 presents a scenario with three players having 4, 4, and 2 bid strategies.

3.1.2 Market 2 Game

Each instance of a Market 2 Game can be defined by a similar tuple to that used to define Market 1 Game. Let $|z_n|$ the cardinality of the total action space for player n in a single instance of Market 2 (as computed in Eq. 2 for the electricity market). The payoffs of the players can be modeled via N payoff matrices of size $|z_1| \times |z_2| \times \ldots \times |z_N|$. Each element of the Market 2 game payoff matrices is derived from the Market 2 revenue together with the payoffs from Market 1.

3.2 *Computation of Payoffs for FTR-Electricity and Allowance-Electricity Games*

In the FTR-electricity case, the payoff for generator n in a single instance of Market 2 game, for each supply bid combination, is computed using Eqs. 3, 10, and 11 as

$$\text{Payoff of Generator } n \text{ from Market 2} = (r_n[z_n] + FP_n - FC_n). \qquad (14)$$

The above represents a joint payoff that includes the FTR and electricity profits.

In the allowance-electricity case, the payoff for generator n in a single instance of Market 2 game, for each supply bid combination, is obtained using Eq. 3

$$\text{Payoff of Generator } n \text{ from Market 2} = r_n\,[z_n] \qquad (15)$$

In the above payoff calculation, the cost of allowances is not considered separately, since the supply bids of the generators account for this cost (see Eq. 12).

The elements of the payoff matrix of the Market 1 game are constructed from the equilibrium payoffs of the corresponding Market 2 games as follows. In the

FTR-electricity case, the payoff for generator *n*, during an FTR bidding period *T*, for each FTR bid combination is the sum of all the equilibrium payoffs for each instance of the day-ahead (or spot) electricity market during *T*. In the allowance-electricity case, the payoff for generator *n*, during an allowance bidding period *T*, for each allowance bid combination is the sum of all the equilibrium payoffs for each instance of the day-ahead (or spot) electricity market during *T* plus the overall revenue/loss X_n (Eq. 13) from trading allowances in the secondary market. The solution procedure presented below further clarifies this approach.

4 A Solution Procedure for the Two-Tier Model

In this subsection we describe the step by step approach (as depicted in Fig. 2) that we have developed for solving the two-tier game-theoretic modeling approach. In Step 0 the action spaces of the players in Markets 1 and 2 are defined. All possible action combinations in Markets 1 and 2 are initialized in Steps 1 and 2, respectively. A specific Market 1 action combination is chosen in Step 3. In Step 4, Market 1 settlement is carried out. In the FTR market case, this step considers solving the FTR allocation problem described by Eqs. 6–9. In the case of the allowance market, this step involves settling the allowance auction (finding the allowance allocations and price). In Step 5 a particular instance of Market 2 is chosen while in Step 6 a particular action combination of the players for the instance is considered. Based on this action combination, Market 2 is settled in Step 7 (i.e., ISO solves the OPF to obtain LMPs and supply allocations for the day-ahead or spot electricity market instance under consideration). In Step 8, the total payoffs for an instance of Market 2, based on settlements in Markets 1 and 2, are computed for each of the bidders ((Eqs. 14 and 15), for the FTR-electricity and allowance-electricity scenarios, respectively). These values are then used to populate the payoff matrix of the current instance of Market 2 game. This loop continues until all the action combinations for the current instance of Market 2 are considered, as shown in Step 9. In Step 10, the current instance of Market 2 game is solved using a value approximation based reinforcement learning (RL) algorithm [29] (that we describe in the next subsection). The loop composed of Steps 5–10 is repeated until all instances of Market 2 are considered, as shown in Step 11. In Step 12, the Market 2 equilibrium payoffs (obtained from Step 10) are used to populate the element corresponding to the current Market 1 action combination in the Market 1 payoff matrix (as described in the final paragraph of the previous section). Steps 3–12 are repeated until all Market 1 action combinations are considered (Step 13). In Step 14, the Market 1 game is solved (again using the value approximation based reinforcement learning (RL) algorithm in [29]). The procedure ends with the selection of the Market 2 equilibrium action combination (for each instance of Market 2) corresponding to the Market 1 equilibrium action combination obtained in Step 14.

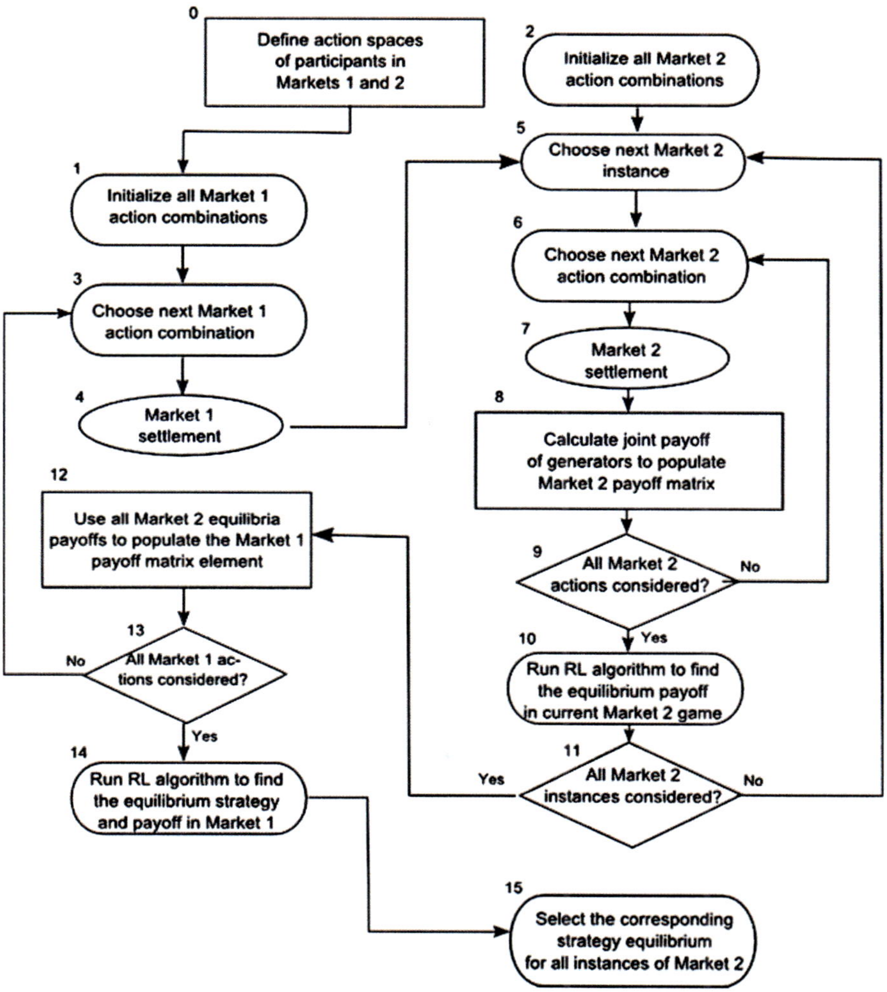

Fig. 2 A schematic of the solution approach for the two-tier game theoretic modeling approach

4.1 Reinforcement Learning Algorithm

We use a reinforcement learning (RL) algorithm, developed in [29], in steps 10 and 14. The RL algorithm uses a value-iteration approach to find a pure Nash equilibrium solution, when one or more exists, otherwise it finds a pure out-of-equilibrium solution [2]. The following are the main steps of the RL algorithm, presented in the context of a single instance of the electricity day-ahead or spot market game (a single instance of Market 2).

- **Step 1.** Let iteration count $j = 0$. Initialize R-values for each generator n with $|z_n|$ supply bid vector choices ($R_0(n,1)$,... $R_0(n,|z_n|)$) to an identical small positive number. Also initialize the learning parameter γ_0, exploration parameter ϕ_0, and parameters γ_τ and ϕ_τ needed to obtain suitable decay rates of learning and exploration. Let *Maxsteps* denote the maximum iteration count.
- **Step 2.** If, $j <$ *Maxsteps* continue learning of the R-values through the following steps:

  – Action Selection

    Greedy action selection:
    Each generator n, with probability $(1-\phi_j)$, chooses a supply bid vector (action) z_n for which $R_j(n, z_n) \geq R_j(n, \overline{z_n})$ where $\overline{z_n}$ stands for all the other supply bid vectors excepting z_n. A tie is broken arbitrarily.
    Exploratory action selection:
    With probability ϕ_j, a generator chooses a supply bid vector (action) z_n from all the possible supply bid vector choices (excluding the greedy action), where each supply bid vector can be chosen with equal probability.

  – R-Value Updating

  Update the specific R-values for each generator n corresponding to the chosen supply bid vector z_n using the learning scheme given below.

$$R_{j+1}(n, z_n) \leftarrow (1 - \gamma_j)R_j(n, z_n) + \gamma_j(r_n(z_n, z_{-n})), \tag{16}$$

where $r_n(z_n,z_{-n})$ is the joint payoff of generator n for choosing supply bid vector z_n when the other generators choose actions z_{-n} (these joint payoffs are computed in Step 7 in Fig. 2).

  - Set $j \leftarrow j + 1$
  - Update the learning parameter γ_j and exploration parameter ϕ_j as in [29].
  - If $j <$ *MaxSteps*, go back to beginning of Step 2, else go to Step 3.

- **Step 3.** From the set of R-values, select the supply bid vector z_n^* for each generator n as follows.

$$z_n^* = \arg\max_{z_n} R_j(n, z_n) \tag{17}$$

The RL algorithm is applied in a similar way to solve the Market 1 (FTR or allowance) game.

It is well known that matrix games, with or without pure strategy Nash equilibrium, always have one or more mixed strategy Nash equilibria. However, in the FTR, allowance, and electricity markets, for any instance of a market that is likely to occur during a period (with expected market conditions), each participant would always play with an equilibrium pure strategy designed for that market instance. Hence, while solving matrix game models for FTR, allowance, and electricity markets, our methodology considers only pure strategy solutions. As shown empirically in [29], the above *value* based reinforcement learning (RL) algorithm finds pure strategy solutions, which for most of the sample games coincide with a respective Nash equilibrium. When multiple pure strategy NE exist, the RL algorithm finds the one with the highest *values* (as computed with Eq. 16) for the bidders. For games without pure strategy NE, an out-of-equilibrium solution [2] provides a practical alternative. For such games, the greedy action selection approach of the RL algorithm (that prevails after the exploration ends) drives each player to choose the highest-*value* action. The resulting action combination of the players and the corresponding payoffs constitute the out-of-equilibrium solution for the game [3]. It may be noted that the RL algorithm is not equipped to determine the uniqueness of a NE solution. However, the nature of the solution (NE or out-of-equilibrium) can be easily determined.

5 Applications

In this section, our main objective is to demonstrate the use of our framework through two example networks. We also draw observations about the bidding behavior of the network participants in the FTR, spot electricity, and allowance markets. It may be noted that the examples do not fully replicate real-life market conditions. In the first example, we obtain joint FTR and electricity bids. In the second example, we obtain joint allowance and electricity bids under a CO_2 cap-and-trade program. We conclude this section with a brief discussion on how the game theoretic modeling framework can be used to develop capacity expansion plans by the generators in restructured electricity markets under CO_2 cap-and-trade programs.

5.1 Example 1: Joint FTR and Electricity Bids

We adopted a three-node sample network, from [19], which is shown in Fig. 3. We assume that the network is lossless and all transmission lines have the same impedance. The generator in node 2 (G2) is considered to have a higher marginal cost than the generator in node 1 (G1). The line joining nodes 1 and 2 is assumed to be congested with maximum line capacity K MW while the other two lines are assumed to have excess capacity.

We assume that the generators sell power at the load node through the following forward contracts: G1 sells 118.4 MWh at $15.602/MWh and G2 sells 53.6 MWh at $15.602/MWh. We assume that the generators do not hold FTRs from previous periods and thus, there are only FTR purchases (and not sells) in the auction. We also assume that the expected order of nodal prices in the network ($E[LMP2] > E[LMP3] > E[LMP1]$) is known to the generators. Therefore, the forward position of

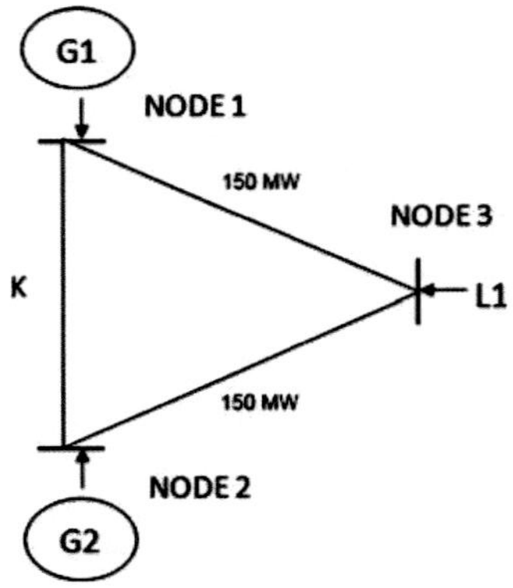

Fig.3 Example 1: A three-node sample network

Table 1 FTR bid parameters (ε,τ) choices for G1 and G2 – Example 1

Generator	Bid choice 1	Bid choice 2	Bid choice 3
G1	(6.0,0.019)	(5.0,0.0035)	–
G2	(4.77,0.0)	(5.0,0.0035)	(5.5,0.011)

G1 entails risk that is hedged by bidding for FTR on path 1–3, whereas G2 faces no risk and enters the FTR market as a speculator. It is considered that the forecasted operating point of the network (based on marginal cost supply bids, load profile, and contingency scenarios presented below) results in an expected congestion revenue equal to \$309.3 and a ΔLMP between nodes 1 and 3 equal to \$4.77/MWh. Thus, G1 should acquire $\frac{309.3}{4.77} = 64.8$ MWh FTR on node-to-node path 1–3 at \$4.77/MWh. We note that this amount of FTR does not leave G1 fully hedged but for the conditions considered in the example, it is the most FTR quantity that G1 can expect to be awarded. An expected full hedge for G1 (which would be achieved if G1 receives 118.4 FTRs) is not possible in our example due to the lack of the counterflow in line 1–2 resulting from G2 neither holding nor selling FTRs in 2–3.

Based on the pivot point ($q = 64.8$, $p = 4.77$), we derive FTR bids for G1 (see Table 1). In the case of G2, a speculator, we consider three non-strategic bid curves that also pivot at ($q = 64.8$, $p = 4.77$) (see Table 1). All the bids that we consider in the example are for obligation type FTRs, since they are more common and more readily analyzed. We note that FTR bids and holdings from previous periods may influence the bids for the current FTR period. In this example, we consider a single bidding period for FTRs free of any influence from previous periods. Therefore, the observations that we draw from the numerical results should be interpreted in that limited context.

The ISO obtains the FTR allocation by solving the optimization problem described by Eqs. 6–9 considering different contingency scenarios (explained below). As presented in Eq. 10, for the computations of FTR profits we consider the FTR costs despite the fact that these costs can be recovered by the generators via bidding for other financial instruments such as Auction Revenue Rights (ARR) [33]. For the electricity market, we consider three choices of supply function bids

Table 2 Supply function bid parameters (a, b) choices for G1 and G2 – Example 1

Generator	Bid choice 1	Bid choice 2	Bid choice 3
G1	(10,0.007)	(12,0.007)	(15,0.007)
G2	(20,0.007)	(24,0.007)	(30,0.007)

for each generator obtained by varying the respective intercept a as presented in Table 2.

The choices for the supply function bid parameters of the generators ensure that G2 is always more expensive than G1, which is a characteristic assumption for the network [19]. It is also assumed that the parameters of the true marginal costs of the generators are $a_1^0 = 10(\$/MWh)$, $\frac{1}{2}b_1^0 = 0.0035(\$/MWh^2)$ and $a_2^0 = 10(\$/MWh)$, $\frac{1}{2}b_2^0 = 0.0035(\$/MWh^2)$ for G1 and G2, respectively. Thus, the cheapest bids of the generators, given by (10,0.007) and (20,0.007) (Bid Choice 1 for both generators), represent the case in which G1 and G2 bid with their true marginal costs. The remaining bid choices for both generators are strategic. The lower and upper generation bounds (0 and 200 MWh, respectively) are assumed to be equal for both generators.

We assume that the transmission line between nodes 1 and 2 is subject to the following contingencies (CS) during the month long (8,640 h) FTR bidding period: 40% of the period K = 30 MW (CS1), 15% of the period K = 25 MW (CS2), 10% of the period K = 20 MW (CS3), 15% of the period K = 15 MW (CS4), and 20% of the time K = 8 MW (CS5). To facilitate analysis, we consider that the generators maintain the same set of supply function bid curves for all congestion scenarios (contingencies). As alluded to earlier, these scenarios are considered by the generators when developing their FTR bids.

In addition to the above, we analyze four different load profiles (LP) as defined by the following demand bids: LP1:p = 50–0.2q, LP2:p = 40–0.142q, LP3: p = 20–0.026q, and LP4:p = 17–0.0081q. These bids are obtained by pivoting at the point (172 MW, \$15.602), which represents an expected market settlement under marginal cost supply bids and the above contingencies. LP1 and LP2 represent conditions where price sensitivity of load is low. LP3 and LP4, on the other hand, represent conditions where the load is somewhat sensitive and very sensitive, respectively. We find joint equilibrium bids for each of the load profiles separately, since they do not represent a source of variation for the network conditions during a bidding period (as the contingencies do). Instead, the LPs are included in the analysis for comparing the resulting bidding strategies under different price sensitivities of the load. The results are presented in Fig. 4 and Table 3. From these results, we draw the following observations.

- **Choice of FTR bids:** It can be seen in Table 3 that the selected FTR bid of G1 remains constant (6.0,0.019) under all 4 load profiles. G2, on the other hand, chooses (5.5,0.011) (which results in an FTR allocation greater than 0) under LP1 and LP2 and (4.77,0.0) (which results in no FTR allocation) under LP3 and LP4. The choice of G1 is simply explained by the fact that G1 needs to obtain FTRs to hedge against the congestion charges. The choice of G2 under LP1 and

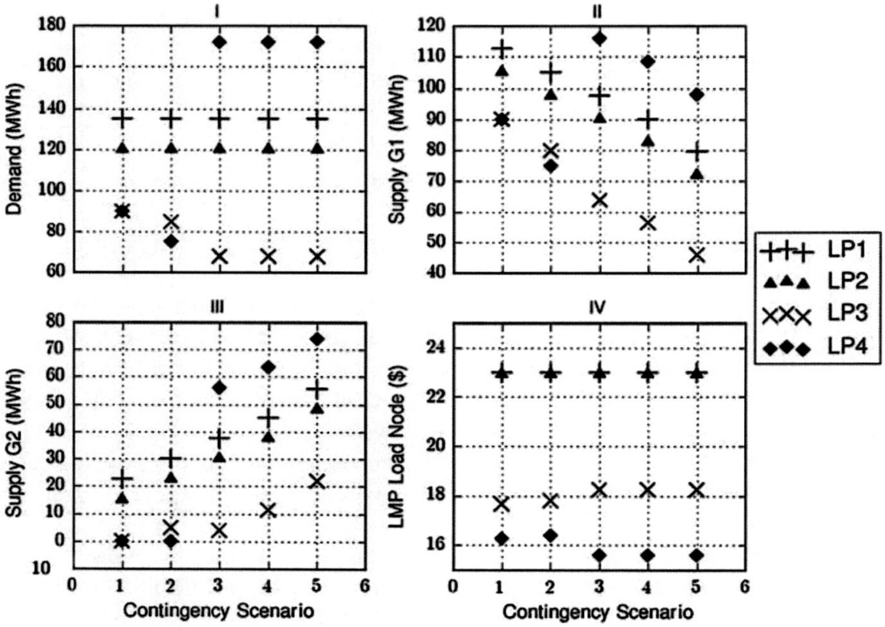

Fig. 4 Contingency scenarios versus quantity demanded, quantities supplied, and load LMP under each load profile

Table 3 Market equilibria under different load profiles

LP	Bids CS 1	Bids CS 2	Bids CS 3	Bids CS 4	Bids CS 5	FTR bids
1	(15,0.007), (30,0.007)	(15,0.007), (30,0.007)	(15,0.007), (30,0.007)	(15,0.007), (30,0.007)	(15,0.007), (30,0.007)	(6.0,0.019), (5.5,0.011)
2	(15,0.007), (30,0.007)	(15,0.007), (30,0.007)	(15,0.007), (30,0.007)	(15,0.007), (30,0.007)	(15,0.007), (30,0.007)	(6.0,0.019), (5.5,0.011)
3	(15,0.007), (20,0.007)	(15,0.007), (20,0.007)	(12,0.007), (24,0.007)	(12,0.007), (24,0.007)	(12,0.007), (24,0.007)	(6.0,0.019), (4.77,0.0)
4	(15,0.007), (20,0.007)	(15,0.007), (20,0.007)	(10,0.007), (20,0.007)	(10,0.007), (20,0.007)	(10,0.007), (20,0.007)	(6.0,0.019), (4.77,0.0)

LP load profile, *CS* contingency scenario

LP2 is based on G2's ability to increase its supply bid price, thus increasing the LMP at the load node and its profits from holding FTRs. Under LP3 and LP4, G2 loses this ability, and thus is better off not holding FTRs.

- **Choice of supply bids:** Under an insensitive load (LP1 and LP2), G2 is able to raise its bid price, "forcing" G1 to follow suit (see first two rows, columns 2–6 in Table 3). As the load becomes more sensitive, G2 is not able to raise its bid price significantly whereas G1 bids its highest supply price only when congestion is low (see last two rows, columns 2–3 in Table 3). When congestion increases, G1 chooses a cheaper supply bid.

- **Impact of contingencies (congestion scenarios):** The higher price generator (G2) generally produces more, as expected, when congestion increases (see plot III in Fig. 4). The opposite is true in general for G1. A notable exception to this occurs when congestion increases under LP4 (see diamonds on plots II and III). Here, G1 is not able to raise its supply bid price (due to both, load sensitivity and congestion) and neither is G2 (due to sensitivity of load). Thus, the equilibrium occurs at the cheapest bids of both generators, increasing the electricity consumed at the load.
- **Impact on load:** As seen from Fig. 4, when the load is less sensitive (LP1 and LP2), the demand and price at the load node remain constant across all contingencies, while the balance of quantity supplied varies with decreasing share of G1 and increasing share of G2. The load LMP is relatively high (around $23), as expected (see plot IV). For high sensitivity profiles (LP3 and LP4), the load is able to significantly reduce demand and price (between $16 and $18.5).

We note that, since there are two generators bidding in the FTR market and we consider two bid choices for the first player and three bid choices for the second player, the payoff of the players for the FTR game has been modeled via two payoff matrices of size 2×3 (there are 6 FTR bid combinations of the players). Similarly, since the same two players have three bid choices in the electricity market, the payoff of the players for each instance of the spot electricity game has been modeled via two payoff matrices of size 3×3 (there are nine supply bid combinations of the players). The element of our solution approach that is a major contributor to the computational needs is the repeated settlement of Market 2 (see Step 7 in Fig. 2). For the joint FTR-electricity example, the settlement of Market 2 corresponds to solving a single OPF. Thus, for this example, we solve 6 (number of FTR bid combinations) $\times$ 9 (number of supply bid combinations) $=$ 54 OPFs for each instance of the electricity spot market. Given than we consider five congestion scenarios (instances of electricity market), the total number of OPFs solved is 270. If we add a new player to the joint game with three bid choices for the FTR market and three supply bid choices, then the total number of OPFs to be solved would increase to 2,430 ($18 \times 27 \times 5$). Consideration of more choices of FTR and supply bids for the generators will also increase the number of OPFs to be solved and thus, the computational need.

In this example, we considered a limited number of choices for the supply and FTR bid strategies by coarsely discretizing the range of possible values of the bid parameters. To further evaluate the equilibrium solutions that we have obtained, one could resolve the games by considering new sets of bids with higher granularity around the current equilibrium solutions. The change, if any, in the new equilibrium solutions should dictate if the problem needs to be resolved again with bids of even higher granularity.

5.2 Example 2: Joint Allowance and Electricity Bids

In this example, we consider a similar three-node network as that from Example 1 but with some changes as presented in Fig. 5. We further assume that G1 and G2 are fossil-fuel generators with G1 emitting more CO_2 /MWh than G2 (emissions factors: for G1 = 1.12 tonnes CO_2/MWh, for G2 = 0.2 tonnes CO_2/MWh). We also consider that the generators procure CO_2 allowances to support electricity production. Allowances are distributed via a uniform market clearing price auction. Generators bid for allowances indicating price ω and quantity ζ. We consider two choices of allowance bid strategies for the generators by modifying the bid price and keeping quantity constant. G2 bids for less allowances given that it is a "cleaner" generator and has a smaller market share. The bid parameter choices of the generators for the allowance and electricity markets are presented in Tables 4 and 5, respectively. Note that the values of the intercept term (a) presented in the tables do not include the allowance cost component. Hence, each supply bid choice is increased by the cost of allowance using Eq. 12 after the allowance auction is cleared (we assume that the generators pass the allowance cost on to the consumers in both supply bid choices).

We consider that the generators submit their bids to the allowance auction which takes place at an earlier time than the competition in the electricity market. We consider only one instance of the electricity day-ahead (or spot) market. In reality, there are multiple instances of the electricity spot market for each instance of the

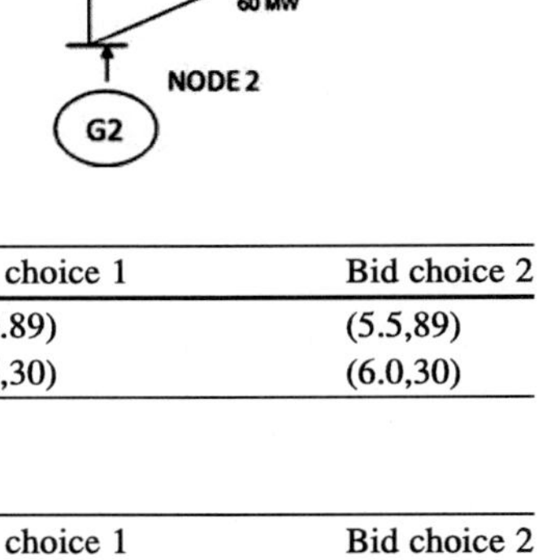

Fig.5 A three-node sample network – Example 2

Table 4 Allowance bid parameters (ω,ζ) choices for G1 and G2 – Example 2

Generator	Bid choice 1	Bid choice 2
G1	(4.0.89)	(5.5,89)
G2	(3.5,30)	(6.0,30)

Table 5 Supply function bid parameters (a,b) choices for G1 and G2 – Example 2

Generator	Bid choice 1	Bid choice 2
G1	(10,0.0025)	(30,0.0025)
G2	(35,0.0025)	(45,0.0025)

allowance market (similar to the situation in the FTR market presented earlier). A total of 89 allowances are offered in the auction. This number is obtained based on the quantity allocations to the generators (obtained by clearing the electricity market without emissions considerations) and the emissions characteristics of their plants. The allowance auction is cleared using the uniform rule. It may be noted that if there are any allowances that remain unused at the end of the electricity market settlement, these can be traded between the generators in the secondary market (provided there is unmet need). The trading price in the secondary market is considered to be an exogenous parameter, and is assumed to be 1.2 times the market clearing price. If a generator does not have enough allowances to surrender even after attempting to procure them in the secondary market, a penalty is imposed by increasing the outstanding allowance balance by a multiple of 3 [36].

In Fig. 6, we present the bid choices and the resulting joint payoffs in the allowance game (the allowance game corresponds to the upper tier game in Fig. 1). Note that the payoffs for each allowance bid combination were obtained by solving the corresponding lower tier electricity games (not shown). The equilibrium of the allowance game occurs at the bid combination, 89–4.0, 30–3.5, for G1 and G2, respectively (shaded box). This results in an allowance clearing price equal to $4.0 with all 89 allowances being allocated to G1. The corresponding equilibrium in the electricity market occurs at the supply bid combination with intercept (a) equal to 30+ cost of allowance for G1 and 45 + cost of allowance for G2 i.e., the most expensive choices for both generators. Note that in the case where no allowance market (no emissions control) is considered, the supply bid chosen by the generators are also the most expensive ones. However, as shown in Table 6, the load sees reduction in its total benefit function in presence of emissions regulation.

It may be noted that a reduction in emissions is achieved primarily due to the reduction in demand and loss of load benefit. As the cost of supply of G1 and

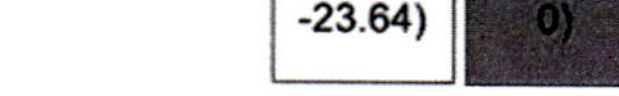

Fig.6 Allowance game – first element of each pair (i,j) is the joint payoff of G1; second element is the joint payoff of G2

Table 6 Comparison of electricity market equilibrium results between scenarios with and without emissions control

Scenario	d (MWh)	q1 (MWh)	q2 (MWh)	LMP load ($/MWh)	Total load benefit ($)	Total emissions (tonnes CO_2)
With emissions control	60	60	0	39.7	$2,391.00	72
No emissions control	100	80	20	39.5	$3,975.00	100

G2 increase due to implementation of emissions control (via allowance price and penalty), the high sensitivity of demand (with slope 0.005) prompts a drastic reduction in equilibrium load and the resulting benefit. The lower equilibrium demand is also influenced by the reduction of G2's profit due to the added penalty when its supply quantity exceeds the allowances acquired.

5.3 *Example 3: Developing Multi-generator Multi-year Generation Expansion Plans (GEP) Under a Cap-and-Trade Program*

Generation expansion planning (GEP) in a restructured market is the challenge of determining *which* type, *where*, and at *what* time periods new generation capacities are likely to be installed by the *competing generators* in response to expected demand growth, changes in network conditions, and market design changes such as the implementation of a CO_2 cap-and-trade program. The matrix game modeling framework can be employed to address this challenge. To that end, we have extended the two tier game theoretic framework by adding an extra tier representing the expansion game where the action of the players are choices of generation expansion plans for a time horizon.

We consider a network with N generators. Each generator owns an array of plants based on different technologies at one or more nodes in the network. We also consider that each generator has a discrete set of expansion plans for the planning horizon, which is derived based on forecasted demand increase and fuel price variations, environmental and societal constraints, technology growth, market share target, and capital availability. We assume that the region served by the power network operates under a CO_2 cap-and-trade program where the generators are required to obtain allowances via auction. The generators bid for allowances with price and quantity, and surrender the acquired allowances at the end of the production period commensurate with their emissions. It is assumed that the generators pass the cost of the allowances on to the consumers in the electricity market.

Figure 7 shows a schematic representation of the generation expansion model. The *expansion game*, in the top tier, represents all possible combinations of the expansion plans of the generators. An expansion plan comprises a set of yearly actions to add generation capacity (or to do nothing) over the planning horizon. The attributes of an action include location of the new capacity and its size, technology, and cost. Each of the combinations in the expansion game (e.g., the shaded portion) represents a specific network generation portfolio (in terms of nodal capacities and technology mix). Thus, an expansion plan combination directly influences the generators' bid strategies in the *allowance game* for each year of a planning horizon H. In turn, each combination of the allowance bid strategies (and the corresponding allowance settlement) influences the supply function bid strategies in the *electricity game*.

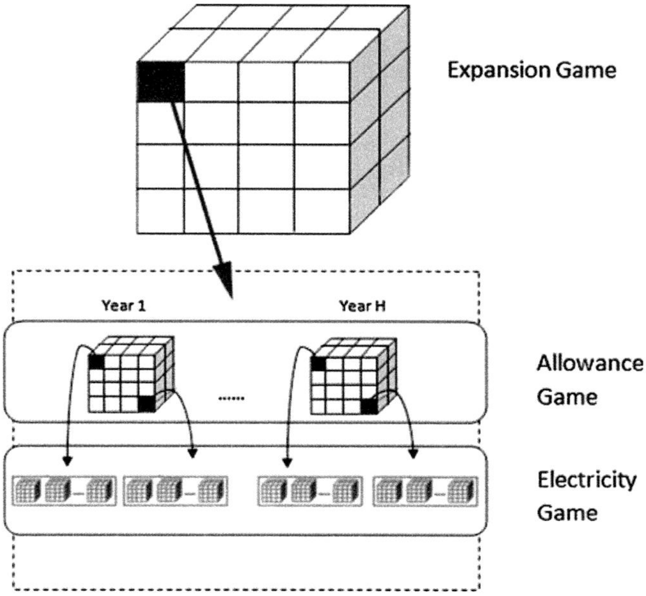

Fig.7 Schematic of a three-tier framework for generation expansion under cap-and-trade

Clearly, the action choices of the generators in the expansion, allowance, and electricity games, collectively determine the equilibrium profit from the electricity market. In our model, this equilibrium profit is used to construct the payoff matrices for the allowance game. Finally, the equilibrium profits from the electricity and the allowance markets are used to construct the payoff matrices of the expansion game, whose solution provides the equilibrium expansion plan combination. Consequently, the solution methodology for the game theoretic framework must begin by solving the electricity game, followed by the allowance and expansion games. A more complete treatment of this model and its application on a power network representing the northern Illinois market of U.S. can be found in [37].

6 Conclusions

The multi-tier modeling framework presented in this chapter provides a means of jointly considering competition among a variety of market participants in the day-ahead (or spot) electricity and other related markets. Since the payoffs of the different markets are interdependent, it is required that the bidding strategies be developed with a joint view of the markets. Though the consideration of competition in multiple correlated markets presents a complex modeling challenge, the tiered modeling framework presented in this chapter allows us to solve one market at a time while still considering the impacts of the competitions in the other markets. This tiered solution approach exploits the natural structure of the markets which are settled in a specific time order. For example, an FTR market is always settled before day-ahead and spot electricity markets. The framework also allows a detailed consideration of many important aspects of the markets such as transmission constraints in both electricity and FTR markets, cap-and-trade features including allowance auction, allowance penalties, and trading in secondary markets.

Our methodology solves matrix games formed by discrete set of bid choices as opposed to continuous sets as considered by methods such as MPEC and EPEC. Thus, our solutions can theoretically be called local, since they are chosen from a limited discrete set of alternative bids for each participant. However, a further improved local solution can be obtained, if desired, by re-solving the game with different sets of bid choices formed around the previous local solution. Using a high level of discretization of the state-spaces, though, which can approach continuous sets, will significantly increase the computational need of our methodology.

As mentioned earlier, our matrix game theoretic framework is well suited to accommodate several parameters/features of the markets considered. Examples of these features in the case of a CO_2 allowance market include penalties for negative allowance balance and allowance trading in secondary markets. To our knowledge, both of these features have not been explicitly addressed by the EPEC and MPEC literature (see Linares et al. [25]) though they have been addressed in other non-game theoretic models such as [32, 38]. Also, in the generation expansion problem (discussed in brief in the chapter), our methodology [37] can handle diverse features of an expansion plan such as location of the new capacity and lead times as well as the transmission constraints in the power network which are not considered in the MPEC model presented in papers such as [10, 26].

Acknowledgement This research was supported in part by a grant from the Florida Energy Systems Consortium (FESC), 2009–2011.

References

1. Alvarado FL, Oren SS (2000) A tutorial on the flowgates versus nodal pricing debate. In: PSERC IAB Meeting, 2000
2. Arthur WB (2006) Out-of-equilibrium economics and agent-based modeling. In: Judd KL, Tesfatsion L (eds) Handbook of computational economics, vol 2. Elsevier/North-Holland, Amsterdam, pp 1551–1564
3. Babayigit C, Rocha P, Das TK (2010) A two-tier matrix game approach for obtaining joint bidding strategies in FTR and energy markets. IEEE Trans Power Syst 25:1211–1219
4. Bautista G, Quintana VH (2005) Screening and mitigation of exacerbated market power due to nancial transmission rights. IEEE Trans Power Syst 20:213–222
5. Berry CA, Hobbs BF, Meroney WA, O'Neill RP, Stewart WR Jr (1999) Understanding how market power can arise in network competition: a game theoretic approach. Utilities Policy 8:139–158

6. Bhattacharya K, Bollen MHJ, Daalder JE (2001) Operation of restructured power systems. Kluwer, Boston
7. Budhraja V, Woolf F (1994) Poolco: an independent power pool company for an efficient market. Electr J 7:42–47
8. Burtraw D (2006) CO_2 allowance allocation in the Regional Greenhouse Gas Initiative and the effect on electricity investors. Electr J 19:79–90
9. Burtraw D, Sweeney R, Walls M (2008) The incidence of US climate policy: where you stand depends on where you sit. DP 08–28. Resources for the Future, Washington, DC
10. Chuang AS, Wu F, Varaiya P (2001) A game-theoretic model for generation expansion planning: problem formulation and numerical comparisons. IEEE Trans Power Syst 16:885–891
11. Das TK, Rocha P, Babayigit C (2010) A matrix game model for analyzing FTR bidding strategies in deregulated electric power markets. Int J Electr Power Energy Syst 32:760–768
12. Department of Energy, EIA. http://www.eia.doe.gov
13. Garber D, Hogan B, Ruff L (1994) Poolco: an efficient electricity market: using a pool to support real competition. Electr J 7:48–60
14. General Accounting Office US (2002) Lessons learned from electricity restructuring: Report to congressional requesters. Technical Report GAO-03-271. General Accounting Office, Washington, DC
15. Hobbs BF (2001) Linear complementarity models of NashCournot competition in bilateral and poolco power markets. IEEE Trans Power Syst 16:194–202
16. Hobbs BF, Metzler CB, Pang JS (2000) Strategic gaming analysis for electric power systems: an MPEC approach. IEEE Trans Power Syst 15:638–645
17. Hogan WW (2002) Financial transmission right formulations. Technical Report. Harvard University, Cambridge, MA
18. Joskow PL (2008) Lessons learned from electricity market liberalization. Energy J 29:9–42
19. Joskow PL, Tirole J (2000) Transmission rights and market power on electric power networks. RAND J Econ 31:450–487
20. Kirby BJ, Van Dyke JW (2002) Congestion management requirements, methods and performance indices. Technical report ORNL/TM-2002/119. Oak Ridge National Laboratory, Oak Ridge
21. Kirschen D, Strbac G (2005) Basic concepts from economics, in fundamentals of power system economics. Wiley, Chichester. doi:10.1002/0470020598.ch2
22. Kirschen D, Strbac G (2005) Transmission networks and electricity markets, in fundamentals of power system economics. Wiley, Chichester. doi:10.1002/0470020598.ch6
23. Lee KH, Baldick R (2003) Tuning of discretization in bimatrix game approach to power system market analysis. IEEE Trans Power Syst 18:830–836
24. Li T, Shahidehpour M (2005) Risk-constrained FTR bidding strategy in transmission markets. IEEE Trans Power Syst 20:1014–1021
25. Linares P, Santos FJ, Ventosa M, Lapiedra L (2008) Incorporating oligopoly, CO_2 emissions trading and green certificates into a power generation expansion model. Automatica 44:1608–1620
26. Murphy F, Smeers Y (2005) Generation capacity expansion in imperfectly competitive restructured electricity markets. Oper Res 53:646–661
27. Nanduri N, Das TK (2007) A reinforcement learning model to assess market power under auction-based energy pricing. IEEE Trans Power Syst 22:85–95
28. Nanduri N, Das TK (2009) A survey of critical research areas in the energy segment of restructured electricity markets. Electr Power Energy Syst 31:181–191
29. Nanduri N, Das TK (2009) A reinforcement learning approach to obtain Nash equilibria of multiplayer matrix games. IIE Trans Oper Eng 41:158–167
30. Nanduri V, Das TK, Rocha P (2009) Generation capacity expansion in energy markets using a two-level game theoretic model. IEEE Trans Power Syst 24:1165–1172

31. O'Neill RP, Helman U, Hobbs BF, Stewart WR Jr, Rothkopf MH (2002) A joint energy and transmission rights auction: proposal and properties. IEEE Trans Power Syst 17:1058–1067
32. Paltsev S, Reilly JM, Jacoby HD, Gurgel AC, Metcalf GE, Sokolov AP, Holak JF (2007) Assessment of US cap-and-trade proposals. No. 13176. National Bureau of Economic Research, Cambridge, MA
33. PJM ISO – ARR analyses. http://pjm.com/planning/arr-analyses.aspx
34. PJM ISO. http://www.pjm.com
35. Ragupathi R, Das TK (2004) Stochastic game approach for modeling wholesale energy bidding in deregulated power markets. IEEE Trans Power Syst 19:849–856
36. Regional Greenhouse Gas Initiative, RGGI. http://www.rggi.org
37. Rocha P, Das TK, Nanduri V, Botterud A (2010) Generation capacity expansion in restructured power markets under a CO_2 cap-and-trade program. In review
38. Ruth M, Gabriel SA, Palmer KL, Burtraw D, Paul A, Chen Y, Hobbs BF, Irani D, Michael J, Ross KM, Conklin R, Miller J (2008) Economic and energy impacts from participation in the regional greenhouse gas initiative: a case study of the state of Maryland. Energy Policy 36:2279–2289
39. Son YS, Baldick R, Lee K, Siddiqi S (2004) Short-term electricity market auction game analysis: uniform and pay-as-bid pricing. IEEE Trans Power Syst 19:1990–1998
40. Su CL (2005) Equilibrium problems with equilibrium constraints: stationarities, algorithms, and applications. Dissertation from the Department of Management Sciences, Stanford University
41. Yao J (2006) Cournot equilibrium in two settlement electricity markets: formulation and computation. Dissertation, University of California, Berkeley

Short-Term Electricity Market Prices: A Review of Characteristics and Forecasting Methods

Hamid Zareipour

Abstract In this chapter, short-term electricity price modeling and forecasting in competitive electricity markets is presented. Dominant characteristics of short-term electricity prices such as, seasonality, non-stationarity, spikes and volatility are discussed and numerical examples from real-life markets are presented. A review of the existing literature on short-term electricity price forecasting is also provided, followed by an overview of the process of building data-driven models for electricity prices. Furthermore, some popular time series models for electricity market price modeling and forecasting, such as ARIMA models, are discussed. A case study is also presented in which Ontario electricity market prices are modeled and 24-h-ahead forecast are generated.

Keywords Artificial intelligence • Competitive prices • Electricity market • Price forecasting • Price volatility • Time series models

1 Introduction

Reforms in electricity markets and introduction of competition in the generation sector started in Chile in 1982. Since then, privatization, deregulation, restructuring or re-regulations of the power sector has been gradually adopted in various countries around the world, including several states in the US, Alberta and Ontario provinces in Canada, the UK, Nordic Countries, most European countries, Australia and New Zealand. In the reformed electricity markets, wholesale prices are determined competitively based on the available supply and the demand for electricity. Retail prices, however, are generally still regulated and small consumers are not

H. Zareipour (✉)
University of Calgary, Calgary, AB, Canada
e-mail: h.zareipour@ucalgary.ca

A. Sorokin et al. (eds.), *Handbook of Networks in Power Systems I*,
Energy Systems, DOI 10.1007/978-3-642-23193-3_4,

directly exposed to the variable wholesale market prices [1]. The focus of this chapter is on wholesale prices in competitive electricity markets.

Future price of electricity is a key factor in optimizing operational and planning activities of wholesale electricity market participants in the short, medium and long terms. While there is no general consensus on a threshold when it comes to the future time horizons, in the short-term, usually in the order of a few hours to a week ahead of operation time, market participants are particularly interested in future electricity prices mainly to schedule their operation and set their bids. In many electricity markets, unit commitment, i.e., deciding which unit should be operating over a certain planning period, is left to generation companies, such as in the Ontario, New England, California, New Zealand, and Australia markets. In those markets, self-scheduling the generation fleet of a generation company combined with the company's bidding strategies are the basis of short-term bids. Forecasts of next-day prices for energy and operating reserves are the key inputs to generation self-scheduling models. In the medium-term, usually a few months to a few years ahead, the trend of future prices is needed for budgeting and setting bilateral energy and fuel contracts, and pricing derivatives. In the long-term, usually between 5 and 15 years, a prediction of future price behavior is factored in major investment plans, such as building new generation facilities [2–4].

Available price models in the literature can generally be classified into predictive and descriptive models. Predictive models are used to forecast future prices, usually hourly prices, and their accuracy are measured against the actual observed prices [5]. Short-term price models fall in this category and usually narrow down to a mathematical function that relates future prices to some known inputs, such as previous prices or electricity demand. Descriptive models, on the other hand, are built to describe the general behavior of price over a longer period of time, and are not usually used for forecasting hour-by-hour prices. A descriptive model can simply be a probability density function or a price duration curve [6, 7]. Unlike predictive models, accuracy of descriptive models is not always easy to assess.

Over the past decade, numerous research works are reported in the literature on short-term electricity market price forecasting. Reviews of the existing methods can be found in [8] and [9]. Compared to short-term electricity market price forecasting, works on medium- and long-term price modeling and forecasting are very limited [10]. This could be attributed to lack of enough data for longer term studies due to the relatively short history of electricity markets, continuous changes in market structures and regulatory regimes which impact the observed price patterns, and various uncertainties involved in formation of prices in longer terms.

This chapter only focuses on short-term electricity market price modeling and forecasting. The goal of this chapter is to provide a fundamental understanding of the main characteristics of short-term electricity prices and the available methodologies for price forecasting. This understanding is important in developing customized models for predicting short-term prices in a specific electricity market.

The rest of this chapter is organized as follow: In Sect. 2, characteristics of electricity market prices are presented. In Sect. 3, an overview of the literature on electricity price forecasting is provided. Also, the process of building a data-driven price model is discussed in this section. Furthermore, available data-driven price forecasting models and forecast accuracy measures are briefly discussed in Sect. 3. In Sect. 4, the case study of forecasting electricity prices in Ontario's market using time series models is presented. Finally, concluding remarks are provided in Sect. 5.

2 Characteristics of Short-Term Electricity Market Prices

Despite the sometimes significant differences in the design and regulations of competitive electricity markets around the world, prices in these markets share a number of common features. Seasonality, non-stationarity, occurrence of unusually high or low price spikes and extreme volatility are the main reported features of electricity prices for several electricity markets [2, 11, 12].

2.1 Seasonality

Driven by human activities, the demand for electricity varies by time in different time scales, i.e., within the hours of a day, weekdays versus weekends and peak and off-peak seasons. In the short-term, seasonality in the demand translates into similar price seasonality. Figure 1 shows the Ontario hourly demand and the Ontario Hourly Energy Prices (HOEPs)[1] for the week of July 7–13, 2008. Despite the sometime severe intra-day or inter-day variations in prices, the daily seasonality of demand and price is obvious in this figure.

Several tools are available to study seasonality in a time series, such as autocorrelation function (ACF) and spectral analysis [14]. Autocorrelation functions basically show the self-similarity in a time series in terms of the linear correlation coefficients. Spectral analysis discovers the cyclic properties of a time series based on Fourier transformation. These techniques are particularly efficient in studying short-term seasonalities. Focusing on ACF, the seasonality in the HOEP are studied here.

If $\{p_1,p_2,\ldots,p_n\}$ are the observations of time series p, sample ACF of time lag $l = 0,1,\ldots n\text{-}1$ is defined as:

$$ACF(l) = \widehat{\rho}(l) = \frac{\widehat{\gamma}(l)}{\widehat{\gamma}(0)}, \tag{1}$$

[1] Electric energy and operating reserves markets in Ontario are based on a joint uniform market clearing pricing scheme. The average of the 12 five-minute energy market clearing prices (MCPs) over an hour is defined as the HOEP. The HOEP uniformly applies to all load zones and generators within the province, and forms the basis of most of the financial settlements. The transmission system is ignored when calculating Ontario MCPs [13].

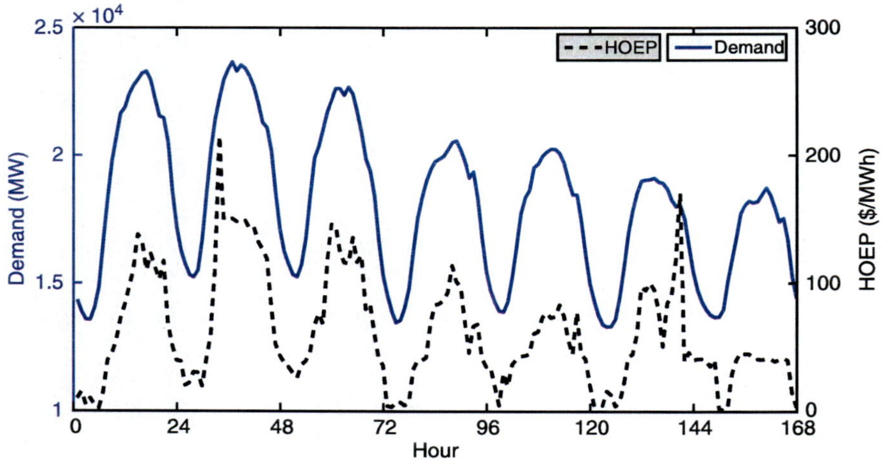

Fig. 1 Ontario hourly demand and electricity prices for the week of July 7–13, 2008

where

$$\widehat{\gamma}(l) = \frac{1}{n} \sum_{t=1}^{n-l} (p_{t+l} - \overline{p})(p_t - \overline{p}), \tag{2}$$

and $\overline{p} = \frac{1}{n} \sum_{t=1}^{n} pt$ is the sample mean.

If the underlying process that generates time series p is completely random, AFCs are statistically insignificant for all time lags l. Otherwise, the self-similarities among the observations are reflected in ACFs. Figure 2 shows the sample ACFs of the HOEPs for year 2008. A daily (i.e., $1 = 24$) and a weekly seasonality ($1 = 168$) is evident in this figure.

For discovering longer term seasonalities, other approaches are usually used. For instance, a sinusoidal function is fit to the time series after removing its short-term (e.g., daily and weekly) seasonalities. Using Wavelet decomposition is another popular method [12].

2.2 Non-stationarity

Non-stationarity in a time series arises when statistical characteristics of the time series change over time [14]. Non-stationarity has been observed both in the mean and in the variance of short-term electricity markets prices [2].

Mean non-stationarity of the electricity prices is generally due to the intra-day and intra-week price fluctuations. Mean non-stationarity can be removed by de-trending the times series over various time scales. A popular and easy to

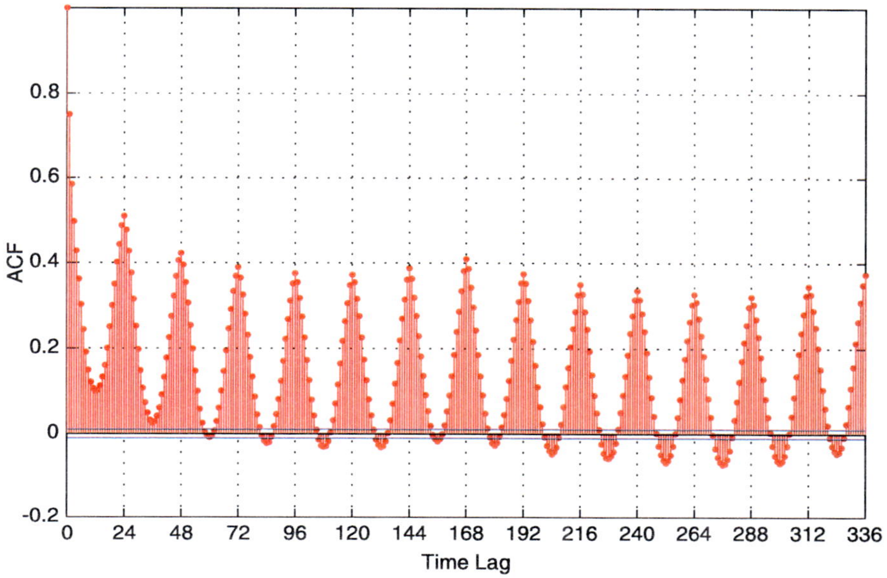

Fig. 2 Sample ACF for the hourly Ontario energy prices (*HOEPs*), year 2008. The *horizontal lines* show the 95% confidence interval for white noise

implement technique in removing short-term seasonalities is differencing, where change in prices is calculated for the cyclic lags. For example, by defining $y_t = x_t - x_{t-24}$, time series x is de-trended for its daily variations. Differencing has been efficiently used in various studies to induce mean-stationarity on electricity price time series [5, 11].

Unusually high or low prices result in an inconstant variance in electricity market price time series, and thus variance non-stationarity. The Box-Cox power transformation, which leads to a natural logarithm transfer of the time series in most cases, is used to reduce the variance non-stationarity of electricity market prices [11, 14]. However, any transformation only results in a weak stationarity where severe variance non-stationarity exists in a time series, which is the case for short-term electricity prices.

Autocorrelation functions are usually used to detect non-stationarity of the mean. Persistent high values of ACF over a large number of time lags are an indication of mean non-stationarity. Figure 3 shows the sample ACFs of de-trended HOEP time series for daily and weekly seasonalities. Compared with Fig. 2, the values of sample ACFs rapidly decay for the de-trended HOEP time series.

Visual inspection of the original or de-trended time series data is the easiest way to detect non-stationarity in the variance. Figure 4 shows the HOEP fluctuations for year 2008. It is evident that the price variations are not homogeneous over time, indicating a non-constant variance.

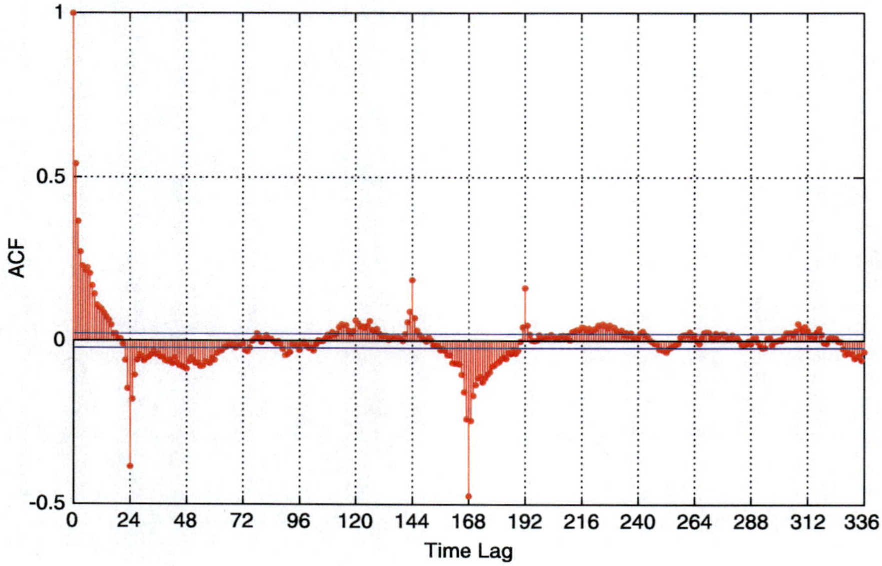

Fig. 3 Sample ACF of the de-trended hourly Ontario energy prices (*HOEPs*) for the weekly and daily seasonalities

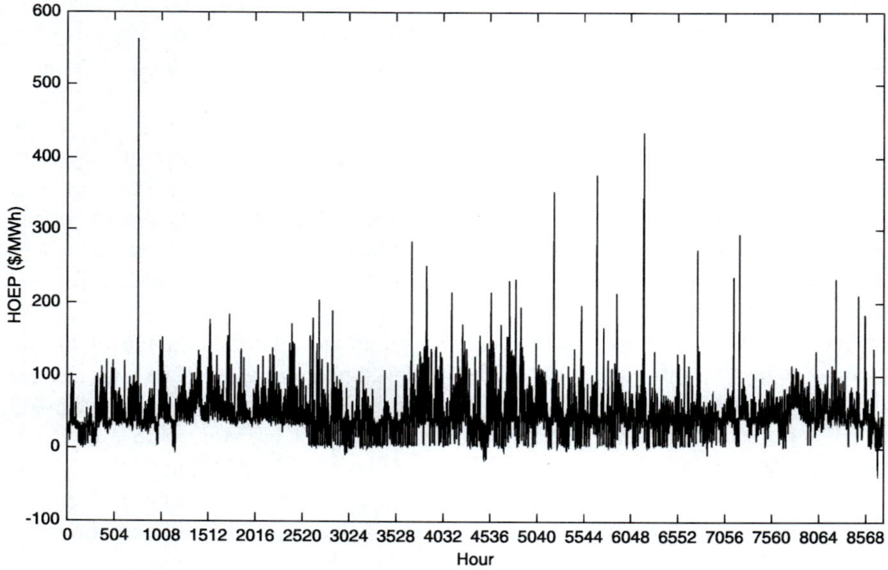

Fig. 4 Fluctuations of hourly Ontario energy prices (*HOEPs*) for year 2008

2.3 Price Spikes

Unusually high or low prices are observed in electricity markets [15]. Although there is no commonly accepted definition of 'unusually high or low', the problem of price spikes has been reported for all markets around the world. Although, the frequency and severeness of price spikes may vary across different electricity markets, the main reason for electricity price spikes is the real-time nature of electricity production and consumption and lack of bulk electricity storage. In addition, market design and structure, generation portfolio and market power, transmission system and connection to other markets, and elasticity of demand are the other factors that may contribute to frequency or severance of price spikes.

The hourly Alberta pool prices (HAPP) [16] for year 2005 are presented in Fig. 5. Alberta has an energy-only single-settlement real-time electricity market where a uniform market price applies to all market participants across the province. The annual average HAPP for year 2005 was \$70.36/MWh. However, for 461 h during this year, the HAPP was above \$200/MWh; this is a threshold that is sometimes used for the analysis of price spikes. Also, for 334 h, the prices were under \$10/MWh. The very steep supply curve in Alberta's market partly explains the problem of price spikes in this market. A typical supply curve, for hour ending 19, Nov. 17, 2005, is shown in Fig. 6. As it can be seen from this figure, after a certain level of supply, a small increment in demand can result in a sharp increase in price. For example, when demand increases by 82 MW from 8,733 to 8,815 MW, the price raises from \$88.19 to \$598.1/MWh. In general, any 'hockey-stick' shape

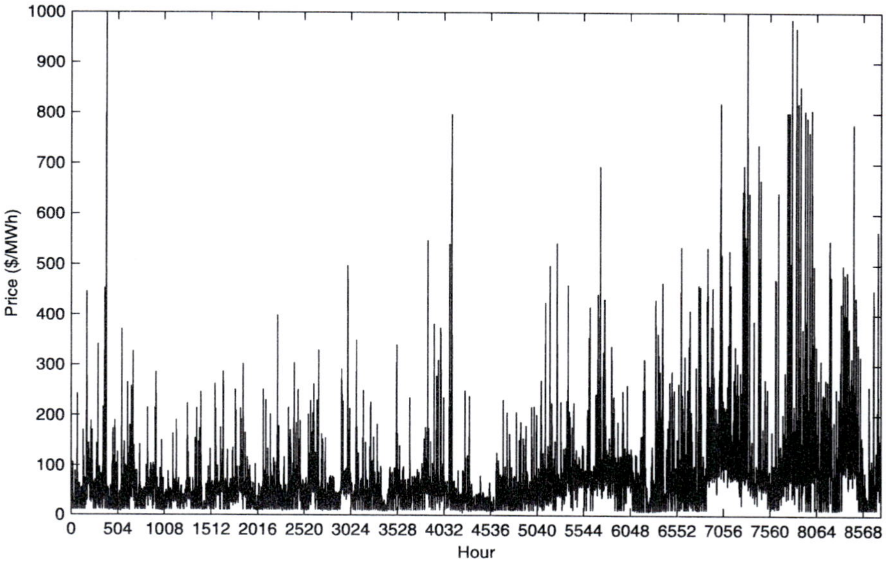

Fig. 5 Hourly Alberta pool prices (*HAPPs*) for year 2005

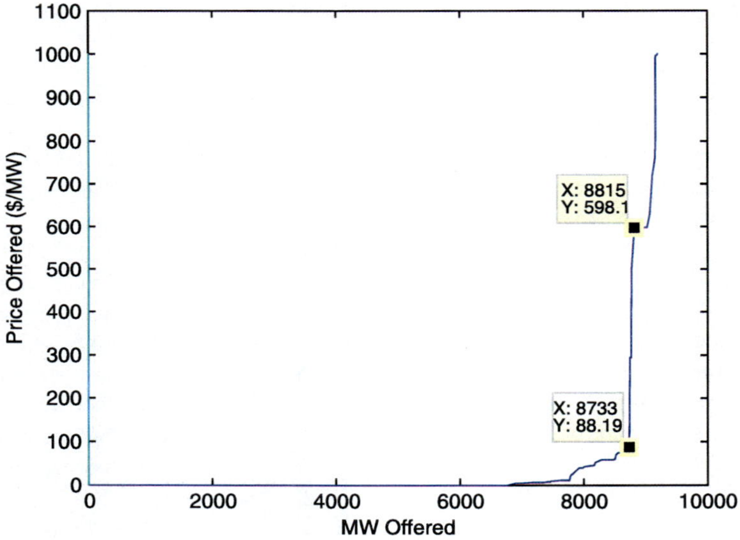

Fig. 6 Supply curve for a typical hour in Alberta's market

supply curve leads to very sharp changes in the market clearing prices if the demand and supply curves intersect in the 'blade' region.

In recent years, predicting the occurrence and severity of price spikes has been the subject of some research work. In particular, price spike prediction models have been developed based on availability of supply, system demand, and weather temperature [13, 32]. The prediction capability of available spike forecasting models is not yet satisfactory. This is not surprising since some of the price spikes simply occur randomly, for example, because of the forced outage of a large generation unit. No model can predict such events and the subsequent price spikes.

2.4 *Price Volatility*

Volatility refers to the unpredictable fluctuations of a process observed over time. Volatility analysis, volatility modeling, and volatility forecasting has been reported in the literature for different commodities, including electricity, and have various applications such as risk management and option valuation [17, 18]. For the time series p_t, say price of electricity, the arithmetic return over a time period h is defined as:

$$R_{t,h} = \frac{p_t - p_{t-h}}{p_{t-h}} \tag{3}$$

and the logarithmic return, over the time period h, is defined as:

$$r_{t,h} = \ln\left(\frac{p_t}{p_{t-h}}\right) = \ln(p_t) - \ln(p_{t-h}) \tag{4}$$

When returns are small, the arithmetic and logarithmic returns are close, given the fact that:

$$r_{t,h} = \ln\left(\frac{p_t}{p_{t-h}}\right) = \ln\left(1 + R_{t,h}\right) \approx R_{t,h} \tag{5}$$

Most volatility analysis studies consider the logarithmic return over arithmetic return for volatility analysis. If the return values are identically and independently distributed (i.i.d.) over a time window T, one can present them as:

$$r_{t,h} = \widehat{\mu}_{h,T} + \widehat{\sigma}_{h,T}\varepsilon_t \tag{6}$$

where $\widehat{\mu}_{h,T}$ is the conditional mean return; $\widehat{\sigma}_{h,T}$ is the conditional return variance; and the random variable ε is a mean zero, unit variance, i.i.d. innovation. $\widehat{\sigma}_{h,T}$ is referred to as the historical volatility over the time window T; in other words, the historical volatility is defined as the standard deviation of arithmetic or logarithmic returns over a time window T. For observed return values, the estimated value of $\widehat{\sigma}_{h,T}$ can be calculated as:

$$\sigma_{h,T}(\%) = 100 \times \sqrt{\frac{\sum_{t=1}^{N_o}\left(r_{t,h} - \overline{r}_{h,T}\right)^2}{N_{o-1}}} \tag{7}$$

where $\sigma_{h,T}$ is the estimated value of historical volatility, N_o is the number of $r_{t,h}$ observations, and $\overline{r}_{h,T}$ is the simple $r_{t,h}$ average, all of them over the time window T.

Considering the daily and weekly seasonality observed in electricity market price behavior, volatility of electricity prices may be defined based on price changes in consecutive hours, i.e., hourly volatility and $h = 1$, or in subsequent days, i.e., daily volatility and $h = 24$, or in subsequent weeks, i.e., weekly volatility and $h = 168$.

Based on several volatility studies reported in the literature, electricity market prices are significantly more volatile than other commodity prices. For example, the 12-year average of daily volatilities, i.e., average of $\sigma_{24,24,}$, of the Nordic electricity market prices has been reported to be about 16% [19], whereas the daily volatility of crude oil and natural gas prices usually fall in the range of 2–3% and 3–4%, respectively.

Using the aforementioned volatility definitions, price volatilities for some of the North American electricity markets for year 2004 are presented in Table 1. Observe

Table 1 Historical volatilities (%) for the Ontario and its neighboring markets prices in 2004

	$\bar{\sigma}_{1,24}$	$\bar{\sigma}_{24,24}$	$\bar{\sigma}_{168,24}$
New England	8.44	6.76	7.22
New York	11.17	8.37	9.07
PJM	16.37	12.94	13.43
Ontario	22.12	28.13	28.05

that Ontario's market prices are significantly more volatile than those of New York, New England and Pennsylvania, Jersey and Maryland (PJM) day-ahead market prices.

Various factors may contribute to the high price volatility in electricity markets. For example, in the case of Ontario [18], errors in predicting the demand and non-dispatchable supply, failure in import/export transactions, and the real-time nature of this market were suggested as the key factors impacting price volatility. In addition, comparing the structure of New England, New York and PJM electricity markets with that of Ontario's, those markets have a multi-settlement structure where electricity trades take place in day-ahead, hour-ahead and real-time markets, as opposed to the real-time only structure of Ontario's market.

3 Forecasting Short-Term Electricity Market Prices

Participants in electricity markets are required to submit price-quantity bids to the market operator well before the real-time operation. In most markets, the "gate closure" is mid-day today for submitting bids for all planning intervals (usually hourly and in some cases half-hourly intervals) of the next day or next week. Although in some cases market participants are allowed to revise their bids up to a few hours before real-time, most markets restrict such bid revisions to unavoidable situations because of unforeseen technical difficulties. Determining optimal bids, however, requires a prediction of short-term future prices. In this section, short-term price forecasting is discussed and some of the most popular models are reviewed.

3.1 Building Short-Term Electricity Price Models

Wide range of models, mainly based on times series analysis and artificial intelligence, have been applied to prices from various markets worldwide. For example, the work presented in [20] describes Neural Networks based models for forecasting prices in the Spanish and Pennsylvania, Jersey and Maryland (PJM) markets with an overall forecasting error of about 5%; however, the model accuracy was reported to collapse when only high price hours were concerned. Weighted nearest neighbors techniques are proposed in [21] and forecasting errors ranging between

5% and 16% were reported for the Spanish market; the variations in forecast accuracy over the studied period were attributed to various unpredictable factors including extreme weather conditions. A dynamic model based on system identification techniques is detailed in [22] with forecasting errors varying between 5% and 36% for the Italian, New England and New York markets. Several statistical parametric and semiparametric models are applied to forecasting in California and Nordic electricity market prices in [23] and errors ranging from 3% to 15% were reported. It was concluded in [23] that no single model could presently be chosen as the single best approach. A hybrid method, composed of support vector machines (SVMs) and a self-organized map, is applied to New England market prices in [24] with resulting forecasting errors of about 10% and 7% were reported for the prices before and after the implementation of the standard market design, respectively. In [11] and [25], time series and neuro-fuzzy models are applied to forecasting Ontario's electricity prices; the forecast errors were reported to vary between 16% and 22% and the high forecast errors were attributed to the high volatility of prices in Ontario [11]. In [26], several linear and non-linear models are employed to forecast Ontario prices for a 3-year study period. The average forecasting errors in [26] vary between 22.98% and 31.86%, with a SVM-based model yielding the lowest errors. In another study [27] on forecasting prices in National Electricity Market of Australia, a SVM-based model is optimized using genetic algorithms and the resulting forecast errors are reported to vary between 16.39% and 23.26% for different time periods.

It can be observed from the existing literature that the reported range of errors in electricity price forecasting is significantly higher when compared to the typical 1–3% error range of short-term electric load forecasting [28]. Also, the price-demand relationship has been mostly considered to be uni-directional and potential impacts of price on demand have been ignored. This might be true and acceptable in the current power markets where elasticity of demand is extremely low. However, under the "smart grid" initiatives [29], consumers of electricity are expected to be enabled to react to electricity prices in the future, leading to higher demand elasticity. This may require new approaches to electricity price and load forecasting.

In addition to publications regarding numerical electricity price forecasting, other alternative approaches have also been explored. For example, several papers have included the estimation of prediction or confidence intervals. Among those, a method based on Neural Networks and Kalman Filters is described in [30] and a different approach based on SVMs is detailed in [31]. Moreover, [15] and [32] focus on the treatment and handling of price spikes and propose hybrid models to predict their occurrence. Furthermore, in [33], the idea of predicting the "class" of future electricity prices with respect to user-defined price thresholds are introduced. In price classification, instead of predicting the exact values of future prices, the range of prices is predicted. Price classification is shown in [33] to be particularly useful when the forecast users do not need to know the exact value of future prices in planning their operation. For example, a large industrial consumer may decide to shut down a certain production line if prices hit a certain value, say $1,000/MWh. Thus, such forecast user only needs to know when such threshold is hit.

While the modeling techniques used vary from case to case, the general premise is to detect price patterns from the historical data and forecast future price accordingly. In general, constructing a model for forecasting purposes involves three main steps: data preprocessing, feature selection and model building.

3.1.1 Data Pre-processing

In this step, relevant data are gathered and processed for descriptive statistics, anomalies, missing values and necessary transformations. The reported studies in electricity price forecasting literature highlight, and deal with, two aspects of price data in this step: non-stationarity of prices and price spikes.

As discussed before, mean and variance non-stationarities of prices are dealt with using power transformations or differencing [5]. Price spikes or outliers disturb the historical price patterns and can mislead or confuse prices models. On the other hand, price spikes are one of the important features of electricity market price time series. In some of the electricity market price forecasting literature, 'pre-processing' of the price spikes are reported. In such cases, the price spikes are usually either removed from the data or capped at a certain value or replaced by a weighted average of previous prices for that particular hour. Some other reported works have build forecasting models based on the original price time series data without manipulating the price spikes. All in all, there are no strong evidences favoring either of the choices in terms of the out-of-sample accuracy of the forecasting models [12].

Decomposition of the price time series can sometime be used to deal with the problems of non-stationarity and price spikes at once. For example, wavelet transform has been reported to decompose price time series into a number of new time series, which have a more stable variance and mean, and do not suffer from unusual spikes [34]. In such cases, independent models are used to forecast the future values of each of the individual new time series. Inverse wavelet transform is then used to compose the forecasted values and form price forecasts.

The pre-processing step in artificial intelligence-based models generally involves data normalization and mapping of the data into a higher dimension space. Data normalization is used especially when the input data have very different ranges and units. For example, in a typical price forecasting model, previous values of price and demand are usually used as model inputs, while the numerical value of demand is much larger than price. If not normalized, such differences in input range can potentially degrade the information value of some input factors against others. Mapping the data into a higher dimension space is generally required in support vector machine-based models. The mapping simplifies the model estimation process and reduces computation times [35].

3.1.2 Feature Selection

In feature selection, a subset of the most informative features is selected from a predetermined pool of potential features [35, 36]. The selected features, also referred to as model inputs or explanatory variables, can potentially explain the fluctuations of the process under study. Two major groups of feature selection techniques are filter and wrapper methods. In filter methods [37–39], the features are assessed for their relevance in explaining the target variable, independent of the prediction model. Relevance criteria such as, the Pearson correlation coefficient, the Fisher criterion and Mutual Information have been often used. Filter methods are fast and simple, but the potential disadvantage of them is that feature selection is isolated from the prediction model.

In wrapper methods [37–39], the complete feature set is searched for a near-optimal sub-set, and the relevance of features is evaluated by the accuracy of the final predictions. One of the main differences in wrapper techniques is in the way they search for the best possible feature sub-set. In forward selection methods, the process starts with an empty sub-set and new features are added step-by-step depending on how they contribute to model accuracy. In backward elimination methods, individual features are eliminated from the complete set of features through the search process if they do not contribute to model accuracy. Classic examples of search methods are hill-climbing and best-first, however, various population-based evolutionary search methods (e.g., based on genetic algorithm) have been proposed for wrapper feature selection. While wrapper methods have been shown to generally provide high prediction accuracy [39], they are computationally expensive for large numbers of features. A third group of feature selection methods is embedded methods in which, finding the near-optimal sub-set for explaining the target variable is left to the model. In other words, all features are fed into the model and based on the final weights of each feature, the final set of features are selected [35, 36].

In the context of electricity markets, the most popular features are previous prices and loads. Other features such as, day and hour indexes; transmission constraints (quantified by shadow prices) [11, 25, 40]; load level of neighboring systems [11, 25, 40]; variants of reserve margin [11, 32]; generator outages and temperature [25]; and availability of different types generation resources [8, 20, 41] have also been reported. However, inclusion of some of the above features in the forecasting models has not always led to better forecast accuracy. For example, clearing prices are determined in the Ontario market based on the unconstrained market simulations, and thus, inclusion of transmission constraints has not enhanced forecast accuracy in [25]; however, in [40] transmission constraints have been reported to improve the accuracy of forecasting locational marginal prices in the PJM market. In addition, although including electricity demand in forecasting electricity prices has improved forecast accuracy to some extent, the achieved improvements have not always been significant. For example, in [5, 11, 33], the forecast accuracy has only improved by 1–2% points when demand data are included in model inputs. The reason behind this observation is that although the general trend in electricity prices follows the same pattern as electricity demand, most of the price fluctuations are driven by other factors, such as scarcity of supply and transmission constraints.

The most frequently reported feature selection method in electricity price forecasting is filter method where linear (e.g., cross-correlation and auto-correlation [11, 22, 23, 25, 40, 42]) or non-linear (e.g., mutual information [20]) measures are used to evaluate the relevance of the input features to the price.

An important factor when constructing the set of potential features that may explain price variations is the timely availability of each particular feature [11]. In other words, although some of the price fluctuations may be explained by certain market variables, those variables cannot be used as inputs to predictive price models if they, or their reliable forecasts, are not know at the time of generating forecasts. For example, some of unusual price spikes may be explained by forced outages of generation units. Those outages, however, are not known until they happen.

3.1.3 Model Building

In this step, a model is built that relates the selected features, i.e., inputs, to the target variable, i.e., price. Following the three-stage procedure for building time series models proposed by Box and Jenkins in their pioneering book of Time series Analysis and Forecasting [14], building a predictive model generally involves three stages, regardless of the type of model employed. The first stage is model identification, in which an initial model structure is identified as the starting point. While the process of model identification may vary depending on the type of model, the initial model is identified based on the characteristics of the process under study, modeler's experience and some analytical tools. An example of those tools is the sample ACF discussed before.

The second stage of building a model is model estimation. In this stage, unknown parameters of the model are estimated using available historical data. Suppose β is the set of model parameters to be estimated. Also, suppose there are N pairs of historical instances of $\{y_t, X_t, t = 1,\ldots,N\}$, where y_t represents the observed value of the target variable (e.g., hourly prices) at time t and X is the set of model inputs (e.g., demand at hour t). If the forecasted value of the target variable for the tth instance, generated using the model, is $\widehat{y}_t$, the goal of model estimation stage is to find optimal values of β that minimize the differences between y_t and $\widehat{y}_t$ across all historical instances of $t = 1,\ldots, N$. In practice, and depending on the type of model, a loss (or likelihood) function is defined as a variant of y_t and $\widehat{y}_t$, and β is estimated through an optimization process that minimizes (maximizes) the loss (likelihood) function. For example, a famous loss function is the sum of square errors, i.e., $\sum_{t=1}^{N} (\widehat{y}_t - y_t)^2$, and its minimization leads to the famous least square parameter estimation that is widely used in regression models [14].

The third stage of building a predictive model is diagnosis and testing. In this stage, model's performance is tested against the fundamental assumptions made in the previous two stages [11, 14, 41]. In addition, accuracy of the initially identified model is evaluated and improved by varying the specifications of the identified model.

Another decision that needs to be made in the model building step is selecting the type of model. A wide range of predictive modeling approaches is available and have been used in several applications. A review of different families of models applied to electricity market price forecasting is presented in Sect. 3.2. Complexity and characteristics of the process under study are generally the main factors influencing the choice of modeling approach. In the case of electricity market prices, the common features of the spot electricity prices such as, seasonality of the prices and price spikes, discussed in Sect. 2, and the nature of market clearing processes are taken into account when selecting a modeling approach. Electricity market prices are determined as the intersection of the supply and demand curves at each market clearing interval. As mentioned in Sect. 2.3, the supply curve in most markets has a 'hockey-stick' shape. On the 'shaft' portion of the curve, price of power raises smoothly and a small change in demand is followed by a small change in price. On the 'blade' portion of the curve, however, a small increase of the demand sometimes lead to significant price jumps. Thus, the price clearing process is highly non-linear with respect to the relationship between price, demand and supply. Ability of a particular modeling approach in capturing such non-linearities makes it a good candidate for electricity market price forecasting. Although non-linear models have been reported to outperform their linear counterparts in some cases, there is no evidence that shows this is always the case [23].

In addition, performance of a certain modeling approach when applied to other similar processes is another driver for choosing a particular modeling technique. For instance, electricity demand and price share some characteristics, such as seasonality and long-term non-stationarity. Hence, a model that performs well in forecasting electric load may be a good starting candidate for forecasting electricity prices. Obviously, other characteristics of electricity prices, such as unusual spikes which do not occur in electricity load, need to be taken into account in bringing the starting model to the next level.

It should be noted that, similar to any other forecasting problem, there is no "best" model for forecasting electricity prices which consistently outperforms other modeling approaches. While a model may perform well in a particular period of time for a certain market, that particular model may not necessarily result in the best forecasting results in a different situation. This fact makes it difficult to select a generalized modeling approach whose performance versus other models can be guaranteed. Thus, a selected number of models should be dynamically trained and tested for best results for forecasting short-term prices in a specific market.

3.2 Data-Driven Models for Short-Term Electricity Price Forecasting

Models applied to electricity market price forecasting may be classified according to various measures, such as: linear versus non-linear, stationary versus non-stationary, single versus hybrid, point-forecasting versus interval-forecasting, and time series versus artificial intelligence-based,.

Linear models are those with that relate the target variable to the explanatory variables through a linear function. A classic example of a linear model is an Autoregressive Moving Average (ARMA) model, explained in Sect. 3.2.1. In a non-linear model, on the other hand, the relationship between the target variable and at least one of the explanatory variables is non-linear. For example, an artificial neural network with a non-linear transfer function applied to the model inputs is a non-linear model [20]. Sometime, a non-linear model is composed of a number of linear models that are activated according to a pre-defined criterion. Such models, sometimes referred to as adaptive or regime-switching models, are also classified as non-linear. The Multivariate Adaptive Regression Splines models [43], or the Threshold Autoregressive models [23], are examples of adaptive models.

Stationary models are strictly applicable to stationary time series, because of the fundamental underlying modeling assumptions. ARMA model, discussed later in this section, is an example. Non-stationary models, on the other hand can be applied to any time series. An artificial neural network is an example of a non-stationary model.

While some of the models applied to electricity price forecasting are built purely based on one single modeling approach, some others are a combination of more than one approach. For example, a combination of fuzzy systems and artificial neural networks or a hybrid of similar days method and artificial neural networks have been reported in the literature, generally presenting more accurate forecasting results than single-model approaches for the particular studied data [25].

Most of the reported price forecasting models are developed for point-forecasting, i.e., finding the exact value of future prices. Some other models, however, also provide the expected range in which the prices are expected to fall with a certain confidence level. These models generally find the estimated variance of the point-forecasts and generate a 95% confidence interval assuming a Normal distribution for model residuals [30, 31]. In most cases, however, the variance of the point-forecasts are relatively high, leading to the forecast of wide confidence intervals for the prices which are hardly useful.

Time series models refer to a group of statistical models that are based on the random nature and the self-similarity of a stochastic process. In time series models, past observations of the process under study are used to model the next values of the process. The Autoregressive Integrated Moving Average (ARIMA), Dynamic Regression (DR) and Transfer Function (TF) are among the most commonly used time series models for electricity market price forecasting which provide reasonably good forecast accuracy despite their simplicity [11, 12, 34, 41, 44].

Artificial intelligence-based models, on the other hand, are 'black box' models that relate the target variable to the explanatory variables through a complex function. Artificial neural networks, fuzzy-logic systems, data mining and machine learning-based approaches (e.g., k-nearest neighbor models, Input/output hidden Markov models, or support vector machines) are the main groups of artificial intelligence-based models

In this section, the ARIMA, DR and TF approaches are reviewed [5, 11, 14]. ARIMA is a univariate model that relates the future values of a time series to its own past values. DR and TF, on the other hand, are multi-variate models which relate the future values of a time series not only to its own past values but also to the values of other explanatory variables. These methods have shown competitive forecasting performance despite their simple structure.

3.2.1 ARIMA Model

Let denote the equally sequenced values of a stationary stochastic process z by $z_{t,}$ z_{t-1}, $z_{t-2,}\ldots$, $z_{t-n.}$ An autoregressive moving average model ARMA (p, q) for this process can be expressed as:

$$z_t = c + \sum_{i=1}^{p} \phi_i z_{t-i} + \varepsilon_t + \sum_{j=1}^{q} \theta_j \varepsilon_{t-j} \tag{8}$$

where c, ϕ_i and θ_j are the model parameters to be estimated. ε_t is assumed to be an independently and identically distributed (i.i.d.) normal random variable (shock) with mean zero and variance σ_ε^2 Using the backward shift operator B, defined as $Bz_t = z_{t-1}$, model (8) can be represented as:

$$\phi(B)z_t = c + \theta(B)\varepsilon_t \tag{9}$$

where $\phi(B) = 1 - \phi_1 B - \ldots - \phi_p B^p$ is the nonseasonal autoregressive operator AR(p), and $\theta(B) = 1 - \theta_1 B - \ldots - \theta_q B^q$ is the nonseasonal moving average operator MA(q).

As discussed before, variance non-stationarity is dealt with by the Box-Cox power transformations, which is defined as $u_t = \left(z_t^\lambda - 1\right)/\lambda$ for $\lambda \neq 0 \in IR$. For a given model, the optimal value of λ is found by minimizing the sum of squares of the residuals of the model. In case λ turns out to be close to or equal to zero, a natural logarithmic transformation $u_t = \ln(z_t)$ is used. If non-stationarity is the result of a variable mean, the dth order differenced process $\upsilon_t = (1 - B)^d z_t$ is modeled; setting $d = 1$ or $d = 2$ usually induces constant mean. The ARMA(p,q) model for the differenced process υ is referred to as the ARIMA(p,d,q) model for the process z.

A time series with potential seasonality, indexed by s, is represented by a general ARIMA(p,d,q) (P,D,Q) model: $\phi_p(B)\Phi_p(B^s)(1-B)^d(1-B^s)^D z_t = c+\theta_q(B)\Theta_Q(B^s)\varepsilon_t$ where $\phi_p(B)$ and $\theta_q(B)$ are nonseasonal AR(p) and MA(q) operators; $\Phi_P(B^s)$ and $\Theta_Q(B^s)$ are seasonal AR(P) and MA(Q) operators; and B^s is the seasonal backward shift operator which is defined as $B^s z_t = z_{t-s}$. For hourly data, $s=24$ and $s=168$ indicate daily and weekly seasonality, respectively.

The three-stage Box-Jenkins model building procedure for building ARIMA models involves sample ACF and sample partial autocorrelation functions (PACF) for model identification. Normalized residuals time domain plots, residuals ACF, the Ljung-Box statistics, residuals probability plots, and plotting residuals against the fitted values are popular tests in the diagnostic checking stage.

3.2.2 Dynamic Regression (DR) Model

A dependent variable y can be related to a set of explanatory variables x_i ($i=1,2,\ldots,m$) by a DR model, as follows:

$$
\begin{aligned}
y_t &= c + \left(\phi_1 y_{t-1} + \phi_2 y_{t-2} + \ldots + \phi_p y_{t-p}\right) + \\
&\left(\omega_{1,0} x_{1,t} + \omega_{1,1} x_{1,t-1} + \ldots + \omega_{1,r_1} x_{1,t-r_1}\right) + \ldots + \\
&\left(\omega_{m,0} x_{m,t} + \omega_{m,1} x_{m,t-1} + \ldots + \omega_{m,r_m} x_{m,t-r_m}\right) + \varepsilon_t
\end{aligned} \tag{10}
$$

where, c, ϕ_is and $\omega_{i,j}$s are model parameters to be estimated. Note that up to p lagged values of the dependent variable y, as well as r_i lagged values of the explanatory variable x_i are included in the model. Using the backward shift operator B, model (10) can be represented as:

$$
\phi(B) y_t = c + \sum_{i=1}^{m} \sum_{j=0}^{r_i} \omega_{i,j} B^j x_{i,t} + \varepsilon_t \tag{11}
$$

3.2.3 Transfer Function (TF) Model

In a more general form than the DR model, the relationship between the dependent variable y and the independent variables x_i, can be defined as a rational transfer function term and a disturbance term N_t as follows:

$$
y_t = c + \sum_{i=1}^{m} \frac{\omega_i(B) B^{b_i}}{\delta_i(B)} x_{i,t} + N_t \tag{12}
$$

where $\omega_i(B) = \sum_{j=0}^{r_i} \omega_{i,j} B^j$; $\delta_i(B) = 1 - \sum_{k=1}^{k_i} \delta_{i,k} B^k$; k_i is the order of the polynomial $\delta_i(B)$; b_i is referred to as the delay time for variable x_i; the disturbance term N_t

is expressed by an ARMA model, $N_t = \theta(B)\varepsilon_t/\phi(B)$; and the polynomial operators $\phi(B)$ and $\theta(B)$ were defined before. The model in (12) is referred to as a TF model.

3.3 Accuracy Assessment of Price Forecasting Models

The most popular accuracy measure for electricity market price forecasts are the mean absolute error (MAE) and the mean absolute percentage error (MAPE), and are defined as:

$$MAE = \frac{1}{N}\sum_{t=1}^{N} |p^f{}_t - p^a{}_t| (\$/MWh) \tag{13}$$

$$MAPE = \frac{100}{N}\sum_{t=1}^{N} \frac{|p^f{}_t - p^a{}_t|}{p^a{}_t} (\%) \tag{14}$$

where N is the number of forecast instances in the assessment period, and $p^f{}_t$ and $p^a{}_t$ are the forecast and actual price of electricity at time t, respectively. Typically, N is 24 for hourly prices to determine the daily MAPE. In some other cases, N is 168 for weekly MAPE. For markets where prices sometime hit zero, the above definition of MAPE is modified by replacing the actual price in the denominator with daily or weekly average prices. Other accuracy measures, such as root mean square error have also been used.

New evidences show that using purely statistical measures, such as MAPE, for choosing between alternative forecasting models may not always efficient in terms of the economic impact of the forecasts on the user [45]. In other words, two forecasting models with significantly different MAPE values may result in the same economic impacts. Instead, cost-based error measures that are defined based on the operational characteristics of forecast users should be employed. As an example, from a consumer's point of view, a Forecast Inaccuracy Economic Impact (FIEI) index can be defined as:

$$FIEI(\%) = \frac{Cost^{fp} - Cost^{ap}}{Cost^{fp}} \times 100 \tag{15}$$

Here, $Cost^{fp}$ represents the cost of operation when forecast prices are used for operation planning. $Cost^{ap}$ is the same cost if perfect price forecasts, i.e., actual prices, were available. Note that FIEI indicates the percentage of the actual cost of operation attributable to price forecasting error. From a supplier's point of view, the Economic Loss Index (ELI) may be defined as:

$$ELI(\%) = \frac{Profit^{ap} - Profit^{fp}}{Profit^{ap}} \times 100 \quad (16)$$

where, $Profit^{ap}$ is the gained profit when generation schedules were determined based on perfect price forecasts. $Profit^{fp}$, on the other hand, represents the profit gained when forecast prices were used for generation scheduling. ELI indicates the percentage of profit loss that is due to price forecasting inaccuracies.

4 Short-Term Price Forecasting: A Case Study

In this section, ARIMA and TF models, are applied to forecasting prices for the Ontario electricity market. The data for year 2004 is used in this study and can be found at *www.ieso.ca*. ARIMA models are also built for the locational marginal prices for certain locations in neighboring markets to Ontario, i.e., the New York, New England and PJM markets. A 24-h forecasting horizon is considered in this section.

4.1 Feature Selection

A wide range of data that are available before real-time operation are evaluated here for their relationship with the HOEP fluctuations and a set of potential features that may explain some of the variations of the HOEP is identified. The potential features are selected based on two main criteria. The first criterion is the consideration of implicit and/or explicit effects of the variables on the Ontario market clearing process. The second criterion is the consideration of linear correlations between current HOEP values and current and past values of the variables, since TF and DR are linear models. While these correlations are measured by the Cross Correlation Functions (CCF) [46], linear correlation coefficients between current HOEP values and current values of each feature, referred to as ρ, are the only correlation coefficients discussed here.

The Ontario Market Surveillance Panel (MSP) reports [47] reveal that coal and gas fired generators are the main price setters in the Ontario electricity market. However, although fuel prices have shown to affect the long-term HOEP trends, no short-term relationship between the HOEP and fuel prices was found in [47]. Therefore, the present study does not consider fuel prices among the explanatory variable candidates.

The Ontario Independent Electricity System Operator (IESO) publishes two sets of system operation data prior to real-time dispatch of energy, which are mined here. The first set consists of conventional forecasts for some of the market variables, and is published as the "System Status Report" (SSR). The second set is referred to as the "Pre-Dispatch Report" (PDR) and it provides the market participants with simulation-based forecasts of market outcomes. Potential features from both sets are discussed and a set of most informative features is eventually selected.

4.1.1 Potential Features from the SSR

The SSR provides forecasts for Ontario demand and supply, energy imports, and capacity excess or shortfall; it also contains total planned transmission and generation outages and other market advisory notices. The SSR is released for each day at least 24 h in advance; it is updated in case of any change in the system status or forecasts.

Demand Forecasts

The SSR 24-h-ahead demand forecasts have annual MAPEs of 2.1% and 4.8% for 2004, when compared with actual Ontario demand and actual market demand (demand plus exports and losses), respectively. The linear correlation coefficient between the SSR demand forecast variable and corresponding HOEP values is 0.68. Since demand is one of the main factors influencing the HOEP, and since the SSR provides the most accurate forecast of actual Ontario demand 24 h before real-time, the SSR demand forecast variable is included in the set of potential features.

Predicted Supply Cushion

Supply cushion (SC), that is basically a variant of available reserve in the system, is defined as follows:

$$SC = \frac{CO - (TD + OR)}{TD + OR} \times 100 \tag{17}$$

where CO is the actual capacity offered, TD is the total market demand, and OR is the operating reserve requirement. Data analyses show that price spikes are more likely when the SC is below 10%. In the present case study, (18) is modified and actual quantities are substituted with respective forecasts from the SSR; the resulting SC is referred here to as the Predicted Supply Cushion (PSC). The PSC is found to be linearly correlated with the HOEP ($\rho = -0.60$), and hence is added to the set of potential features.

Planned Outages

Outage of cheap generation facilities can result in higher energy prices, especially during low-demand hours. However, the total outages reported in the SSR are the aggregation of various planned generation and transmission system outages, and this total is found to be not meaningfully correlated with the HOEP ($\rho = 0.18$). Hence, the SSR planed outages variable is not considered in the model building process.

Capacity Excess or Shortfall

The SSR capacity excess or shortfall variable is found to be linearly correlated with the HOEP ($\rho = -0.65$), and hence, is included in the set of potential features. However, it should be noted that when demand is low, capacity excess is high and vice versa, a fact confirmed by the high negative correlation between the SSR capacity excess or shortfall and demand ($\rho = -0.75$). Therefore, the SSR capacity excess or shortfall variable is highly collinear with the SSR demand forecast variable and should be included in the model only after the possible effects of demand have been modeled.

Imports

The SSR import forecast variable deviates significantly from actual values. Therefore, the import forecasts are not considered in the set of explanatory variable candidates. No export forecasts are published in the SSR.

4.1.2 Potential Features from the PDR

The Ontario market clearing algorithm is run in two time frames, namely, the pre-dispatch and the real-time (dispatch) [13]. The pre-dispatch run provides the market participants with the "projected" schedules and prices, based on the most recent available market information. Outcomes of the pre-dispatch run are published by the IESO as the PDR for a variety of variables, including energy and operating reserves prices, total load, dispatchable load not served, system losses, and some of the system security constraints. From the PDR variables, the PDP and Pre-Dispatch Demand (PDD) variables carry the latest information about demand and price in the coming hours; hence, they are examined here for their role in improving accuracy of HOEP forecasting.

The PDP/PDD values that correspond to hour t and that are published k hours before real-time are called k-hour-ahead PDPs/PDDs. Linear correlation coefficients between the HOEP and k-hour-ahead PDPs ($\rho_{HOEP,PDP}$), and between the HOEP and k-hour-ahead PDDs ($\rho_{HOEP,PDD}$), for $k = \{1, 2, 3, 24\}$, are presented in

Table 2 Correlation between hourly Ontario energy price (*HOEP*), *k*-hour-ahead pre-dispatch prices (*PDPs*), and pre-dispatch demands (*PDDs*)

k	24	3	2	1
$\rho_{HOEP,\ PDP}$	0.16	0.74	0.77	0.78
$\rho_{HOEP,\ PDD}$	0.62	0.63	0.63	0.64
$\rho_{PDD,\ Demand}$	0.97	0.98	0.98	0.98

Table 2. Linear correlation coefficients between actual Ontario market demand and k-hour-ahead PDDs ($\rho_{PDD,\ Demand}$) are also presented in Table 2.

k-Hour-Ahead PDPs

It can be inferred from the correlation coefficients presented in Table 2 that when k is small, the k-hour-ahead PDPs are closer to the HOEP. Hence, considering k-hour-ahead PDPs as explanatory variables depends on the forecasting horizon. For 24-h-ahead forecasting, the 24-h-ahead PDP variable is clearly not useful and hence is not considered as an explanatory variable candidate.

k-Hour-Ahead PDDs

It can be observed from Table 2 that the k-hour-ahead PDDs do not deviate significantly from actual market demand, thus these should be included in the set of potential features. However, since the accuracy levels of the 24-h-ahead PDDs and the SSR demand forecasts are very close, only one of them, i.e., the SSR demand forecast, is included in the set of potential features.

4.1.3 Demand and Energy Price in the Neighboring Areas

The Ontario electricity market is interconnected with the New York electricity market, and Quebec, Michigan, Manitoba, and Minnesota control areas. The last three control areas are now part of the Midwest market. The New York electricity market is also interconnected with the PJM and New England electricity markets, and New England and PJM trade energy with Quebec and Michigan. With such a complex interconnection between neighboring areas, it is difficult to assess the effects of energy price and demand of the neighboring areas on the HOEP. Furthermore, lack of publicly available information on quantity and price of energy transactions between Ontario and Quebec, Michigan, Manitoba, and Minnesota constrained the authors to consider only the effects of demand and price of the New York, New England and PJM electricity markets on the HOEP. The data employed are available at www.nyiso.com, www.iso-ne.com and www.pjm.com.

Table 3 Correlation between demand in the neighboring markets, Ontario Hourly Energy Price (*HOEP*) and Ontario demand, year 2004

	New York demand	New England demand	PJM demand
HOEP	0.54	0.63	0.37
Ontario demand	0.83	0.89	0.52

Demand

To evaluate the possible effects of New England and PJM market demands on HOEP, actual demand data from these markets are considered. But, since these data are not available before real-time, they cannot be considered in the final models even if they turn out to be significant. For the New York market, historical demand forecasts are available and hence are used in this study.

Linear correlation coefficients between demand in the neighboring markets and demand and price in Ontario market are presented in Table 3. Climatic conditions that are similar across New York, Ontario, and New England could be a reason for the collinearity in demand between these markets. Consequently, demand collinearity could be the reason for high correlation between the HOEP and the New York and New England demands. On the other hand, the low correlation between the Ontario and PJM demands can be attributed to variations in the residential and industrial load distribution pattern across these markets, plus climatic differences between the two.

In this study, the New York and New England markets demands are considered as potential features. However, due to the collinearity between the Ontario demand and the other demands, they should be included in the model building process only after the effects of the Ontario demand on the HOEP are modeled. Given its small correlation with the HOEP, the PJM market demand is not considered an explanatory variable candidate.

It was also observed that actual quantities of power transactions through the Ontario-New York transmission link (intertie) had no meaningful correlation with demand or price in the neighboring markets. This lack of correlation is due to the fact that much of the overall transactions constitute power wheeling transactions from different parties taking place through this intertie.

Price

Only day-ahead prices in the neighboring markets are examined for their possible effects on the HOEP, because they are known before real-time. Note that the main components of the costs of any energy transactions between Ontario and the neighboring markets are the HOEP and the LMPs at the pricing points in those markets involved in the trade. These LMPs are denoted as LMP_{NYON} for the New York to Ontario interface in the New York market, LMP_{NENY} for the New England to New York interface in the New England market, and LMP_{PJMON} for the PJM to Ontario interface in the PJM market. Thus, only these three LMPs are studied here.

Table 4 The final set of potential features

Variable	ρ
x_1: Predicted supply cushion (PSC)	−0.60
x_2: The SSR Ontario demand forecast	0.68
x_3: New England market demand	0.63
x_4: New York market demand	0.56
x_5: LMP_{NENY}	0.67
x_6: LMP_{NYON}	0.69
x_7: LMP_{PJMON}	0.67
x_8: The SSR capacity excess	−0.65

The HOEP is correlated to LMP_{NYON}, LMP_{NENY} and LMP_{PJMON}, with ρ values of 0.69, 0.67, and 0.67, respectively. Hence they are considered as explanatory variable candidates. Since market prices are influenced mainly by demand; however, the high correlations may be due to the similar demand patterns in the neighboring areas. Therefore, the mentioned LMPs need to be included in the HOEP models only after the possible effects of demand in Ontario and other markets on the HOEP are properly modeled.

The final set of potential features are denoted by x_1 to x_8 and are summarized in Table 4.

4.2 *Time Series Models for the HOEP*

4.2.1 General Considerations

Three time periods, each of 2 weeks duration, are selected for building the time series models and generating HOEP forecasts. The first period comprises two consecutive weeks from April 26 to May 9, 2004, referred to as Week_1 and Week_2; during this period, the Ontario market demand reached its spring low point. The second period comprises two consecutive summer peak-demand weeks from July 26 to August 8, 2004, and are referred to as Week_3 and Week_4. The last period includes two high-demand winter weeks in 2004, spanning December 13–26, and are referred to as Week_5 and Week_6.

Models for each of the 6 weeks have been individually identified, estimated and checked. The ARIMA models are built using 4 weeks of historical data, while the TF and DR models are developed based on 10 weeks of historical data. The main criteria for identifying the final models are as follows: diagnostic checking tests (see Sect. 3.2.1); the principle of parsimony, i.e., a simpler model is generally better; t-value of the estimated model parameters; Akaike Information Criterion; out-of-sample forecasts accuracy; and reality of the identified models.

To illustrate some results of the diagnostic checking stage, the residuals ACFs of the 24-h-ahead forecasts by the TF model developed for Week_3 (in Sect. 3) are presented in Fig. 7; the horizontal bands in this figure represent the significance

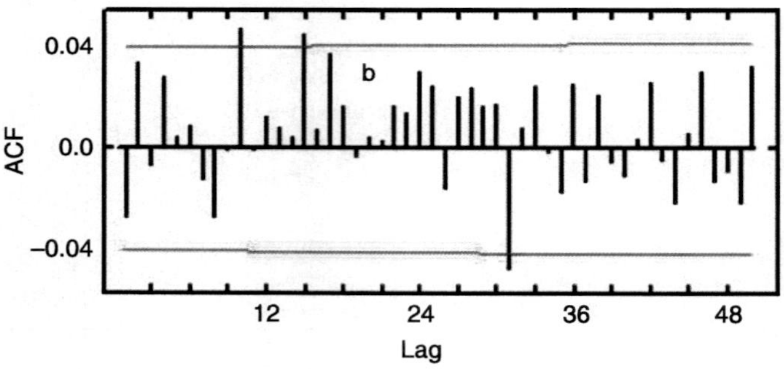

Fig. 7 Residuals ACF of the TF model

Table 5 The ARIMA models for the hourly Ontario energy price (*HOEP*)

$Week_1$	(1) (24, 25, 72, 119) (168, 169) Z_t = (1) (24) ε_t
$Week_2$	(1) (24, 25, 72) (168, 169, 336) Z_t = (1) (24) ε_t
$Week_3$	(1, 2) (24, 25) (168) Z_t = (1, 2, 3) (24) ε_t
$Week_4$	(1, 2) (24) Z_t = (1, 2) (24) ε_t
$Week_{5,6}$	(1) (23) Z_t = (1, 2, 3, 4) (24) ε_t

limits of the ACFs. Observe that no significant correlations for the first few lags and the relevant seasonal lags (24, 48) exist.

The Scientific Computing Associates (SCA) statistical system is used to build the proposed models. To deal with outliers, the Chen-Liu algorithm for joint estimation of model parameters and outliers, implemented in the SCA system, is employed. Natural logarithmic transformation is found to be the optimal Box-Cox transformation for variance stability in this study, given the historical data and the identified models. Furthermore, a seasonal differencing with $s = 24$ was enough to induce mean stationarity in all models. The transformed differenced HOEP time series, $(1–B^{24})$ ln $(HOEP_t)$, is referred to as Z_t here onward.

4.2.2 ARIMA Models for the HOEP

ARIMA is a univariate model. Thus, none of the potential features, discussed in the previous section, are considered for model building. A shorthand convention is employed here for simplicity to show the developed ARIMA models. According to this convention, an AR or an MA operator is represented by the orders of the respective backward shift operator. For example, the ARIMA model $(1 - \phi_1 B) \times (1 - \Phi_{24}B^{24} - \Phi_{47}B^{47})z_t = (1 - \Theta_2 B^2)(1 - \Theta_{24}B^{24})\varepsilon_t$ is shown as (1) (24, 47) z_t = (2) (24) ε_t. The ARIMA models developed for each of the six studied weeks are listed in Table 5.

4.2.3 TF Models for the HOEP

TF is a multivariate model. Building such models involves a careful selection of final features that efficiently explain price fluctuations. Two important factors should be kept in mind when building multivariate parametric models, namely, model simplicity and model reality. A model should be as simple as possible to minimize the number of model parameters to be estimated. Estimating a large number of parameters can result in biased estimations that make the model invalid. Model reality is necessary to avoid any meaningless relationship between the target variable and the explanatory variables. As an example, electricity price at hour t may be dependent upon electricity demand at hour $t-4$, but not at hour $t-17$.

An initial analysis of the potential feature, selected in Sect. 4.1, reveals that some of the variables in that set are collinear. Multicollinearity arises in a regression problem if there is a linear dependency among the explanatory variables. In this study, a two-step procedure is designed for building the TF models in the presence of multicollinearity among the potential features, as follows:

1. In the first step, market knowledge, theoretical justifications, and linear correlation between the HOEP and the explanatory variable candidates are used to choose the most influential explanatory variable, referred to as the "first-step variable". TF models for the HOEP are fully built assuming that the first-step variable is the only explanatory variable. In this step, the power transformation and the differencing order, which are needed for stabilizing the variance and the mean of the time series are identified. Given the critical effect of Ontario demand on the HOEP, the SSR demand forecast variable, namely x_2, is selected as the first-step variable.
2. In the second step, the general form of the transfer function term associated with the first-step variable is kept constant and other variables are added to the model in a step-wise manner; variables with collinearity with the first-step variable are considered first. The performance of the new models is monitored using the identification criteria mentioned in Sect. 1. The transfer function terms associated with each significant variable, as well as the disturbance terms, are modified appropriately in this step, and the final model is identified by adding other explanatory variable candidates and repeating this step.

Note that the set of potential features was selected based on a filter paradigm. However, the final models in this study are built according to the wrapper filter selection paradigm.

Lags 1, 2, 3, 4, 23, 24, 25, 47, 48, 49, 71, 72, 73, 95, 96, 97, 119, 120, 121, 143, 144, 145, 167, 168, 169, 335, and 336, of each candidate feature are considered for building TF models. The inclusion of trading-day effects, i.e., adding the index of the day as a dummy feature, in the TF models was not found to improve overall forecast accuracy; this can be attributed to the fact that demand forecasts are already used as model inputs, carrying the corresponding trading-day information.

Using this procedure, the following TF models are finally identified:

Table 6 Disturbance terms for the TF models

Week_1	(1) (24, 25, 72) N_t^{24} = (1, 2, 3, 4) (24) ε_t
Week_2	(1) (23, 24, 25, 72) N_t^{24} = (24) ε_t
$\text{Week}_{3,4}$	(1) (24, 25) (168, 169) N_t^{24} = (1, 2, 3) (24) ε_t
$\text{Week}_{5,6}$	(1) (24) N_t^{24} = (2) (24) ε_t

$$Z_t = \sum_{j=0}^{2} \left(\left(\omega_{3,j} B^j \right) \left(1 - B^{24} \right) x_{3,t} + \left(\omega_{4,j} B^j + \right) \left(1 - B^{24} \right) \ln \left(x_{4,t} \right) \right) + N_t \tag{18}$$

where the disturbance terms corresponding to model (19) is presented in Table 6 for each of the study periods.

4.2.4 ARIMA Models for the Neighboring Markets' LMPs

In order to compare the price behavior in Ontario with that in the neighboring markets, ARIMA models are also developed for LMP_{NENY}, LMP_{NYON}, and LMP_{PJMON} (x_5, x_6 and x_7). Ten weeks of historical data are used to identify and estimate the following models for Week_1:

1. New England: (1) (23, 24, 25, 48, 72, 96, 120, 144) (167, 168, 169, 170) × $(1 - B^{24}) \ln (x_{5,t})$ = (1, 2, 3, 4, 5) (24, 25) ε_t
2. New York: (1, 2) (24, 48, 49, 72, 96) (168, 169, 336, 337, 504) × $(1 - B^{24}) \ln (x_{6,t})$ = (1) (24, 48, 72, 96) ε_t
3. PJM: (1, 2, 3) (24, 25, 26, 47, 72) (167, 168, 169) × $(1 - B^{24}) \ln (x_{7,t})$ = (24) (167, 168, 169) ε_t

It is observed that the studied LMPs exhibit a stable behavior; in other words, models (20), (21), and (22) fit the data well for Week_2, and even for Week_3, and Week_4. However, the final identified ARIMA, TF, and DR models for the HOEP are different for the studied weeks. The unstable behavior of the HOEP models highlights the fact that these models have to be re-identified and re-estimated after new observations are available. The need for model re-identification implies that market participants cannot count on a single model in order to produce HOEP forecasts in a non-supervised automatic manner, a fact that must be taken into account for commercialization of the forecasting models.

4.3 Discussions and Numerical Results

Observe from the final TF models identified for the HOEP that, PSC and New England demand are identified as significant variables in the second step. New England electricity market prices are generally higher than the HOEP, a factor affecting exports from Ontario, which in turn affects HOEP behavior. This would explain why New England demand appears in the developed TF and DR models. Also, demand and price from other markets, and the SSR capacity excess or shortfall variable are insignificant variables in the developed models, thereby implying that they do not carry additional information once the effects of the Ontario and New England demands are modeled.

4.3.1 Forecasting Results for Ontario

The weekly MAPEs and MAEs of the generated HOEP forecasts for the six studied weeks are presented in Table 3. The accuracy measures of the IESO-generated PDPs are also presented in this table for reference. The forecasting origin for a set of 24-h-ahead forecasts is the last hour of the previous day; for example, for the 24-h-ahead forecast horizon spanning from today midnight to tomorrow midnight, the forecasting origin is 11 PM today. It is to be noted that the PDP values used in the present work have the same forecasting origins as the generated HOEP forecasts.

The results presented in Table 3 show that the accuracy of the generated HOEP forecasts is significantly higher than that of the IESO-generated PDPs. In addition, these results clearly show that for the high-demand period ($Week_3$ to $Week_6$), multivariate models outperform the univariate models; the MAPEs of the 24-h-ahead forecasts generated by the TF models improves by 2.1%. In terms of MAE, the improvement achieved over the ARIMA models by the TF models is \$1/MWh. However, for the low-demand period ($Week_1$ and $Week_2$), inclusion of the market data in the multivariate models does not improve forecast accuracy.

Although inclusion of the "before-the-fact" market data into the forecasting models improves forecast accuracy to some extent, this improvement is not significant. The small improvements in accuracy of the multivariate HOEP models can be attributed to the real-time nature of the Ontario market. In Ontario, the entire demand obligation has to be cleared in real-time; thus, given the "hockey stick-shape" generation offer curve in Ontario, unpredictable events such as demand over-forecasting, demand under-forecasting, and import/export failures oblige the market operator to commit expensive units on the "blade" portion of the offer curve, or de-commit some of the already committed units and move back on the "shaft" portion of the offer curve. This requirement puts upward or downward pressures on the HOEP, leading to price spikes. Furthermore, out-of-market actions by which the market operator manipulates the market clearing procedure affect the patterns behind the price behavior. Hence, the HOEP is highly volatile, and the information contents of the before-the-fact market data have a high level of uncertainty.

The 24-h-ahead HOEP forecasts generated by the ARIMA and the TF models are plotted against the corresponding actual HOEP values for $Week_4$ in Fig. 8, which presents the poorest forecasting results. During $Week_4$, the prices are unusually high for the first 2 days, and relatively low for the rest of the week. Note that none of the models can reasonably forecast the unusually high or low prices,

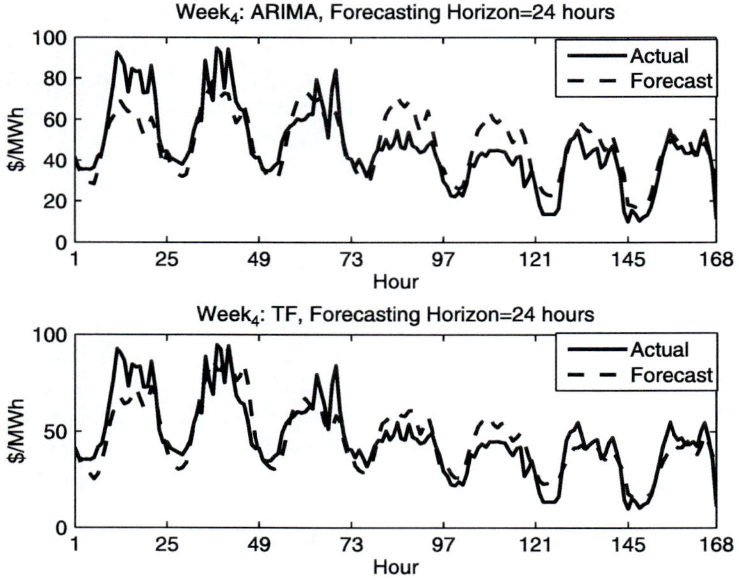

Fig. 8 Twenty-four-hour-ahead HOEP forecasts for Week_4 by the ARIMA and TF models

Table 7 Weekly MAPEs (%) of the neighboring LMP forecasts, and, the volatilities

	New England	New York	PJM	Ontario
Week_1	6.2	6.9	10.1	15.9
Week_2	5.4	7.1	11.4	18.6
Week_3	3.7	6.1	8.7	13.6
Week_4	7.1	8.1	17.3	21.5
Average	**5.6**	**7.1**	**11.9**	**17.4**
$\overline{\sigma}_{24,24}$	6.76	8.37	12.94	28.05

although the TF models predict high/low prices relatively better than ARIMA (Table 7).

4.3.2 Forecasting Results for the Neighboring Markets

The models developed in Sect. 4 are used to generate 24-h-ahead forecasts for the studied LMPs for Week_1 to Week_4. The time duration of these weeks is the same as that defined in Sect. 1, and correspond to two typical low-demand and two typical high-demand weeks in these markets. For comparison, the calculated weekly MAPEs of the forecasts are presented in Table 8 along with the respective results for the HOEP forecasts. Observe that the accuracy of the forecasts generated for the New England, New York, and PJM day-ahead market LMPs, x_7, x_8, and x_9, is higher than the accuracy of the HOEP forecasts. To partly explain the relatively low

Table 8 Weekly MAPEs (%), and weekly MAEs ($/MWh) for the hourly Ontario energy price (*HOEP*) models

	ARIMA		TF		PDP	
	MAPE	MAE	MAPE	MAE	MAPE	MAE
$Week_1$	15.9	7.2	15.6	7.1	39.7	17.5
$Week_2$	18.6	8.2	18.0	8.2	30.3	12.0
Average	**17.2**	**7.7**	**16.8**	**7.7**	**35**	**14.7**
$Week_3$	13.6	6.9	12.3	6.4	36.9	20.6
$Week_4$	21.5	8.7	18.3	7.3	31.6	12.3
$Week_5$	15.4	9.6	14.8	9.2	60.2	34.3
$Week_6$	20.8	12.0	17.5	10.1	37.3	22.8
Average	**17.8**	**9.3**	**15.7**	**8.2**	**41.5**	**22.5**
6-week Average	**17.6**	**8.8**	**16.1**	**8.1**	**40**	**19.9**

accuracy level of the HOEP forecasts, the daily volatilities of prices in the four studied markets, calculated in Sect. 2.4, are also presented in Table 8. Note that price predictability for prices with lower volatility is higher. In general, high price volatility indicates that of a high portion of the price is unpredictable, and thus, hard to be predicted by any model.

5 Concluding Remarks

Modeling and forecasting short-term electricity market prices helps market participants in optimizing their operational strategies. However, those prices follow non-stationary, non-linear and highly volatile processes, and hence difficult to predict. The complexity in the price process arises from the facts that electricity must be generated and consumed instantaneously and a electricity market is not only governed by economic principles but also but physics laws.

Over the past decade, several numerical price forecasting approaches have been proposed in the literature that predict the exact value of future prices. The reported errors range between 5% and 36% for different markets, with the highest errors reported for real-time markets. This range of error is significantly higher than the typical 1–3% error range of electricity load forecasting. Alternatively, some other research works have focused on predicting a confidence interval for prices, forecasting price spikes, or classifying the future prices with respect to user-defined thresholds.

It can be observed from the literature that the price-demand relationship has been generally considered unidirectional, i.e., the demand does not respond to prices. However, with the advent of "smart grid" technologies and new market place initiatives, the consumers of electricity are expected, and will be enabled, to be more active in participating in electricity markets by responding to high prices. Such a price-responsive environment would make the price formation dynamics more complicated than before. Thus, new forecasting approaches would be required to capture the interrelationship between price and demand and forecast them in a smart grid environment.

References

1. Bhattacharya K, Bollen MH, Dallder JE (2001) Operation of restructured power systems. Kluwer, Boston
2. Conejo AJ, Nogales FJ, Arroyo JM (2002) Price-taker bidding strategy under price uncertainty. IEEE Trans Power Syst 17(4):1081–1088
3. Vehvilinen I, Keppo J (2003) Managing electricity market price risk. Eur J Oper Res 145(1):136–147
4. Conejo A, Garcia-Bertrand R, Diaz-Salazar M (2005) Generation maintenance scheduling in restructured power systems. IEEE Trans Power Syst 20(2):984–992
5. Nogales FJ, Contreras J, Conejo AJ, Espinola R (2002) Forecasting next-day electricity prices by time series models. IEEE Trans Power Syst 17(2):342–348
6. Huisman R, Mahieu RJ (2003) Regime jumps in electricity prices. Energ Econ 25(5):425–434
7. Ruibal C, Mazumdar M (2008) Forecasting the mean and the variance of electricity prices in deregulated markets. IEEE Trans Power Syst 23(1):25–32
8. Gonzalez AM, Roque AMS, Garcia-Gonzalez J (2005) Modeling and forecasting electricity prices with input/output hidden Markov models. IEEE Trans Power Syst 20(1):13–24
9. Niimura T (2006) Forecasting techniques for deregulated electricity market prices - extended survey. In: 2006 IEEE PES power systems conference and exposition PSCE'06, Atlanta, 2006, pp 51–56
10. Vehviläinen I, Pyykkönen T (2005) Stochastic factor model for electricity spot price–the case of the nordic market. Energ Econ 27(2):351–367, special Issue on Electricity Markets
11. Zareipour H, Canizares C, Bhattacharya K, Thomson J (2006) Application of public-domain market information to forecast Ontario wholesale electricity prices. IEEE Trans Power Syst 21(4):1707–1717
12. Weron R (2006) Modeling and forecasting electricity loads and prices: a statistical approach. Wiley, Chichester
13. Zareipour H, Canizares CA, Bhattacharya K (2007) The operation of Ontario's competitive electricity market: overview, experiences, and lessons. IEEE Trans Power Syst 22(4): 1782–1793
14. Box GEP, Jenkins GM, Reinsel GC (1994) Time series analysis, forecasting and control. Prentice Hall, Englewood Cliffs
15. Zhao JH, Dong ZY, Li X, Wong KP (2007) A framework for electricity price spike analysis with advanced data mining methods. IEEE Trans Power Syst 22(1):376–385
16. AESO (2009) The Alberta electric system operator (AESO). http://www.aeso.ca/
17. Li Y, Flynn PC (2004) Deregulated power prices: comparison of volatility. Energ Policy 32 (14):1591–1601
18. Zareipour H, Bhattacharya K, Canizares C (2007) Electricity market price volatility: the case of Ontario. Energ Policy 35(9):4739–4748
19. Simonsen I (2005) Volatility of power markets. Phys A Stat Mech Appl 335(1):10–20
20. Amjady N, Keynia F (2009) Day-ahead price forecasting of electricity markets by mutual information technique and cascaded neuro-evolutionary algorithm. IEEE Trans Power Syst 24 (1):306–318
21. Lora A, Santos J, Exposito A, Ramos J, Santos J (2007) Electricity market price forecasting based on weighted nearest neighbors techniques. IEEE Trans Power Syst 22(3):1294–1301

22. Bompard E, Ciwei G, Napoli R, Torelli F (2007) Dynamic price forecast in a competitive electricity market. IET Gener Transm Distrib 1(5):776–783
23. Weron R, Misiorek A (2008) Forecasting spot electricity prices: a comparison of parametric and semiparametric time series models. Int J Forecast 24(4):744–763
24. Fan S, Mao C, Chen L (2007) Next-day electricity-price forecasting using a hybrid network. IET Gener Transm Distrib 1(1):176–182
25. Rodriguez CP, Anders GJ (2004) Energy price forecasting in the Ontario competitive power system market. IEEE Trans Power Syst 19(1):366–374
26. Aggarwal S, Saini L, Kumar A (2009) Day-ahead price forecasting in Ontario electricity market using variable-segmented support vector machine-based model. Electr Power Compon Syst 37(5):495–516
27. Saini L, Aggarwal S, Kumar A (2010) Parameter optimisation using genetic algorithm for support vector machine-based price-forecasting model in national electricity market. IET Gener Transm Distrib 4(1):36–49
28. Taylor J, McSharry P (2007) Short-term load forecasting methods: an evaluation based on European data. IEEE Trans Power Syst 22(4):2213–2219
29. Ipakchi A, Albuyeh F (2009) Grid of the future. IEEE Power Energ Mag 7(2):52–62
30. Zhang L, Luh P (2005) Neural network-based market clearing price prediction and confidence interval estimation with an improved extended Kalman filter method. IEEE Trans Power Syst 20(1):59–66
31. Zhao JH, Dong ZY, Xu Z, Wong KP (2008) A statistical approach for interval forecasting of the electricity price. IEEE Trans Power Syst 23(2):267–276
32. Boogert A, Dupont D (2008) When supply meets demand: the case of hourly spot electricity prices. IEEE Trans Power Syst 23(2):389–398
33. Zareipour H, Janjani A, Leung H, Motamedi A, Schellenberg A (2011) Classification of future electricity market prices. IEEE Transactions on Power System 26(1):165–173
34. Conejo AJ, Plazas MA, Espinola R, Molina AB (2005) Day-ahead electricity price forecasting using the wavelet transform and ARIMA models. IEEE Trans Power Syst 20(2):1035–1042
35. Han J, Kamber M (2006) Data mining: concepts and techniques, The Morgan Kaufmann series in data management systems. Morgan Kaufmann, Amsterdam
36. Witten IH, Frank E (2005) Data mining: practical machine learning tools and techniques. Morgan Kaufmann, Amsterdam
37. Saeys Y, Inza I, Larranaga P (2007) A review of feature selection techniques in bioinformatics. Bioinformatics 23(19):2507–2517
38. Guyon I, Elisseeff A (2003) An introduction to variable and feature selection. J Mach Learn Res 3:1157–1182
39. Dash M, Liu H (1997) Feature selection for classification. Intell Data Anal 1(3):131–156
40. Li G, Liu CC, Mattson C, Lawarree J (2007) Day-ahead electricity price forecasting in a grid environment. IEEE Trans Power Syst 22(1):266–274
41. Contreras J, Espinola R, Nogales F, Conejo A (2003) ARIMA models to predict next-day electricity prices. IEEE Trans Power Syst 18(3):1014–1020
42. Pindoriya N, Singh S, Singh S (2008) An adaptive wavelet neural network-based energy price forecasting in electricity markets. IEEE Trans Power Syst 23(3):1423–1432
43. Zareipour H, Bhattacharya K, Canizares C(2006) Forecasting the hourly Ontario energy price by multivariate adaptive regression splines. In: Proceedings of the IEEE PES Annual General Meeting, San Francisco, 2006, p 7
44. Conejo AJ, Contreras J, Esanola R, Plazas MA (2005) Forecasting electricity prices for a day-ahead pool-based electric energy market. Int J Forecast 21(3):435–462
45. Zareipour H, Canizares C, Bhattacharya K (2010) Economic impact of electricity market price forecasting errors: a demand-side analysis. IEEE Trans Power Syst 25(1):254–262
46. Pankratz A (1991) Forecasting with dynamic regression models. Wiley, New York
47. MSP (2011) Monitoring reports on the IESO-administered electricity markets, The Market Surveillance Panel, Ontario, 2002–2005. www.ieso.ca/imoweb/marketSurveil/mspReports.asp

Forecasting Prices in Electricity Markets: Needs, Tools and Limitations

H.A. Gil, C. Gómez-Quiles, A. Gómez-Expósito, and J. Riquelme Santos

Abstract Electricity is a fundamental good for society. The price at which it is sold as a commodity influences all levels of economic activity and determines the profits and benefits that generators and consumers reap from participating in the electricity markets. Forecasting the electricity prices at different time-frames, namely in the short-run (daily), medium-term (seasons) or long-term (years), is of foremost importance for all industry stakeholders for cash flow analysis, capital budgeting and financial procurement as well as regulatory rule-making and integrated resource planning, among others. On the other hand, the process of price formation in competitive electricity markets is unique in terms of the different factors that come into play in the settlement process. These factors, which may be endogenous or exogenous to the market, bring about uncertainty and volatility to the electricity prices. This uncertainty hinders the forecast user's ability to estimate the prices with accuracy at the different time-frames. This chapter explores the different reasons why forecasting electricity prices is necessary in electricity markets, the most widely used methodologies for short-term electricity price forecasting and their fundamental common limitations. This analysis is carried out using actual electricity price datasets.

Keywords Electricity markets • forecasting • heuristic and statistical models • prices • uncertainty • volatility

---

H.A. Gil • C. Gómez-Quiles • A. Gómez-Expósito (✉) • J.R. Santos
Department of Electrical Engineering, University of Seville, Seville, Spain
e-mail: hgil@us.es; catalinagq@us.es; age@us.es; jsantos@us.es

A. Sorokin et al. (eds.), *Handbook of Networks in Power Systems I*,
Energy Systems, DOI 10.1007/978-3-642-23193-3_5,

1 The Need for Electricity Price Forecasting

The notions of electricity market and power industry deregulation were first put into practice by Chile in the early 1980s. The UK followed suit in 1990, an example adopted in the subsequent years by a number of Commonwealth countries, namely Australia, New Zealand and Canada (Ontario and Alberta). Several jurisdictions in the United States and Europe have also embarked in the industry deregulation process, such as the PJM, New York, New England, California, Spain, Italy, etc.

All these jurisdictions have laid out different priorities and objectives for the deregulation of their electricity industries and there has not been a single recipe that guarantees success in all settings. These deregulation strategies have been, however, initiated by a industry unbundling through which the activities of generation, transmission, distribution and, generally, commercialization, are separated into different independently owned and operated companies. These independent companies may or may not be privatized, according to the needs of the regulators or governments. The most common goal in this process is the establishment of a competitive electricity market in the supply side, generally combined with the introduction of competition in the demand side as well. Although the implementation details for each of these markets varies a great deal according to the jurisdiction, the most important outcome of an electricity market is the formation of a price at which all power is traded, at least on a daily basis, by way of the so-called ‘Spot’ market.

In most electricity markets, independent bilateral trading between suppliers and consumers is permitted, even encouraged and sometimes mandatory. The resulting bilateral contracts can be physical (with the option of generation self-scheduling) or financial and electricity is sold and purchased at pre-negotiated fixed price, unknown to the operator or the other market participants. Nonetheless, not all available supply and demand can be sold and purchased through bilateral contracts for obvious reasons. The daily (spot) electricity market serves as a marketplace of last resort for generators and demands to trade their remaining available not-contracted power. Almost all spot electricity markets currently under operation have implemented a mandatory day-ahead bidding framework, which may or may not be complemented with intra-day and real-time (balancing) markets. In terms of the amount of energy being traded, the day-ahead market is the most significant one among all spot (intraday or real-time) markets. Even if most of the electricity is traded through bilateral contracts (at a price only known to the two parties involved), the electricity prices cleared at the spot market have a substantial influence on the contract prices since any shortfall or surplus in generation or consumption with respect to the contracted positions must be traded at the spot prices.

As a matter of fact, the economics of the whole electricity industry hinges a great deal on the electricity prices cleared at the market. There are, therefore, a number of reasons why forecasting future electricity prices is crucial for all stakeholders, depending on the time frame under which this forecasting is carried out. In the short-run (from 3 to 24 h) electricity price forecasting is especially important in electricity markets in which participants must optimize their positions (bidding price and quantity for the various markets, namely day-ahead and intraday) based on their perception of what the future hourly prices and incremental costs will be over the bidding period. Moreover, some agents, especially large consumers with self power production, are able to decide which portion of their consumption is to be supplied by the market or by their own production and the corresponding timing. Nonetheless, the driving force behind the decision-making of all market participants is the maximization of profit, which is given by the difference between price and their own incremental cost. Estimating the profits for future bidding exercises calls for an estimation of the electricity prices at which the electricity will be traded over the bidding period. In the medium and long-term, generators and demands must work out estimates of future market clearing prices in order to put together optimal spot versus bilateral-contract supply portfolios. When it comes to investment in new generating capacity, technological upgrades or retrofits, estimating future electricity prices at different time frames (usually in the long-term) is also determinant for the calculation of future cash flows necessary for capital budgeting analysis and financing procurement (capital or equity). Particularly to investments in renewable energy technologies, government incentives are common, which are usually tied up to the evolution of typical technological costs. Profitability in this type of investments is determined not only by how the incentives are rolled out in time but also on the expected prices at which the output will be sold over the lifetime of the project.

Electricity price forecasting is also important for regulators and policy-makers. Electricity is a fundamental commodity and its price affects all levels of the economic activity. In fact, one of the main goals of regulators is to ensure that electricity is traded at fair prices, by making sure that the market remains competitive, that open access is guaranteed to new participants and, through anti-trust regulation, that no generators take advantage of their potential dominant positions. It is desirable that regulatory actions anticipate the evolution of the market, that is, that regulation be preventive and not corrective. Forecasting electricity prices in the long-run considering indicative generation capacity expansion plans, integrated resource planning, demand growth, etc., is crucial for the development of optimal regulatory actions put together to indirectly steer the market toward the government's basic objectives with minimal intervention.

With the previous notions in mind, Fig. 1 summarizes the main applications for which electricity price forecasting is necessary in the short-run or long-term, according to the stakeholder involved. There are a number of challenges, however, that get on the way of an accurate price forecasting process at the different time-frames. This chapter will examine in detail the different factors that come into play against this process. A description of the most common electricity price forecasting methodologies is also included. The main reason why electricity price forecasting is challenging is simply because prices are uncertain. The following section will discuss the different sources of uncertainty that characterize the price formation process in electricity markets and how this process is particularly different from other economic market-based activities.

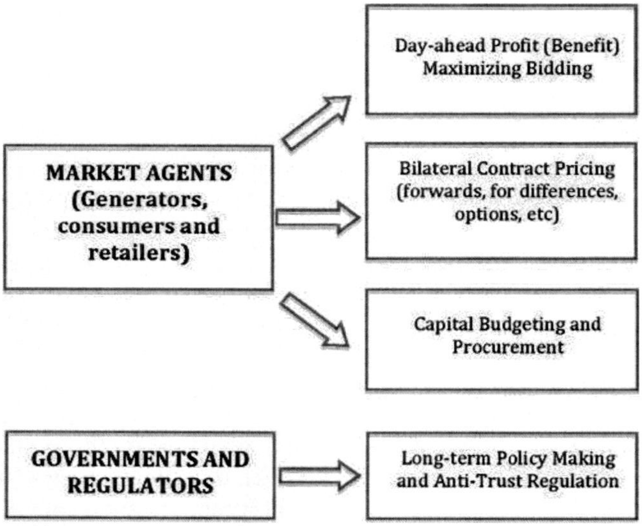

Fig. 1 Main applications for electricity price forecasting

2 Uncertainty in Electricity Markets

Forecasting electricity prices is a challenging task not only because the prices are uncertain but, most importantly, because of the particularities of how these prices are brought into being. The process of price formation in electricity markets follows in essence the basic rule of microeconomic theory by which the price of the underlying commodity in a competitive market should reflect the relative scarcity of the supply for a given demand level. If the demand for a commodity is low, those suppliers with higher incremental costs must step out of competition (or make negative profits) and give way to suppliers with the lowest incremental costs. This process results in relatively low equilibrium prices. On the other hand, as the demand increases, those suppliers with the lowest incremental costs are the first ones to enter the market and use up their production capacity so more and more expensive suppliers have to come in to supply the increasingly scarce commodity, rising the equilibrium price. This process is observed in electricity markets on a regular basis. The market clearing prices tend to follow closely the daily and seasonal swings in consumption. If consumption and price were determined by a one-to-one deterministic relationship, anticipating the electricity prices would boil down to forecasting accurately the demand, which is one of the most investigated problems in power systems operation and planning. The influence of the demand on the electricity prices is, however, far from being deterministic. There are a series of factors that bring about uncertainty to the price formation process even if the demand is known with certainty. This will be discussed next.

2.1 *Sources of Uncertainty in the Electricity Price Formation Process*

The different sources of uncertainty in the price formation process may be classified according to whether these sources are *endogenous* or *exogenous* to the process. The most important endogenous factors are:

- *Demand*: As mentioned above, the demand is the counterpart of the supply (generation) in the market clearing mechanism, with the price being its most important outcome. If the incremental costs of the suppliers are assumed constant, the equilibrium price is therefore defined by the level of consumption. The demand is therefore the most determinant factor outside the supply side in the market clearing mechanism. Uncertainty is added to this process since the total demand is a mix of consumption patterns that vary across different types of consumers (residential, commercial, industrial, public service, etc.). These patterns are also influenced by external factors such as the day-by-day weather fluctuations, seasons, economic cycles (prosperity or recession), etc. The uncertain characteristic of the demand is carried over into the electricity prices through the market settlement process.
- *The exercise of market power by dominant generators*: According to the fundamentals of Game Theory, the price of a commodity in perfectly competitive electricity markets is set in such a way that no generator will increase profits by increasing its output unilaterally, resulting in a market equilibrium [1]. This process is substantially influenced by the number of competing generators [2]. Dominant generators influence the clearing electricity prices and have the ability to break down the perfectly competitive conditions by intentionally withdrawing needed capacity or artificially inflating (or understating) their true incremental costs. This behavior cannot be anticipated so the resulting market clearing prices are uncertain.
- *Generation outages*: This issue is closely related to the previous one. The competitiveness of an electricity market (as in any market) depends largely on the number of suppliers available for competition. The way generation outages influence the competitive process and therefore the electricity prices depends on the timeframe under which these outages occur. If the outages are unexpected, last-minute actions must be taken by the system operator in order to avoid or minimize supply interruptions to consumers. Ancillary (reserve) or real-time markets have been developed for this purpose but the interaction of these markets with the day-ahead market adds another level of complexity to the definition of the final prices under which electricity will be sold and purchased. These prices can only be determined ex-post.
- *Transmission congestion*: Electricity cannot be stored in large amounts and must be produced and transported instantaneously to the consumers over a grid dominated by physical laws and sometimes congested. This particular characteristic influences the competitive process since some generators may be

prevented from accessing parts of the network located 'downstream' of the congested paths. The way transmission congestion affects the electricity prices depends on how the market is settled. In uniform (or single) price electricity markets, the effects of transmission congestion in actual sell or purchase prices for all agents are only known ex-post since the grid is not taken into account in the day-ahead market clearing process. In nodal electricity markets, the capacity of the transmission lines is considered in the clearing process but the price may display a large 'granularity', that is, there are as many clearing prices as there are buses in the network. In any case, the availability of the transmission network is determined by a series of environmental and technical conditions only known to the transmission operator, incorporating another level of uncertainty to the electricity prices that will clear in the market.

- *Considerable penetration of renewable energy generation*: This type of generation has grown dramatically in recent years due to the numerous incentives that governments and regulators have put on to encourage investments in clean technologies. Among these technologies, wind power is in the clear front. The availability of wind power depends mostly on the resource (the winds) and the accuracy of the day-ahead bids from wind farms depends largely on the accuracy of the wind speed forecasts. Since forecasting wind speeds is in general more complex than forecasting the output from conventional generation, the presence of large amounts of wind power may impact the operator's ability to forecast accurately the residual demand to be met by conventional generation (wind power is given priority in the day-ahead scheduling over the remaining technologies by assigning a zero price to its output). Large wind power forecast errors may lead to relatively large real-time imbalances of wind power output with respect to the day-ahead bids and may therefore introduce uncertainty in the market clearing mechanism. Large amounts of wind power may also drive electricity prices to extremely low values (even negative) especially during load demand periods ([3], among others).

Alongside the previous endogenous sources of uncertainty, there are a number of exogenous factors that may introduce uncertainty into the price formation process as well. These are listed next:

- *Price of fuels*: The uncertainty in the price of fuels is perhaps the most important source of exogenous risk for conventional generators. Although most generators guarantee their fuel supply with long term contracts (take-or-pay, options, forwards, etc.), a volatile fuel market in the short-run tends to raise the costs of long-term fuel supply contracts in a way that cannot be easily determined. Since the bids that generators submit to the electricity market are tied up with their incremental costs, which are primarily determined by their fuel costs, uncertainty in the fuel supply side will be passed on to the bids and, subsequently, to the resulting market clearing price.
- *Volumetric uncertainty*: It is related to the uncertainty in the amount of prime-mover fuel that generators will have available for the production of power. This type of uncertainty is particularly evident in weather-based renewable

generation technologies (hydro plants, wind and solar farms, among others) although conventional power plants may also be subject to uncertainty in the fuel supply chain. This exogenous uncertainty may translate in resource forecast errors, an endogenous source of uncertainty as was previously explained.

- *Regulatory uncertainty*: Since the onset of the deregulation of the electricity industry in the early 1980s, different restructuring strategies have been put forward by regulators following some basic principles (industry unbundling, privatization, etc.). The learning process has been turbulent in some cases. Some electricity markets have failed to fulfill some fundamental requirements (competitiveness, fairness, liquidity or others; [4, 5] for instance). As regulators and governments have improved their knowledge on the proper conditions needed to foster the success of an electricity market, generators, consumers and other market agents have been subject in many cases to frequent changes and reviews in the adopted regulation. These actions bring about what is called regulatory uncertainty. Although regulators and government agencies have devoted substantial efforts to curb this type of uncertainty in the power industry [6], it remains one of the major sources of risk in electricity markets in the medium and long-term since it is sometimes exogenous not only to the market but to the industry itself (sometimes the result of political interests).

The previous endogenous and exogenous factors lie primarily upon the supply side (generation). Due to the simultaneous, unforeseeable interaction of these factors, the electricity prices display a random behavior even if the demand, as the counterpart of the supply side in the competitive market clearing process, is known with certainty. This randomness is closely related with volatility. The following subsection will describe briefly how econometric theory can be used to characterize the demand-depend entvolatility of the electricity prices.

2.2 Characterization of Electricity Price Volatility

There are different ways of characterizing the volatility of the electricity prices. In fact, there is no formal universally-accepted measure for it. Some works have quantified the time-dependent volatility of the electricity prices based on time-series representation of actual electricity price datasets and compared across different markets [7–9]. These approaches, however, by using exclusively price data, internalize the effect of all endogenous and exogenous variables in the analysis and provide a single volatility measure over a predetermined time span (1 year, for instance). It is therefore interesting, not only to measure volatility according to a pre-established criterion, but also to quantify the effect of some explanatory variables (either endogenous or exogenous) in the volatility of the electricity prices. As was mentioned before, however, since the electricity price is the result of an equilibrium between supply and consumption forces, the demand, once known, becomes the most important explanatory variable in the price formation process.

Volatility can therefore be measured relative not only to time but also to the state of the system (particularly, the demand). Econometric theory can be used for this purpose as detailed next.

2.2.1 Demand-Dependent Electricity Price Volatility

According to the statistics theory, a random variable is fully characterized by its infinite moments (if they exist). Perhaps the most important moments are the first two: the expectation and variance. In terms of uncertainty and risk, the variance (the second central moment of a random variable) provides the most useful information on how all possible realizations of the random variable are spread around the expectation: the higher the variance relative to the expectation the larger the spread of realizations and the larger the uncertainty in determining its actual values. A random variable with relatively low variance indicates that many realizations lie close to the expectation. If this expectation can be determined with accuracy, the uncertainty of the estimation of any realization is therefore relatively low. In electricity markets, characterizing the variance of the market clearing prices is therefore a crucial step towards the characterization of the uncertainty of the electricity prices. This uncertainty is often referred to a *volatility* although there is no a single formal measure for it. It is at least evident that volatility is proportional to variance: the larger the 'cross-sectional' or time-dependent variance of the electricity price the larger its volatility.

According to the econometric theory, the causal relationship between the electricity price cleared at the market and the corresponding system demand may be represented by the fundamental linear regression model:

$$\lambda = \mathrm{X}\beta + \mathrm{u} \tag{1}$$

Where λ is a vector of observations, that is, the hourly electricity prices (or prices corresponding to any market settlement timeframe) over a predetermined period of study, **X** is a matrix of explanatory variables (regressors), made up in this case of the hourly demands represented by d_h (or **d** in vector form) or any non-linear functions of it, provided that the model remains linear in the parameters and **u** the vector of residuals. Several issues must be considered for the proper specification of the residuals, namely the daily and seasonal demand oscillations and trends. Regression models account for this by way of the so-called dummy variables which are added to matrix **X** as columns (refer to [10] for a detailed description on the implementation of regression models with seasonality). The key issue is that in order for the least-square (ordinary or generalized) estimates of the parameters β in (1) to be unbiased, it can be shown that a necessary condition is that the explanatory variables which form the columns of **X** are *predetermined*, meaning that the residual term at hour h, u_h does not depend on the regressors at the same hour. This is to say that:

$$E[u_h|X_h] = 0 \tag{2}$$

Where E[.] is the expectation operator. Since $\mathbf{X}_h$ (row h of $\mathbf{X}$) is made up of the hourly demands and possibly any functions of it as only explanatory variables, (2) is simplified into:

$$E[u_h|d_h] = 0 \tag{3}$$

Without discussing in detail the issues of estimator efficiency, the final outcome of the fundamental regression (1) is a function of the conditional expectation of the electricity price on the demand, which may be written as:

$$E[\lambda_h|d_h] = X\beta \tag{4}$$

Model (4) represents the fundamental causal relationship between the demand (or linear-in-the-parameters of it) and the corresponding electricity price cleared at the market. Once this model is identified, the residuals $\mathbf{u}$ in (1) contain key information on the variability of the electricity prices with respect to the demand. This issue hinges on the notion of 'Heteroskedasticity' in regression models.

Heteroskedasticity refers to the condition that the variance of the error terms in a regression such as (1) is not constant with respect to the explanatory variables. Assuming that the demand is predetermined to the model, from (3) it can be proven that:

$$\text{var}[u_h|d_h] = E[u_h^2|d_h] \tag{5}$$

Where [.] is the variance operator. This result indicates that the conditional expectation of the square of the error terms is equal to their conditional variance on the explanatory variable (the demand). The nature of this conditional variance indicates the influence of the demand on the relative volatility of the electricity prices. Now, assume that this conditional variance can be expressed as:

$$\text{var}[u_h|d_h] = E[u_h^2|d_h] = \phi(d_h) \tag{6}$$

Where $\phi(.)$ is a positive-valued function (linear or non-linear). This function is generally referred to as *skedastic function*. This function is a quantitative measure of how volatile the electricity market prices may become as the demand grows, based on the historic behavior of the market. In [11] it is shown that the most representative non-linear least squares estimation for $\phi(d_h)$ using actual data from two large-scale electricity markets is an exponential function of the form:

$$\phi(d_h) = b_0 e^{b_1 d_h} \tag{7}$$

Since the errors, u_h, result from the difference between the price observation, λ_h, and their conditional expectation, $E[\lambda_h, | d_h]$ it can be shown that the conditional variance of the electricity prices on the demand also grows exponentially. The derivative of an exponential function is also exponential, meaning that the volatility of the electricity markets is substantially dependent on demand fluctuations, especially near the peak. This considerable demand-dependent, heteroscedastic price volatility influences considerably the performance of any electricity price forecasting technique especially at high demand periods. This will be explored in detail later on using actual electricity price data and a well established forecasting technique as a sample method. The following section will introduce and describe the most widely used methods for electricity price forecasting.

3 Models for Electricity Price Forecasting

A number of electricity price forecasting models and techniques have arisen over the past years based on the long-standing mathematical and heuristic load-forecasting techniques [12–16]. Electricity price forecasting techniques may be classified according to three major basic approaches, namely *Production-Cost*, *Statistical* and *Heuristic* or *Data Mining* models. It is not the purpose of this study to describe the details of implementation for each of these models but to provide basic guidelines toward their fundamental characteristics and differences.

3.1 Production-Cost Models

These models rely on the detailed simulation of the electricity market and the strategic behavior of all agents, their costs and expected bids. A network-constrained (or security-constrained) optimization problem is set up so as to minimize overall operating costs or maximize expected profits and benefits subject to a set of technical constraints. The Lagrange multipliers (shadow prices) associated with the power balance constraints are assumed to be the simulated electricity prices. The success of this approach hinges on the quality of the estimates of the numerous inputs. This process requires a considerable amount of information (in many cases confidential) and the development of customized large-scale optimization algorithms. Several worldwide vendors offer high-performance commercial computer packages for this purpose [17] although it is not rare for power companies to develop their own 'in-house' models for power system optimization and simulation [18].

3.2 Statistical Models

These models are based on the assumption of an implicit time-dependent and recurrent relationship between the current and previously-realized electricity market prices without explicitly modeling the underlying physical-economic market settlement process. This model category comprises a diversity of methods, ranging from the simplest 'black box' time-series models, which are based exclusively on previous price realizations as input data, to more complex structural forecasting models that include explanatory (causal) variables such as the demand for electricity, fuel prices and generator availability. This category of models was first made popular by the introduction of the so-called "Box-Jenkins" (BJ) methods [19] which comprise Auto-Regressive (AR), Moving-Average (MA) and Auto-Regressive, Moving-Average (ARMA) methods. Exogenous variables can be incorporated into these models, typically referred to as ARX, MAX and ARMAX models. Classical methods have been used for electricity price forecasting [20] and even including Markov chains [21].

The main goal of time-series analysis is not limited to forecasting (i.e., to predict future realizations of a given random variable). There are some attributes that shed light on the nature and underlying mechanisms that define the physical process being simulated. These attributes must be modeled in the process of system identification and are detailed next:

- *Serial autocorrelation*: It is perhaps the most common attribute to physical or economic time-series, indicating that a realization of the random variable at a given time is correlated with immediately previous realizations. For instance, it is likely that the electricity price will be high in the next hour if the current price is high. A similar reasoning applies to other physical random variables such as electricity consumption, air temperature, etc. Properly modeling this serial autocorrelation is crucial to the success of the time-series in representing the underlying process.
- *Heteroskedasticity*: Refers to the fact that the conditional variance of the random variable that defines the time-series is not constant. This phenomenon is common to economic time-series. For instance, electricity prices tend to be more volatile during periods of high demand, as will be explained in detail later on. The classical ARMA models rely upon the assumption that the time-series is stationary. Techniques have been developed to convert (normalize) non-stationary time-series into stationary ones so the classic theory can be applied. As an alternative, the Auto-Regressive Integrated Moving-Average (ARIMA) has been developed as a generalization of the classical ARMA models when there is evidence of non-stationarity [22]. ARIMA models have been incorporated into the BJ set of time-series analysis models. Generalized Auto-Regressive Conditional Heteroskedasticity (GARCH) models for time-series with stochastic, auto-regressive variance have also been used for electricity price forecasting [23]. The performance of GARCH models relies on the validity of the various assumptions, which are sometimes simplistic (e.g., normality of random innovations).

- *Seasonality*: It is given by the daily and seasonal cycles in all the physical explanatory or instrumental variables that define the time-series. The most representative example of such variable in power systems is the demand. Any demand-dependent random variable and its corresponding time-series (such as the electricity prices) will display a seasonal pattern, which must be properly characterized. Seasonality must be checked for in time-series models and removed, using the so-called seasonal dummy variables in regression-based models [10] or frequency-domain methods (Fourier transforms [24, 25] or spectral and wavelet analysis [26]).
- *Trend*: Refers to the steady increase or decline of the expectation of the underlying random variable in the time-series over a predetermined period of time. In economic time-series, these trends are given, for instance, by the economic growth or recession periods that influence consumption patterns for goods and services. Electricity is one example of such goods. Periods with steady investment in more efficient generation capacity may also reduce the long-term electricity marginal cost and may result in a steady decline in the average electricity prices.

The following subsection will provide specific details of how statistical models are set up and fine-tuned for forecasting purposes.

3.2.1 Details of Statistical Time-Series Models for Forecasting

Among the different statistical time-series models for forecasting, one of the simplest forms is to represent the observation of the forecasted variable at time t, y_t, as a linear combination of a number of p past realizations. This is the case of an AR(p) model, which is formulated as:

$$\left(1 - \sum_{i=1}^{p} \phi_i L^i\right) y_t = \varepsilon_t \tag{8}$$

where ϕ_i are the linear coefficients of the so-called AR polynomial, L^i is the lag operator (so that $L^i y_t = y_{t-i}$) and ε is a random variable representing zero-mean white noise (not necessarily normally-distributed).

An alternative formulation to an statistical time-series forecasting model is to assume that the forecasted variable y_t is a linear combination of the past q forecasting errors. This is the so-called MA(q) model, which can be formulated as:

$$y_t = \left(1 + \sum_{i=1}^{q} \theta_i L^i\right) \varepsilon_t \tag{9}$$

where θ_i are the linear coefficients of the so-called MA polynomial.

A generalized model which incorporates both AR and MA components in the linear combination of observation y_t based on past information is the ARMA(p, q) model, formulated as:

$$\left(1-\sum_{i=1}^{p}\phi_i L^i\right)y_t=\left(1+\sum_{i=1}^{q}\theta_i L^i\right)\varepsilon_t \qquad (10)$$

As mentioned before, when heteroskedasticity is detected in the underlying simulated process, differentiation of the original time-series up to d realizations back may be needed until stationarity in the transformed series is achieved. ARIMA (p,q,d) models are used for this purpose, which are formulated as:

$$\left(1-\sum_{i=1}^{p}\phi_i L^i\right)(1-L)^d y_t=\left(1+\sum_{i=1}^{q}\theta_i L^i\right)\varepsilon_t \qquad (11)$$

where it is assumed that due to the non-stationarity of the original time-series, the polynomial

$$1-\sum_{i=1}^{p}\phi_i L^i \qquad (12)$$

has a unitary root of multiplicity d.

Exogenous inputs can also be incorporated into a generalized ARMA(p,q) model as explanatory variables. For instance, when electricity prices are forecasted, the expected hourly demand may offer substantial explanatory power to the process. An ARMA(p,q) model in which an exogenous time-series e_t of which up to b observations are incorporated becomes an ARMAX(p,q,b) model. Such model is written as:

$$\left(1-\sum_{i=1}^{p}\phi_i L^i\right)y_t=\left(1+\sum_{i=1}^{q}\theta_i L^i\right)\varepsilon_t+\left(1-\sum_{i=1}^{b}\rho_i L^i\right)e_t \qquad (13)$$

In order to set up a statistical time-series forecasting model, four steps are usually followed:

- *Identification*: In this step, the AR and MA orders (p and q, respectively) are estimated using the Partial Auto-Correlation Function (PACF) (for p) and the Auto-Correlation Function (for q). These functions are found by regressing successively historical observations at time t on previous observations t-k for different lags k. The optimal order p is determined by the last significant estimated parameter in the auto-regressive regression using a given t-statistic. The MA order q is usually obtained by estimating the Durbin-Watson statistic to the error time-series obtained in the previous regression under the null hypothesis of no auto-correlation. Similar regression-based estimations are used to determine the optimal orders for the ARIMA and ARMAX models (d and b, respectively).

- *Estimation*: Once the optimal orders for the AR and MA models are estimated, the next step corresponds to the estimation of the polynomial coefficients ϕ, θ and ρ. This is done by minimizing the sum of squared prediction errors using part of the available historical dataset.
- *Validation*: This step consists of applying the estimated model to the remaining available historical dataset. The key issue is that the obtained errors must display no auto-correlation (basically behave as white noise). The Durbin-Watson statistic may be used for this purpose under the null hypothesis of no auto-correlation. The estimated statistic should not be significantly different from zero if error white noise is detected (the null hypothesis should hold true).
- *Verification*: Once the chosen model is systematically used for forecasting, its performance is continuously verified using different metrics such as the forecast mean-square or mean average error, error variance, and others. This step is what determines the optimal pool of forecasting models to be used according to their recurrent performance.

3.3 Heuristic or Data-Mining based Models

The last family of forecasting methods are the Heuristic or based on Data-Mining (DM). These methods were first developed at the onset of digital computers and massive data storage hardware. The main advantage of these models is that no assumptions have to be made about the underlying phenomena behind the price formation process considering its substantial non-linearity. Among the several techniques that fall into this category, the most widely used are Decision Trees, Artificial Neural Networks (ANN), Clustering, Fuzzy Logic-based and Weighted Nearest Neighbours (WNN). The latter is essentially based on pattern recognition but can be used for price forecasting under the assumption that the similarities in price behavior seen in the past will persist for the future. The main advantage of this model is its simplicity and the relatively few amount of parameters upon which it relies.

Forecasting methods based on ANN have received a great deal of attention due to their relative simplicity and intuitiveness. They are structured to resemble the functioning of the brain and the learning process by interconnecting basic functional units (neurons) through synapses. ANN are trained by feeding past information into a series of inputs and the desired output, minimizing some sort of 'learning error' by way of mathematical or heuristic optimization techniques ([27] and [28] among others). More complex arrangements of heuristic methods have also been developed, combining different features within each family or across different families. For instance, ANN with Fuzzy-Logic [29] or ARIMA models combined with the Wavelet transform [30] or with ANN [31].

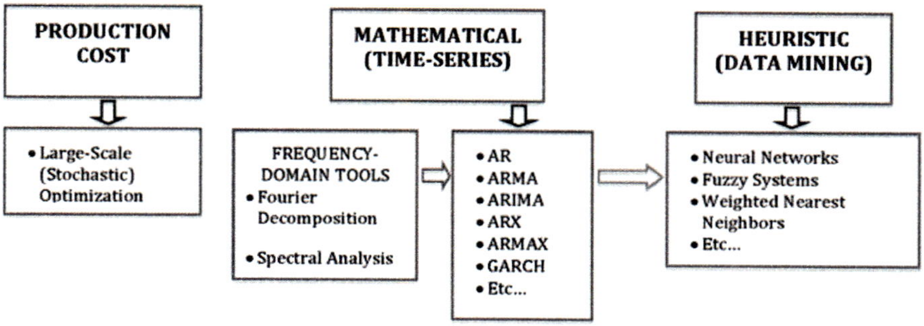

Fig. 2 Classification of electricity price forecasting methods according to fundamental structure

Figure 2 summarizes a general classification of the most common price forecasting models according to their fundamental structure. Price forecasting models, especially those based on heuristic techniques, are highly customized according to the electricity market and user (generating companies, retailers, utilities, etc.). The portability of these customized models across different settings is therefore limited. In addition, price forecast users barely use one single general model, but a suite of mathematical and heuristic techniques, which are continuously fine-tuned according to its recurring performance.

It is not the purpose of this study to implement and compare the performance of all or several of the previous forecasting techniques. This type of cross-model performance analysis can be found in different sources already available in the specialized literature ([32] or [33] for instance). Selecting a sample forecasting technique is useful to illustrate the different issues and challenges discussed so far on price forecasting, which is the main objective of this work. With this in mind, the WNN will be used as sample forecasting technique (among the several others that could be selected) for the subsequent analysis using actual electricity price data. The WNN method has displayed excellent performance compared to other forecasting methods given its intuitiveness and relative ease of implementation. This method will be explained in detail next.

3.3.1 The Weighted Nearest Neighbours Forecasting Method

The details of this method are provided in [32], which are summarized here. Given the historic price records up to day d, the problem consists on predicting the 24 hourly prices corresponding to day $d + 1$. Let $Pi \in \mathbf{R}^{24}$ be a vector made up by the 24 hourly electricity prices corresponding to an arbitrary day i:

$$P_i = [p_1, p_2, \ldots, p_{24}] \tag{14}$$

The associated vector $WP_i \in \mathbf{R}^{24\mathrm{m}}$ is then defined by collecting the prices contained in a window of m consecutive days, from day i backward, as follows:

$$WP_i = [p_{i-m+1}, p_{i-m+2}, \ldots, p_{i-1}, p_i] \tag{15}$$

where m is a parameter to be determined. Note that when $m = 1$ vector WP reduces to P. For any couple of days, i and j, a distance can be defined:

$$\text{dist}(i,j) = \|WP_i - WP_j\| \tag{16}$$

where ||·|| represents a suitable vector norm (the Euclidean norm is generally used). The WNN method first identifies the k nearest neighbors of day d, where k is a number to be determined and 'neighborhood' in this context is measured according to (16) above. This leads to the neighbor set:

$$NS = \{\text{set of } k \text{ days}, q_1, \ldots, q_k, \text{closest to day } d\} \tag{17}$$

in which q_1 and q_k refer to the first and k-th neighbors respectively, ordered by distance. According to the WNN methodology, the 24 hourly prices of day $d + 1$ are predicted by linearly combining the prices of the k days succeeding those in NS, that is:

$$P_{d+1} = \frac{1}{\sum_{i \in NS} \alpha_i} \cdot \sum_{i \in Ns} \alpha_i P_{i+1} \tag{18}$$

Where the weighting factors α_i are obtained from:

$$\alpha_i = \frac{\frac{1}{dist(1,d)}}{\sum_{j=i}^{k} \frac{1}{dist(j,d)}} \tag{19}$$

Note that, although the 24×m prices contained in WP_i are used to determine whether i is a good neighbor of d, only the 24 prices of P_{i+1} are relevant in determining P_{d+1}. Fig. 3 illustrates the basic idea behind the WNN methodology. It considers that, if WP_i is close WP_d to, then P_{d+1}, already known, should be also similar to P_{d+1}. In order to find candidate neighbors, a window of m days is simply slid along the entire list of hourly prices contained in the data base.

Before applying the WNN method, a training phase is necessary in order to find suitable values for m and k. Generally, after using the resulting model for a certain period, prediction errors tend to increase slightly, particularly when applied to a relatively volatile electricity. This may call for new training processes. As will be seen below, the training phase may be relatively costly and, in any case, even if the model were trained in a daily basis, the prediction accuracy would be always limited by the intrinsic uncertainty of prices. Therefore, a compromise should be found between cost and accuracy. In this paper, unless otherwise indicated, monthly training periods will be considered, but yearly training provides almost as accurate results.

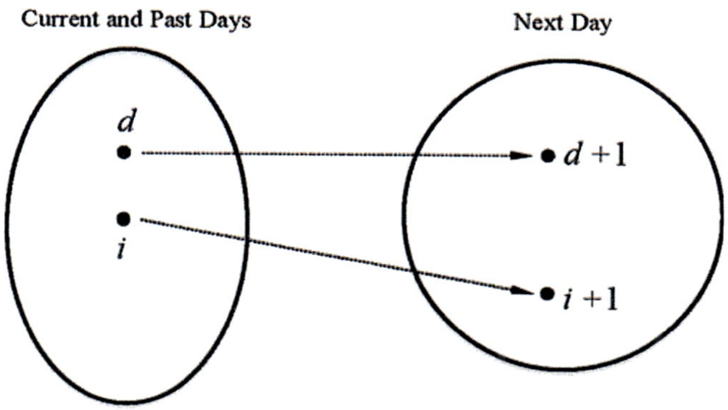

Fig. 3 Illustration of the WNN approach

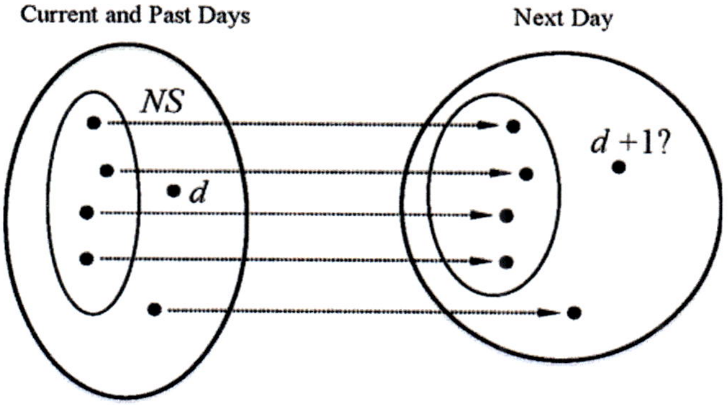

Fig. 4 False nearest neighbors

Optimal Window Length: The number of days contained in the window that will be used to find candidate neighbors (parameter m) is determined in advance by resorting to the so-called False Nearest Neighbors (FNN) method [34]. This method compares the distance between a day d and a candidate neighbor i with that between $d + 1$ and $i + 1$. If the second distance is larger, as illustrated in Fig. 4, it is said that d and i are false neighbors, because the trajectory of the associated prices tend to diverge. Note that, according to (16), such distances depend on m in an implicit manner. By trying all days contained in the training set, m can be chosen so as to minimize the number of false neighbors.

In practice, as m increases, the cost of the training process also increases and the number of candidate neighbors gets significantly reduced (in the limit, if m approached the size of the training set, a single candidate would remain). Hence, the suboptimal but cheaper scheme adopted here consists of choosing the minimum value of m leading to a percentage of false neighbors not exceeding a given threshold (e.g., 10%). Frequently, but not necessarily, $m = 1$ leads to forecasting errors that are close to that of the optimal value.

Optimal Number of Nearest Neighbors: The optimal number of nearest neighbors (parameter k) is the one that minimizes the forecasting error when the WNN methodology is applied to the training set. Mathematically, this is equivalent to finding the value of k that minimizes the following quadratic function,

$$\sum_{d \in TS} \left\| \hat{P}_{d+1} - P_{d+1} \right\| \tag{20}$$

where $\widehat{P}_{d+1}$ are forecasted prices for day $d + 1$, according to the WNN method, P_{d+1} are actual recorded prices and TS refers to the training set. Note that, according to (18), $\widehat{P}_{d+1}$ is an implicit function of the discrete variable k, which prevents application of standard mathematical programming methods when searching for k. In practice, k is assigned successive integer numbers (k = 2,3,...) until a local minimum is found.

Performance Indicators: Among the different indices that can be used to measure the performance of a forecasting technique, the following two are used here, namely the Mean Daily Absolute Error (MDAE) and Mean Daily Total Error (MDTE). These are defined as:

$$\begin{aligned} MDAE &= \frac{1}{24_{\bar{p}}} \sum_{h=1}^{24} |\hat{p}_h - p_h| \\ MDTE &= \frac{1}{24_{\bar{p}}} \sum_{h=1}^{24} (\hat{p}_h - p_h) \end{aligned} \tag{21}$$

Where $\overline{p}$ is the average price for the corresponding day. As the name indicates, these normalized indices are calculated on a daily basis (although any other timeframe may be used). Note that in (21), the MDAE is defined as the average absolute error (any deviation of the forecast, $\widehat{p}_h$, with respect to the actual realization, p_h, is added in absolute value), whereas in the MDTE negative deviations around the realization may offset positive deviations. This measure is of particular value when the analyst is more concerned about estimating daily trading balances rather than the performance of the forecasting tool hour by hour.

4 Performance and Limitations of Price Forecasting in Actual Electricity Markets

The performance of any electricity price forecasting methodology is affected by the different sources of uncertainty common to the price formation process, as described in Sect. 2. In this section, an analysis of the influence of these factors in an actual electricity market is carried out. To that end, hourly day-ahead electricity price and demand data were obtained from the Spanish electricity market operator [35] for the period between 2002 and 2009. The peak demand in 2009 was 39.73 GW. This electricity market is an adequate illustration of how generators, consumers and the market in general may respond to the different sources of uncertainty and how price forecasting may be influenced by those factors.

Consider Fig. 5, which shows a typical daily and weekly market demand pattern from the dataset, together with the corresponding electricity prices cleared at the market. It can be seen how the prices follow closely the fluctuations in demand (the peaks and valleys in demand and price tend to coincide with one another) with some random excursions as the footprint of the different sources of uncertainty in the price formation process. This is a typical week in the sense that there are five business days (with no holidays) and two weekend days. These can be noticed by a clear decline in demand from hour 120 onwards. This example is also illustrative with respect to the sudden drop in the electricity price observed at about hour 26th. These large price drops are due in part to the presence of large amounts of wind power generation in Spain and also to the operation with large base-load, relatively inflexible nuclear power plants. This effect can also be seen in other power systems not necessarily with large wind power penetrations, such as in Ontario, Canada, which has a considerable amount of nuclear and coal base-load generation [3]. Figure 6 shows as illustration the same electricity prices shown in Fig. 5 together with the corresponding forecast using the WNN method, using price data from 2007 through the day before the forecasted day.

Following with the analysis, electricity prices were forecasted for all days in 2009 using the WNN method, the results of which are shown in Fig. 7. The left-hand side plot shows a histogram of the MDAEs obtained for each day as defined in (21) with bins of 1%. The plot on the right shows a similar histogram for the MDTE. It is clear that the MDAE yields higher errors since any deviation around the actual realization during the day is added in absolute value. The median MDAE for 2009 was 6.5%. The MDTE can be useful when forecast error fluctuations (above or below the actual realization) are netted at the end of the bidding exercise (a day for instance). It can be seen how almost half of the MDTEs were less than 1%, an excellent result for forecasting revenues for generation with relatively constant outputs throughout the day.

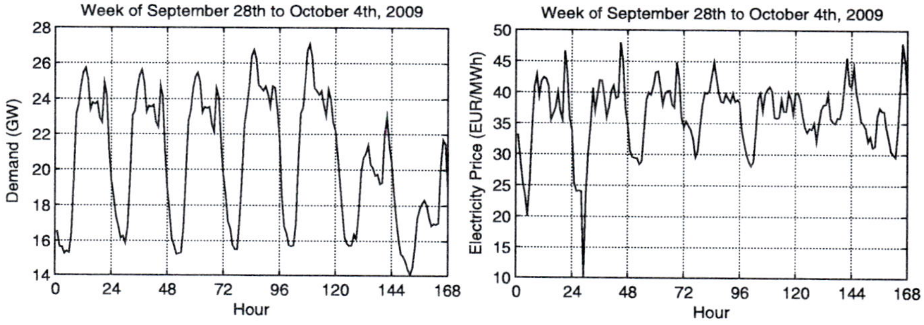

Fig. 5 Typical weekly demand and electricity price pattern from the dataset

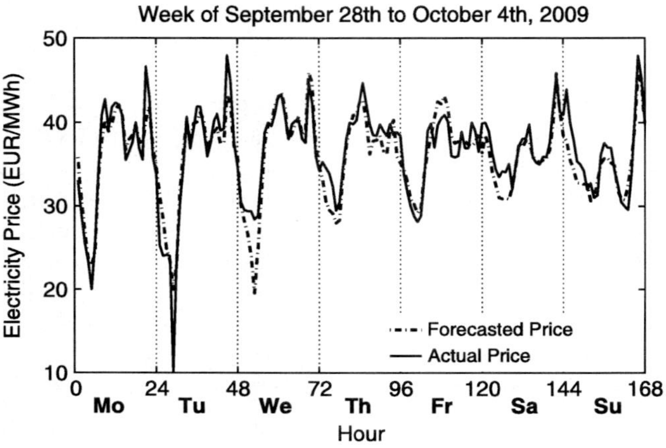

Fig. 6 Sample 7-day forecast by the WNN method

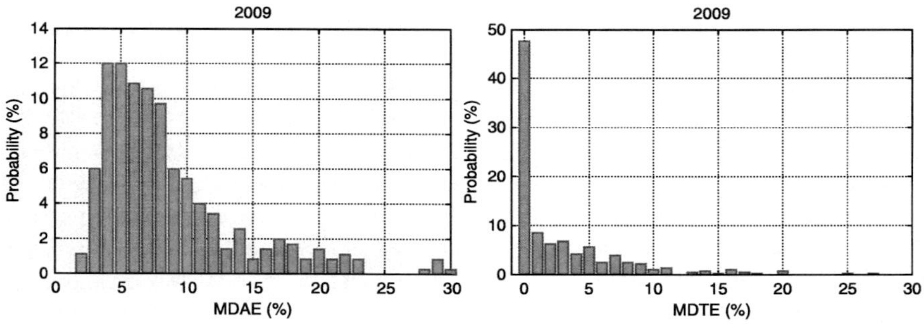

Fig. 7 Histograms of MDAE and MDTE errors from WNN method for 2009

Weekends must be treated with special care in any price forecasting methodology due to the sudden change in consumption pattern with respect to the more common business-day activities. All information or learning on price and consumption patterns gathered during the business days cannot be used directly for holiday or weekend forecasting. Forecasting models are therefore 'trained' so as to look for the best cross-correlations among previous weekend days in order to maximize the accuracy of the forecast on those days. It can be seen from Fig. 6 that the performance of the WNN method is similar during the weekend when compared to the business days for that particular sample week. This is regularly observed for any other week.

Unlike business days and weekends, holidays represent, on the other hand, a particular challenge when it comes to electricity price forecasting. The reason is the few amount of holidays from which forecasting models can learn and their rather non-uniform distribution across the calendar. In fact, Fig. 8 shows on the left hand side the hourly demand for a business-day weekly cycle but with Monday as holiday (October 12, 2009). The one-of-a-kind consumption pattern for this holiday is clearly seen. In fact, this pattern depends even on the day of the week on which the holiday falls (consumption patterns are different if the holiday falls on Monday, resulting in a long weekend, or Wednesday, for instance). The effect of this special pattern on the performance of the forecasting tool is seen on the right hand side plot. The largest errors were observed for the holiday (Monday). These relatively high errors may propagate to the next day as seen in the plot. Heuristic rules can be implemented to forecast those business days preceded by holidays. One simple rule is, for instance, to consider this particular Monday as a business day in order to forecast for Tuesday (which has not been done for this particular example). Table 1 summarizes the information provided by Fig. 8. It shows the daily forecast MDAE (in %) for each day of the sample week together with the maximum forecast error for each day (in EUR/MWh). The relatively high forecasting errors for Monday as holiday are confirmed in numeric terms.

Continuing with the analysis, a greater picture of the nature of the relationship between the market demand and the prices can be observed in Fig. 9, which shows scatter plots of the hourly day-ahead electricity prices cleared in the Spanish electricity market against the corresponding hourly demands. These plots are shown for each year from 2002 through 2009, each one containing 8,760 data-points.

Table 1 Forecasting errors for a week with holiday

Day	MDAE (%)	Maximum error (EUR/MWh)
Monday[a]	19.54	15.37
Tuesday	13.88	10.50
Wednesday	3.27	4.17
Thursday	5.00	3.97
Friday	6.02	7.17

[a]A holiday in this particular week

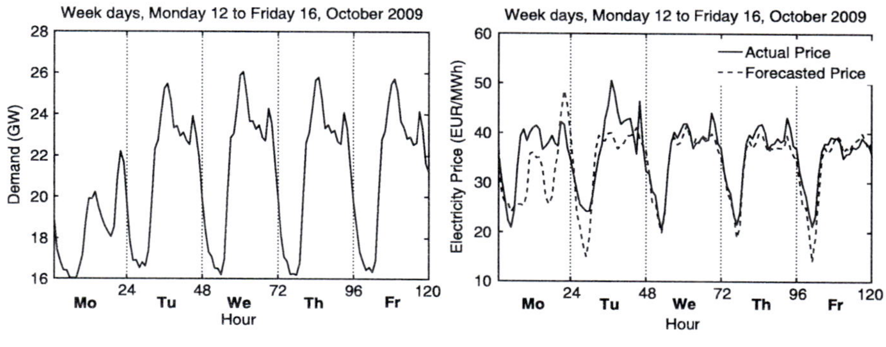

Fig. 8 Weekly demand and electricity price pattern with a holiday (Monday) and price forecast

From Fig. 9, it can be seen that typical price-demand scatters were observed for years 2002–2005. These scatters are determined by a single cluster of data points with some evident correlation. Moreover, in line with was discussed in Sect. 1, these scatters indicate from within that electricity markets tend to become more volatile as the demand grows toward the peak. This growth in volatility is in fact exponential, if the demand-dependent price heteroscedasticity is characterized using econometric theory as explained in Sect. 1. It is therefore natural to assume that forecasting electricity prices at higher demand levels may become more challenging due to this conditional volatility. In order to visualize this effect, day-ahead electricity prices have been forecasted for year 2009 using the WNN method, which has been fine-tuned following different tests. In particular, $m = 1$ and $k = 4$ (refer to the description of the WNN in Sect. 1). The relative performance of the price forecasting method for the whole year can be observed in Fig. 10. The plot has two ordinate axes: the left axis shows Spain's average monthly demand for 2009 while the right axis shows the average monthly MDAE in percent. The abscissa corresponds to the month. It can be seen how the forecasting accuracy is clearly influenced by the seasonal fluctuations in demand. As was discussed above, the higher the demand the more volatile the market and the prices become. The effect of this conditional volatility on the relative performance of the price forecasting methodology is evident.

On the other hand, the influence of regulatory uncertainty on the behavior of the electricity market can also be observed by looking at the scatter plots for year 2006 in Fig. 9. An special rule for generation-owning distribution companies was put in place in mid-2006 in Spain, resulting in the withdrawal of part of the demand from the market from that date on. The consequence is indicated by the two clear clusters of points, one for lower demands and one for higher demands for that year. The most important outcome is, however, the high price volatility that resulted from these actions. The range of demands for both clusters is relatively narrow but the range of prices under which these demands were settled in the market is relatively high for all demand levels. This situation remained as such until the first quarter of 2007, which can also be seen in the price-demand plot for that year.

The effects of the regulatory rule-making on the behavior of the market can also be seen for year 2009. Starting July that year, the minimum demand threshold for open electricity market access to consumers was lowered, resulting in an important increase in market demand. A transitional period followed in which the electricity price was capped by the regulator until new long-term contracts were signed by consumers, aggregators and distribution companies, which led to a subsequent reduction of the day-ahead market demand back to pre-transition levels. The consequences of this process can be seen by the unusual ramification of high demand values and relatively low prices that branches off the bottom-right of the scatter for 2009. The previous situations illustrate clearly how the market agents may respond to regulatory actions and how this change in behavior may lead to an increased price volatility. Forecasting the electricity prices in those situations becomes more challenging.

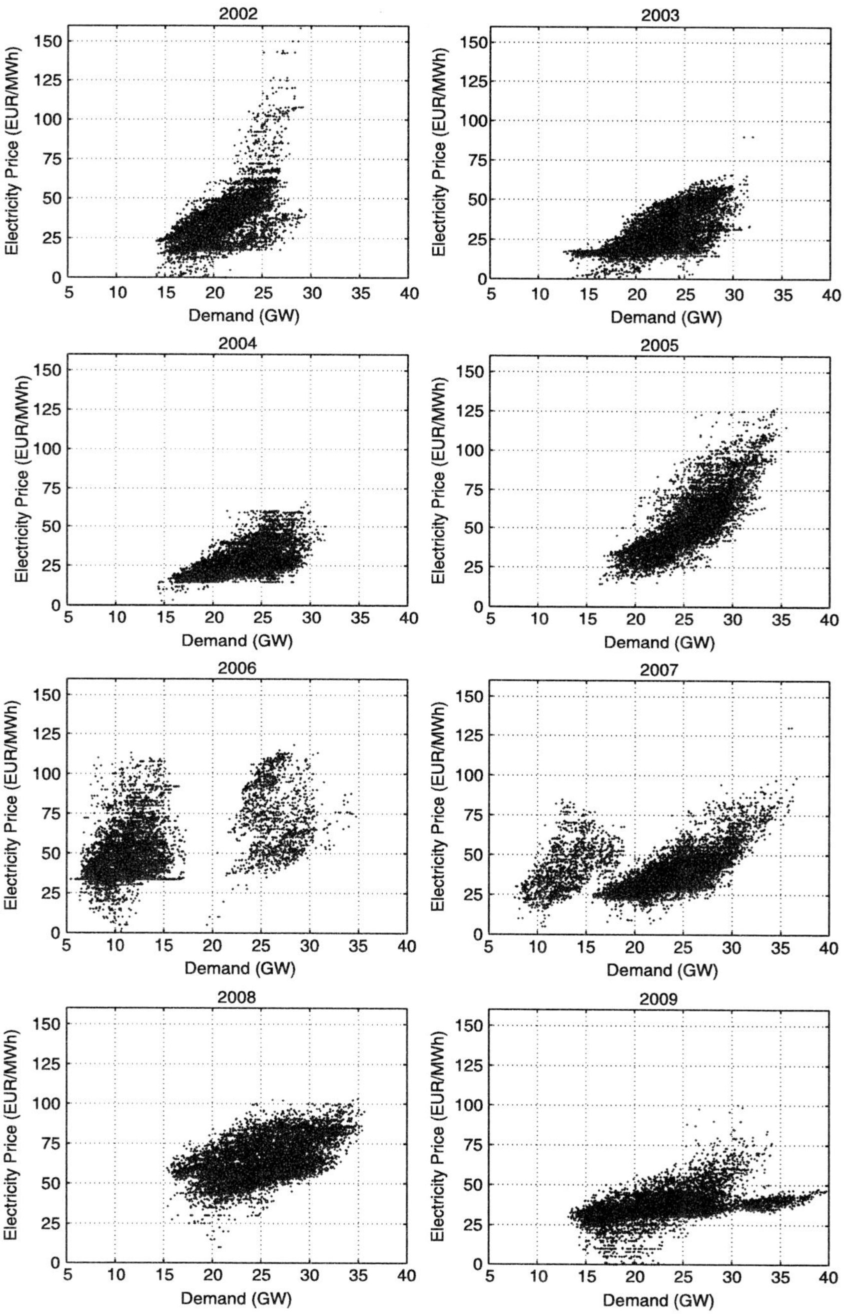

Fig. 9 Scatter plots of electricity price versus demand in Spain from 2002 through 2009

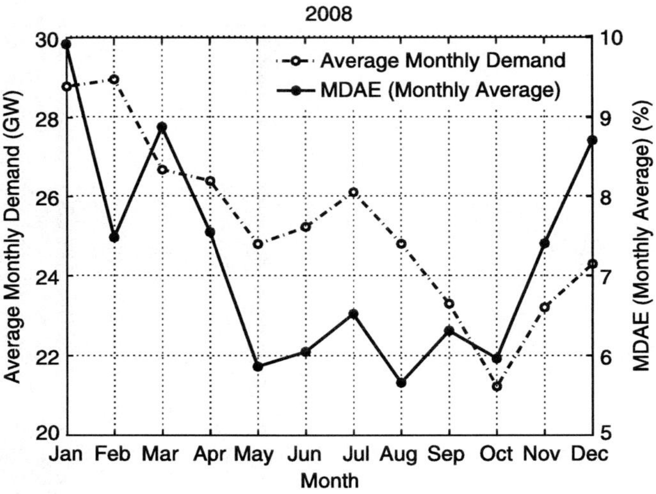

Fig. 10 Indirect influence of the demand on the relative performance of the WNN method

The example of the Spanish electricity market is also of great value when it comes to assessing the relative influence of large amounts of renewable energy on the uncertainty of the electricity prices. This case is of particular interest due to the presence of large amounts of wind power in the system, which have been increasing steadily over the last years. Total wind power generation in the country accounts sometimes for more than half of the total country's electricity consumption (for instance, wind power penetration reached 54.1% with respect to the demand on December 30, 2009, at 3:00 am, [36]). As was discussed above, the variable nature of the wind resource as a weather-related phenomenon influences the generators and operator's ability to forecast the residual demand to be met by conventional generation at the day-ahead timeframe when bids must be submitted to the market. For instance, if generators with some market power forecast a large amount of wind power for the next day, they may adjust their positions according to their own operational priorities and interests (for instance, if high amounts of wind are expected, they could bid low prices if a shutdown is not desirable, or in turn, bid higher prices if small amounts of wind are expected and they realize that system will be in substantial need of conventional power). If the foreseen weather conditions change considerably according to the forecast, unexpected results may arise in terms of clearing prices, which may oscillate from hour to hour or day by day according to each generator's perception of the market conditions, their own bidding strategies and their ability to forecast the wind power output and the system's residual demand.

It is natural to observe that this added volatility to the electricity prices influences the relative performance of any price forecasting methodology. One way of visualizing these effects is to measure the relative performance of the WNN method in forecasting the electricity prices subject to different country-wide wind power regimes. To that end, electricity prices for year 2008 have been

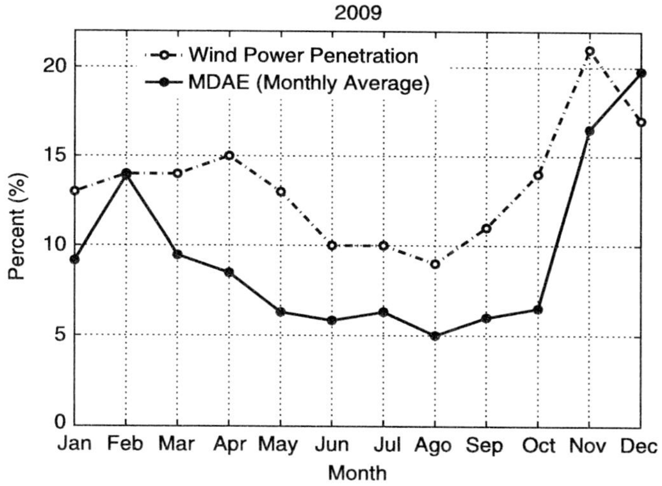

Fig. 11 Indirect influence of large wind power penetrations on the relative performance of the WNN method

forecasted using historic data from 2007 until the day before the forecasted day (year 2009 has not been examined due to the market abnormalities that took place that year as described above, basically a large summer market demand settled under a capped price). Figure 11 shows the average monthly MDAE for the WNN day-ahead (24 h) price forecast for Spain versus the country's average daily wind power penetration for 2008. Figure 11, confirms that there is a strong evidence that large amounts of wind power lead to relatively larger price forecasting errors due to the increased market volatility. There is evidence, however, from the same price dataset, that large amounts of wind power in Spain have caused a sustained decline in the electricity market prices. The resulting economic benefits accrued by the bulk of the consumers, on top of the considerable environmental benefits brought about to the society in general, may have served to offset the negative consequences of the relatively added price volatility and the impacts that lower prices have on the financial performance of fossil-fueled power plants.

5 Conclusion

In electricity markets, the prices cleared on a daily basis are a major determinant of the economic performance of any power industry-related investment. Although in many electricity markets most of the power may be traded by way of bilateral contracts, the prices cleared at the daily spot markets, as a marketplace of last resort, have an indirect influence in the contracted prices set independently by the parties involved. Forecasting the electricity market prices is, therefore, fundamental to all industry stakeholders due to a variety of reasons, ranging from cash flow and capital budgeting analysis to financial procurement and regulatory rule-making, among others.

The process of price formation in electricity markets follows the microeconomic principle in which competitors submit bids and offers for production and consumption and an equilibrium price and quantity is reached in a coordinated manner. If this process were deterministic, there would be a one-to-one relationship between consumption and price and determining the prices would boil down to obtaining adequate demand forecasts, which is one of the most studied problems in the power industry. There are, however, numerous endogenous and exogenous factors that bring about uncertainty to the settlement process and, therefore, to the electricity prices cleared at the market. The most important endogenous source of uncertainty in the prices is the demand, as the counterpart of the supply side in the competitive process. In addition, the electricity markets tend to become substantially more volatile as the demand grows towards the peak. Forecasting electricity prices in volatile situations is obviously more challenging.

Different price forecasting methodologies have been proposed over the last decades based on optimization, time-series models and heuristic techniques. These models are usually customized to account for the particular needs and experience of the user in each setting. All electricity markets are different in a variety of ways, ranging from the number of participants and the fuel mix to the regulatory framework and the climate region, to name a few. This reduces the portability of any price forecasting model across utilities and jurisdictions. Mathematical models are the most general but their practical implementation and performance hinges on the compliance of a series of fundamental assumptions, which have to be carefully examined by the analyst. Heuristic techniques are extensive in data handling but are simpler to implement and are open to the user's intervention, customization and continuous learning. These techniques have become popular since the onset of high performance computing and massive digital storage. Price forecasting users, however, barely rely on a single method or technique but on a suite of models that are continuously compared and improved.

References

1. Frank R, Parker I (2002) Microeconomics and behavior (Canadian Edition), 1st edn. McGraw-Hill, New York
2. Hasan E, Galiana FD (2008) Electricity markets cleared by merit order-part II: strategic offers and Market Power. IEEE Trans Power Sys 23(2):372–379
3. Ontario Power Authority (Canada) (2006) Ontario wind integration study. Schenectady
4. Zhou H, Tu Z, Talukdar S, Marshall KC (2005) Wholesale electricity market failure and the new market design. In: IEEE power engineering society general meeting, San Francisco, 12–16 June 2005, pp 503–508
5. Wooa CK, Lloyd D, Tishlerd A (2003) Electricity market reform failures: UK, Norway, Alberta and California. Energy Policy 31:1103–1115

6. Chao HP, Oren S, Wilson R (2006) Alternative pathway to electricity market reform: a risk-management approach. In: Proceedings of the 39th Hawaii international conference on system sciences, January 4–7, 2006
7. Alvarado FL, Rajaraman R (2000) Understanding price volatility in electricity markets. In: Proceedings of the 33rd Hawaii international conference on system sciences, January 4–7, 2000
8. Zareipour H et al (2007) Electricity market price volatility: the case of Ontario. Energy Policy. doi:10.1016/j.enpol.2007.04.006
9. Hadsell L, Marathe A, Shawky HA (2004) Estimating the volatility of wholesale electricity spot prices in the US. Energy J 25(4):23–40
10. Davidson R, MacKinnon JG (2004) Econometric theory and methods. Oxford University Press, New York
11. Gómez-Quiles C, Gil HA (2010) Regression-based estimation of the electricity market price demand-dependent volatility. In: 11th international conference on probabilistic methods applied to power systems (PMAPS), Singapore, 14–17 June 2010
12. Gross G, Galiana FD (1987) Short-term load forecasting. Proc IEEE 75(12):1558–1573
13. Mohammed O, Park D, Merchant DR, Dinh T, Tong C, Azeem A, Farah J, Drake C (1995) Practical experiences with an adaptive neural network short-term load forecasting system. IEEE Trans Power Syst 10(1):254–265
14. Douglas P, Breipohl AM, Lee FN, Adapa R (1998) The impacts of temperature forecast uncertainty on bayesian load forecasting. IEEE Trans Power Syst 13(4):1507–1513
15. Papalexopoulos A, Hesterberg TC (1990) A regression-based approach to short-term load forecasting. IEEE Trans Power Syst 5(4):1535–1550
16. Troncoso A, Riquelme JM, Riquelme JC, Gómez-Expósito A, Martínez JL (2004) Time-series prediction: application to the short-term electric energy demand. Lect Notes Artif Intell 3040:577–586
17. Bam L, Jewell W (2005) Review: power system analysis software tools. In: Proceedings of the 2005 IEEE power engineering society general meeting, San Francisco, June 12–16 2005
18. Queiroz AR, Oliveira FA, Marangon Lima JW, Balestrassi PP (2007) Simulating electricity spot prices in Brazil using neural network and design of experiments. In: Proceedings of the 2007 power tech conference, Lausanne, July 1–5 2007
19. Box G, Jenkins G (1976) Time series analysis: forecasting and control, Revised Edth edn. Holden-Day, San Francisco
20. Nogales FJ, Contreras J, Conejo AJ, Espínola R (2002) Forecasting next-day electricity prices by time series models. IEEE Trans Power Syst 17:342–348
21. Mateo González A, Muñoz San Roque A, García-González J (2005) Modeling and forecasting electricity prices with input/output hidden Markov models. IEEE Trans Power Syst 20 (1):13–24
22. Contreras J, Espínola R, Nogales FJ, Conejo AJ (2003) ARIMA models to predict next-day electricity prices. IEEE Trans Power Syst 18(3):1014–1020
23. García RC, Contreras J, van Akkeren M, Batista J, García C (2005) A GARCH forecasting model to predict day-ahead electricity prices. IEEE Trans Power Syst 20(2):867–874
24. Bodger PS, Brooks DRD, Moutter SP (1987) Spectral decomposition of variations in electricity loading using mixed radix fast Fourier transform. IEE Proc Gen Transm Distrib 134(3):197–202
25. Saito M, Kakemoto Y (2004) Demand forecasting by the neural network with discrete Fourier transform. Proc IEEE Int Joint Conf Data Min 4:2759–2763
26. Vahabie H, Yousefi A, Araabi MMR, Lucas BN, Barghinia C. Combination of singular spectrum analysis and autoregressive model for short term load forecasting. In: IEEE power tech conference, Lausanne, 1–5 July 2007, pp 1090–1093
27. Ramsay B, Wang AJ (1998) A neural network based estimator for electricity spot-pricing with particular reference to weekend and public holidays. Neurocomputing 23:47–57

28. Zhang L, Luh PB, Kasiviswanathan K (2003) Energy clearing price prediction and confidence interval estimation with cascaded neural networks. IEEE Trans Power Syst 18 (1):99–105
29. Rodriguez CP, Anders GJ (2004) Energy price forecasting in the Ontario competitive power system market. IEEE Trans Power Syst 19(1):366–374
30. Conejo AJ, Plazas MA, Espínola R, Molina AB (2005) Day-ahead electricity price forecasting using the wavelet transform and ARIMA models. IEEE Trans Power Syst 20 (2):1035–1042
31. Areekul P, Senjyu T, Toyama H, Yona A (2010) A hybrid ARIMA and neural network model for short-term price forecasting in deregulated Market. IEEE Trans Power Syst 25 (1):524–530
32. Troncoso Lora A, Riquelme Santos JM, Gómez-Expósito A, Martínez Ramos JL, Riquelme Santos JC (2007) Electricity market price forecasting based on weighted nearest neighbors techniques. IEEE Trans Power Syst 22(3):1294–1301
33. Gómez-Expósito A, Troncoso A, Gómez-Quiles C, Riquelme JM, Martínez JL, Riquelme JC Application of the weighted nearest neighbor method to power system forecasting problems. In El-Hawary M (ed) Forecasting in power systems. IEEE Press (in press)
34. Kennel MB, Brown R, Abrabanel HDI (1992) ÒDetermining embedding dimension for phase-space reconstruction using a geometrical constructionÓ. Phys Rev A 45(6):3403–3411
35. Spanish Electricity Market Operator (OMEL) Information System. http://www.omel.es/. Accessed Dec 2009
36. Spanish Wind Energy Association Information System. http://www.aeeolica.es/. Accessed Dec 2009

ECOTOOL: A general MATLAB Forecasting Toolbox with Applications to Electricity Markets

Diego J. Pedregal, Javier Contreras, and Agustín A. Sánchez de la Nieta

Abstract Electricity markets are composed of different agents that make their offers to sell and/or buy energy. These agents need forecasting tools to have an accurate prediction of the prices that they will face either in the day-ahead or long-term time spans. This work presents the ECOnometrics TOOLbox (ECOTOOL), a new MATLAB forecasting toolbox that embodies several tools for identification, validation and forecasting models based on time series analysis, among them, ARIMA, Exponential Smoothing, Unobserved Components, ARX, ARMAX, Transfer Function, Dynamic Regression and Distributed Lag models. The toolbox is presented in all its potentiality and several real case studies, both on the short and medium term, are shown to illustrate its applicability.

Keywords ARIMA • ARMAX • ARX • Dynamic regression • Exponential smoothing • MATLAB forecasting toolbox • Transfer function

D.J. Pedregal (✉)
University of Castilla – La Mancha, E.T.S. de Ingenieros Industriales and Institute of Applied Mathematics to Science and Engineering (IMACI), Ciudad Real, Spain
e-mail: Diego.Pedregal@uclm.es

J. Contreras
University of Castilla – La Mancha, E.T.S. de Ingenieros Industriales and Institute of Energy Research and Industrial Applications (INEI), Ciudad Real, Spain
e-mail: Javier.Contreras@uclm.es

A.A. Sánchez de la Nieta
University of Castilla – La Mancha, E.T.S. de Ingenieros Industriales, Ciudad Real, Spain
e-mail: Agustin.Sanchez@uclm.es

A. Sorokin et al. (eds.), *Handbook of Networks in Power Systems I*,
Energy Systems, DOI 10.1007/978-3-642-23193-3_6,

1 Introduction

Price and load demand forecasting has become increasingly important in electricity markets all over the world. It has been applied to short-term day-ahead markets and also to long-term markets. In day-ahead markets, the producers submit offers consisting of a set of quantities at certain prices. The market operator matches offers and bids setting the 24 hourly clearing prices for the next day. Market clearing prices are publicly available in many markets, as it is the case of the day-ahead pool of mainland Spain and Portugal [27]. The prediction of these prices is crucial for electric companies in order to construct their daily offers. On the other hand, in long-term markets, companies wish to decide their future contract prices and compare them to the prices in day-ahead markets.

Recent years have seen a number of prediction methods applied to price forecasting in electric markets [1]. A brief review of the most important references on forecasting methods follows.

Artificial Neural Network (ANN) are techniques used by e.g. [2], who propose a hybrid approach based on neural networks and fuzzy logic with examples from the England-Wales market. Other authors, like Amjadi [3] presents a fuzzy neural network that has inter-layer and feed-forward architecture for the Spanish and Californian markets, and Szkuta [4] propose a three-layered ANN with backpropagation for the Victorian electricity market in Australia.

Other forecasting methods are based on times series, for example, the Auto Regressive (AR) models. They have been used to predict weekly prices in the Norwegian system [5]. In addition, Auto Regressive Integrated Moving Average (ARIMA) models have also been applied to predict day-ahead hourly prices in California and Spain [6].

Different time series models are based on transfer function and dynamic regression [7], with applications to the Spanish and Californian day-ahead markets.

To take into account the ever changing volatility of prices, GARCH models are commonly used methods [8], for example to study the hedging performance of electricity futures in the Nordic market [9].

To conclude, Kalman filters stand as control devices able to forecast prices [10–12].

For a more complete review of state-of-the-art forecasting methods interested readers are referred to in [13, 14].

Many commercial tools have embedded forecasting capabilities to predict time series [26, 28, 29] among them, MATLAB [30] is very good to visualize results graphically. In particular, there is an official MATLAB GARCH toolbox, a recent Econometrics toolbox and several other time series functions in the System Identification Toolbox that implements some time series models. The main problem is that none of the official toolboxes in MATLAB offers the flexibility often required by real applications in certain contexts. This is mainly due to the fact that most of them are developed by and for engineers, making such tools too rigid or cumbersome to use for researchers or practitioners in other areas of science. One typical case is related to time series analysis in electricity markets, where analysts miss typical models (for example Exponential Smoothing, Unobserved Components, etc.) or find the ones implemented in MATLAB too rigid.

One example of the aforementioned rigidity is the expression of multiplicative seasonal polynomials in ARIMA models. For example, it is not obvious how the product of polynomials is implemented in the System Identification Toolbox, whilst this is straight away in ECOTOOL. In addition, we allow for the possibility of having more than two multiplicative seasonal polynomials. Please see modelTF function in Sects. 3 and 4.

Another example is the flexibility and easiness in writing models in our toolbox, because the code closely mimics the way in which mathematical models are written "on paper". This is particularly useful when considering parameter constraints, because the user can choose arbitrarily the names of the parameters.

Due to these limitations, we present a forecasting MATLAB toolbox (ECOTOOL[1]) to analyze time series of common use in electricity markets by means of several forecasting methods. Its visual capabilities and user-friendly interface can be very useful for students, professors and forecasting professionals dealing with price and demand forecasts in very wide contexts, and especially in electric markets.

The remaining sections are organized as follows. Section 2 presents the mathematical models of some of the forecasting models available in the toolbox. In Sect. 3, a general description of the toolbox is provided. Section 4 presents the graphical and numerical results of the forecasts of real-life examples of the Iberian electricity market, and Sect. 5 provides several conclusions.

2 Forecasting Models

The toolbox offers the possibility of estimating and forecasting time series using models with different degrees of complexity, either Single-Input-Single-Output (SISO), Multiple-Input-Single-Output (MISO) and Multiple-Input-Multiple-Output (MIMO) systems. In addition, it includes user friendly tools for presenting results related to a wide range of statistical tests, identification, and validation (see Sect. 3).

2.1 SISO Models

Three main families of SISO systems (or univariate models) are available in ECOTOOL, namely ARIMA, Exponential Smoothing (ES) and Unobserved Components models (UC).

[1] The toolbox is available upon request to the authors via email.

2.1.1 ARIMA Models

Following the standard ARIMA jargon [15], the models implemented on the toolbox are of the $ARIMA(p_0, d_0, q_0) \times (p_1, d_1, q_1)_{s_1} \times \cdots \times (p_k, d_k, q_k)_{s_k}$ class. Its specific formulation corresponds to Eq. 1.

$$y_t = \frac{1}{(1-B)^{d_0}(1-B^{s_1})^{d_1}\cdots(1-B^{s_k})^{d_k}} \frac{\vartheta_{q_0}(B)}{\varphi_{p_0}(B)} \frac{\vartheta_{q_1}(B^{s_1})}{\varphi_{p_1}(B^{s_1})} \cdots \frac{\vartheta_{q_k}(B^{s_k})}{\varphi_{p_k}(B^{s_k})} e_t \quad (1)$$

where y_t is the observed time series; e_t is a zero mean and constant variance Gaussian white noise; s_j, $(j = 0, 1, \ldots, k)$ are a set of seasonal periods, with $s_0 = 1$; $(1 - B^{s_j})^{d_j}$, $(j = 0, 1, \ldots, k)$ are the $k + 1$ differencing operators necessary to reduce the time series to mean stationarity; $\vartheta_{q_j}(B^{s_j})$ and $\phi_{p_j}(B^{s_j})$, $(j = 0, 1, \ldots, k)$, are invertible and stationary polynomials in the backshift operator $B : B^l y_t = y_{t-l}$ of the type $\vartheta_{q_j}(B^{s_j}) = \left(1 + \vartheta_1 B^{s_j} + \vartheta_2 B^{2s_j} + \cdots + \vartheta_{q_j} B^{q_j s_j}\right)$.

It is clear that model (1) is particularly general in its formulation and includes multiple ARIMA models as particular cases. As an example, consider the well known *airline model* in Eq. 2, which, for monthly data is an $ARIMA(0, 1, 1) \times (0, 1, 1)_{12}$.

$$y_t = \frac{(1 + \vartheta_1 B)(1 + \vartheta_{12} B^{12})}{(1 - B)(1 - B^{12})} e_t \quad (2)$$

Estimation of this type of models may be performed by a number of methods, being the conditional sum of squares and exact maximum likelihood [15] the ones implemented in ECOTOOL.

2.1.2 Exponential Smoothing

Six different cases of additive Exponential Smoothing models may be used,[2] depending on the number of components included. Table 1 summarizes such models.

In Table 1, l_t, b_t, s_t and e_t are the level, slope, seasonal and irregular components, respectively; and α, β, γ and ϕ are unknown parameters that should be estimated.

Each of the six models has to be used for the appropriate data, indicated by the respective code (second column of Table 1) and the first equation in each model. The first letter in the code indicates the model for the trend, while the second letter is reserved for the seasonal model. A ‘N’ indicates that the component is not present; ‘A’ implies an additive component; ‘D’ implies a damped component, reserved for the trend solely.

[2] Multiplicative options may be implemented by taking logarithms to the data.

Table 1 Models for common linear forms of Exponential Smoothing. Taken from [16]

Case	Code	Model	Description
1	NN	$y_t = l_{t-1} + e_t$ $l_t = l_{t-1} + \alpha e_t$	Simple exponential smoothing [22]
2	AN	$y_t = l_{t-1} + b_{t-1} + e_t$ $l_t = l_{t-1} + b_{t-1} + \alpha e_t$ $b_t = b_{t-1} + \beta e_t$	Trend-corrected exponential smoothing [23]
3	DN	$y_t = l_{t-1} + b_{t-1} + e_t$ $l_t = l_{t-1} + b_{t-1} + \alpha e_t$ $b_t = \varphi b_{t-1} + \beta e_t$	Damped trend [24]
4	NA	$y_t = s_{t-m} + e_t$ $s_t = s_{t-m} + \gamma e_t$	Elementary seasonal case
5	AA	$y_t = l_{t-1} + b_{t-1} + s_{t-m} + e_t$ $l_t = l_{t-1} + b_{t-1} + \alpha e_t$ $b_t = b_{t-1} + \beta e_t$ $s_t = s_{t-m} + \gamma e_t$	Winters additive method
6	DA	$y_t = l_{t-1} + b_{t-1} + s_{t-m} + e_t$ $l_t = l_{t-1} + b_{t-1} + \alpha e_t$ $b_t = \varphi b_{t-1} + \beta e_t$ $s_t = s_{t-m} + \gamma e_t$	Damped trend with seasonal effects

Models 1–3 are used for time series without a seasonal component, either considering that the slope of such level or trend is zero (model 1, i.e. the series is considered just as a time varying mean) or different from zero. Models 4–6 are appropriate when a seasonal component is present.

One classical issue considered in the literature is the so-called *reduced form* or ARIMA form of ES models, which is particularly easy to obtain for the models above (see [16], Chap. 11). Any Exponential Smoothing model may be expressed as a constrained ARIMA model. As an illustration, Eq. 3 shows the reduced form for case 1 (or NN), obtained by substitution of the second equation into the first one. The process is equivalent to an $ARIMA(0, 1, 1)$ in which the parameter of the MA(1) term is $\alpha - 1$.

$$y_t = \frac{\alpha B}{1 - B} e_t + e_t = \frac{1 + (\alpha - 1)B}{1 - B} e_t \tag{3}$$

Exponential Smoothing models are observationally equivalent to their reduced form counterparts, and produce the same forecasts and prediction intervals [16]. Therefore, Exponential Smoothing estimation in the toolbox is based on the reduced form, based on the MATLAB code developed for ARIMA estimation. The main advantages are: (1) either exact or conditional likelihood is possible, (2) further procedures for outlier automatic detection and estimation would be applicable (see next section), and (3) further extensions of the model with inputs may be used (see MISO models below).

2.1.3 Unobserved Components Models

In a univariate Unobserved Components model (UC), the time series are assumed to be the addition of several components, usually a trend, seasonal and irregular, each one with its own physical interpretation. Using the same notation than for ES models the UC model implemented in the toolbox may be written as:

$$y_t = l_t + s_t + e_t \tag{4}$$

One successful way to handle this type of model is by means of a State Space (SS) framework, in which Eq. 4 plays the role of the observation equation, while the transition equations (i.e. the dynamic behaviour of the components) has to be explicitly defined. The SS representation of each of them used in this chapter is the typical of the so called Basic Structural Model (BSM) of Harvey [17] and is given in Eq. 5.

$$\mathbf{x}_t = \begin{pmatrix} l_t \\ b_t \\ crs_t \\ x_t^1 \\ x_t^2 \\ \vdots \\ x_t^{s-1} \end{pmatrix} = \begin{pmatrix} \alpha & 1 & 0 & 0 & 0 & \cdots & 0 \\ 0 & 1 & 0 & 0 & 0 & \cdots & 0 \\ 0 & 0 & -1 & -1 & -1 & \cdots & -1 \\ 0 & 0 & 1 & 0 & 0 & \cdots & 0 \\ 0 & 0 & 0 & 1 & 0 & \cdots & 0 \\ \vdots & \vdots & \vdots & \vdots & \vdots & \ddots & \vdots \\ 0 & 0 & 0 & 0 & 0 & \cdots & 0 \end{pmatrix} \begin{pmatrix} l_{t-1} \\ b_{t-1} \\ s_{t-1} \\ x_{t-1}^1 \\ x_{t-1}^2 \\ \vdots \\ x_{t-1}^{s-1} \end{pmatrix} + \begin{pmatrix} \varepsilon_t \\ \eta_t \\ \omega_t \\ 0 \\ 0 \\ \vdots \\ 0 \end{pmatrix}$$

$$y_t = (1 \quad 0 \qquad 1 \quad 0 \quad \cdots \quad 0)\mathbf{x}_t + e_t \tag{5}$$

Here the trend noises, σ_η^2 and η_t, and the seasonal component noise, ω_t, are random Gaussian noises, independent of each other with zero mean and certain variances, σ_ε^2, σ_η^2, and σ_ω^2, respectively. System (5) is divided visually with dotted lines into two independent subsystems block-concatenated, the first one corresponding to define the dynamics of the trend component and a second one corresponding to the seasonal component. Only three elements in the state vector are meaningful, i.e. the trend, the slope and the seasonal, l_t, b_t and s_t, respectively. Four possible models of trend may be implemented, Local Linear Trend (LLT) [17] with $\alpha = 1$; Random Walk (RW) by eliminating the first state equation and taking b_t as the trend; Integrated Random Walk (IRW) [18] with $\alpha = 1$ and $\varepsilon_t = 0$; or Smoothed Trend (SRW) with $0<\alpha<1$.

Given the full SS system, the Kalman Filter (KF) [19] and the Fixed Interval Smoothing algorithms (FIS) [25] provide the optimal estimation of the first and second order moments of the state vector in the sense of minimizing the Mean Squared Error. They also produce automatically a number of important operations in time series analysis, like interpolation of missing observations, forecasting, backasting (if necessary), etc.

The application of the recursive KF and FIS algorithms requires the knowledge of all the system matrices, something that is not known in general. In the system above, the unknown parameters are all the noise variances, i.e., σ_ε^2, σ_η^2, σ_ω^2 and σ^2, the irregular/innovations variance. The estimation method usually preferred in the literature, due to its general good statistical properties is Maximum Likelihood (ML) based on the *prediction error decomposition*. Further details may be found in [17] and [18].

As it happened to ES models, each UC model have their reduced form counterpart, i.e. ARIMA model with specific constraints on the parameters. Then, when outlier automatic detection is required by the user, a preliminary estimation is carried out based on the reduced form.

2.2 MISO Models

The basic MISO model implemented is the linear Transfer Function [15]. The general formulation may be expressed as in Eq. 6.

$$y_t = \sum_{i=1}^{h} \frac{\omega_{n_i}(B)}{\delta_{m_i}(B)} u_{it} + N(B)e_t \tag{6}$$

where $\omega_{n_i}(B) = (\omega_0 + \omega_1 B + \cdots + \omega_{n_i} B^{n_i})$, $(i = 1, \ldots, h)$, are polynomials in the backshift operator that may have leading zero coefficients when a pure time delay is necessary; and $\delta_{m_i}(B) = (1 + \delta_1 B + \cdots + \delta_{m_i} B^{m_i})$, $(i = 1, \ldots, h)$, are stationary or stable polynomials. The general representation of the noise model $N(B)e_t$ may be any of the SISO alternatives shown in the previous subsection. Looking at Eq. 6 in a different way, we may see it as the previous SISO models (ARIMA, ES or UC) extended with additional linear dynamical terms.

It is important to note that, by imposing specific constraints in model (6), we may obtain other well-known alternatives in the literature, such is the case of regression with correlated noises ($n_i = 0$, $m_i = 0$, $i = 1, \ldots, h$); ARX, dynamic regression, or distributed lag models ($\delta_{m_i}(B) = \delta(B)$, $i = 1, \ldots, h$ and $N(B) = 1/\delta(B)$); or ARMAX models ($\delta_{m_i}(B) = \delta(B)$, $i = 1, \ldots, h$ and $N(B) = \omega(B)/\delta(B)$).

2.3 MIMO Models

VARX models according to Eq. 7 are also included in the toolbox:

$$\left(\mathbf{I} + \mathbf{\Phi}_1 B + \mathbf{\Phi}_2 B^2 + \cdots + \mathbf{\Phi}_p B^p\right)\mathbf{y}_t = \left(\mathbf{\Gamma}_0 + \mathbf{\Gamma}_1 B + \cdots + \mathbf{\Gamma}_k B^k\right)\mathbf{u}_t + \mathbf{e}_t \tag{7}$$

where boldface letters indicate either matrices or vectors; $\mathbf{I}$ is a $n \times n$ identity matrix; $\mathbf{y}_t$ and $\mathbf{u}_t$ are a set of n and m output and input variables, respectively; $\mathbf{\Phi}_j$ are a set of p, $n \times n$ squared matrices of coefficients to estimate; $\mathbf{\Gamma}_j$ are another set of $n \times m$ squared matrices of unknown coefficients; and $\mathbf{e}_t$ are a vector of n Gaussian white noises serially independent with contemporaneous covariance matrix Ω.

Estimation of unrestricted VARX models is particularly simple, since least squares of each equation separately produce consistent and efficient estimates. However, when constraints are imposed onto some of the elements of the coefficients matrices, iterated generalized least squares may to be used, in order to avoid inefficient estimation [20].

3 ECOTOOL MATLAB Forecasting Toolbox

ECOTOOL is intended for a general use in time series analysis, though in this chapter it will only be used for electricity forecasting examples. As it has already be shown in the previous section, several modeling options (from SISO to MIMO models) are available to the user, so that he/she may choose the one that results more appropriate in each particular case.

One main advantage of the toolbox is the flexibility with which models may be specified. Several properties are the salient features of the toolbox: (1) it is user-oriented, i.e. a full time series analysis may be performed with just a few MATLAB functions, so the user should keep in mind just a few function names; (2) thorough help is available for the toolbox as a whole and for each function in particular, as it happens with standard MATLAB toolboxes; (3) specification of models is rather simple and flexible, e.g. in the case of TF models they are written in MATLAB code as they are written analytically; (4) imposing constraints in the parameters is straightforward; (5) estimation methods for some models include both conditional and exact Maximum Likelihood; (6) automatic detection and estimation of four types of outliers is implemented for some models: additive, innovative, level shift and transitory change, see [21]; (7) functions to deal with moving festivals, trading day, and other calendar effects are also included; (8) additional identification and diagnostic tools are included by means of the toolTEST function, as well as a forecast function to compute forecast errors and other statistics related to the appropriateness of the models (toolFORECAST).

Table 2 shows a simplified table of contents that may be obtained by the command help ecotool.

The main functions are in the first block of Table 2, where all the model estimation functions are shown, together with a demo file prepared to show the properties of the toolbox and two Graphical User Interfaces (GUI) for descriptive and diagnostic checking of time series (toolTEST) and for presenting forecasting results and calculating forecast errors (toolFORECAST). The second block offers several functions for outlier intervention that often are important in time series

Table 2 Simplified table of contents of ECOTOOL toolbox

GUIs and main functions	
toolTEST	Exploratory, descriptive and diagnostic checking tool
toolFORECAST	Forecasting tool
modelES	Exponential Smoothing models
modelTF	MISO transfer function analysis
modelUC	Unobserved components models
modelVARX	VAR model with eXogenous variables analysis
ECOTOOLdemos	ECOTOOL demos
Time series intervention functions	
days	Dummy variable for number of days in months or quarters
easter	General dummy Easter variables on monthly or quarterly data
leapyear	Dummy variable for leap year intervention
trading	Trading day variables on monthly or quarterly frequency
General purpose functions	
adjust	Subtracts two time series adjusting the size of the matrices
corrmatrix	Builds correlation matrix from covariance matrix
lag	Lags a vector of variables
meda	Median Absolute deviation with respect to the median
nanprod	Product of matrices with NaN values
printwf	Print table output with format
transform	Standardization or de-standardization of a vector of variables
varstep	Var impulse and step function analysis
vboxcoxinv	Inverse of Box-Cox transformation
vconv	Multiplication of vector polynomials
vdif	Differentiation of a vector of variables
vfilter	Filters a vector of inputs with a vector digital filter
vroots	Calculates the roots of a vector polynomial

analysis. The final block includes a number of general purpose functions, thoroughly used in this toolbox, though any user would find useful in many different contexts.

Due to space constraints it is not possible to fully explain each one of the functions, but in the remaining of this section the capabilities of the two GUI included and the use of the most important modeling functions are briefly described.

Figure 1 shows a typical caption of the toolTEST GUI for a case study discussed in the next section. In addition to the standard figure menus, the tool offers three additional possibilities: (1) *Tests* to perform a long list of statistical tests (see below); (2) *Series* to select the time series (or all available) on which the text will be executed; and (3) *Options* to choose particular options for the current test. The most relevant menu in this GUI is shown open in Fig. 1 so that the reader may check the number of graphical and statistical tests available:

- Descriptive information: time plots, box plots to detect seasonal behavior, scatter plots, descriptive statistics, histograms with hypothetical Gaussian distribution, quantile plots, gaussianity standard tests.

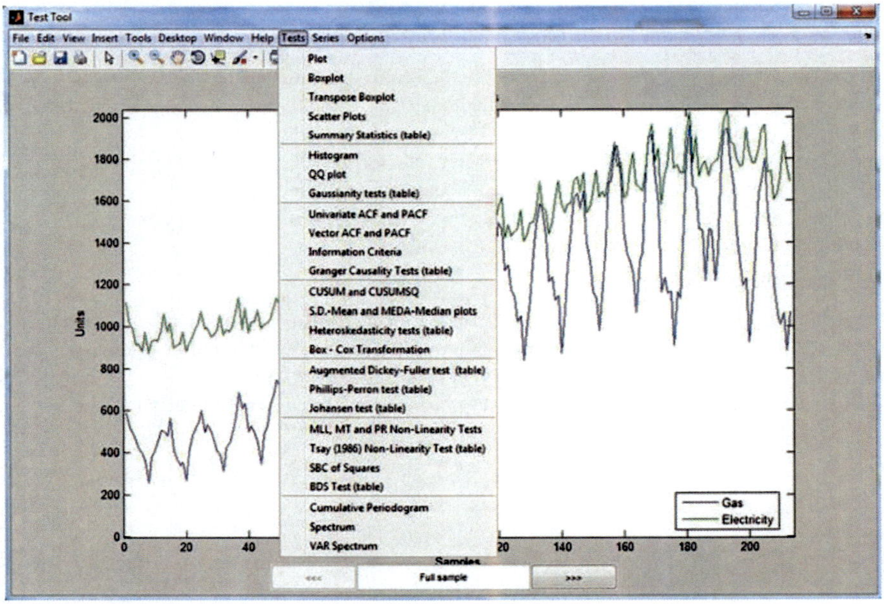

Fig. 1 Typical aspect of the toolTEST window with the Tests menu open

- Identification tools: Univariate Autocorrelation and Partial Autocorrelation Functions, Vector Autocorrelation and Partial Autocorrelation Functions for multivariate time series, several information criteria applied to increasing VAR models, Granger causality tests based on VAR models.
- Heteroskedasticity tests: CUSUM and CUSUMSQ tests, range-mean plot (standard deviation mean or meda-median plots), variance ratio tests, estimation of univariate or multivariate Box-Cox homoskedasticity transformation.
- Unit root and cointegration tests: Dickey-Fuller, Phillips Perron and Johansen tests.
- Non-linearity tests: Tsay, Schwarz criterion on squares, Brock-Dechert-Scheinkman test, etc.
- Spectral tools: cumulative periodogram, smoothed or raw periodogram, AR-spectrum.

Regarding the second GUI, toolFORECAST, Fig. 3 in next section shows a caption on a particular case study. In a similar way to toolTEST, three additional menus may appear: (1) *Series* becomes visible only for multivariate systems to select the specific time series on which the plot or the error calculations are performed; (2) *Forecasts* appears when more than one forecast is available for each time series; (3) *Options* to control the graphical and output general options. The pushbutton labeled as *Show table* prints out the actual data together with forecasts and different forecast error measures.

Commands modelES, modelTF, modelUC, modelVARX are MATLAB functions to estimate the different types of models shown analytically in the previous section. All of them have been written keeping a common nomenclature in order to make their use easier. The typical MATLAB call, common to all of them is:

```
>> [e, pz, stdpz]= model??(z, u, model, nofs);
```

where model?? stands for any of the functions; z and u are matrices of output and input observations, respectively; model is the model specification; nofs are the forecasting information necessary to estimate forecasts and confidence intervals. There are other inputs and outputs specific to the functions that may be consulted on each function individual help.

One of the most comfortable features of the ECOTOOL toolbox is the flexibility with which models are specified, i.e. the variable model above. This input is a cell of strings for MISO models or may be just a string for SISO models. Each string in the cell is a ratio of polynomials as they are written analytically on a paper, reserving the letter 'B' for the backshift operator and using any names (formed just by letters) to identify the parameters. Valid analytical expressions in MATLAB may also be used to define the parameters and constraints may be imposed based on such names. For example, any MISO model with one TF input term may be specified as:

```
>> model= {Noise_model, '(w0+w1*B+w2*B12)/(1+d1*B)'};
```

The values for the first element in the cell (Noise_model) depend on the particular model to estimate. Such values are:

- modelTF: valid cases of ARIMA models are, for example, '(1 + ma1*B) (1 + ma12*B12)/(1-B)(1-B12)', or '(1 + ma1*B)(1-(2*ma1)*B12)'. Beware than in the latter case the model is estimated in a way such that the second parameter is minus twice the first one (only one parameter would be estimated).
- modelES: Noise_model is a string variable followed by the seasonal period. Some valid cases are 'NN', 'NA12', 'AA12' (see Table 1).
- modelUC: Noise_model is a string variable indicating the type of trend to estimate (either 'LLT', 'IRW', 'RW', 'SRW'), followed by the seasonal period, if any. Some valid cases are 'LLT', 'IRW12', 'SRW4'.

The case of VARX models are necessarily different to the previous ones. When modelVARX function is used, model is a cell with two elements, corresponding to the polynomials in Eq. 7. Each polynomial may be either a row vector of positive values for lags in ascending order of powers of the backshift operator, or vector matrices with the block parameters in the backshift operator of appropriate dimensions, where NaN values are interpreted as parameters to estimate, and any number is taken as a constraint. Valid values are, for example, [1 2 3] for an unconstrained polynomial of order 3; [2 4] for a constrained polynomial of order 4 in which block parameters for lags 1 and 3 are zero; [1 0 NaN 0; 0 1 NaN NaN] would represent a polynomial of order 1 in a model with two outputs and selected constraints.

4 Case Studies

In order to show the versatility of ECOTOOL, we explain the MATLAB code that can be used to implement a number of time series of different complexity step by step. This section may be regarded as a tutorial and results may be replicated by running the code on MATLAB.

4.1 Electricity and Gas Monthly Data Forecasts

Spanish Electricity final consumption and *natural and manufactured gas* monthly series are available from the Spanish National Statistics Agency (www.ine.es), from January 1992 to September 2009 (213 months).

The data may be loaded on the MATLAB workspace writing load energy at the command prompt. Two variables are loaded, t (time) and y (gas and electricity consumption in a 213 $\times$ 2 matrix).

Several graphical and statistical tests may be carried out by means of the toolTEST function. Let's consider the following code:

```
>> toolTEST(y);
>> z= log(y);
>> dz= vdif(z, [1 1], [1 12]);
>> toolTEST(dz);
```

With the first command, Fig. 3.1 above will appear on the screen. With this tool, detailed graphical analysis of stationarity, seasonality, heteroskedasticity, etc. may be done. It is clear, as detected from a simple plot of the series, that they are not stationary in mean and variance and they have a strong seasonal pattern (with a period of 12 observations per year).

Stabilizing variance and mean transformations may be taken in with the log and vdif functions and stored in a variable called dz (where a regular and seasonal difference has been taken in the third line of code). Variable dz is assumed now to be stationary in mean and variance, and therefore an ARIMA model for each of them individually may be identified with the help of the toolTEST function again, now applied on variable dz and selecting the *Univariate ACF* and *PACF* option on the *Tests* menu (last line of code).

Monthly series of this type are often influenced by moving festivals (typically Easter) and trading day effects (number of working days in each month). In order to test for these effects in our time series, we generate two artificial variables with the help of easter and trading functions that will be considered two inputs to the models.

```
>> u=[easter([1992 01], 213) trading([1992 01],
   213)];
>> model1={'(1+ma1*B+ma2*B2)(1+ma12*B12)/
   (1-B)(1- B12)','w0', 'w1'};
>> y1= y(1:201, :); u1= u(1:201, :);
>> yf= y(202:end, 2); uf= u(202:end, :);
>> [e1, py1]=
   modelTF(y1(:, 1),u1,model1,uf,'eml',3.2);
>> [e2, py2]=
   modelTF(y1(:, 2),u1,model1,uf,'eml',3.2);
>> toolTEST([e1 e2]);
```

The model is specified in the second line. It may be easily seen that the analytical model is:

$$y_t = \omega_0 u_{1,t} + \omega_1 u_{2,t} + \frac{(1+\vartheta_1 B + \vartheta_2 B^2)(1+\vartheta_{12}B^{12})}{(1-B)(1-B^{12})} e_t$$

i.e., an $ARIMA(0,1,2) \times (0,1,1)_{12}$ with two linear regression terms added to take into account the effects of the dummy variables. Intermediate variables y1, u1, yf and uf are defined to make the difference between the estimating and forecasting periods, in which one year of monthly observations have been kept for checking forecast accuracy of different models. Given the inputs to the function modelTF in the fifth and sixth lines of code, the models are estimated by exact maximum likelihood (this is indicated by the 'eml' input) and with automatic detection of outliers (in this case residuals outside a ± 3.2 standard deviations band will be checked). The residuals and forecasts are stored on variables e1, e2, py1, and py2, respectively. Finally, in the last line of code the estimated residuals are checked in order to assess the adequacy of the model.

The estimation of the model above for the electricity series is shown in Fig. 2, in which in addition to all the parameters specified initially, some other appear, as a result of the automatic outlier searching algorithm. The names of such variables follow certain rules, i.e. two letters that identify the type of outlier (AO- additive; IO- innovative; TC- transitory change; LS- level shift) and the numbers after signal the observation at which the outliers occur.

Other model options, instead of ARIMA may be Exponential Smoothing or Unobserved Components models. These may be run replacing part of the code above by the following:

```
>> model2={'AA12', 'w0', 'w1'};
>> [e3, py3]=
   modelES(y1(:, 1),u1,model2,uf,'eml',3.2);
>> [e4, py4]=
   modelES(y1(:, 2),u1,model2,uf,'eml',3.2);
>> model3={'LLT12', 'w0', 'w1'};
>> [e5, py5]=
   modelUC(y1(:, 1),u1,model3,uf,'eml',3.2);
>> [e6, py6]=
   modelUC(y1(:, 2),u1,model3,uf,'eml',3.2);
```

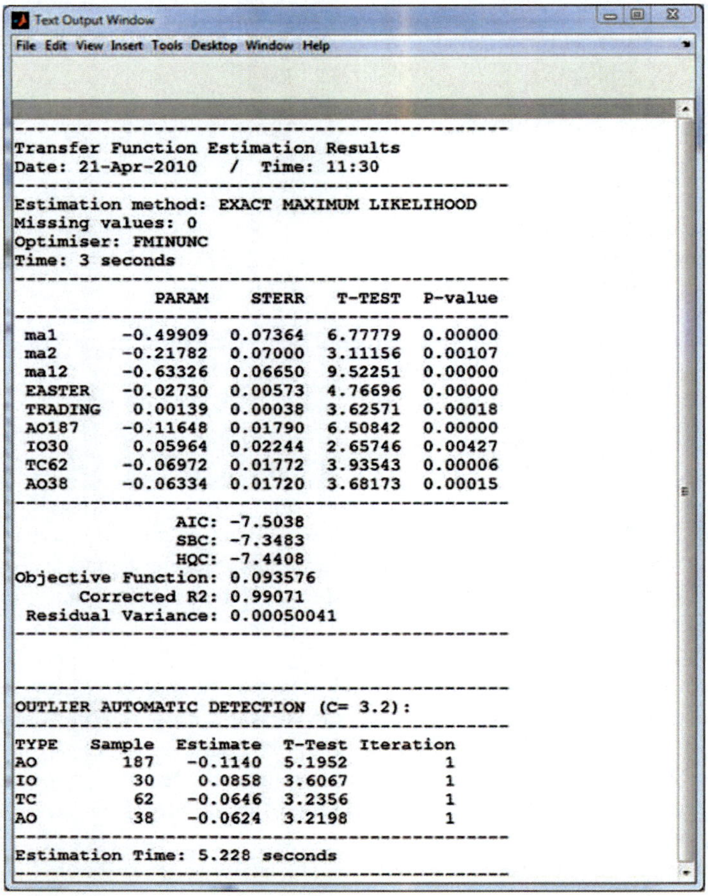

```
Text Output Window
File Edit View Insert Tools Desktop Window Help
---------------------------------------------
Transfer Function Estimation Results
Date: 21-Apr-2010    /  Time: 11:30
---------------------------------------------
Estimation method: EXACT MAXIMUM LIKELIHOOD
Missing values: 0
Optimiser: FMINUNC
Time: 3 seconds
---------------------------------------------
              PARAM     STERR    T-TEST  P-value
---------------------------------------------
 ma1       -0.49909  0.07364  6.77779  0.00000
 ma2       -0.21782  0.07000  3.11156  0.00107
 ma12      -0.63326  0.06650  9.52251  0.00000
 EASTER    -0.02730  0.00573  4.76696  0.00000
 TRADING    0.00139  0.00038  3.62571  0.00018
 AO187     -0.11648  0.01790  6.50842  0.00000
 IO30       0.05964  0.02244  2.65746  0.00427
 TC62      -0.06972  0.01772  3.93543  0.00006
 AO38      -0.06334  0.01720  3.68173  0.00015
---------------------------------------------
                AIC: -7.5038
                SBC: -7.3483
                HQC: -7.4408
Objective Function: 0.093576
      Corrected R2: 0.99071
 Residual Variance: 0.00050041
---------------------------------------------


---------------------------------------------
OUTLIER AUTOMATIC DETECTION (C= 3.2):
---------------------------------------------
TYPE   Sample  Estimate  T-Test Iteration
AO       187   -0.1140  5.1952         1
IO        30    0.0858  3.6067         1
TC        62   -0.0646  3.2356         1
AO        38   -0.0624  3.2198         1
---------------------------------------------
Estimation Time: 5.228 seconds
---------------------------------------------
```

Fig. 2 Typical output of modelTF function

A final option available in ECOTOOL is the VARX model for both time series simultaneously. One possible code in this case would be:

```
>> model4={[1:4 12:13 24:25], 0};
>> [e7, py7]= modelVARX(y1,u1,model2,uf,[1 1],
   [1  12]);
>> toolTEST(e7);
```

The model specified in this case is:

$$\left(\mathbf{I}+\mathbf{\Phi}_1 B+\cdots+\mathbf{\Phi}_4 B^4+\mathbf{\Phi}_{12} B^{12}+\mathbf{\Phi}_{13} B^{13}+\mathbf{\Phi}_{24} B^{24}+\mathbf{\Phi}_{25} B^{25}\right)\times \\ \times\left(\mathbf{I}-\mathbf{I}B\right)\left(\mathbf{I}-\mathbf{I}B^{12}\right)\mathbf{y}_t=\mathbf{\Gamma}_0\mathbf{u}_t+\mathbf{e}_t$$

where all the system matrices are squared of dimension 2. Beware that the differencing operators are included explicitly in the last two inputs to the function modelVARX, ([1 1] and[1 12]). This is quite convenient, since the forecasts are produced directly on the levels of the series.

One final interesting check is the forecast comparisons by means of the toolFORECAST function. One possible call, if we want to compare all the forecasts done so far is:

```
>> toolFORECAST(yf, [py1 py2 py3 py4 py5 py6 py7]);
```

The first input to this function indicates the actual values, while the second input is the collection of the four sets of forecasts produced, i.e. ARIMA, Exponential Smoothing, Unobserved Components, and VARX. Figure 3 shows the output where all the forecasts are compared with the actual values for the electricity consumption function. It is interesting to note the optimistic forecasts of the models with respect to the actual values, due to the fact that models have not yet incorporated the effect of the economic crisis.

4.2 *Day-Ahead Electricity Price Forecasting*

The capabilities of ECOTOOL are stretched in this section, where the performance of a transfer function model is shown with a realistic example that uses hourly data,

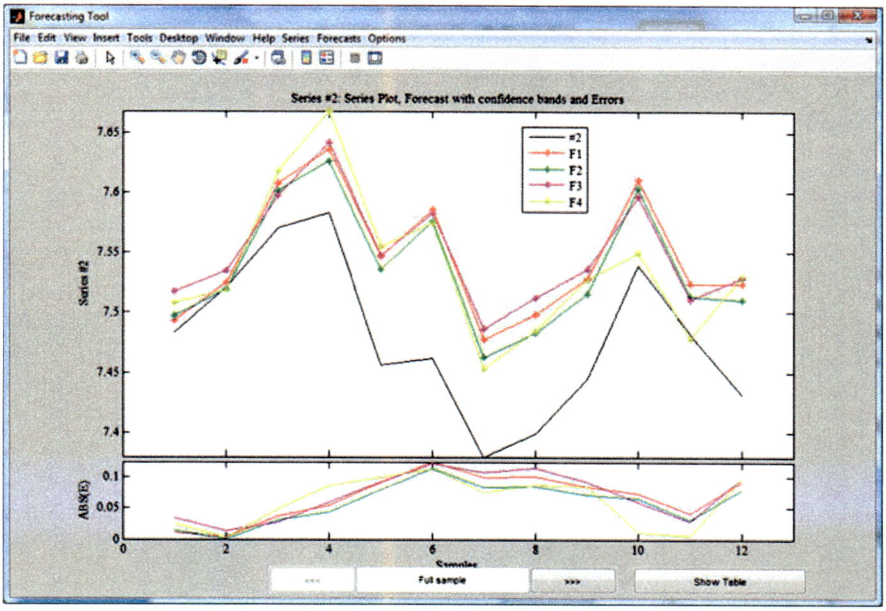

Fig. 3 Graphical output of toolFORECAST function

taken from the Spanish day-ahead electricity market. In particular, the week spanning from October 5th to October 11th, 2009 is selected. Prediction data spans from July 1st, 2009, till the forecasted day. It is assumed that once the day is forecasted, actual data, and not forecasts, are used to predict the following day.

Below, we explain the code implemented in the ECOTOOL toolbox step by step assuming that the price data are known. To begin with, the sample data required for prices needs to be a column vector format; it is graphically depicted through the toolTEST function and logarithms are taken to stabilized the variance, the code is given by:

```
>> toolTEST(price);
>> y=log(price);
```

Figure 4 depicts the temporal shape of price data. By the combination of this plot with the Autocorrelation functions and spectral tools available in the toolTEST function, it is possible to see two different types of seasonal factors, daily and weekly. It is also confirmed by a wide range of publications in the field, see [12, 13].

Similarly, we consider the demand data as shown in Fig. 5:

```
>> toolTEST(demand);
>> d=log(demand);
```

The transfer function model has to take into account three multiplicative factors, regular and seasonal, of 24 and 168 h (1 day and 1 week, respectively). Analytically the model is:

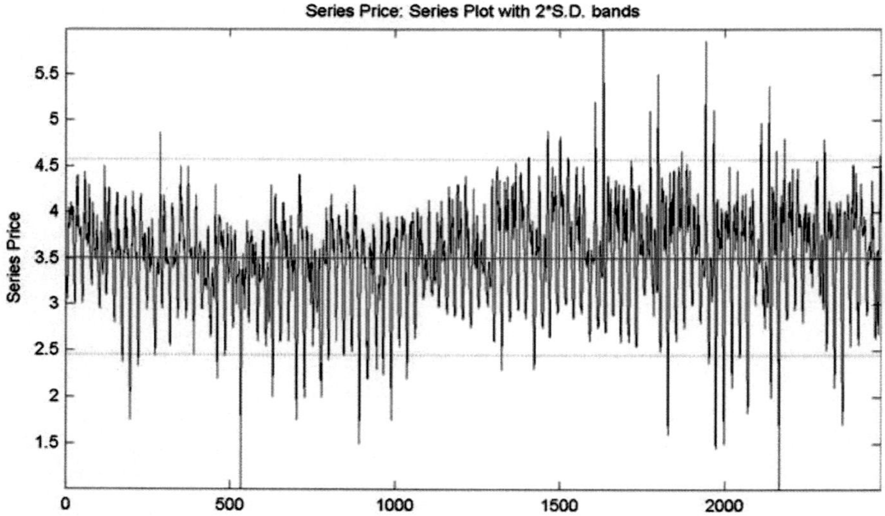

Fig. 4 Price series data of the day-ahead Spanish electricity pool market measured in c€/kWh

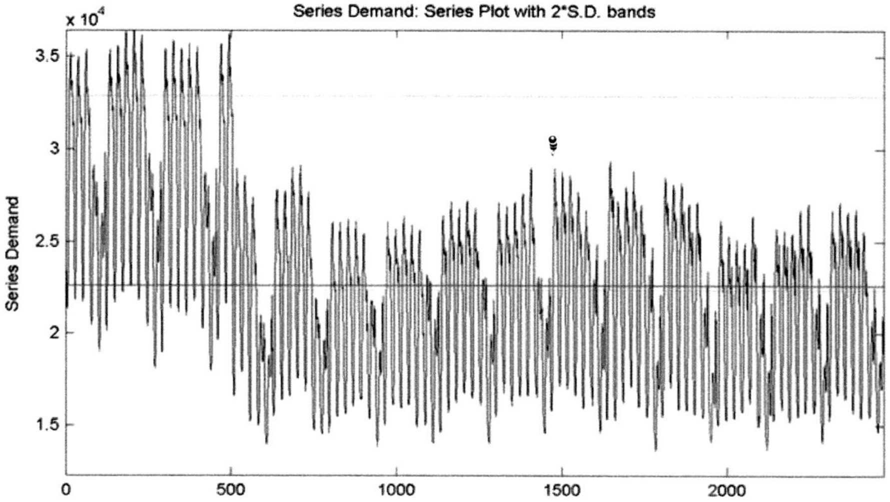

Fig. 5 Hourly demand of the day-ahead Spanish electricity pool market measured in kWh

$$y_t = c + \omega^d(B)d_t + \frac{1}{(1-B)(1-B^{24})(1-B^{168})}\frac{\vartheta_{q_0}(B)}{\phi_{p_1}(B)}\frac{\vartheta_{q_1}(B^{24})}{\phi_{p_1}(B^{24})}\frac{\vartheta_{q_k}(B^{168})}{\phi_{p_k}(B^{168})}e_t$$

where c is a constant, and the polynomial $\omega^d(B)d_t$ expresses the impact of the demand in the price forecast. To forecast the demand alone we use an ARIMA model of the type shown in Eq. 1 whose MATLAB code is:

```
>> modeld1={'(1-ma23*B23-ma24*B24-ma25*B25-ma48*B48-
ma96*B96)(1-ma168*B168)/(1-ar1*B-ar2*B2-ar3*B3-
ar4*B4-ar5*B5)(1-ar24*B24-ar48*B48-ar96*B96)(1-
ar168*B168-ar336*B336-ar504*B504)(1-B)(1-B24)(1-
B168)','constant'};
```

The identification of the transfer function model is rather complex in this case and follows a trial and error process until an appropriate model is found. The final model is specified by the following code:

```
>> model1={'(1-ma24*B24-ma48*B48-ma96*B96)
(1-ma168*B168)/(1-ar1*B-ar2*B2-ar3*B3-ar4*B4-ar5*B5-
ar6*B6-ar7*B7-ar8*B8)(1-ar23*B23-ar24*B24-ar25*B25-
ar48*B48-ar96*B96)(1-ar168*B168-ar336*B336-
ar504*B504)(1-B)(1-B24)(1-B168)','constant','(1-d1*B-
d2*B2-d3*B3-d4*B4-d5*B5)(1-d24*B24-d25*B25-d48*B48-
d49*B49-d72*B72-d73*B73-d96*B96-d97*B97-d120*B120-
d121*B121-d144*B144-d145*B145)(1-d168*B168-d169*B169-
d192*B192-d193*B193-d336*B336-d504*B504)'};
```

Note that in model1, the first terms until constant correspond to the ARMA model of the residual term, e_t, next, the three differentiations appear, 1, 24 and

168 (hourly, daily, and weekly), and, finally, the $\omega^d(B)$polynomial terms of the demand.

In order to produce 24 h ahead forecasts for the last week of data available, 168 hourly prices at the end of the dataset are removed. The next step eliminates the final week of the data, reserves in variable yf the actual future values of prices for the last day to be used later, and produce a vector of ones necessary to estimate the constant c in the model:

```
>> data1=y(1:end-168);
>> yf= y(end-168:end-145);
>> u1=ones(size(data1));
```

To produce the forecasts, the function modelTF has these inputs: the price data sample, a matrix made up of the vector of ones and the demand data sample, the transfer function model itself, and a matrix made up of the number of periods to forecast and the ARIMA model demand forecasts, pyd1, resulting from modeld1. The outputs are: the residuals (errors), the forecasted values and their standard deviation.

```
  >> [E1,py1,stdpy1]=modelTF(data1,[u1 d(1:end-
168)],model1,[ones(24,1) pyd1(end-168:end-145)]);
```

The function toolTEST run all necessary validation tests on the residuals of the model, checking if they can be assimilated to white noise by looking at the Autocorrelation Function (ACF) and Partial Autocorrelation Function (PACF):

```
>> toolTEST(E1);
```

Finally, the toolFORECAST function shows the sampling data, the model adjustment results, and the forecasts and their 95% confidence bands, together with the sampling data in a single screen. The forecasts, forecasting errors and confidence intervals are shown in Fig. 6 correspond to 1 week of hourly forecasts. Finally, Fig. 7 depicts the Auto Correlation Function (ACF) and the Partial Auto Correlation Function (PACF) of the residuals (errors). Note that the daily average forecasting errors in Table 3 are defined as:

$$\frac{1}{24}\sum_{t=1}^{24}\left|\frac{p_t-\widehat{p}_t}{\bar{p}}\right|100;\ \bar{p}=\frac{1}{24}\sum_{t=1}^{24}p_t$$

where p_t and $\hat{p}_t$ are the actual hourly prices and hourly predictions, respectively.

```
  >> toolFORECAST(exp(yf),exp(PY1),[exp(py1-2*stdpy1)
exp(py1+2*stdpy1)],2336);
```

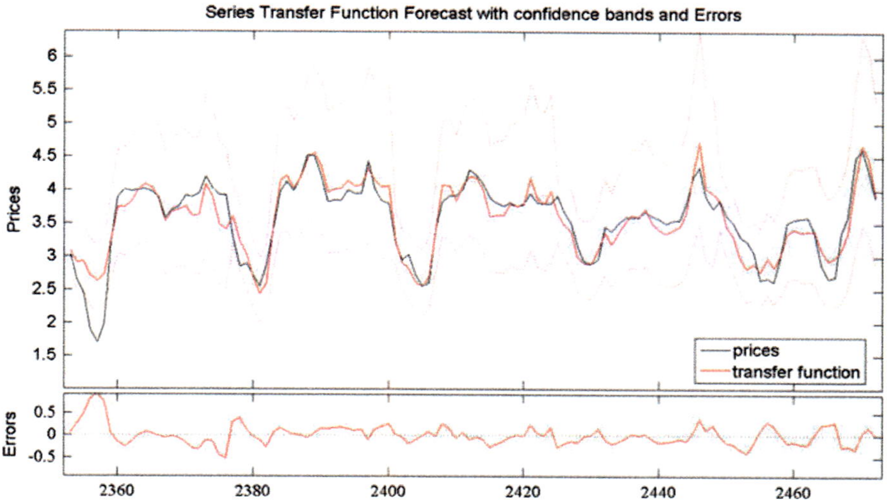

Fig. 6 Transfer function forecasting results for the day-ahead Spanish electricity pool market measured in kWh

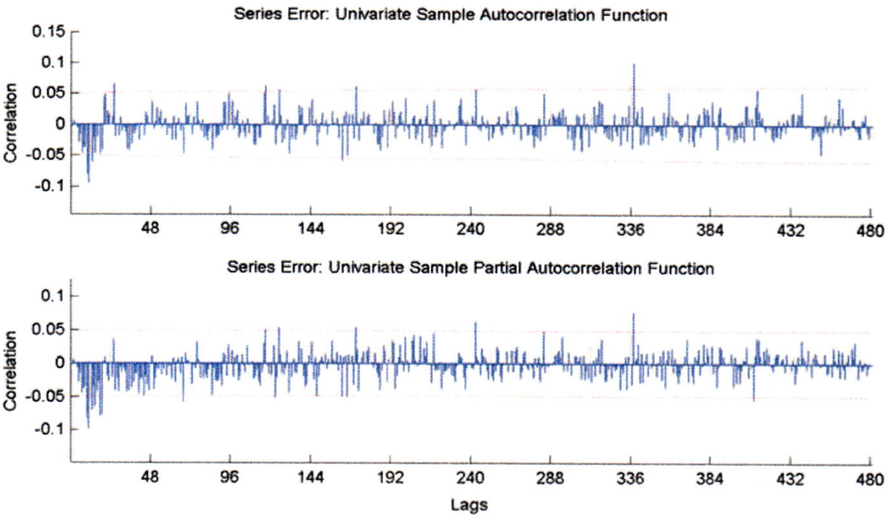

Fig. 7 ACF and PACF of the residuals of the day-ahead Spanish electricity pool market measured in kWh

Table 3 Daily average forecasting errors of the day-ahead Spanish electricity pool market

Days	1	2	3	4	5	6	7
Error (%)	7.51	6.66	7.45	3.75	2.87	3.69	5.63

5 Conclusions

This work has presented the new general MATLAB ECOnometrics TOOLbox (ECOTOOL). The toolbox is intended mainly for professional practitioners, academic researchers, students, and anyone involved in the analysis of time series, forecasting or signal processing.

ECOTOOL is composed of a number of powerful functions to estimate a wide range of models; user friendly tools of identification, validation and graphical representation of results; and other general functions for showing results, building moving festival dummy variables and other calendar effects, etc.

The main types of models implemented belong to the classes of SISO, MISO and MIMO systems. Methodologically, the models are of Box-Jenkins type, Exponential Smoothing, Unobserved Components and VARX.

Several properties are the salient features of the toolbox, e.g. it is user-oriented, just a few MATLAB functions are enough to do an exhaustive analysis; specification of models is rather simple and flexible; several estimation methods are implemented, making the toolbox very robust; automatic detection and estimation of four types of outliers is implemented; etc.

The toolbox also provides a wide range of descriptive information of the data, both graphically and in tabular format; standard and not so standard identification tools; formal and visual tests for gaussianity, independence, causality, heteroskedasticity, non-linearity, unit root and cointegration; spectral tools; tests on forecasting performance; etc.

The first case study included is a medium term forecasting exercise based on monthly Spanish data on global demand of electricity and gas. It is shown and compared the diversity of possible forecasts that may be produced with all the models available.

The second case study challenges the use of ECOTOOL with a rather complicated example, in which short term forecasts of hourly prices recorded at the Spanish day-ahead electricity market are run. The complexity of the ARIMA and Transfer Function models in this case are remarkable, as well as the results found.

As a summary, ECOTOOL is a powerful and user friendly toolbox intended to satisfy the needs of a wide audience with different requirements, with the main advantages of flexibility and easiness to use, perfectly comparable to other existing commercial tools.

Acknowledgments This work was supported in part by the Spanish Ministry of Education grant ENE2009-09541 and the Junta de Comunidades de Castilla – La Mancha grants PII2I09-0154-7984 and PII1I09-0209-6050.

References

1. Weron R (2006) Modeling and forecasting electricity loads and prices: a statistical approach. Wiley, Chichester
2. Wang AJ, Ramsay B (1998) A neural network based estimator for electricity spot-pricing with particular reference to weekend and public holidays. Neurocomputing 23:47–57

3. Amjadi N (2006) Day-ahead price forecasting of electricity markets by a new fuzzy neural network. IEEE Trans Power Syst 21:887–896
4. Szkuta BR, Sanabria LA, Dillon TS (1999) Electricity price short-term forecasting using artificial neural networks. IEEE Trans Power Syst 14:851–857
5. Fosso OB, Gjelsvik A, Haugstad A, Birger M, Wangensteen I (1999) Generation scheduling in a deregulated system. The Norwegian case. IEEE Trans Power Syst 14:75–81
6. Contreras J, Espínola R, Nogales FJ, Conejo AJ (2003) ARIMA models to predict next-day electricity prices. IEEE Trans Power Syst 18:1014–1020
7. Nogales FJ, Contreras J, Conejo AJ, Espínola R (2002) Forecasting next-day electricity prices by time series models. IEEE Trans Power Syst 17:342–348
8. Garcia RC, Contreras J, van Akkeren M, Garcia JBC (2005) A GARCH forecasting model to predict day-ahead electricity prices. IEEE Trans Power Syst 20:867–874
9. Byström H (2003) The hedging performance of electricity futures on the Nordic power exchange. Appl Econ 35:1–11
10. Pedregal DJ, Young PC (2008) Development of improved adaptive approaches to electricity demand forecasting. J Oper Res Soc 59:1066–1076
11. Trapero JR, Pedregal DJ (2009) Frequency domain methods applied to forecasting electricity markets. Energy Econ 31:727–735
12. Pedregal DJ, Trapero JR (2010) Mid-term hourly electricity forecasting based on a multi-rate approach. Energy Convers Manage 51:105–111
13. Conejo AJ, Contreras J, Espínola R, Plazas MA (2005) Forecasting electricity prices for a day-ahead pool-based electric energy market. Int J Forecasting 21:435–462
14. Weron R, Misiorek A (2008) Forecasting electricity prices: a comparison of parametric and semiparametric time series models. Int J Forecasting 24:744–763
15. Box GEP, Jenkins GM, Reinsel GC (1994) Time series analysis forecasting and control, 3rd edn. Prentice-Hall, Englewood Cliffs
16. Hyndman RJ, Koehler AB, Ord JK, Snyder RD (2008) Forecasting with exponential smoothing. Springer, Berlin
17. Harvey AC (1989) Forecasting structural time series models and the Kalman filter. Cambridge University Press, Cambridge
18. Pedregal DJ, Young PC (2002) Statistical approaches to modelling and forecasting time series. In: Clements M, Hendry D (eds) Companion to economic forecasting. Blackwell, London
19. Kalman RE (1960) A new approach to linear filtering and prediction problems. ASME Trans J Basic Eng 83-D:95–108
20. Lütkepohl H (1991) Introduction to multiple time series analysis. Springer, Heidelberg
21. Tsay RS (1986) Time series model specification in the presence of outliers. J Am Stat Assoc 81:132–141
22. Brown RG (1959) Statistical forecasting for inventory control. McGraw-Hill, New York
23. Holt CC (1957) Forecasting seasonals and trends by exponentially weighted moving averages, ONR Research Memorandum 52, Pittsburgh, Carnegie Institute of Technology
24. Gardner ES Jr, McKenzie E (1985) Forecasting trends in time series. Manag Sci 31:1237–1246
25. Bryson AE, Ho YC (1969) Applied optimal control, optimization, estimation and control. Blaisdell Publishing, Waltham
26. EViews. http://www.eviews.com
27. Iberian market energy operator, OMEL. http://www.omel.es
28. SAS. http://www.sas.com
29. Scientific Computing Associates, SCA, http://www.scausa.com
30. The Mathworks: MATLAB. http://www.mathworks.com

Electricity Markets Simulation: MASCEM Contributions to the Challenging Reality

Zita A. Vale, Hugo Morais, Tiago Pinto, Isabel Praça, and Carlos Ramos

Abstract Electricity Markets are not only a new reality but an evolving one as the involved players and rules change at a relatively high rate. Multi-agent simulation combined with Artificial Intelligence techniques may result in sophisticated tools very helpful under this context.

Some simulation tools have already been developed, some of them very interesting. However, at the present state it is important to go a step forward in Electricity Markets simulators as this is crucial for facing changes in Power Systems. This paper explains the context and needs of electricity market simulation, describing the most important characteristics of available simulators. We present our work concerning MASCEM simulator, presenting its features as well as the improvements being made to accomplish the change and challenging reality of Electricity Markets.

Keywords Agents negotiation • Distributed generation • Electricity markets • MASCEM • Multi-agent systems • Virtual power players

1 Introduction

Simulation is an important tool to gain knowledge about systems, whether they already exist or not. Multi-agent simulation has several advantages, besides those related with "traditional" simulation, such as the possibility of analyzing the system as a whole while also analyzing each entity individual behavior and how that behavior influences, or is influenced, by system behavior. Multi-agent models also have the advantage of being easily enlarged to deal with systems evolution.

Z.A. Vale • H. Morais (✉) • T. Pinto • I. Praça • C. Ramos
GECAD – Knowledge Engineering and Decision-Support Research Group, Electrical Engineering Institute of Porto – Polytechnic Institute of Porto (ISEP/IPP), Porto, Portugal
e-mail: zav@isep.ipp.pt; hgvm@isep.ipp.pt; tmp@isep.ipp.pt; icp@isep.ipp.pt; csr@isep.ipp.pt

A. Sorokin et al. (eds.), *Handbook of Networks in Power Systems I*,
Energy Systems, DOI 10.1007/978-3-642-23193-3_7,

MASCEM is a tool that combines agent based-modeling and simulation, with utility function characterization of agent objectives, dynamic strategies, and game theory for scenario analysis. Although MASCEM purpose is not to explicitly search for equilibrium points, but to help understand the complex and aggregate system behavior that emerges from the interactions of heterogeneous individuals, thus possibly converging toward equilibrium. MASCEM has evolved into considering Distributed Generation of electrical energy increasing use by means of Virtual Power Players. Virtual Power Players can reinforce the importance of these generation technologies making them valuable and profitable in electricity markets.

In this paper we start by a brief introduction to multi-agent simulation, and its applicability to study Electricity Markets. We also address some of the most relevant tools developed for this type of markets. Then, we describe MASCEM multi-agent model, agents' strategic behavior and Virtual Power Players. We finish by presenting a case-study that illustrates MASCEM capabilities.

1.1 Multi-agent Simulation

Simulation is generally recognized as an important tool in the strategic and tactical decision making process. Simulation is not a decision making tool but a decision support tool. A simulation model consists of a set of rules that define how a system changes over time. When developing a simulation model, it is important to have a good knowledge about systems behavior and to keep in mind the objectives of the simulation study and that the art of simulation modeling is abstraction and simplification.

In agent-base simulation a multi-agent system is used as a model that describes simulated actors and their decision processes. The systems behavior emerges as a result of the actions of the agents, and their interactions with other agents and the environment.

Agents are knowledge entities characterized by independence and autonomy that have the ability to plan and to establish their actions ahead of time, to develop appropriated problem's solving strategies, to communicate, or to share resources [1, 2]. Agents have the possibility to follow events as they occur in the environment and to learn from past experiences. Agents typically have the following properties:

- Autonomy: agents operate without others having direct control of their actions and internal state;
- Social ability: agents interact with other agents (and possibly humans) through some kind of language (message passing, ...);
- Reactivity: agents are able to perceive their environment (which may be the physical world, a virtual world of electronic networks, or a simulated world including other agents) and respond to it;
- Goal-oriented: an agent does not simply act in response to the environment;
- Proactivity: as well as reacting to their environment, agents are also able to take the initiative, engaging in goal-directed behavior.

Multi-Agent Systems refer to the algorithmic solutions of problems dealing with agents; how agents should interact, avoid conflicts or organize cooperative behavior in order to fulfill common goals. They provide a powerful computational tool, for which dynamic aspects are based on interactions between agents, rather than centralized control. To do their functions the agents have the capacity to communicate with others and with the system. Agents are particularly adapted to complex systems modeling, in which environments are unpredictable [3] and simulation required [4–6].

Agent based simulation allows modeling a system with heterogeneous agents, each having personal motivations and incentives, and to represent groups and group interactions. It is appropriate to study systems driven by interactions among their entities [7] and might reveal complex emergent behavior at the systems level even when the agents involved are fairly simple.

2 Simulation and Multi-agent Simulation of Electricity Markets

All over the world electricity restructuring placed several challenges to governments and to the companies that are involved in the area of generation, transmission, distribution and retail of electrical energy. One of the most significant features is the introduction of electricity markets, aimed at providing competitive electricity service at consumers. Potential benefits, however, depend on the efficient operation of the market. Definition of the market structure implies a set of complicated rules and regulations that should not encourage strategic behaviors that might reduce market performance.

Electricity Markets are not only a new reality but an evolving one as the involved players and rules change at a relatively high rate [8, 9]. The emergence of a diversity of new players (e.g. aggregators) and new ways of participating in the market (Distributed Generation and demand side are gaining a more active role) are signs of this [10].

The restructuring turned electricity markets into an attractive domain for developers of software tools. Simulation and Artificial Intelligence techniques may be very helpful under this context.

Real-world restructured electricity markets are sequential open-ended games with multiple participants trading for electric power on a daily basis. Market players and regulators are very interested in foreseeing market behavior: regulators to test rules before they are implemented and detect market inefficiencies; market players to understand market behavior and operate in order to maximize profits.

Power Systems Generation is a distributed problem by its own nature. In the past Electricity was based on a reduced number of installations (e.g. thermo, nuclear, and hydro power plants). However, guaranteeing sustainable development is an enormous challenge for Power Systems. This requires a significant increasing in Distributed Generation, mainly based on renewable sources. However, this leads to a system that is much more complex to control, since we have many more power generation plants, and the generation is more unpredictable than before, due to the difficulty in forecasting the energy production originated by some renewable sources (e.g. wind and photovoltaic). We are in the presence of a problem that is clearly a distributed problem and where intelligent behavior needs to be assigned to the operation of power plants. Thus Multi-Agent Systems appear as the natural solution to model power systems generation (even in a cyber-physical systems perspective).

Electricity Markets introduced a new dimension in the problem complexity: the economic dimension. However the infrastructure, namely the power system network, has a real physical nature, with specific limitations. The introduction of Electricity Markets, show us the fragility of power systems infrastructures. Several severe incidents, including blackouts, occurred (e.g. the 14th August 2003 Blackout in USA, and the 4th October 2006 quasi-blackout affecting nine European countries). For environmental reasons a lot of the new generation plants are based on renewable sources of energy, for which generation forecasting is more difficult to perform, and generation is much more constrained than in traditional plants (and with higher prices per kWh). In order to overcome these constraints, coalition formation between power producers appears as a natural answer for the problem.

Electricity Market simulators must be able to cope with this evolving complex dynamic reality and provide electricity market players with adequate tools to adapt themselves to the new reality, gaining experience to act in the frame of a changing economic, financial, and regulatory environment. With a multi-agent simulation tool the model may be easily enlarged and future evolution of markets may be accomplished.

Multi-agent simulation combined with other Artificial Intelligence techniques may result in sophisticated tools, namely in what concerns players modeling and simulation, strategic bidding and decision-support [11–14]. For example, consumers' role has significantly changed in this competitive context, making load analysis, consumer profiling and consumer classification very important [15]. The data generated during simulations and by real Electricity Markets operation can be used for knowledge discovery and machine learning, using data mining techniques [15–17] in order to provide Electricity Markets players with simulation tools able to overcome the little experience they have in Electricity Markets operation. Some of the existent Electricity Markets simulators have machine learning abilities [13] but huge advances are required so they are of added value for real Electricity Markets players.

Each player acting in an Electricity Market has his own goals and should use adequate strategies in order to pursuit those goals, its strategic behavior [18, 19] being determinant for its success. A player behavior exhibits changes in response to new information and knowledge that he may have; this may refer to his self knowledge, to knowledge coming from the exterior and from the dynamic complex interactions of the heterogeneous individual entities. Each agent has only partial knowledge of other agents and makes his own decisions based on his partial knowledge of the system. Methodologies for strategic bidding in Electricity

Markets (e.g. [20–24]) can help players to take more successful decisions but they must be combined together with dynamic behavior strategies [13, 18, 19] able to take advantage from the knowledge concerning past experience and other players.

There are several experiences that sustain that a Multi Agent System with adequate simulation abilities is suitable to simulate Electricity Markets [13, 25–27], considered the complex interactions between the involved players. It is important to note that a Multi Agent System is not necessarily a simulation platform but simulation may be of crucial importance for Electricity Markets study, namely concerning scenarios comparison, future evolution study and sensitive analysis.

2.1 *Electricity Markets Simulators*

Several Electricity Markets simulators have been developed [13, 25–28] with very interesting results but they present limitations concerning the application field. In fact several tools and references are available at the "Agent-Based Computational Economics Research Area: Restructured Electricity Markets" maintained by L. Tesfatsion at www.econ.iastate.edu/tesfatsi, hosted by the Department of Economics, Iowa State University.

In general, Electricity Market simulators are intended for some specific application, implementing a specific market model and/or a specific auction type. Some of them are particularly interesting.

The Electricity Market Complex Adaptive System (EMCAS) [25, 26] uses an agent based approach with agents' strategies based on learning and adaptation. Different agents are used to capture the restructured markets heterogeneity, including Generation Companies, Demand Companies, Transmission Companies, Distribution Companies, Independent System Operators, Consumers and Regulators. It allows undertaking Electricity Markets simulations in a time continuum ranging from hours to decades, including several Pool and Bilateral Contracts markets.

Agent-based Modeling of Electricity Systems (AMES) [27] is an open-source computational laboratory (www.econ.iastate.edu/tesfatsi/AMESMarketHome.htm) for the experimental study of wholesale power markets restructured in accordance with U.S. Federal Energy Regulatory Commission (FERC)'s market design. It uses an agent-based model with strategically learning electric power traders to experimentally test the extent to which commonly used seller market power and market efficiency measures are informative for restructured wholesale power markets. The wholesale power market includes Independent System Operator, Load-Serving Entities and Generation Companies, distributed across the busses of the transmission grid. Each Generation Company agent uses stochastic reinforcement learning to update the action choice probabilities currently assigned to the supply offers in its action domain.

The Short – Medium run Electricity Market Simulator (SREMS) [28] is based on game theory and is able to support scenario analysis in the short-medium term and to evaluate market power, in some situations. Some of its main features are: short-medium run (a time horizon of 1 month or multiples of it) simulation of electricity markets based on game theory, calculating price-makers optimal hourly bids; inelastic load, defined hour by hour and zone by zone; tree-like network with interzonal transit limits; monthly scheduling of reservoir hydro pumping storage plants; highly realistic representation of thermal plants; possible quota appointed to physical bilateral contracts, depending on producers share and risk attitude. It is particularly suitable to study the Italian electricity market.

These are important contributions but, in general, lack flexibility as they adopt a limited number of market models and of players' methodologies. The evolution of some of these simulators is difficult to track but some of them, as is the case of AMES, are evolving in a very dynamic way.

At the present state, it is important to go a step forward in Electricity Markets simulators as this is crucial for facing the changes in Power Systems. Increasing number and diversity of active players (due to high penetration of Distributed Generation and demand side participation) are a huge challenge [15, 29].

In MASCEM we have a complex simulation infrastructure, able to cope with discrete events and with the diverse time scales of the supported negotiation mechanisms, considering several players with strategic behavior competing and cooperating with each other. MASCEM combines agent based-modeling and simulation, with utility function characterization of agent objectives, dynamic strategies, and game theory for scenario analysis. MASCEM flexibility concerning the type of market to simulate, from Asymmetric to Symmetric Pools along with the possibility of establishing bilateral contracts, is very important to make it a real Decision Support Tool where different scenarios may be tested, but also where the same scenario may be tested under different market conditions. An important issue, not found on any of the related simulators, is MASCEM inclusion of an agent that represents the new concept of Virtual Power Players (VPP) that represents the aggregation of a set of producers, mainly based on Distributed Generation concerning renewable sources. They can provide the means to adequately support Distributed Generation increasing use, and its participation in the context of competitive electricity markets [10, 30].

The next section describes the MASCEM simulator in detail, pointing out the improvements and model enlargement already planned and being implemented.

3 MASCEM

The Multi-Agent Simulator for Electricity Markets (MASCEM) [13] is based on a Multi-agent System where players can use several dynamic strategies and a scenario analysis algorithm based on game-theory [31].

MASCEM is implemented on the top of OAA (www.ai.sri.com/~oaa/), using OAA AgentLib library, and Java Virtual Machine 1.6.0. The OAA's Interagent Communication Language is the interface and communication language shared by all agents, no matter which machine they are running on or which programming language they are programmed in, which allows integrating a variety of software modules.

Communication and cooperation between agents are brokered by one or more facilitators, which are responsible for matching requests, from users and agents, with descriptions of the capabilities of other agents.

OAA is not a framework specifically devoted to develop simulations; some extensions were made to make it suitable to deal with the energy markets that MASCEM presently supports, namely to introduce the time evolution mechanism of the simulation.

3.1 Multi-agent Model

MASCEM multi-agent model includes: a Market Facilitator Agent, Seller Agents, Buyer Agents, a Market Operator Agent and a System Operator Agent. Three types of markets are simulated: Pool Markets, Bilateral Contracts and Hybrid Markets. Figure 1 illustrates the multi-agent model.

The System Operator Agent is the responsible for the transmission grid and all the involved technical constraints. Every established contract, either through Bilateral Contracts or through the Pool, must first be communicated to the System Operator, who analyses its technical feasibility from the Power System point of view (e.g. analyzing power system flows leading to congestion situations). This agent is indeed another simulator [32], able to simulate the physical system (generators, loads and transmission grid) and its technical and operational constraints.

The Market Operator Agent is responsible for the Pool mechanism. This agent is only present in simulations of Pool or Mixed markets. The Market Operator will receive bids from Sellers and Buyers, analyze them and determine the market clearing price and accepted bids.

Virtual Power Producers agents represent an aggregation of producers, under the Distributed Generation model inclusion, and are described in detail in a later section.

Seller and Buyer Agents are the key players in the market, involved in Bilateral Contracts negotiations and presenting selling/buying bids to the Pool. The structure of these types of agents comprises three functional modules: Events Handler, Negotiation Management and Strategic Decision Making, plus one knowledge-based module: the Market and Individual Knowledge module. Figure 2 illustrates this structure.

The Events Handler Module is responsible for all processes related with messages handling. Incoming messages are ordered by degree of importance and time of arrival. Outgoing messages are sent only to those agents that are known to be possibly interested in that particular piece of information. Agents use ICL – Interagent Communication Language – to exchange messages between themselves.

The Market and Individual Knowledge module contains information about the organizational and operational rules of the market, as well as other agent

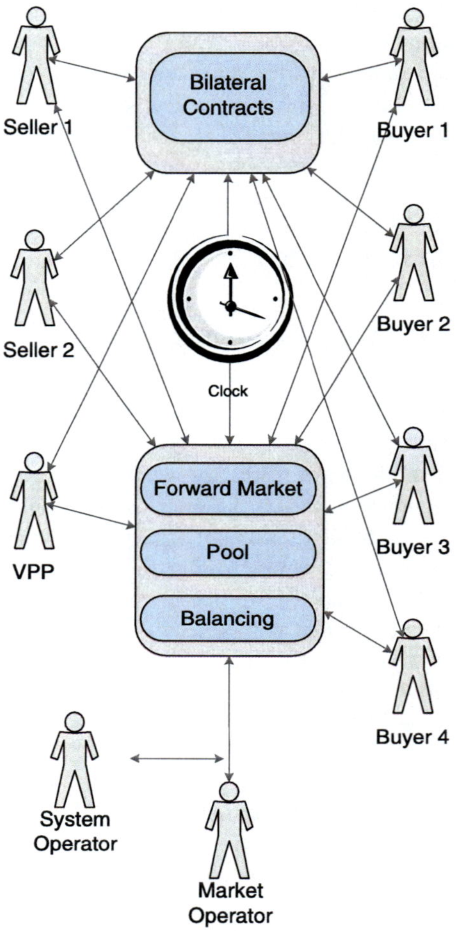

Fig. 1 MASCEM negotiation framework

Fig. 2 Seller and buyer agents structure

Fig. 3 ICL message structure for bilateral contracts negotiations

call_for_proposals (AgentId, Params)
proposal (AgentId, PId, Params)
request(AgentId, PId)
reply_proposal(AgentId, PId, not_accepted, Reason)

call_for_participation (PoolTp, Params)
bid (AgentId, BId, Params)
reply_bid(AgentId, BId, Mprice, [accepted/not_accepted], Reason)

Fig. 4 ICL message structure for pool negotiations

commitments and capabilities, and about the agent itself: agent own capabilities, current availability, past experiences and strategies. Through the analysis of historical market results, this module constructs the profile of each agent in the market, particularly in what concerns their capabilities, limit prices and expected prices.

During a negotiation period agents analyze and formulate several proposals. The Negotiation Management module contains all the processes related to this subject. The proposals received are analyzed taking into account issues such as the price, quantity of energy and viability of the transaction (based on the technical analysis made by the System Operator). The process of formulating proposals results from the interaction of this module with the Strategic Decision Making module. Negotiation is based on a set of messages being exchanged between the involved agents. Figures 3 and 4 give an idea of the type and structure of messages.

MASCEM simulator facilitates agent meeting and matching, besides supporting the negotiation model. In order to have results and feedback to improve the negotiation models and consequently the behavior of user agents, we simulate a series of negotiation periods, $D = \{1,2,\ldots,n\}$, where each one is composed by a fixed interval of time $T = \{0, 1, \ldots, m\}$. At a particular negotiation period, each agent has an objective that specifies its intention to buy or sell and on what conditions. The available agents can establish their own objectives and decision rules. Moreover, they can adapt their strategies as the simulation progresses on the basis of previous effort's successes or failures. The simulator probes the conditions and the effects of market rules, by simulating the participant's strategic behavior.

3.2 Negotiation

As a decision support tool, MASCEM includes several types of negotiation mechanisms to let the user test them and learn the best way to negotiate in each one. So, we include bilateral contracts and a Pool, centralized mechanism based on an auction, and regulated by a market operator. Both types of negotiation may exist at the same time: Mixed Market. These implies each agent must decide whether to, and how to, participate in each market type.

Let Agt_b denote the buyer agent, Agt_s the seller agent and let $[Pi_{min}, Pi_{max}]$ denote the range of values for price that are acceptable for agents. A seller agent has the range $[Psi_{min}, Psi_{max}]$, which denotes the scale of values that are comprised of the minimum value that the seller is disposed to sell to the optimal value.

A buyer agent has the range $[Pbi_{min}, Pbi_{max}]$, which denotes the scale of values that are comprised of the optimal value to buy to the maximum value.

4 Bilateral Contracts

In bilateral contracting buyer agents are looking for sellers that can provide them the desired products at the best price. We adopt what is basically an alternating protocol [18].

Negotiation starts when a buyer agent sends a request for proposal. In response, a seller agent analyses its own capabilities, current availability, and past experiences and formulates a proposal. Sellers can formulate two kinds of proposals: a proposal for the product requested; or a proposal for a related product, according to the buyer preference model.

$PPgi^{DT}_{Agts \to Agtb}$ represents the proposal offered by the seller agent $Agts$ to the buyer agent $Agtb$ at time T, at the negotiation period D for a specific product. The buyer agent evaluates the proposals received with an algorithm that calculates the utility for each one, U^{Agtb}_{PPgi} ; if the value of U^{Agtb}_{PPgi} for $PPgi^{DT}_{Agts \to Agtb}$ at time T is greater than the value of the counter-proposal that the buyer agent will formulate for the next time T, in the same negotiation period D, then the buyer agent accepts the offer and negotiation ends successfully in an agreement; otherwise a counter-proposal $CPgi^{DT}_{Agtb \to Agts}$ is made by the buyer agent to the next time T. The seller agent will accept a buyer counter-proposal if the value of U^{Agts}_{CPgi} is greater than the value of the counter-proposal that the seller agent will formulate for the next timeT; otherwise the seller agent rejects the counter-proposal.

On the basis of the bilateral agreements made among market players and lessons learned from previous bid rounds, both agents revise their strategies for the next negotiation rounds and update their individual knowledge module.

5 Pool

In MASCEM, agents also have the possibility of negotiating through a Pool, which is a centralized mechanism that functions according to an auction mechanism, and is regulated by a market operator. We have two different auction mechanisms: a double and a single uniform auction.

The process starts at the market operator, who sends a request for participation. The *call_for_participation* message triggers the negotiation process and is delivered to all agents in the simulated market. If the agent is interested, or capable, of participating in the Pool, it will formulate a bid and send it to the market operator, specifying for each requested parameter the value of its proposal.

The process of formulating bids, by buyer and seller agents, is related to agent strategies, addressed in detail in the next section. The market operator evaluates all the received bids, analyses them through the pool auction mechanism, defines the market price and accepted bids. Then a *reply_bid* message is sent to all pool participants, specifying the settled market price and if the bid was or not accepted and why.

6 Mixed Markets

The Mixed model combines features of Pools and Bilateral Contracts. Agents decide whether to negotiate only through the Pool, to keep bilateral negotiations simultaneously with Pool negotiations, or to wait for the Pool results and start bilateral negotiations only if bids were not accepted. Agents use their past experiences, market knowledge and their own negotiation strategies to support decisions.

On the basis of the results obtained in a period, Sellers and Buyers revise their strategies for the next day. Seller and Buyer Agents have strategic behavior to define their desired price. These agents have time-dependent strategies [33], and behavior-dependent strategies, to define next period price according to previous results. Time-dependent strategies are very important [34] since they allow the simulation of important issues such as: emotional aspects and different risk behaviors. To adjust price between days, also referred as behavior-dependent strategies, MASCEM provides basic strategies based on composed goals, "Composed Goal Directed" and on previous efforts, "Adapted Derivative Following" [35].

6.1 *Virtual Power Players*

Due to environmental and fossil fuels shortage concerns, renewable energy resources are being more used. The advantages of using renewable are clear from the environment point of view. From the technical and economical point of view there are problems that must be overcome to take advantage of an intensive use of Distributed Generation. An aggregating strategy can enable owners of renewable generation to gain technical and commercial advantages, making profit of the specific advantages of a mix of several generation technologies and overcoming serious disadvantages of some technologies. The aggregation of Distributed Generation plants gives place to the new concept of Virtual Power Player (VPP). VPPs integration into Electricity Markets is a very challenging domain that has being motivating our work regarding MASCEM evolution.

VPPs are multi-technology and multi-site heterogeneous entities, being relationships among aggregated producers and among VPPs and the remaining Electricity Market agents a key factor for their success [36]. Agent coalitions

[37–40] are especially important to address VPPs issue as these can be seen as a coalition of agents that represent the aggregated players [41].

Coalition formation is the coming together of a number of distinct, autonomous agents that agree to coordinate and cooperate, acting as a coherent grouping, in the performance of a specific task. Such coalitions can improve the performance of the individual agents and/or the system as a whole. It is an important form of interaction in multi-agent systems. The coalition formation process comprises several phases: Coalition Structure Generation, Optimization of the Value of the Coalition and Payoff Distribution.

Regarding the coalition formation process, for VPP modeling, the three main activities of Coalition Structure Generation, Optimization of the Value of the Coalition and Payoff Distribution should be considered under a scenario where agents operate in a dynamic and time dependent environment. This entails significant changes on MASCEM core model and communications infrastructure.

To sell energy in the market VPP must forecast the generation of aggregated producers and "save" some power capacity to assure a reserve to compensate a generation oscillation of producers with natural resources technologies dependent.

The VPP can use different market strategies, considering specific aspects such as producers established contracts and range of generation forecast. The prediction errors increase with the distance between the forecasting and the forecast times. The standard error figures are given as percent of the installed capacity, since this is what the utilities are most interested in (installed capacity is easy to measure); sometimes they are given as percent of the mean production or in absolute numbers.

Considering an example of Spanish market (OMEL), the spot market session closes at 11:00 AM, therefore the time slice between the predictions and real day is 13–37 h. In this context, the VPP can change its market strategy during the day to manage the risk. These strategies are also depending of reserves, in other words, VPP can change the reserve to maintain the risk, however, if VPP has a bigger reserve the costs is higher.

Another important factor to the VPP market strategy is the buy energy price to the aggregated producers. The price considered for each producer must be agreed with the VPP so that competitive prices can be obtained, to allow the producers to have revenues from their investments in reasonable periods of time.

If subsidies exist, these will have to be included in the calculation of the prices considered for the producers. The price of the reserve will also have to be previously agreed between the VPP and the producers.

7 MASCEM Evolution

The Introduction of VPP models in MASCEM required to re-think MASCEM architecture, namely in what concerns agent communication. From a conceptual point of view, each VPP is seen as an agent coalition. Figure 5 illustrates this idea.

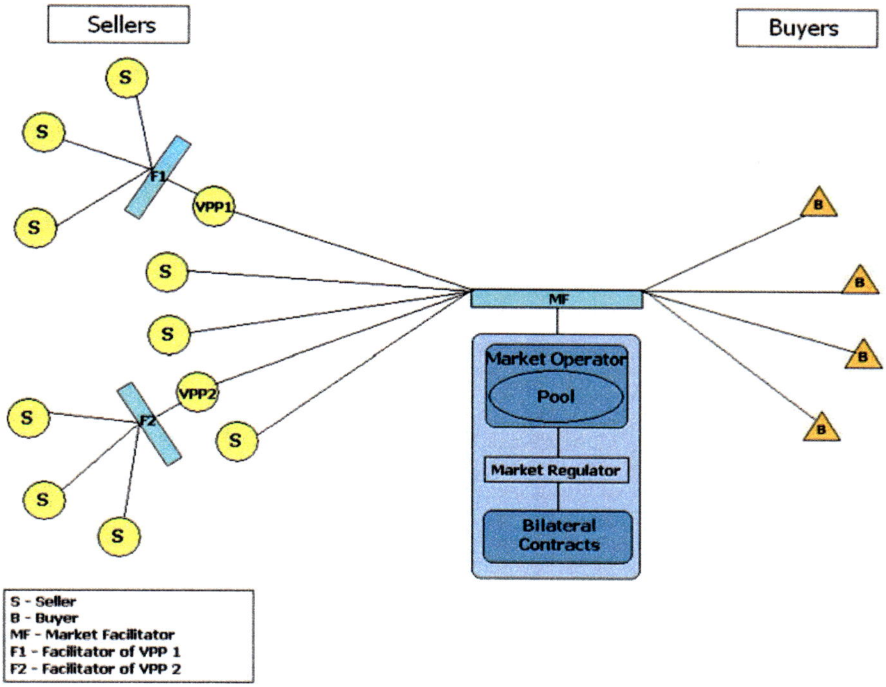

Fig. 5 MASCEM multi-agent architecture

Modeling one agent coalition for each VPP requires not only the modeling of the agents that take part in each coalition but also their interactions. These include interactions during several distinct periods:

- In the pre-bidding period each VPP has to prepare the bids. This means the VPP needs to obtain information from each aggregated producer, namely concerning its generation forecasting and envisaged price;
- After the clearance of the market, the VPP has to internally dispatch the sold energy (considering both the market and bi-lateral contracts) and inform each producer of its dispatch;
- Aggregated producers' operation has to be supervised by the VPP and adequate internal control and management measures implemented. This includes internal reserve management to overcome shortage or surplus generation, which considers the generation resources of all aggregated producers and internal and external available/contracted reserve means.

As the overall performance of the market simulator must be optimized, these VPP internal interactions should only overload the whole simulation in the exact required measure. Moreover, in order to make VPP coalitions act at their best performance the first step was to determine how to integrate them in the market negotiations with minimum degradation of the previous implementation performance.

This led us to face each VPP as a multi-agent system, operating in the scope of the overall multi-agent system that simulates the electricity market. Considering each VPP as a multi-agent system allows an interesting approach from both the performance and the conceptual point of view. In order to implement this conceptual architecture, each VPP has to have its own facilitator. The answer was to add specific facilitators, with no relation to the market facilitator, to coordinate the negotiations between the members of each individual VPP. This means that each VPP has now its own facilitator that allows it to communicate with all the producers that are part of its coalition or intend to join it, independently from the rest of the simulation.

8 Communication

The communication between the VPPs and its aggregated members takes place in two distinct stages. The initial negotiations concern coalition formation. These negotiations take place when individual agents make efforts to aggregate themselves within a new or an existing VPP and when an existing VPP wants to aggregate more producers. The agents that are interested in joining one aggregation can make its request to the VPP, which will accept, refuse or negotiate the proposal, based on the VPP's individual operating strategies and necessities at the moment. VPPs can also propose individual agents to take part in their coalition.

The second stage concerns VPP operation being more complex than the first stage. This includes the communication required when preparing the bids to be presented to the market operator. For the day-ahead market, this is done for all the 24 periods of the next day (or 48 periods if half an hour periods are considered). This begins with gathering all the capacity available from the different aggregated generation resources and establishing the electricity amount that the VPP can trade on the market. Then the market negotiation price is calculated according to VPP strategies, always with the goal of maximizing the profits, and taking into consideration the individual envisaged prices of each coalition member. Finally, after the market session, according to the contracts that were established in the first stage, the VPP calculates each producer's remuneration, and informs them of the results for each period, their individual revenue, the total amount of energy sold and the amount of energy sold in the pool and/or in the bilateral contracts.

All the VPPs acting in the simulated market establish these communications at the same time, independently from each other, each using its own facilitator, and so increasing the efficiency of the simulation.

9 Classification

The producer's selection criteria are different for each VPP, depending on the dimension and on the already aggregated producers. In MASCEM VPPs are classified according to the following five different types:

- Parallel VPP (PVPP) – include different producers with distinct generation capacities, typically upper to 1 MW and lower than 20 MW. The common characteristic is the participation in parallel markets;
- Large Scale VPP (LSVPP) – include producers with large generation capacity, typically higher than 10 MW;
- Micro VPP (μVPP) – these are constituted by many producers with small capacity, typically lower than 2 MW;
- Global VPP (GVPP) – this type of VPP aggregates both producers and consumers, assuming the function of a trader;
- Several VPP (SVPP) – this VPP type does not have a priori defined characteristics so that it allows users to create more specific VPPs.

Decision making for VPP formation and subsequent aggregation of more producers takes into account a large set of producers' characteristics (listed in the first column of Table 1). The weight of each of these characteristics depends on the VPP type, as shown in Table 1.

These weights are based on economic criteria and on VPP market strategies. The characteristics weight ranges from 0 to 10.

These values have been determined based on a set of a priori analyzed cases, considering possible VPP strategies and are used by MASCEM as default values. However MASCEM users can modify these values to adjust the VPP strategy according to their own needs.

The user also has the possibility of developing and simulating scenarios in which VPPs change their aggregated producers, in order to improve VPP strategy in function of market evolution.

The classification structure has been integrated with MASCEM simulator with the purpose of being tested in an actual market simulator, and so allowing to draw conclusions about the efficiency of this procedure, and about the development of the behavior of the VPPs when including such mechanisms. We expect this new

Table 1 Producers' characteristics weights

Characteristics	PVPP	LSVPP	μVPP	GVPP	SVPP
Speculative energy cost	10	10	9	9	10
Dispatchability	7	9	7	10	7
Reliability	7	8	2	8	7
Use of installed power	5	7	2	5	5
Lifespan	3	3	1	3	5
Volatility of prices	7	8	3	7	7
2nd market	9	4	4	6	5
GHG emissions	7	6	5	5	5
Location	4	2	8	6	5
Dimension	4	3	8	5	5
Technology type	5	5	6	6	5
Social impact	5	5	5	4	5
Maturity of technology	4	5	2	4	5
Commercial behavior	5	6	3	5	5

feature to prove to be a real added value to the intelligence of the VPP agents, by increasing their ability to take the best decisions when confronted with particular situations (in this case, the election of the producers who would be a greater asset to the coalition in the present and future, and contribute the most to the achievement of its objectives).

The integration in MASCEM has been done through the implementation of this mechanism in LPA-Prolog, and its inclusion in the private facilitator of each VPP. This facilitator is also implemented in Prolog and has the responsibility of managing the communications with the agents that form the coalition.

The features of this new mechanism are:

- The registration of new VPP types – There are five standard VPP types defined at first. They present static factor weights, for an easier choice when a new VPP is created. Additionally, at the time of creation of a new VPP, we have the choice of defining a new VPP type, naming it, and attributing the desired weights for the factors presented before. This allows the new VPP to present the best possible suiting to the objectives that it ought to have;
- The classification of producers that intend to join the aggregation – A new producer that desires to be a part of a particular VPP must at a first instance provide all the data necessary for its classification concerning the VPP's goals, characteristics, and constituents. Each producer that sends its request for entering the VPP is awarded a classification that can vary from −2 to 2;
- The acceptance or refusal of a producer application – Depending on the strategy adopted by each VPP, the applicant producers will be accepted or refused in the coalition. This acceptance process can be based on a priori defined limit for the minimum classification for which the coalition will allow the new producer to be aggregated. Moreover, the VPP can also define a maximum number of members to be part of the coalition, refusing the entrance of all that apply when that number is reached. Alternatively, a VPP can accept all the applicants, assuming that each one can be an asset for the growth of the coalition.

When a new VPP is created in the simulator, it is required to define the various aspects that will characterize its objectives and desired orientation in the market. These factors and preferences will be the basis for the classification of each producer that intends to join this aggregation, together with the producer's individual characteristics. These characteristics are also required when a new producer is created.

Once a producer makes its application to join a certain VPP, all the information is sent to that VPP's facilitator, which will be responsible for its classification, and for the acceptance/refusal of the application. If it is accepted, it will from that point on be a member of the coalition.

Throughout each day the revenues from all the periods of negotiation must be distributed amongst the members of the aggregation. To manage those transactions, the facilitator is also equipped with a profits' distribution mechanism, which determines the amounts of payoff that the members are going to receive.

This algorithm is based on the total amount of energy that the VPP was able to sell in each period; the market price for that period; and the amount of energy that each producer provided individually, along with its value of classification awarded by the VPP at the time of entrance.

The use of this mechanism ensures that the payoffs reward the producers that were better classified, and those that produced the most in each period.

10 Implementation

VPPs are designed as a type of sellers in the scope of MASCEM. They are seen by the market and by all other agents just like any regular seller. This means that they communicate the same way as the other sellers do; the requests made to the sellers are understood and answered by the VPPs just the same way. The same applies when the requests come from the VPP to others; they are viewed and treated just like if it was a regular seller.

For the VPPs individual facilitators we decided to develop our own version of the OAA facilitator (which is used as the market facilitator). VPPs' facilitators are implemented in LPA Prolog, as it guarantees a level of efficiency and speed of processing that JAVA can never give. VPP facilitators have been implemented so they seem as an adaptation of the OAA facilitator's basic functionalities. For this purpose, it has been decided to maintain as long as possible the name of its methods, so that it would grant an easier adaptation from people who are used to the OAA. So, a VPP facilitator allows the agents to:

- Connect to the facilitator by sending the list of their capabilities;
- Disconnect from the facilitator, and so, from that point on not being asked to solve requests from this facilitator;
- Send "solve" requests;
- Send "solve" requests for a specific agent;
- Get the agents' answers to a solvable;
- Check for requests to solve;
- Respond to a "solve" request;
- Get information about:
  - The list of capabilities of each agent that is connected to this facilitator;
  - The agents that already replied to a current request;
  - The agents that did not yet give their response;
  - All the actions that are being solved;
  - The historic of all the actions that were requested, the agents that replied to each of them, and respective answers.

VPP facilitators are implemented in a Prolog file named facilitator.pl (Fig. 6), which is integrated in the JAVA program by using the LPA Win-Prolog Intelligence Server. Intelligence Server provides a DLL interface between Win-Prolog and other applications, and so it allows the LPA-based components to be embedded in applications written in many Windows programming languages.

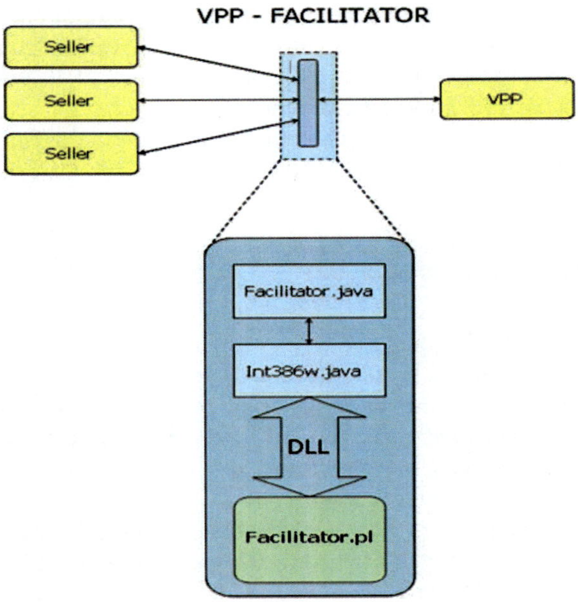

Fig. 6 VPP facilitator communication

Adding the Intelligence Server functionality is obtained by importing the provided lib file and defines the Intelligence Server function prototypes for the DLL. Then, just calling prolog goals as a string, retrieves the results. These are also provided in the form of a string. The first parameter is the return value of the goal (true or false), after this are the results that would be printed in the console. So, the way to send back results other than true/false information is to write the results to the console, and they will be returned together with the result of the goal.

The native code interface to the intelligent server is represented in the class Int386w (Fig. 7). The functions are accessed through an instance of this class.

Additionally, we developed an extra class that provides an easier call to the prolog facilitator goals. This facilitator class contains the methods that allow the agents to:

- Connect to an existing instance of the prolog facilitator;
- Create a new instance of the prolog facilitator;
- Simplify the calls o0f the goals. As mentioned before, the usage of the Int386w class to establish the connection to the IS requires to use the methods of this class to access the prolog (facilitator) goals. To make it easier to use and similar to the OAA facilitator, the facilitator class provides the methods to make the calls more intuitive. So, to solve a goal, instead of having to make the calls:

  – Int386w.initGoal("solve(goal)") ;
  – Int386w.callGoal() ;
  – Int386w.exitGoal() ;

now it is just needed to call:

– Facilitator.solve("goal").

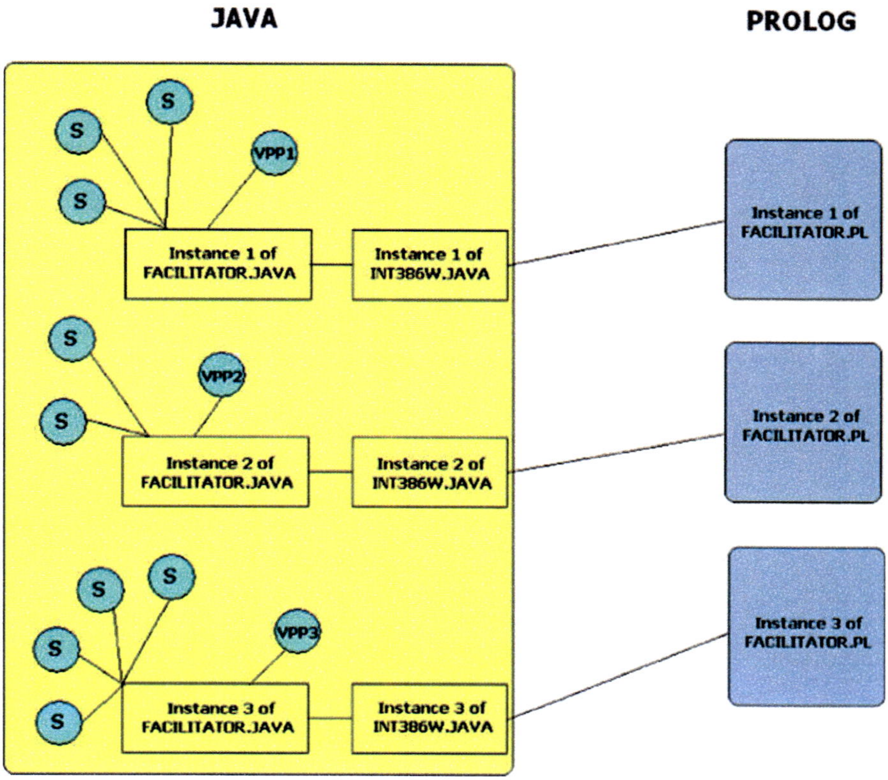

Fig. 7 JAVA/prolog interface

Still with the intention of simplifying, this class also contains the necessary methods for the agent to get the results of an asked goal in the most appropriate way, by transforming the returned string into variables that can be easily understood by the agent.

We can have a better look at this communication process in Figs. 8 and 9. The first one shows how the requests to solve a certain action are processed. The VPP sends its request to the JAVA facilitator class, which through the calls of the necessary methods of the Int386w class – initializing the goal, calling it, and then exiting the call – makes the request get to the prolog facilitator. When the facilitator receives it request processes, checks the agents that are connected and that have the ability to solve it, and then sends to each ones the request to be solved, via the same way – Int386w class, and java facilitator class.

Once the agents get the request to solve the action, the reverse process occurs (Fig. 9), they solve it and return each individual result to the java facilitator class, which makes sure it gets to the prolog facilitator, through the necessary calls of the Int386w class once more.

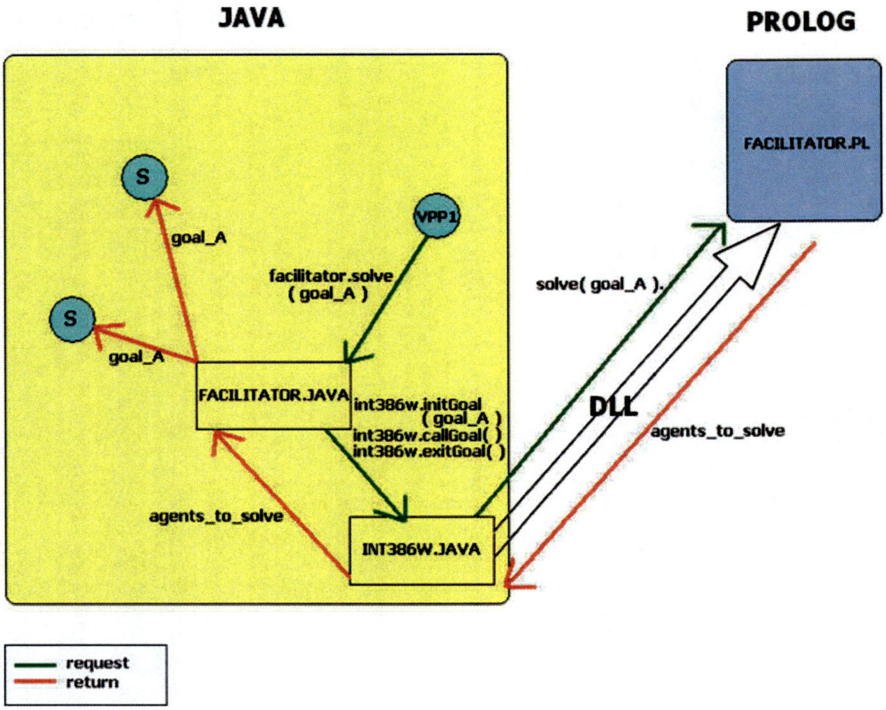

Fig. 8 Solve request architecture

When the prolog facilitator receives the answers, it combines them and sends the final answer back to the VPP that made the request.

10.1 Case Study

This case study considers 11 seller agents, based on real producers of OMEL (Spanish electricity market) with several technologies, together with 3 VPPs with distinct characteristics and strategies.

The main goal is that each VPP chooses the best producers to aggregate, according to its initial objectives. After this process is concluded, the agents will start the negotiation in the market, using distinct strategies, which allow studying their performance and taking some conclusions on those strategies which are better to be used by the agents in the future.

The results of VPP remuneration after each period of the market negotiations will also be presented to show how the profits distribution mechanism works,

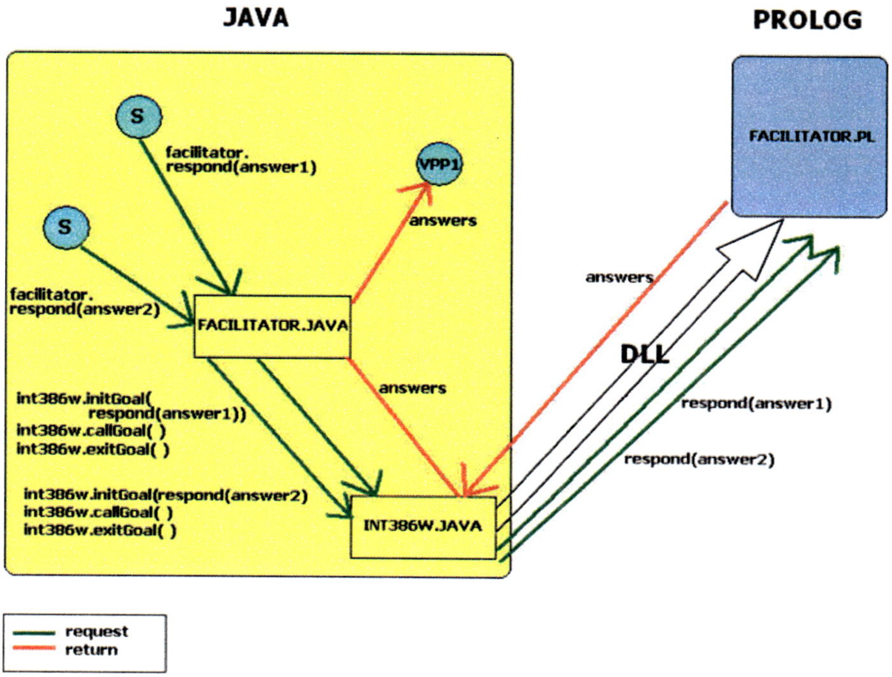

Fig. 9 Solve answer architecture

considering the members value of classification, along with their individual production as main factors to the determination of the individual revenues.

11 Classification

For better understanding, and to simplify tables drawing, we first present the legend used on the tables that illustrate the VPP agents' characteristics and tolerance factors.

These factors combined with the characteristics of each individual candidate producer determine whether that producer will or not be accepted by the VPP (Table 2).

Tables 3–5 characterize the three VPPs, showing their objectives and tolerance factors.

Along with the factors presented in Tables 3–5 the producers' data must be gathered in order to allow the classification to take place. Table 6 presents the data respective to each producer considered for this simulation.

Using the proposed method for the classification with the inputs being the historical values from the various producers, combined with the VPP's data

Table 2 VPP's legend related to agents characteristics and tolerance factors

Legend	
E. C.	Expected cost
A.F.	Average failure
I.P.U.	Installed power use
Reli.	Reliability
L.C.	Life cycle
Volat.	Volatility
Emis.	Emissions
Dim.	Dimension
P.M.B.V	Parallel market business value
E.M.B.V.	Electric market business value
P.M.R.	Parallel market remuneration
A.C.E.	Average cost of electricity
T. Tec.	Type of technology
Tec.M.	Technology maturity
Local.	Localization
S.Imp.	Social impact
C. Com.	Commercial comportment
S. M.	Second market
*	Non real value
**	Not used values

Table 3 VPP1 characteristics

VPP1			Choosing factors/investments			
Type	Local		Dim.	T. Tec.	Tec.M.	Local.
Large	Barcelona		1	0	1	1
Objectives						
E.C.	A.F.	I.P.U.	Reli..	L.C.	Volat.	
0.045	110*	8,000	97	10	0.003	
Emis.	Dim.	P.M.B.V.	E.M.B.V.	P.M.R.	A.C.E.	
0.3	20,000	5,000*	95,000*	0.009*	0.05*	
Formula Changing Factors/Tolerance						
E.C.	A.F.	I.P.U.	Reli..	L.C	Volat.	Emis.
20	10	5	5	5	5	10
Dim.	T.Tec	S.Imp.	Tec.M.	Local.	C.Com.	S. M.
10	0**	0**	0**	20	0**	10

shown before, originates the individual classification for each of the producers for this VPP, and consequent proposal results (entrance in the coalition accepted or not accepted). Next tables show the classification assigned by the VPPs to the producers that proposed to enter each aggregation.

It can be seen from Tables 7–9 that all the producers would have the entrance in the various aggregations guaranteed because the VPPs are initially empty and have positively classified all of them (although the profit a producer can provide is little, it is always higher than the null profit the VPP gets while being empty). As there are

Table 4 VPP2 characteristics

VPP2			Choosing factors/investments			
Type	Local		Dim.	T. Tec.	Tec.M.	Local.
Large	*Madrid*		1	1	0	1
Objectives						
E.C.	A.F.	I.P.U.	Reli..	L.C.	Volat.	
0.039	110*	7,500	90	15	0.1	
Emis.	Dim.	P.M.B.V.	E.M.B.V.	P.M.R.	A.C.E.	
0.1	5,000	5,000*	95,000*	0.009*	0.05*	
Formula changing factors/tolerance						
E.C.	A.F.	I.P.U.	Reli..	L.C	Volat.	Emis.
10	10	10	10	10	10	10
Dim.	T.Tec	S.Imp.	Tec.M.	Local.	C.Com.	S. M.
10	0**	0**	0**	20	0**	10

Table 5 VPP3 characteristics

VPP3			Choosing factors/investments			
Type	Local		Dim.	T. Tec.	Tec.M.	Local.
Large	*Irun*		1	0	1	1
Objectives						
E.C.	A.F.	I.P.U.	Reli..	L.C.	Volat.	
0.043	110*	8,500	85	20	0.1	
Emis.	Dim.	P.M.B.V.	E.M.B.V.	P.M.R.	A.C.E.	
0.1	200	5,000*	95,000*	0.009*	0.05*	
Formula changing factors/tolerance						
E.C.	A.F.	I.P.U.	Reli..	L.C	Volat.	Emis.
10	10	15	15	15	10	10
Dim.	T.Tec	S.Imp.	Tec.M.	Local.	C.Com.	S. M.
10	0**	0**	0**	20	0**	10

some producers that proposed entrance to more than one VPP, they will have to pass through a negotiation stage, to determine which producers will be assigned to each VPP.

12 Entrance Negotiation

Once the classification process is finished, it is time to determine which of the producers will enter which of the VPPs. As all the producers obtained positive classifications for entrance in the three VPPs, there were selected the ones with the higher values for each of the three aggregations. The next figures show the producers that obtained the higher scores for each of the VPPs, and consequently were selected to join them.

In Fig. 10 it can be seen the diagram for the initial constitution of VPP 1, with its four members, the ones that obtained the higher scores of classification of all that

Table 6 Producers' data

Producers		Historical values			
Name	*Producer 1*	C.E.	I.M.	U.P.I.	Fiab.
Technology	*SOLAR*	0.06	120*	1839.6	98
Local	*Barcelona*	V. Util	Volat.	Emis.	Dim.
		24	0	0	4.2
		V.N.Mp	V.N.Me.	R.Mp.	
		5,110*	100,000*	0.0089*	
Name	*Producer 2*	C.E.	I.M.	U.P.I.	Fiab.
Technology	*SOLAR*	0.055	120*	1664.4	99
Local	*Barcelona*	V. Util	Volat.	Emis.	Dim.
		23	0	0	41.8
		V.N.Mp	V.N.Me.	R.Mp.	
		5,110*	100,000*	0.0089*	
Name	*Producer 3*	C.E.	I.M.	U.P.I.	Fiab.
Technology	*WIND*	0.03	120*	3328.8	96
Local	*Ourense*	V. Util	Volat.	Emis.	Dim.
		14	0	0	40,000
		V.N.Mp	V.N.Me.	R.Mp.	
		5,110*	100,000*	0.0089*	
Name	*Producer 4*	C.E.	I.M.	U.P.I.	Fiab.
Technology	*WIND*	0.034	120*	2890.8	92
Local	*Cadiz*	V. Util	Volat.	Emis.	Dim.
		19	0	0	39,000
		V.N.Mp	V.N.Me.	R.Mp.	
		5,110*	100,000*	0.0089*	
Name	*Producer 5*	C.E.	I.M.	U.P.I.	Fiab.
Technology	*BIOMASS*	0.04	120*	8146.8	86
Local	*Cordoba*	V. Util	Volat.	Emis.	Dim.
		17	0.005	0	6,619
		V.N.Mp	V.N.Me.	R.Mp.	
		5,110*	100,000*	0.0089*	
Name	*Producer 6*	C.E.	I.M.	U.P.I.	Fiab.
Technology	*WOOD*	0.039	120*	8497.2	90
Local	*Vizcaya*	V. Util	Volat.	Emis.	Dim.
		16	0	0	3,871
		V.N.Mp	V.N.Me.	R.Mp.	
		5,110*	100,000*	0.0089*	
Name	*Producer 7*	C.E.	I.M.	U.P.I.	Fiab.
Technology	*ALMOND*	0.032	120*	7971.6	87
Local	*Tarragona*	V. Util	Volat.	Emis.	Dim.
		7	0	0	500
		V.N.Mp	V.N.Me.	R.Mp.	
		5,110*	100,000*	0.0089*	
Name	*Producer 8*	C.E.	I.M.	U.P.I.	Fiab.
Technology	*ORUJILLO*	0.036	120*	7,884	84
Local	*Malaga*	V. Util	Volat.	Emis.	Dim.
		16	0	0	9,150

(continued)

Table 6 (continued)

Producers		Historical values			
		V.N.Mp	V.N.Me.	R.Mp.	
		5,110*	100,000*	0.0089*	
Name	*Producer 9*	C.E.	I.M.	U.P.I.	Fiab.
Technology	*ANIMALS*	0.025	120*	7,884	91
Local	*Lerida*	V. Util	Volat.	Emis.	Dim.
		21	0	−0,2	6,000
		V.N.Mp	V.N.Me.	R.Mp.	
		5,110*	100,000*	0.0089*	
Name	*Producer 10*	C.E.	I.M.	U.P.I.	Fiab.
Technology	*OLIVE*	0.037	120*	8,322	78
Local	*Cordoba*	V. Util	Volat.	Emis.	Dim.
		17	0	0	300
		V.N.Mp	V.N.Me.	R.Mp.	
		5,110*	100,000*	0.0089*	
Name	*Producer 11*	C.E.	I.M.	U.P.I.	Fiab.
Technology	*USED*	0.045	120*	7708.8	88
Local	*Ciudad Real*	V. Util	Volat.	Emis.	Dim.
		23	0	0.457	10,000
		V.N.Mp	V.N.Me.	R.Mp.	
		5,110*	100,000*	0.0089*	

Table 7 Producers' classification in VPP 1

VPP1						
	Assigned values					
Candidates	T.Tec.	S.Imp.	Tec.M.	C.Com.	Classif.	Accepted
Producer 1	2	2	0	1	*9.09*	*Yes*
Producer 2	2	2	0	2	*9.21*	*Yes*
Producer 3	1	2	−1	0	*18.11*	*Yes*
Producer 4	1	2	−1	2	*21.96*	*Yes*
Producer 5	1	0	0	0	*6.89*	*Yes*

Table 8 Producers' classification in VPP 2

VPP2						
	Assigned values					
Candidates	T.Tec.	S.Imp.	Tec.M.	C.Com.	Classif.	Accepted
Producer 2	2	2	0	1	*6.97*	*Yes*
Producer 4	1	1	−1	1	*75.59*	*Yes*
Producer 6	0	−2	1	−2	*13.18*	*Yes*
Producer 7	0	1	−2	−1	*7.55*	*Yes*
Producer 8	2	−2	2	−1	*23.08*	*Yes*

proposed for entrance in this VPP. Those four are: Producer 4, Producer 3, Producer 2 and Producer 1. Figures 11 and 12 present the producers with higher classification in VPPs 2 and 3, excluding the ones already assigned to VPP1.

Table 9 Producers' classification in VPP 3

VPP3						
	Assigned values					
Candidates	T.Tec.	S.Imp.	Tec.M.	C.Com.	Classif.	Accepted
Producer 3	1	2	−1	1	*1152.65*	*Yes*
Producer 6	0	−2	1	−1	*122*	*Yes*
Producer 9	1	−2	1	1	*200.91*	*Yes*
Producer 10	0	−1	−1	0	*19.87*	*Yes*
Producer 11	−1	−1	−1	0	*266.43*	*Yes*

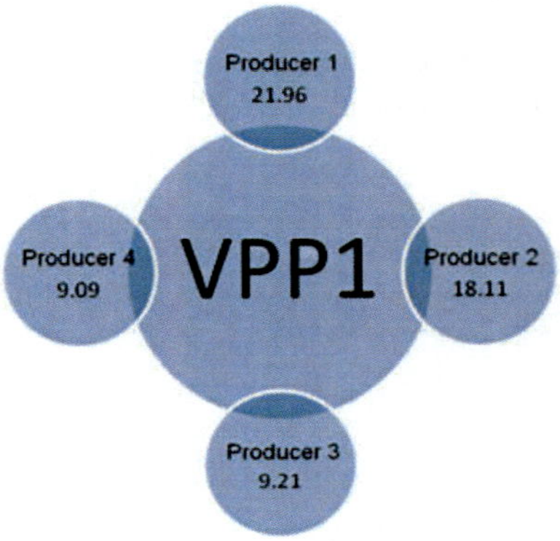

Fig. 10 VPP1 selected members

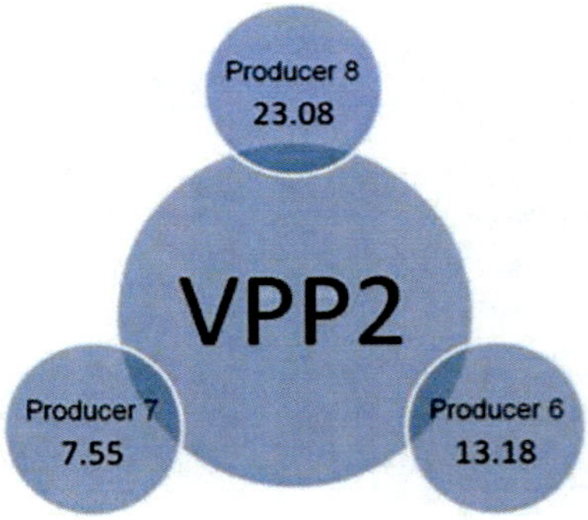

Fig. 11 VPP2 selecetd members

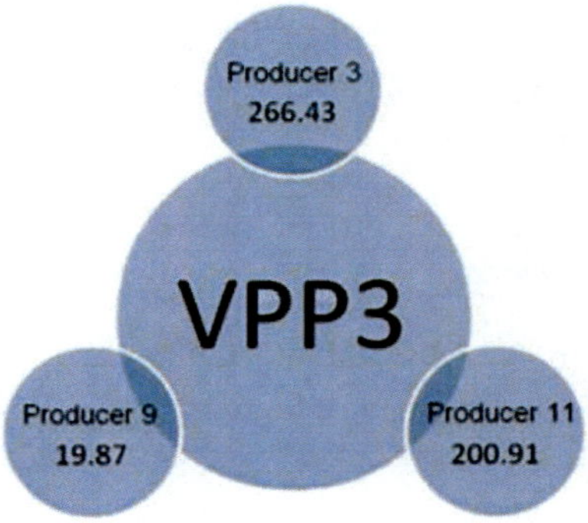

Fig. 12 VPP3 selected members

13 Market Negotiation

Being the classification and aggregation processes completed, and so the VPPs aggregated players defined, the agents are ready for entrance in the market. We present two simulations, related to two consecutive days, Wednesday, 29th October, 2008 and Thursday, 30th October, 2008. The data used in this case study has been based on real data from the Spanish market, extracted from OMEL.

This simulation involves seven buyers and five sellers (three "normal" sellers and two of the VPPs considered before – VPP1 and VPP2). This group of agents was created with the intention of representing the Spanish reality, reduced to a smaller summarized group, containing the essential aspects of different parts of the market, in order to allow a better individual analysis and study the interactions and potentiality of each of those actors.

Each agent has a different strategy for the **first day:**

- Buyer 1 – This buyer buys energy independently of the market price. The offer price is 18.30 c€/kWh (this value is much higher than average market price)
- Buyer 2 – This buyer bid price varies between two fix prices, depending on the periods where it really needs to buy, and the ones in which the need is lower. The two variations are 10.00 and 8.00 c€/kWh
- Buyer 3 – This buyer bid price is fixed at 4.90 c€/kWh
- Buyer 4 – This buyer bid considers the average prices of last 4 Wednesdays
- Buyer 5 – This buyer bid considers the average prices of last 4 months
- Buyer 6 – This buyer bid considers the average prices of last week (considering only business days)
- Buyer 7 – This buyer only buys energy if market prices are low (this value is lower than average market price)
- Seller 1 – This seller needs to sell all the energy that he produces. The offer price is 0.00 c€/kWh
- Seller 2 – This seller bid considers the average between average prices of last 4 months and average prices of last week (considering only business days)
- Seller 3 – This seller bid considers the average prices of last 4 months with an increment of 0.5 c€/kWh
- VPP 1 – The VPP 1 includes four wind farms and offers a fix value along the day. The offer price is 3.50 c€/kWh
- VPP 2 – The VPP 2 includes one photovoltaic plant, one co-generation and one mini-hydro; the offer price is based on generation costs of co-generation and on the total forecasted production.

The average prices have been computed based on July, August, September and October OMEL market prices. Buyers and Sellers offers are represented in Figs. 13 and 14 respectively.

Figures 15 and 16 present Buyers' and Sellers' transactions respectively. Analyzing Figs. 15 and 16 it is possible to see that the market prices are imposed

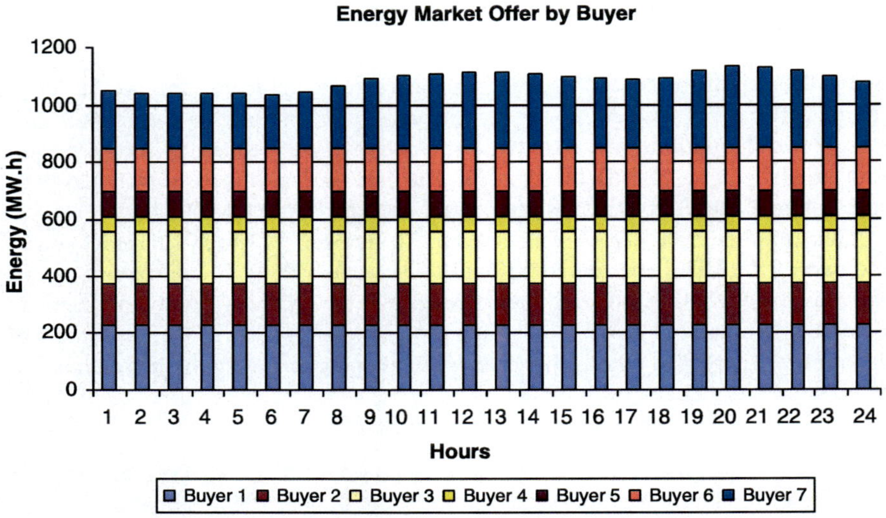

Fig. 13 Energy market offer by buyer

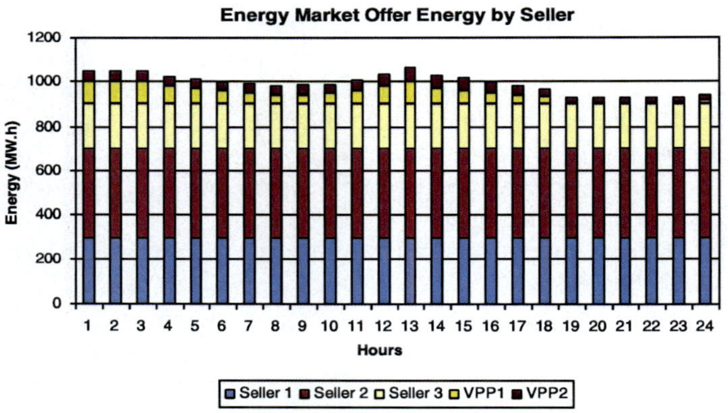

Fig. 14 Energy market offer by seller

by the agents who do variable offers along the day. In the present case, Seller 2 is the agent who most influences the price.

Analyzing Fig. 15 it is possible to observe that Buyers 4, 5 and 6 have success with the use of their strategies but this success depends on the negotiation periods. This and other case studies results point to the need to use complex strategies, combining the presented basic strategies, according to the characteristics of each negotiation period.

We can get a better understanding of the price factor by analyzing the bid prices independently of the negotiated energy.

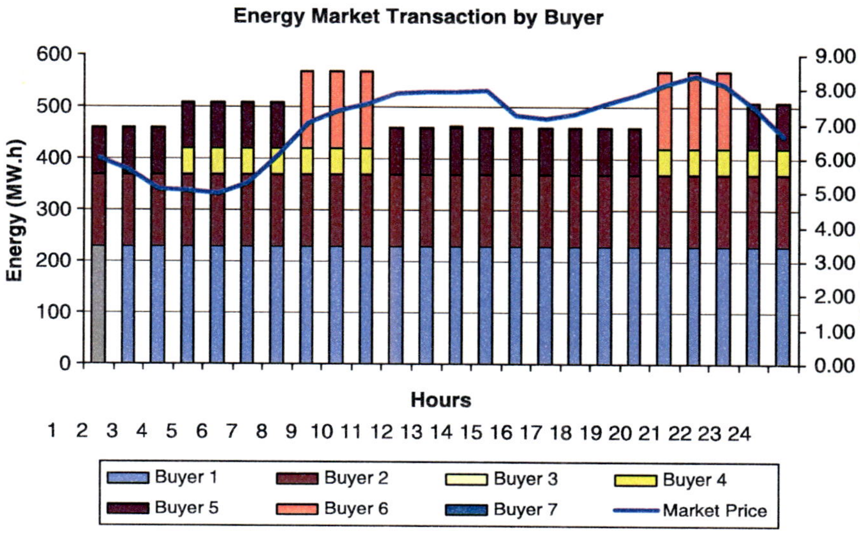

Fig. 15 Energy market transaction by buyer

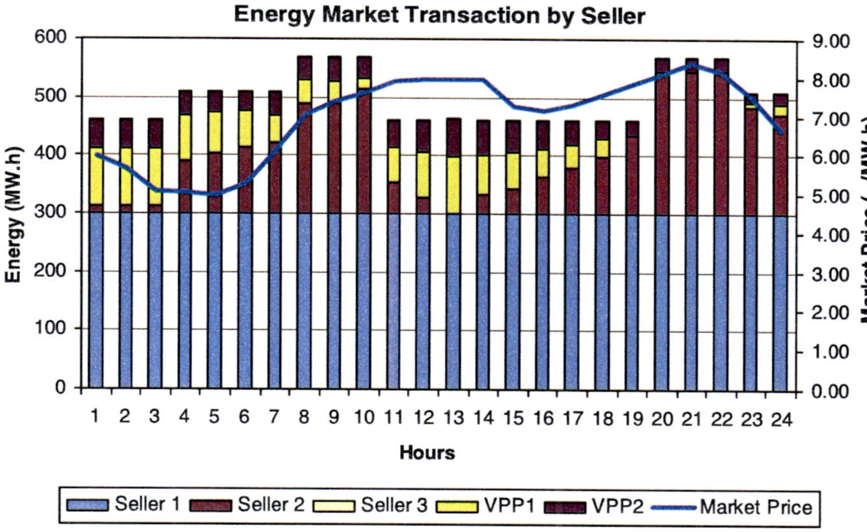

Fig. 16 Energy market transaction by seller

So in Figs. 17 and 18 we present a comparison between the agents' bid prices and the market price. Figure 17 also discriminates the individual prices of the VPPs' aggregated agents. Sellers 4, 5, 6 and 7 represent each of the four wind farms that are part of VPP1, and Sellers 8, 9 and 10 represent the photovoltaic plant, the co-generation and the mini-hydro that form VPP2.

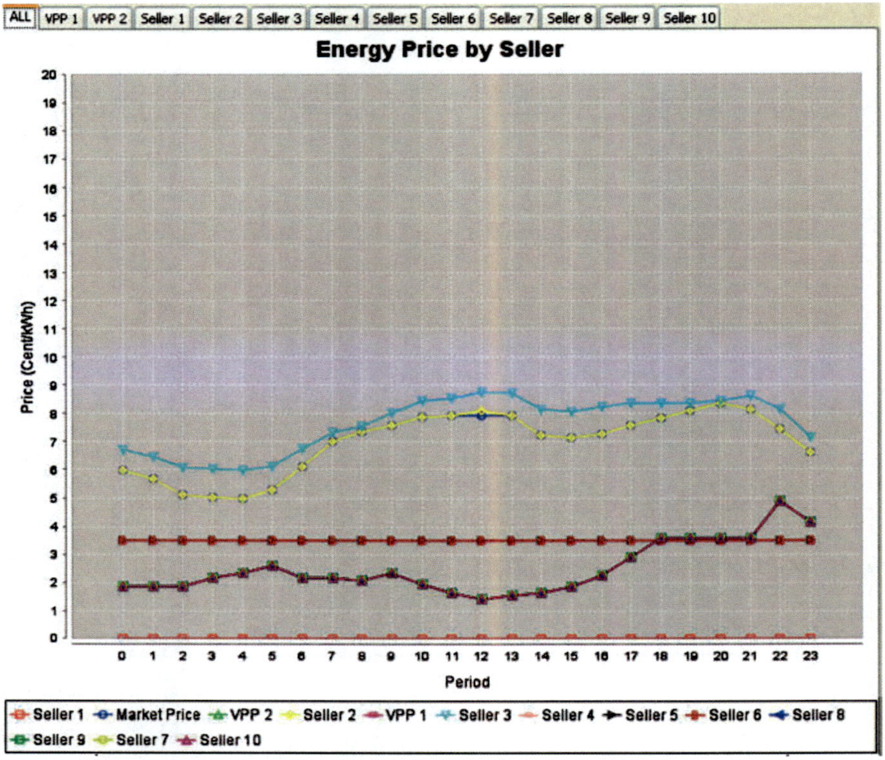

Fig. 17 Bid price by seller vs. market price

In Fig. 17 we can see that, as mentioned before, Seller 2 is the agent with the higher influence on the market price. Its graphic curve is, in practically all periods, overlapped with the market price, in a way that we almost aren't able to see the market price's curve, except for period 12. We can also watch the behavior of Seller 3, which wasn't able to sell at any period of this day. Even though that was verified, its bid prices were always following the market price, what indicates that its strategy, with some adjustments to scale the prices a bit lower, could probably be a very successful one. As for the other sellers, their prices are low comparing to the market price, and so they faced no problem in selling all of their available energy in all periods.

In what refers to the Buyers, Fig. 18, we can see that the market price varies accordingly to three agents, buyers 4, 5 and 6, the three agents that have their strategies based on the historic market prices from the previous months. Although the three base their strategies on the same data, they adopt different strategies when treating it, and so we can see that each gets its periods of success and failure, depending on the period. This fact suggests that a combination of such strategies, taking into account the characteristics of each period could prove to undertake the weaknesses of each separate one.

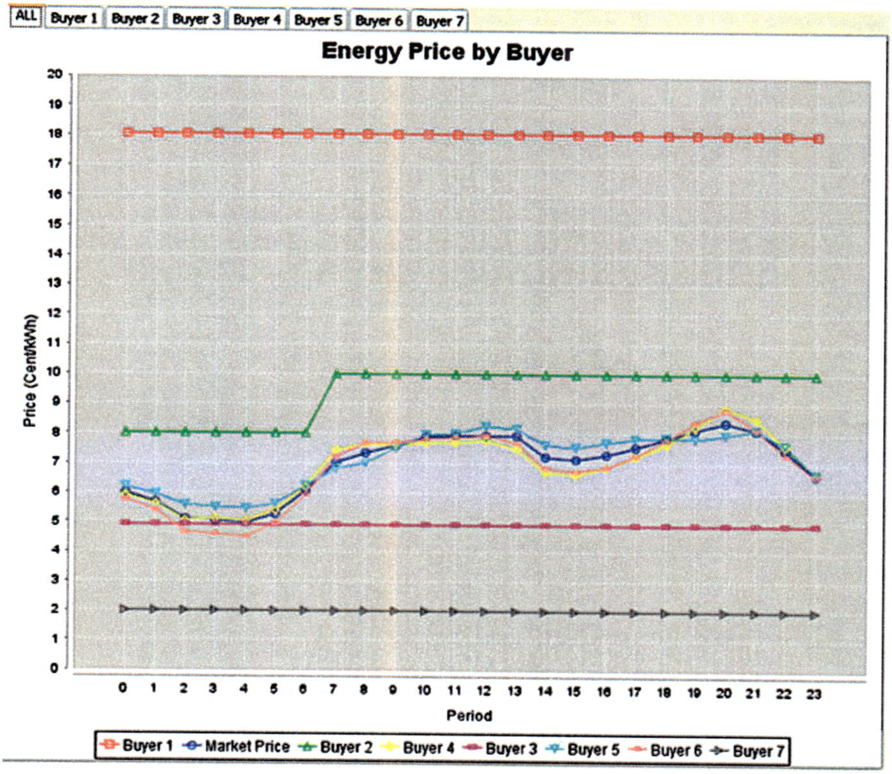

Fig. 18 Bid price by buyer vs. market price

As for the two VPPs, their prices are below market price in all periods, so they always sell the full amount of energy available.

Figures 19 and 20 show that the bid prices from both VPPs are always bellow the market price, making it possible that their energy is sold in its totality in all periods.

Second day

Based on the results of the first day we set the simulation for the next day. Considering the same agents and simulation scenario, this time we attributed two different strategies to only two seller agents in order to compare them directly, and independently from the rest of the participants. In the analysis of the first case study results we got to the conclusion that sellers 2 and 3 were the ones that had the higher influence over the market, and so, those were the agents chosen to test these new approaches. The rest of the agents considered in this case study remain with the same strategies as in the first day. The new strategies for the two agents are:

- Seller 2 – Bidding price based on **linear regression** over the historic OMEL data considering the price of all previous business days of October;
- Seller 3 – Price based on the results of a **Neural Network** with an input layer of eight units, regarding the prices and powers of the same period of the previous

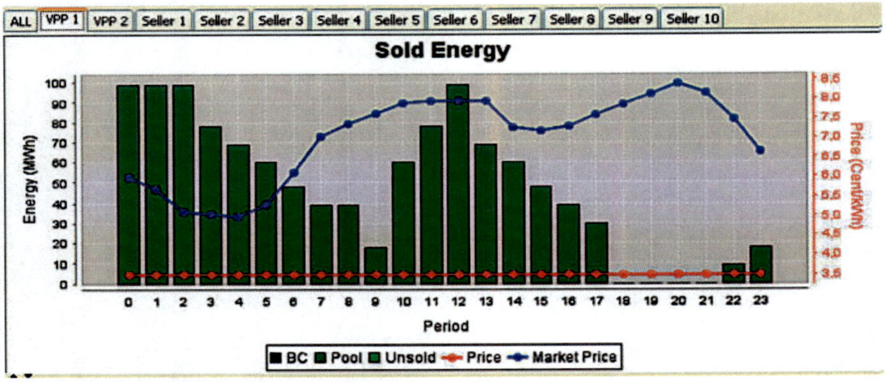

Fig. 19 VPP1 results for the first day

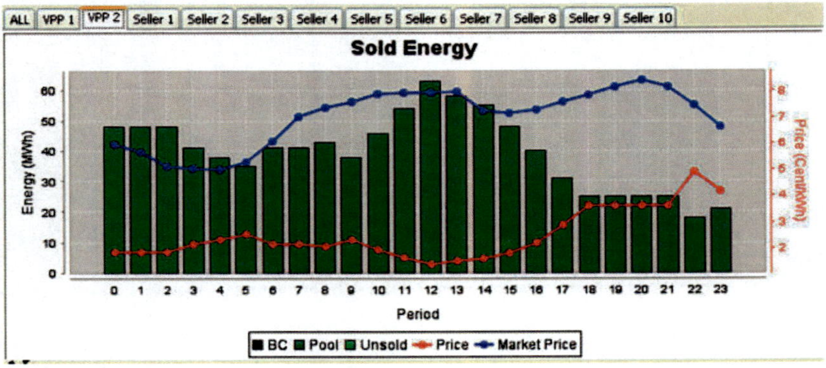

Fig. 20 VPP2 results for the first day

day, and the same week days of the previous 3 weeks. Also containing an intermediate hidden layer of four units, and an output of one unit – the predicted market price for the period in question. For example, the first training set for period 1 is (remembering that the study day is now Thursday, 30th October, 2008):

- The price of the 1st period of the previous day – Wednesday, 29th October, 2008;
- The total amount of power traded in the 1st period of the previous day – Wednesday, 29th October, 2008;
- The price of the 1st period of 1 week before – Thursday, 21st October, 2008;
- Total power traded in the same period of the same day- Thursday, 21st October, 2008;
- The price of the 1st period of 2 weeks before – Thursday, 16th October, 2008;
- Total power traded in the same period of the same day – Thursday, 16th October, 2008;

- The price of the 1st period of 1 week before – Thursday, 9th October, 2008;
- Total power traded in the same period of the same day – Thursday, 9th October, 2008;

The NN structure can be seen in Fig. 21. The input layer consists in 8 units, which correspond to the prices and powers of each period in the previous day, 1 week ago, 2 weeks, and 3 weeks. The second layer is a hidden, intermediate layer of four units, originating the final layer, the output – the forecasted price. Simulation results for the 2nd day are shown on Figs. 22–24.

Analyzing Fig. 22 we can see that the two Sellers that are the object of our study – Sellers 2 and 3, are still the ones responsible for the definition of the market price, as it is always defined around the values of their bids. That is an important issue, to verify that the changes in both their strategies didn't affect their influence over the market price definition.

Looking more closely to the two agents, and comparing them through Figs. 23 and 24, we can see that Seller 2 was the one that determined the market price more often, in ten periods. That also meant that it was never able to sell the complete amount of energy. Seller 3 determined the market price in four periods, but in the rest it was more successful than Seller 2, for it sold the complete amount of available energy in nine occasions.

These results show that in spite of the strategies used, their success varies from period to period. This and other case studies results point out to the need to use

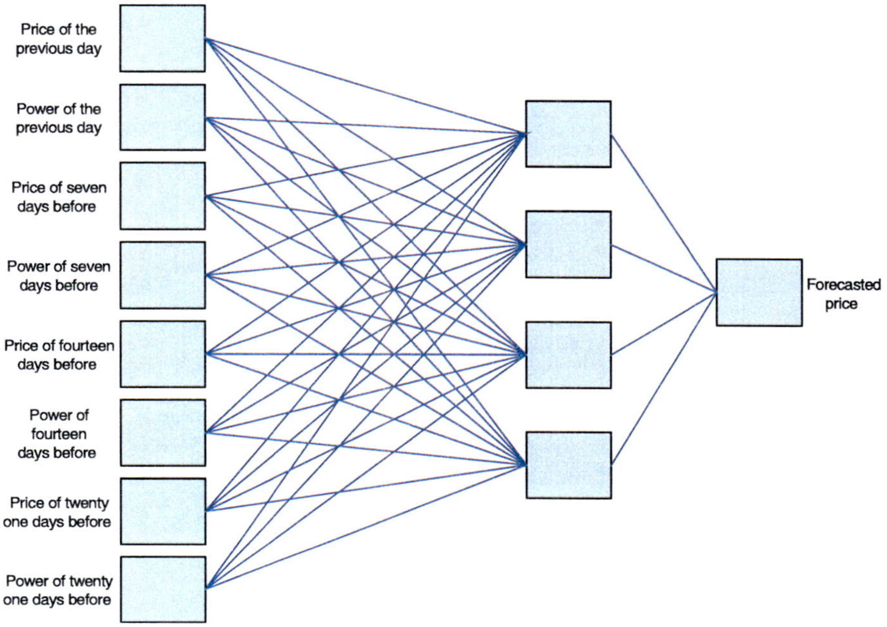

Fig. 21 Neural Network used to forecast the market price for one period

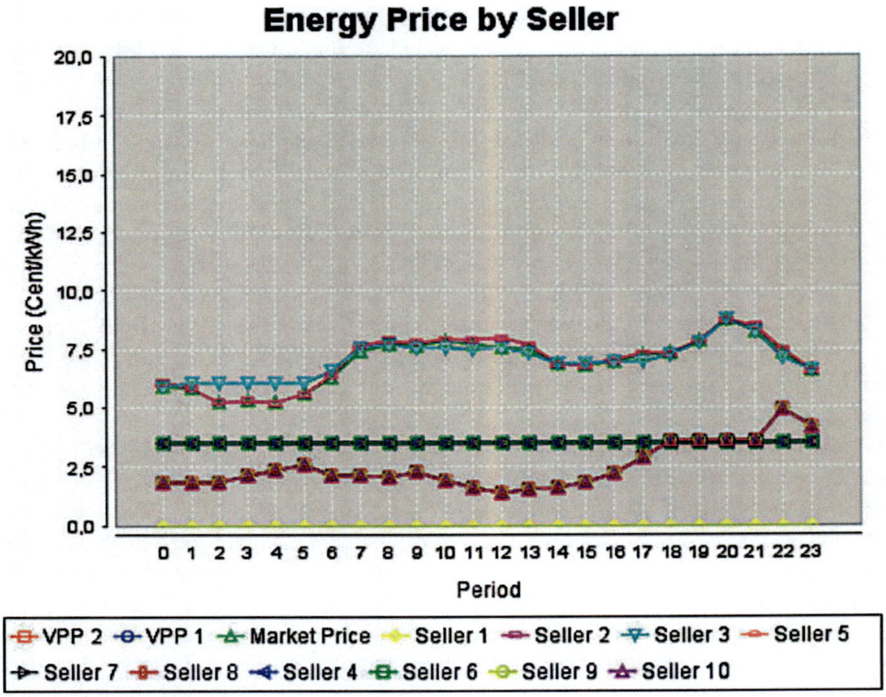

Fig. 22 Sellers results for day 2

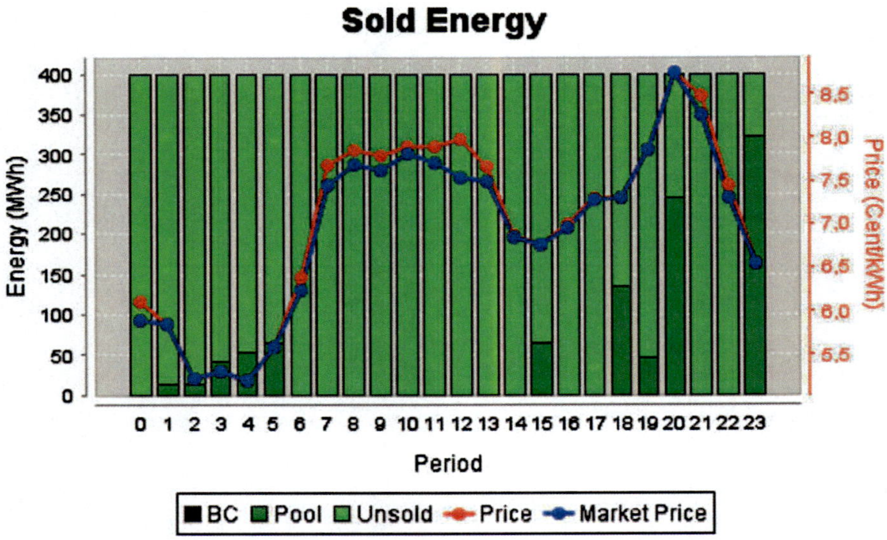

Fig. 23 Seller 2 results for Scenario 1

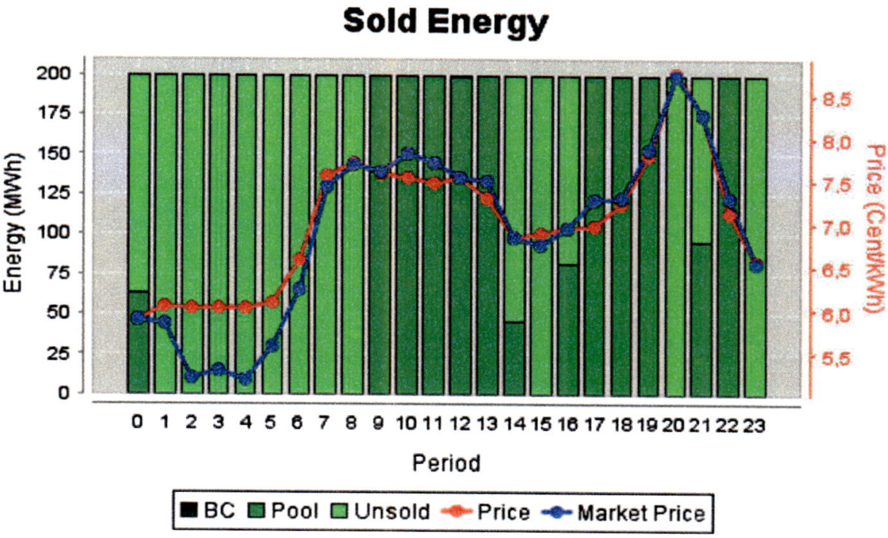

Fig. 24 Seller 3 results for Scenario 1

complex strategies, combining the presented basic strategies, according to the characteristics of each negotiation period.

14 Profits Division

Throughout the simulation, based on the classification that each producer obtained, it is needed for the VPP to divide its profits among its members, having the classification along with the amounts of energy sold as main factors to consider. In Fig. 25 is shown the comparison of the distribution of the profits among two of the members of VPP 2.

We can see in Fig. 25 that, despite having sold the same amounts of energy, Producer eight presents a superior profit than Producer 7. That happens due to the much higher classification awarded to it. This proves the satisfaction of the need of having the profits division depending on the classification.

15 Conclusions and Future Developments

With MASCEM several experiences have been made which gives us the experience, the know-how and the motivation to go forward and overcome the limitations as well as enlarge the scope of the simulator. In this section we point out some of the issues that are already previewed and being started, such as those related to the improvement of the agents learning process, the improvement of negotiations

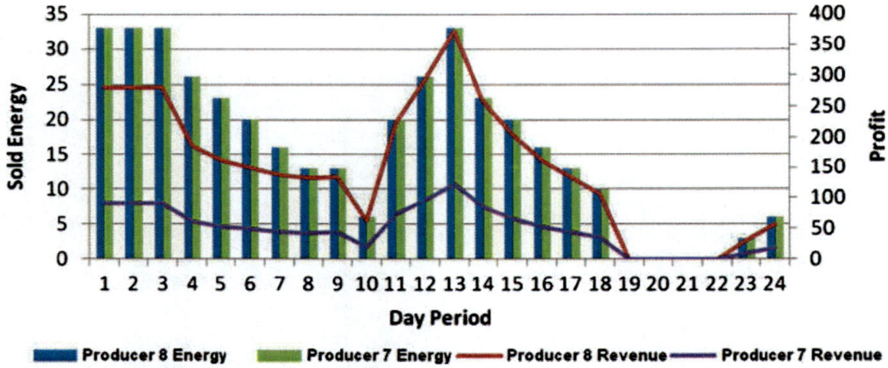

Fig. 25 Producer 7 and Producer 8 produced energy and remuneration

inside coalitions for VPP and model enlargement to deal with ancillary services and settlements handling.

One of the main goals of MASCEM electricity market simulator is to study the market players' interactions, providing the means to understand how their actions influence the states of the market. This knowledge can prove to be a great advantage when trying to get the most profits out of the market, and even decrease the energy prices through competition.

Therefore, a very important aspect to consider in the development of this simulator is the capability of market agents to make intelligent decisions, being able to adapt their actions at each moment by analyzing the environment around them.

There is work in progress to asset the agents with such capabilities. This new adapting mechanism will provide the agents with several learning technologies and data mining analysis, offering the agents a means to analyze the data that each one is able to perceive. Through the gathering of several distinct mechanisms that offer different approaches in the dealing with the data, agents will have on their hands various proposals of strategies to use in the market, and will be able to choose the ones that offer more value in each moment, according to their previous results when dealing with similar situations.

The learning algorithms go from simple strategies like averages of the previous market prices, and linear regressions on that data, to more complex procedures as neural networks, strategies based on data mining mechanisms, application of the game theory and other prediction and negotiation mechanisms. The system will provide an interface with several technologies, such as Matlab, Clementine, and Win-Prolog. This will also facilitate the process of a continuous evolution, through the inclusion of more strategies.

This mechanism will not only allow the agents to adapt and learn, but will also provide a good mean for comparison between learning algorithms and adaptive strategies of various types.

Negotiations inside coalition's structure and coalition negotiations on the market require sophisticated protocols, where other Artificial Intelligence techniques, such as argumentation and multiple criteria based negotiations may be combined. Inside the context of multi-agent negotiation, an argument is seen as a piece of information able to influence other agents negotiation stance and/or justify the own agent negotiation stance [42–44]. Argumentation can be excellent to justify possible choices and to convince other elements of the coalition that one alternative is better or worse than another. Since it is adequate to decision making inside groups, one of our future goals is to identify how the argumentation based negotiation can be used inside the coalition structure, and which types of arguments will be the most adequate. We intend to consider negotiations based on both multiple criteria and argumentation. One of the approaches will be to consider multiple criteria to set up the agent intentions and proposals definition, and use argumentation to justify them.

Electricity Markets have as the underlying negotiated product the electrical energy required by the demand side; the energy transactions occur in the scope of the energy market. Additionally, services required for assuring security and reliability, known as ancillary services, must be obtained [36, 45–47]. Ancillary Services include active power (load/frequency control, including primary control, automatic generation control, tertiary control, balancing service and black start provision) and reactive power (voltage/reactive power control).

In Electricity Markets ancillary services can be seen as a set of products separated from the energy production. A competitive environment requires adequate procedures and methodologies for determining Ancillary Services needs, the ways they are obtained and priced. These must assure that the Ancillary Service required for assuring Power Systems security and reliability are permanently available and these services are fairly remunerated. Ancillary Services provision can be mandatory or remunerated under market-driven mechanisms [9, 48–51].

Ancillary Services can be treated jointly with the energy market or on a separated basis; a hybrid approach, treating a part of ancillary service (e.g. frequency control) jointly with the energy markets and the remaining ancillary service separately is also possible.

Traditionally ancillary services are provided by large power plants; presently there is the need for mechanisms able to deal with smaller generation resources and loads participation in ancillary services provision.

Most energy transactions are decided in an anticipated basis, based on forecasted loads, resulting from bi-lateral contracts or day-ahead markets. Due to forecasting errors and to unpredicted outages, additional mechanisms are required to undertake balancing (adjustments/settlements) to assure the required load–generation balance. Traditionally settlements are undertaken trough generation adjustments (increase or decrease generation bids); in the present context, both supply and demand bids (increase or decrease loads) should be accepted. Settlements handling requires an efficient approach able to deal with high volumes of data in a short time [52].

Under the new vision of MASCEM, Electricity Markets will be simulated in a much more complete way what requires to implement a more complex simulation infrastructure, able to cope with discrete events and with the diverse time scales of the supported negotiation mechanisms (bi-lateral contracts, pool, derivatives market, ancillary services and settlement mechanisms) as well as with the time constraints inherent to the physics of the Power System.

References

1. Weiss G (ed) (1999) Multiagent system: a modern approach to distributed artificial intelligence. MIT Press, Cambridge, MA
2. Wooldridge MJ (2002) An introduction to multiagent system. Wiley, New York
3. Brooks R J, Shi D (2006) The calibration of agent-based simulation models and their use for prediction. In: Robinson S, Taylor S, Garnett J (eds) Proceedings of the 2006 OR society simulation workshop, UK
4. King A, Streltchenko O, Yesha Y (2005) Using multi-agent simulation to understand trading dynamics of a derivatives market. Ann Math Artif Intel 44(3):233–253. Springer, Netherlands
5. Law A, Kelton W (2000) Simulation modeling and analysis, 3rd edn. McGraw-Hill International, New York
6. Streltchenko O, Finin T, Yesha Y (2005) Multi-agent simulation of financial markets. In: Kimbrough SO, Wu DJ (eds) Formal modeling in electronic commerce. Springer, New York
7. Helleboogh A, Vizzari G, Uhrmacher A, Michel F (2007) Modeling dynamic environments in multi-agent simulation. JAAMAS 14(1):87–116
8. Ilic M, Galiana F (1998) Power syst restructuring: engineering and economics. Kluwer Academic, Dordrecht
9. Meeus L, Purchala K, Belmans R (2005) Development of the internal electricity market in Europe. Electr J 18(6):25–35
10. Hatziargyrious N, Meliopoulos S (2002) Distributed energy sources: technical challenges. In: IEEE power engineering society winter meeting, vol 2, pp 1017–1022
11. Azevedo F, Vale ZA, Oliveira PBM (2007) A decision-support system based on particle swarm optimization for multiperiod hedging in electricity markets. IEEE T Power Syst 22 (3):995–1003
12. McArthur S et al (2007) Multi-agent system for power engineering applications – part I: concepts, approaches, and technical challenges; part II: technologies, standards, and tools for building multi-agent system. IEEE T Power Syst 22(4):1743–1759
13. Praça I, Ramos C, Vale Z A, Cordeiro M (2003) MASCEM: A multi-agent system that simulates competitive electricity markets. IEEE Intell Syst 18(6):54–60. Special issue on Agents and Markets
14. Vale ZA (2003) Knowledge-based system techniques and applications in power system control centers. In: Cornelius T Leondes (ed) Intelligent system technologies and applications, vol 6. CRC Press, pp 61–110, USA
15. Figueiredo V, Rodrigues F, Vale Z, Gouveia JB (2005) An electric energy consumer characterization framework based on data mining techniques. IEEE T Power Syst 20(2):596–602
16. Han J, Kamber M (2006) Data mining, concepts and techniques, 2nd edn. Morgan Kaufmann, San Francisco
17. Witten I, Frank E (2005) Data mining, practical machine learning tools and techniques with java implementations, 2nd edn. Morgan Kaufmann Series - Elsevier, San Francisco, USA
18. Fatima SS, Wooldridge M, Jennings NR (2006) Multi-issue negotiation with deadlines. J of AI Res 27:381–417
19. Wellman MP, Greenwald A, Stone P (2007) Autonomous bidding agents: strategies and lessons from the trading agent competition. MIT Press, Cambridge
20. Hatami AR, Seifi H, Sheikh-El-Eslami MK (2009) Optimal selling price and energy procurement strategies for a retailer in electricity markets. Electr Power Syst Res 79:246–254

21. Hortaçsu A, Puller SL (2007) Understanding strategic bidding in multi-unit auctions: a case study of the Texas electricity spot market. RAND J Econ 39(1):86–114, Wiley InterScience
22. Li T, Shahidehpour M (2005) Strategic bidding of transmission-constrained GENCOs with incomplete information. IEEE T Power Syst 20(1):437–447
23. Wu Z, Ilic M (2008) The effects of multi-temporal electricity markets on short- and long-term bidding. IEEE PES general meeting 2008, Pittsburgh
24. Yucekayaa AD, Valenzuelaa J, Dozierb G (2009) Strategic bidding in electricity market using particle swarm optimization. Electr Pow Syst Res 79:335–345
25. Koritarov V (2004) Real-world market representation with agents. IEEE Power Eng Mag 2:39–46
26. North MJ, Macal CM (2007) Managing business complexity: discovering strategic solutions with agent-based modelling and simulation. Oxford University Press, New York
27. Somani A, Tesfatsion L (2008) An agent-based test bed study of wholesale power market performance measures. IEEE Comput Intel Mag 3(4)
28. Migliavacca MG (2007) SREMS-electricity market simulator based Game Theory and incorporating network constraints. IEEE Power Tech 2007, Lausanne
29. Khodr H, Vale ZA, Ramos C (2008) A benders decomposition and fuzzy multicriteria approach for distribution networks remuneration considering DG. IEEE T Power Syst (TPWRS-00683-2008, accepted for publication)
30. Praça I, Ramos C, Vale Z et al (2005) Intelligent agents for negotiation and game-based decision support in electricity markets. Intern J Eng Intell Syst 13(2):147–154. CRL Publishing
31. Praça I, Morais H, Ramos C et al (2008) Multi-agent electricity market simulation with dynamic strategies & virtual power producers. IEEE Power & Energy Society -2008 PES general meeting, Pittsburgh
32. Ferreira J, Vale Z, Cardoso J, Puga R (2008) Transmission price simulator in a liberalized electricity market. In: 5th international conference on European electricity market, pp 1–6, Lisbon, Portugal
33. Morris J, Greenwald A, Maes P (2003) Learning curve: a simulation-based approach to dynamic pricing. Electronic Commerce Res: Special Issue on Aspects of Internet Agent-based E-Business Syst 3(3–4):245–276. Kluwer Academic
34. Faratin P, Sierra C, Jennings N (1998) Negotiation decision functions for autonomous agents. Int J Robot Auton Syst 24(3):159–182
35. Greenwald A, Kephart J (1999) Shopbots and Pricebots. In: Proceedings of the sixteenth international joint conference on artificial intelligence. IJCAI, Stockholm
36. Vale Z, Morais H, Cardoso M et al (2008) Distributed generation producers' reserve management. IEEE PES general meeting 2008, Pittsburgh
37. Dang VD, Jennings N (2004) Generating coalition structures with finite bound from the optimal guarantees. In: Proceedings of the 3 rd international conference on autonomous agents and multi-agent systems, New York, pp 564–571
38. Norman TJ, Preece A, Chalmers S et al (2004) Agent-based formation of virtual organisations. Int J Knowl-Based Syst 17:103–111
39. Rahwan T, Ramchurn SD, Dang VD, Giovannucci A, Jennings NR (2007) Anytime optimal coalition structure generation. In: Proceedings of the 22nd conference. on artificial intelligence (AAAI), Vancouver, pp 1184–1190, 2007
40. Rahwan T, Jennings NR (2008) Coalition structure generation: dynamic programming meets anytime optimisation.In: Proceedings 23 rd conference on AI (AAAI), Chicago, pp 156–161
41. Morais H, Cardoso M, Castanheira L et al (2007) VPPs information needs for effective operation in competitive electricity markets. In: Proceedings of the 5th international conference on industrial informatics, Indin, Vienna
42. Jennings NR, Parsons S, Noriega P et al (1998) On argumentation-based negotiation. In: Proceedings of international workshop on multi-agent system, Boston

43. Marreiros G, Santos R, Ramos C, Neves J (2010) Context-Aware Emotion-Based Model for Group Decision Making. IEEE Intel Syst 25(2):31–39
44. Ramchurn SD, Sierra C, Godo L, Jennings NR (2007) Negotiating using rewards. Artif Intel J 171(10–15):805–837
45. Li Y, Jiang JN (2007) Experience with operating the ancillary-service markets in ERCOT. Power Engineering Society general meeting, 24-28 June, Tampa, Florida, USA
46. Miguélez EL, Cortésa IE, Rodríguez LR et al (2008) An overview of ancillary services in Spain. Electr Pow Syst Res 78(3):515–523
47. Pereira A, Vale ZA, Moura AM et al (2004) Provision and costs of ancillary services in a restructured electricity market. In: International conference on renewable energy and power quality (ICREPQ'04)
48. Amundsen ES, Bergman L (2007) Provision of operating reserve capacity: principles and practices on the Nordic electricity market. Compet Regul Networ Ind (Intersentia) 1(2):73–98
49. Cheung KW (2008)Ancillary service market design and implementation in North America: from theory to practice. DRPT2008. Nanjing
50. Raineri R, Ríos S, Schiele D (2006) Technical and economic aspects of ancillary services markets in the electric power industry: an international comparison. Energ Policy 34 (13):1540–1555
51. Thorncraft S R, Outhred HR (2007) Experience with market-based ancillary services in the Australian national electricity market. Power engineering society general meeting, 2007, IEEE
52. Chen D (2008) Post-market computation engine toward CAISO market settlements. In: IEEE 2008 Transmission and distribution conference and exposition, Chicago

Differentiated Reliability Pricing Model for Customers of Distribution Grids

Arturas Klementavicius and Virginijus Radziukynas

Abstract The paper addresses the idea of reliability differentiation for peer electricity customers willing to choose a standardized reliability level of electricity supply for the respective tariff.

The authors suggest the reliability differentiation concept based on standardized reliability levels (categories). As a partial prototype, the existing 3-grade reliability differentiation system in Lithuania is referred to. According to the concept, customers with higher reliability category pay a higher distribution reliability tariff, in proportion to the incurred grid operation, management and amortization cost.

The operator provides a contracted higher category mainly through the parallel supply path in the grid, i.e. by the switch-over of customer's load to the 2nd independent supply point after the failure of electrical path from the 1st independent supply point. Accordingly, the supply notion is split to major, joint and reserving supply where major supply for a customer is provided in normal operation situations (i.e. from the 1st independent supply point), reserving supply – in unusual situations (from 2nd independent supply point) while joint supply denotes situation when the same electrical path serves as both a major supply route for one customer and a reserve supply route for another.

The suggested concept is supported by a new mathematical reliability pricing model for 2-grade reliability system. The model is cost-reflective and applicable to the distribution grid area controlled by one operator. Its rationale is fair allocation of grid cost between customer groups with different reliability categories, and subsequent derivation of distribution tariffs for reliability categories. The grid cost is split to the components related to four types of grid equipment – distribution transformers, medium voltage lines, local transformers and low voltage lines. These cost components are allocated between customer classes with different reliability

A. Klementavicius • V. Radziukynas (✉)
Lithuanian Energy Institute, Kaunas, Lithuania
e-mail: arklemen@mail.lei.lt; virginijus@mail.lei.lt

A. Sorokin et al. (eds.), *Handbook of Networks in Power Systems I*,
Energy Systems, DOI 10.1007/978-3-642-23193-3_8,

categories and connection-to-grid voltages. An allocation criterion is the scope of usage of grid equipment type.

The applicability of the presented model is illustrated by numerical setup for the sample grid.

Keywords Customers • Distribution grid • Pricing model • Reliability • Supply

1 Introduction

The differentiated reliability pricing is yet neither a challenge for grid operators and other power sector's stakeholders nor a focus of attention for energy research and regulation framework. Apparently, the dominating view may be defined as that expressed by European regulators in [1], where doubts concerning regulated prices of reliability are expressed. Practically, the authors conclude that currently it is impossible to differentiate the delivered electricity in terms of reliability. In general, the idea of differentiated reliability is promoted to moderate extent and without consolidated efforts.

Historically, utilities and customers addressed reliability pricing of electricity service for decades. In the early period of power market liberalization, the outlooks, concerns and approaches related to differentiated reliability were reviewed and investigated in study [2].

The majority of contributions to differentiated reliability research is done in transmission reliability. The differentiation principles are based either on additional reserve generation cost or forced outages of customers or transmission line reserving cost. The combination of those principles is also applied.

The commonly known reliability index – the loss of load probability (LOLP) – may be applied for load service probability differentiated nodal pricing [3]. This index is re-defined as locational LOLP. Using the concept of domains and commons and generators' contributions to the loads, the locational load service probability (LSP) indices can be determined from the locational LOLP. Further LSP are incorporated into the locational marginal prices (LMPs). This way, the customers located at a load bus are fairly priced, taking into account the LSP they receive from the system operator. The traditional LMPs are thereby modified to include the LSP of a customer at a bus.

Similarly, authors in [4] propose a pricing model featuring both reliability pricing and real-time pricing based mainly on generation costs. It accounts for customers' reliability preferences as well as the randomness of outages in the power supply system. The expected total and marginal production cost is estimated on an hourly basis, as well as system reliability measured by the loss-of-load probability. These estimates are inputs to a welfare-maximization model with revenue constraints differentiated by customer class. The authors state that the proposed pricing approach is significantly superior to spot and Ramsey pricing in terms of economic efficiency, energy conservation, and generation-capacity requirements.

Authors [5] link the real-time pricing with the reliability differentiation. The proposed pricing scheme combines real-time pricing and priority pricing with reliability differentiation based on consumer outage costs. The scheme has a particular emphasis on consumer behaviour and welfare effects. The authors [6] developed a consistent transmission reliability allocation method. They suppose it seems plausible to compute the reliability contributions of all market participants based on the probabilistic approach which takes notice of the forced outage rate for each transmission line as well as the line outage impact factor and then to allocate the transmission reliability cost among all the system users in proportion to their "extent of use" of reliability reserves in transmission facilities.

Interesting differentiation concept is outlined in [7]. It is underlain by the transmission reliability margin (TRM) and forced outage rate. The paper explores the probabilistic approach to the calculating of all market participants' contributions to the TRM by taking into account the forced outage rate of each circuit across the entire network. With those reliability contributions, the reliability cost can be allocated to the users in a more reasonable way.

The differentiation of transmission reliability could be undertaken as an instrument to solve transmission congestion problem [8]. The capacity reservation tariff is proposed to induce the users to self-select their preferred levels of reliability. Based on these self-selected reliability levels, the Independent System Operator can efficiently allocate limited transmission capacity. Such a differentiation was suggested as an opportunity to Californian Independent System Operator.

A comprehensive approach was outlined in [9]. It refers to the principle that the design of pricing menu should consist of the price and the reliability indices. The authors determine the price ratio of high reliability menu (menu-H) to standard reliability menu (menu-L) and assign each customer to one of them. The reliability cost is derived from the outage cost. The assignment allows for minimization of total outage cost. Taking sample small system, the authors evaluate how the reliability differentiated pricing menu is affected by the local characteristics in power systems.

Most optimistic view on advantages and possibilities of differentiated reliability prices is expressed in recent paper [10]. It finds out that end-customers have different needs and preferences, such that choosing the level of availability that they are willing to pay for becomes a key issue. This paper elaborates the concept by differentiating electricity in multiple availability types and multiple types of customers classes.

Anyway, on the distribution level, different reliability levels could be determined from network topology and number of independent supply points (backup buses and feeders). Feeder automation enables automatic load transfer from failed feeder to a backup feeder (i.e. second independent supply point). This transfer is carried out by circuit breaker or another change-over switch located either in a substation or in remote feeder section. The automation at feeder level in North America is assessed to be less than 20% [11].

2 Reliability Differentiation for Peer Customers and Peer Customer Groups

Axiomatically, to differentiate a parameter, say, a tariff or a quality indicator would mean to fix (standardize) a set of different parameter values. Clearly, if there is only one fixed value, such a parameter is non-differentiated. The reliability differentiation in electricity supply might be regarded as existing, if several reliability levels for customers (their connection-to-grid points) were promulgated and applied.

The characteristic feature of reliability framework in EU countries is absence of any interruption standard for a single customer in vast majority of countries. As exceptions of it, the Belgium stands out with one standard for all customer classes – 4 h maximum duration of one long unplanned interruption. The same French standard is 6 h, and the Finnish – 12 h [12].

Nevertheless, among those exceptions, a few isolated cases exist when reliability is differentiated for peer customer groups, i.e. different peer groups are assigned to different reliability standards. Here "peer customers" is understood as those being similar or identical with respect to connection voltage and geographic-territorial conditions (as they predetermine the similar structure and configuration of distribution grids).

Reliability differentiation for peer customer groups

The existing frameworks of reliability differentiation deal with either interruption duration standards or multiple interruption standards [12].

In case of duration standards, the duration limits are laid down either for a single unplanned long interruption (under operator's responsibility) or for annual duration of such interruptions. There is only one single-duration framework in Europe – in Czech Republic: 12 h for medium voltage (MV) customers (1st peer group) and 18 h for low voltage (LV) customers (2nd peer group). As for annual values, the case of Spain represents a paradigm of nice differentiation:

- LV customers (4 peer groups distinguished): urban customers – 6 h/year, suburban – 10 h/year, concentrated rural – 15 h/year, scattered rural – 20 h/year;
- MV customers (4 peer groups): urban customers – 4 h/year, suburban – 8 h/year, concentrated rural – 12 h/year, scattered rural – 16 h/year;
- High voltage (>36 kV) customers (1 peer group) – 6 h/year.

In case of multiple interruption standards, limits for yearly number of unplanned long interruptions are laid down. Specifically in Italian framework, the following limits have been applied:

- LV customers: no standard;
- MV customers (3 peer groups): in high density areas – 3 interr./year, in medium density areas – 4 interr./year, in low density areas – 5 interr./year;
- High voltage (HV) customers (1 peer group) – 1 interr./year.

Reliability differentiation for peer customers

Nevertheless, the reliability differentiation between peer customer groups is only a partial differentiation because one typical value is assigned to an entire peer group. A substantial differentiation would take place, if peer customers within a group would be entitled a right to request the higher reliability level than typical for this peer group (for the adequate price).

To illustrate the difference between both types of differentiation, a hypothetical case could be assumed when reliability standard for Italian HV customers (1 peer group), 1 interr./year, is differentiated, say, into two reliability categories:

- Typical reliability – 1 interr./year (the same as referred to above),
- Higher reliability – 0.5 interr./year (i.e. 1 interr./2 years).

Then the reliability would be differentiated namely for peer customers.

Reliability differentiation for peer customers is neither applied nor discussed in EU [12]. Herein only Lithuania can be regarded as some exception (see Sect. 2).

It could be noted that the most comprehensive studies like [13] do not refer to reliability differentiation, neither for peer customer groups nor for peer customers.

As concerns European future grid vision [14], it would be rather complicated to find arguments that the reliability differentiation is envisaged in the smart grids in the horizon to 2030. The reliability of electricity supply is supposed to be increased, inter alia, by means of distributed generation and energy storage devices, including a large proportion of customer generators and storage devices. Nonetheless, the commercial distributed generators could be contracted by grid operator to serve as additional independent supply sources and provide basis to ensure the higher reliability levels for requesting customers.

For the purpose of thorough evaluation of reliability differentiation idea, the reliability differentiation models and differentiated reliability pricing models could be helpful and contribute to the better quality of power supply in future distribution grids.

3 Grid-Provided Reliability Differentiation in Lithuania

Historically, Lithuanian electricity customers have been categorized to three different levels of grid-provided reliability of supply (continuity of supply). The recent situation in reliability differentiation is outlined in Table 1 using data presented in [15]. A standard interruption in Lithuanian regulations is defined as a legally established maximum duration of a single interruption for a single customer when the cause of interruption is attributable to the liability of grid operator.

In order to obtain the 2nd or 1st reliability category, a customer is obligated to motivate the need for higher reliability against grid operator (for environmental, safety, technology reasons).

Referring to concept of peer group, it might be considered that there is one peer customer group in Lithuania with three differentiated reliability levels applied.

Table 1 The breakdown of Lithuanian customers into reliability categories in 2006

Category	Standard for unplanned single long interruption, h		Customers		Declared capacity	
	Before October 11 2005	From October 11 2005	Number	Percentage	MW	Percentage
3rd	24	24	~1,420,000	99.7	~12,000	79.2
2nd	2.5	6	3,181	0.22	1,740	14.5
1st	~0[a]	2.5	79	0.01	760	6.3

[a]Practically no-interruption supply

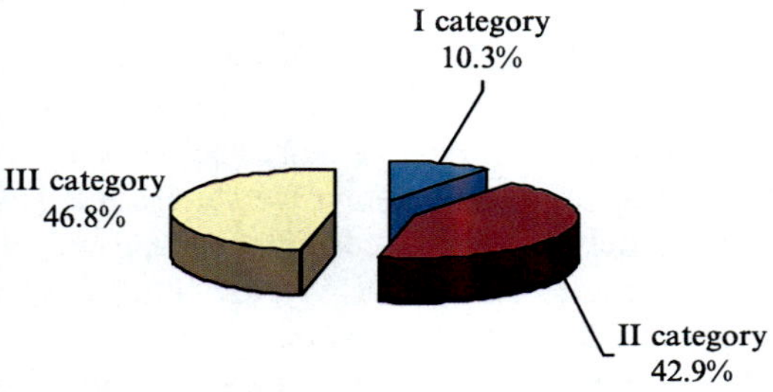

Fig. 1 Proportions of Lithuanian customers by reliability categories (2006) in medium voltage networks (6–10-kV) in terms of declared capacities

By the typical standard of 24 h, majority of Lithuanian customers (as holders of 3rd category) lag behind the European countries leading in reliability regulation. On the other hand, they do better than a group of countries without such a standard at all.

It should be noted that there is a specific voltage level on which the connected customers with 2nd and 1st categories occupy jointly as large market share as those with 3rd category. It is MV level covering 284 customers with 1,190 MW of declared load (capacity). This capacity accounts for 53.2% of total loads (including 3rd category loads) connected to 6–10-kV distribution grids (Fig. 1).

The current Lithuanian framework for pricing of grid services seems to be not fair because the privileged customers (provided with 2nd and 1st reliability categories) do not pay more for higher quality service. Therefore the cross-subsidization by customers provided with typical reliability level (3rd) is taking place.

4 Preliminary Concept of Differentiated Reliability Pricing

As proposed in [15], the higher reliability levels should be additionally charged by grid operator for the recipient customers. The idealized system of reliability categories was assumed to associate the categories with the number of independent supply points (feeding points) in operator's grid:

- 3rd category customer (load) is connected to 1 point;
- 2nd category – to 2 points (i.e. 1 point can serve for reserving);
- 1st – to 3 points (i.e. double reserving is ensured).

The independent supply point is conceived to be a top point of a separate path (circuit) in power grid from a distribution transformer (HV/MV, i.e. connecting transmission and distribution networks) to the boundary of customer's internal grid (i.e. customer's connection-to-grid point).

If a customer has availability of the 2nd independent grid point in case of failure of major 1st point and even to the 3rd point (in case of failure of two points), he should reimburse extra grid cost related to such an availability. Namely, he should share that cost with other customers fed from them as from their 1st supply points.

The preliminary conceptual model (1) as presented in [15], was intended to reflect the operator's operation and management cost (operating expenses) of grid components serving as the connecting links to the 2nd and 3rd independent grid points. The idea as how to estimate this cost was to find weights (proportions) of those connecting links in the total structure of grid. Subsequently, the additional tariff for higher reliability would be derived from this cost. A bit modified, this preliminary model (1) looks like that:

$$T_{I,II} = \frac{\left(OM_{transf} \cdot \frac{S_{I,II}}{S_{transf}} + OM_L \cdot \frac{L_{I,II}}{L} \cdot k_{joint}\right)}{P_{I,II}}, \tag{1}$$

where $T_{I,II}$ – additional tariff for 1 kW of customer's declared load provided with 1st or 2nd reliability category, currency/year;

OM_{transf} – total operation and management cost of all transformers and related substation equipment in operator's grid;

OM_L – total operation and management cost of all MV and LV electric lines in operator's grid;

S_{transf} – total installed transformer capacities in operator's grid (kVA);

$S_{I,II}$ – total installed transformer capacities in operator's grid in the connecting links to the 2nd and 3rd independent supply points for the customers with 2nd and 1st category (kVA). Here only parts of transformer capacities count that are adequate to the declared loads of those customers;

L – total length of electric lines in operator's grid (km);

$L_{I,II}$ – total length of electric lines in connecting links to the 2nd and 3rd supply points in operator's grid (km);

k_{joint} – co-feeding factor taking into account the 3rd category customers receiving major supply through the same electrical lines connected to 2nd and 3rd supply points for 2nd and 1st category ($k_{joint} < 1$);

$P_{I,II}$ – total declared capacity (load) of all privileged customers, kW.

5 Sample Grid for Development of Differentiated Reliability Pricing Model

The conceptual model (1) should be extended to a detailed model in order to identify:

- The role of major types of grid equipment in differentiated reliability provision;
- The respective grid cost components related to those equipment;
- How the different customer classes participate in the joint use of grid infrastructure;
- The tariff structure for different reliability categories with respect to different customer classes (voltage levels).

Such a detailed pricing model for 2-grade reliability system (of 3rd and 2nd categories) is presented below in Sects. 5–9. It was developed on the basis of a sample grid. The sample grid and the description of grid-provided reserving with the necessary denotations is the matter of this section.

5.1 *Configuration and Structure*

The sample grid as presented in Fig. 2 is assumed to have a configuration and structure sufficient for the purposes of the model. Concurrently, it seems to be representative for a real grid.

5.2 *Denotations and Terms*

- *Major supply* – supply of a customer (load) in normal grid configuration from 1st independent supply point
- *Reserving supply* – supply of a customer (load) in provisional grid configuration from 2nd independent supply point (following the failure of major supply)
- *Joint supply* – use of the same grid element to provide major supply for some customers and, in addition to it, reserving supply for other customers in unusual situations

Customer classes

A	Class of 10-kV customers (their loads) provided with 2nd reliability category
$A1$	A-class customer No1 (its load)
$A1(1)$	1st load of customer $A1$
B	Class of 10-kV customers (their loads) provided with 3rd reliability category
$B1$	B-class customer No1 (its load)
a	Class of 0.4-kV customers (their loads) provided with 2nd reliability category
$a1$	a-class customer No1 (its load)
$a2(2)$	2nd load of customer $a2$
b	Class of 0.4-kV customers (their loads) provided with 3rd reliability category
$b1$	b-class customer No1 (its load)
⬚	Boundaries of customer's grid

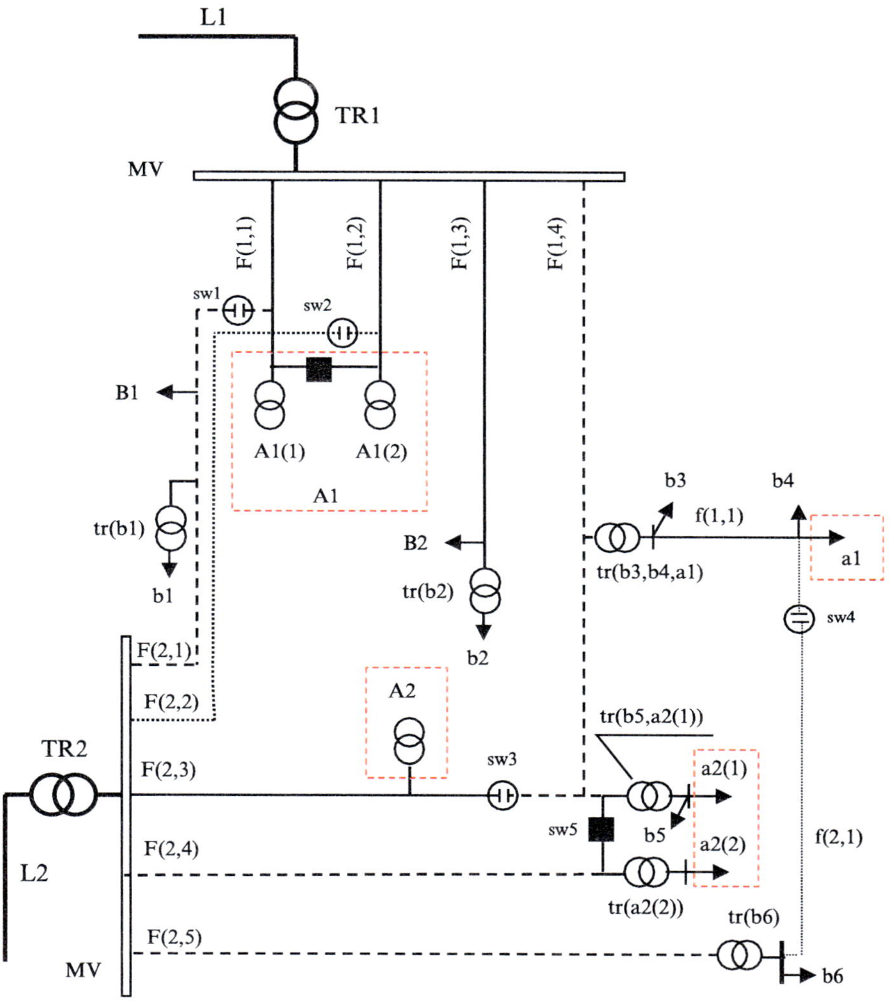

Fig. 2 Configuration of sample distribution grid with 3rd reliability category customers (*B* and *b*) and 2nd reliability category customers (*A* and *a*)

Operator's grid

LV, *MV*	Low and medium voltage, respectively
L	High voltage (e.g. 110-kV) transmission line
TR	High voltage (e.g. 110-kV) distribution transformer
tr	Local transformer (e.g. 10/0.4-kV)
tr(*b*3, *b*4, *a*1)	Local transformer feeding customers *b*3, *b*4, *a*1 in normal grid configuration
F	Medium voltage (e.g. 10-kV) distribution line (feeder)

(continued)

$F(1,4)$	F-type line No4 connected directly to $TR1$
f	Low voltage (0.4-kV) distribution line (feeder)
$f(2,1)$	f-type line No1 connected to $TR2$ in normal grid configuration
sw	Switching device
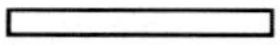	Medium voltage bus of TR (independent supply point)
————————	F-type line providing only major supply for customer of A, a, B and b classes
....	F-type line providing only reserving supply for customers of A and a classes
— — — ·	F-type line providing joint supply
———————	f-type line providing only major supply for customers of a and b classes
....	f-type line providing only reserving supply for customers of a class
	Switching device, switched off in normal grid configuration
	Medium voltage circuit breaker, switched off in normal grid configuration
L_F	Total length of F-type lines
L_F^m, L_F^j, L_F^r	Total lengths of F-type lines applied for major, joint and reserving supply, respectively
S_{tr}^m, S_{tr}^j, s_{tr}^r	Total installed capacities of local transformers applied for major, joint and reserving supply, respectively

Capacities (loads) and energy

P_A, P_a, P_B, P_b	Total declared capacities (loads) of A-class, a-class, B-class and b-class customers, respectively
W_A, W_a, W_B, W_b	Annual consumptions of A-class, a-class, B-class and b-class customers, respectively
P_{Aa}	Total declared capacity of $A\&a$ customers
P_{Bb}	Total declared capacity of $B\&b$ customers
W_{Aa}	Total annual consumption of $A\&a$ customers
W_{Bb}	Total annual consumption of $B\&b$ customers
P_m, P_r	Total declared capacities (loads) of customers provided with major supply and reserving supply, respectively
k_{Aa}	Weight of A-class and a-class in the total capacity of all the customers
k_a	Weight of a-class in the total capacity of a-class and b-class customers

Herein $A\&B$ classes are regarded to form one peer customer group, and $a\&b$ classes – another peer group.

Cost

C	Grid-related cost experienced by the grid operator
C_{TR}	Distribution transformers related cost
C_F	F-type lines related cost
C_{tr}	Local transformers related cost
C_f	f-type line related cost
C_F^m, C_F^j, C_F^r	Cost components for using F-type lines for major, joint and reserving supply, respectively
C_{tr}^m, C_{tr}^j, C_{tr}^r	Cost components for using local transformers for major, joint and reserving supply, respectively
C_f^m, C_f^j, C_f^r	Cost components for using f-type lines for major, joint and reserving supply, respectively

Tariffs

$T(TR)$	Single capacity tariff for the use of distribution transformers
$T'(TR)$	Basic capacity tariff for the use of distribution transformers in the double-tariff differentiation scheme
$T'_{Bb}(TR)$, $T'_{Aa}(TR)$	Differentiated capacity tariffs for the use of distribution transformers in the double-tariff differentiation scheme, assigned to *B&b* customers and *A&a* customers, respectively
$T''(TR)$	Basic capacity tariff for the use of distribution transformers in the capacity-proportional differentiation scheme
$T''_{Bb}(TR)$, $T''_{Aa}(TR)$	Differentiated capacity tariffs for the use of distribution transformers in the capacity-proportional differentiation scheme, assigned to *B&b* and *A&a* customers, respectively
$t(TR)$	Single energy tariff for the use of distribution transformers
$t'(TR)$	Basic energy tariff for the use of distribution transformers in the double-tariff differentiation scheme
$t'_{Bb}(TR)$, $t'_{Aa}(TR)$	Differentiated energy tariffs for the use of distribution transformers in the double-tariff differentiation scheme, assigned to *B&b* and *A&a* customers, respectively
$t''(TR)$	Basic energy tariff for the use of distribution transformers in the capacity-proportional differentiation scheme
$t''_{Bb}(TR)$, $t''_{Aa}(TR)$	Differentiated energy tariffs for the use of distribution transformers in the capacity-proportional differentiation scheme, assigned to *B&b* and *A&a* customers, respectively
$T(F)$	Single capacity tariff for the use of *F*-type lines
$T^m(F)$, $T^j(F)$, $T^r(F)$	Capacity tariff components for using *F*-type lines for major, joint and reserving supply, respectively
$T_{Bb}(F)$, $T_{Aa}(F)$	Differentiated capacity tariffs for the use of *F*-type lines, assigned to *B&b* customers and *A&a* customers, respectively
$t(F)$	Single energy tariff for the use of *F*-type lines
$t^m(F)$, $t^j(F)$, $t^r(F)$	Energy tariff components for using *F*-type lines for major, joint and reserving supply, respectively
$t_{Bb}(F)$, $t_{Aa}(F)$	Differentiated energy tariffs for the use of *F*-type lines, assigned to *B&b* customers and *A&a* customers, respectively
$T(tr)$	Single capacity tariff for the use of local transformers
$t(tr)$	Single energy tariff for the use of local transformers
$T^m(tr)$, $T^j(tr)$, $T^r(tr)$	Capacity tariff components for using the local transformers for major, joint and reserving supply, respectively
$T_b(tr)$, $T_a(tr)$	Differentiated capacity tariffs for the use of local transformers, assigned to *b*-class and *a*-class customers, respectively
$t^m(tr)$, $t^j(tr)$, $t^r(tr)$	Energy tariff components for using the local transformers for major, joint and reserving supply, respectively
$t_b(tr)$, $t_a(tr)$	Differentiated energy tariffs for the use of local transformers, assigned to *b*-class and *a*-class customers, respectively
$T(f)$	Single capacity tariff for the use of *f*-type lines
$t(f)$	Single energy tariff for the use of *f*-type lines
$T^m(f)$, $T^j(f)$, $T^r(f)$	Capacity tariff components for using *f*-type lines for major, joint and reserving supply, respectively
$T_b(f)$, $T_a(f)$	Differentiated capacity tariffs for the use of *f*-type lines, assigned to *b*-class and *a*-class customers, respectively

(continued)

$t^m(f)$, $t^j(f)$, $t^r(f)$	Energy tariff components for using *f*-type lines for major, joint and reserving supply, respectively
$t_b(f)$, $t_a(f)$	Differentiated energy tariffs for the use of *f*-type lines, assigned to *b*-class and *a*-class customers, respectively
ET_A, ET_a, ET_B, ET_b,	End capacity tariffs for *A*-class, *a*-class, *B*-class and *b*-class customers, respectively
et_A, et_a, et_B, et_b	End energy tariffs for *A*-class, *a*-class, *B*-class and *b*-class customers, respectively

5.3 Grid Reserving Mechanism for Customers with 2nd Reliability Category

The customers of *A&a* classes are provided with parallel grid path that ensures the reserving supply, i.e. supply from 2nd independent supply point. This parallel path might be used exclusively by *A&a* customers or jointly with *B&b* customers. Theoretically, the parallel grid infrastructure will incur cost for grid operator. This cost should be either entirely reimbursed or shared by the *A&a* customers.

The total grid cost *C* will include, first and foremost, operation and management cost and amortization cost:

$$C = C(OM) + C(AM). \tag{2}$$

The reserving mechanism, as it should theoretically function in the sample grid configuration (see Fig. 2), might be set out as follows.

Customer *A*1 receives major supply from *TR*1 as from 1st independent point, through lines *F*(1,1), *F*(1,2). If this path fails, the 10-kV switches *sw*1 and *sw*2 are switched in and the reserving supply is provided by *TR*2 as 2nd independent point, through lines *F*(2,1) and *F*(2,2).

Customer *A*2 receives major supply from *TR*2 as 1st independent point, through *F*(2,3). If this path fails, the 10-kV switch *sw3* is switched in and the reserving supply starts from *TR*1 as 2nd independent point, through line *F*(1,4).

Customer *a*1 receives major supply from *TR*1 as from 1st independent point, through line *F*(1,4), local transformer *tr*(*b*3,*b4*,*a*1) and 0.4-kV line *f*(1,1). If this path fails, the 0.4-kV switch *sw*4 is switched in and the reserve supply starts from *TR*2 as 2nd independent point, through line *F*(2,5), local transformer *tr*(*b*6) and 0.4-kV line *f*(2,1).

In case of major supply, the load *a*2(1) of customer *a*2 is supplied from *TR*1 as from 1st independent point, through the line *F*(1,4) and local transformer *tr*(*b*5, *a*2(1)). If this path fails, the reserving supply starts by switching the 0.4-kV links between loads *a*2(1) and *a*2(2) in the internal grid of customer *a*2, and, if appropriate, by switching in the 10-kV circuit breaker *sw*5. In the latter case, the power flows from *TR*2 (2nd independent point), through the line *F*(2,4) and transformer *tr*(*b*5,*a*2(1)).

The load $a2(2)$ of this customer is supplied from $TR2$ in major supply case (1st independent point), through the line $F(2,4)$ and local transformer $tr(a2(2))$. If this path fails, the reserving supply also starts by switching 0.4-kV links between the loads $a2(1)$ and $a2(2)$ in the internal grid of customer $a2$, and, if appropriate, by switching the same device $sw5$. In the latter case, the power flows from $TR1$ (2nd independent point), through the line $F(1,4)$ and transformer $tr(a2(2))$.

The grid operator might issue the higher reliability categories (levels) for customers that need (or wish) to be provided with. Such categories would be some standardized sets of continuity-of-supply indices.

6 Allocation of Distribution Transformer Related Costs

To allocate the cost generated by distribution transformers between the classes of customers, the application of transformers for the functions "only major supply", "only reserving supply" and "joint supply" should be identified with respect to the individual customers. The results of such identification for the sample grid is presented in Table 2.

As seen from Table 2, each customer of $A\&a$ classes needs twice as much of transformer capacity as that of $B\&b$ classes (compare P_m and P_r). Both transformers $TR1$ and $TR2$ are applied for joint supply and it could be reasonably presumed that in real grids the majority of TR transformers provide the joint supply.

Single tariff case

If there is no tariff differentiation for the use of distribution transformers, the customers of $B\&b$ classes will cross-subsidize the customers of $A\&a$ classes through a single tariff. Such a tariff might be opted either as energy or capacity tariff.

Single energy tariff would be:

$$t(TR) = \frac{C_{TR}}{W_{Bb} + W_{Aa}}, \text{currency/kWh}, \tag{3}$$

Table 2 Application of distribution transformers in the sample grid (Fig. 2)

TR	Major supply loads	Reserving supply loads	Transformer application form
$TR1$	$A1, B2, b2, b3, b4, a1, b5, a2(1)$	$A2, a2(2)$	$TR1$ serves for joint supply
$TR2$	$b1, B1, A2, a2(2), b6$	$A1, a1, a2(1)$	$TR2$ serves for joint supply
$TR1 + TR2$	$P_m = P_A + P_a + P_B + P_b$ $W_A + W_a + W_B + W_b$	$P_r = P_A + P_a$ Negligible for $A\&a$	Negligible – because the reserving supply is needed in contingent cases

while single capacity tariff:

$$T(TR) = \frac{C_{TR}}{P_{Bb} + P_{Aa}}, \text{currency/kW/year.} \quad (4)$$

Differentiation to double tariffs

Since total demand of transformer capacities consist of terms P_m and P_r (see Table 2), the basic capacity tariff $T'(TR)$ would be derived as:

$$T'(TR) = \frac{C_{TR}}{P_m + P_r} = \frac{C_{TR}}{P_{Bb} + 2P_{Aa}}. \quad (5)$$

If (5) and (4) are compared, it is clear that $T'(TR) < T(TR)$.
Now the C_{TR} can be expressed from (5) as:

$$C_{TR} = T'(TR) \cdot P_{Bb} + 2T'(TR) \cdot P_{Aa}. \quad (6)$$

Consequently, the basic tariff $T'(TR)$ should be imposed on customers of $B\&b$ classes and when doubled – on customers of $A\&a$ classes:

$$T'_{Bb}(TR) = T'(TR), \quad (7)$$

$$T'_{Aa}(TR) = 2T'(TR). \quad (8)$$

If (6)–(8) are expressed in equivalent energy-related terms:

$$C_{TR} = t'(TR) \cdot W_{Bb} + 2t'(TR) \cdot W_{Aa}, \quad (9)$$

$$t'_{Bb}(TR) = t'(TR), \quad (10)$$

$$t'_{Aa}(TR) = 2t'(TR). \quad (11)$$

Differentiation to capacity-proportional tariffs

Since the capacity P_{Aa} does not entail the construction of a new transformer for reserving supply, it might be assumed that the contribution of $A\&a$ customers to the reimbursement of C_{TR} could be not twice that of $B\&b$ customers. It could be in proportion to the weight k_{Aa} of $A\&a$ customers in total capacity P_{Bb} + P_{Aa}:

$$k_{Aa} = \frac{P_{Aa}}{P_{Bb} + P_{Aa}}. \quad (12)$$

When

$$C_{TR} = T''(TR) \cdot P_{Bb} + (1+k_{Aa})T''(TR) \cdot P_{Aa}, \quad (13)$$

and the differentiated capacity tariffs will be:

$$T''_{Bb}(TR) = T''(TR). \tag{14}$$

$$T''_{Aa}(TR) = (1 + k_{Aa}) \cdot T''(TR). \tag{15}$$

If (13)–(15) are transformed to energy-related terms:

$$C_{TR} = t''(TR) \cdot W_{Bb} + (1+k_{Aa})\ t''(TR) \cdot W_{Aa}, \tag{16}$$

$$t''_{Bb}(TR) = t''(TR), \tag{17}$$

$$t''_{Aa}(TR) = (1 + k_{Aa}) \cdot t''(TR). \tag{18}$$

7 Allocation of MV Lines Related Costs

As outlined in Sect. 5, any distributing transformer is applied jointly for major supply and reserving supply purposes. Referring to MV lines, their application is more varied and covers all three forms: "only major supply", "joint supply" and "only reserving supply". The distribution of customers by such forms is determined in Table 3.

Single tariff case

Without a tariff differentiation for the use of MV lines, the *B&b* customers will cross-subsidize the *A&a* customers through a single tariff. Alike the case of distribution transformers, such a tariff might be opted either as energy or capacity tariff.

The single capacity tariff will be:

$$T(F) = \frac{C_F}{P_{Bb} + P_{Aa}} \tag{19}$$

Table 3 Application of MV lines in the sample grid (Fig. 2)

MV line	Major supply loads	Reserving supply loads	Line application form
$F(1,1)$	$A1(1)$	–	Only major supply ("m")
$F(1,2)$	$A1(2)$	–	Only major supply ("m")
$F(1,3)$	$B2$, $b1$	–	Only major supply ("m")
$F(1,4)$	$b3$, $b4$, $a1$, $b5$, $a2(1)$	$A2$, $a2(2)$	Joint supply ("j")
$F(2,1)$	$b1$, $B1$	$A1(1)$	Joint supply ("j")
$F(2,2)$	–	$A1(2)$	Only reserving supply ("r")
$F(2,3)$	$A2$	–	Only major supply ("m")
$F(2,4)$	$a2(2)$	$a2(1)$	Joint supply ("j")
$F(2,5)$	$b6$	$a1$	Joint supply ("j")

and the single energy tariff:

$$t(F) = \frac{C_F}{W_{Bb} + W_{Aa}} \tag{20}$$

Differentiation of tariffs according to MV lines application

Hence, the total length of MV lines L_F in any real grid can be split to three components by the application forms:

$$L_F = L_F^m + L_F^j + L_F^r, \tag{21}$$

where L_F^m, L_F^j, L_F^r – lengths of lines applied for "only major supply", "joint supply" and "only reserving supply", respectively.

Accordingly, the total cost C_F can be split to components C_F^m, C_F^j, C_F^r. Each of them could be determined by the weight of its lines in the total length L_F, e.g.:

$$C_F^m = C_F \frac{L_F^m}{L_F^m + L_F^j + L_F^r}. \tag{22}$$

To allocate these cost components due and fairly between customer classes, the following assumptions should be made:

1. C_F^m is reimbursed by all customers through a single tariff,
2. C_F^j is shared by *B&b* and *A&a* customers through differentiated tariffs,
3. C_F^r is imposed exclusively on *A&a* customers and reimbursed through a special tariff.

The cost components might be expressed through the respective capacity tariff components as follows:

$$C_F^m = T^m(F) \cdot (P_{Aa} + P_{Bb}), \tag{23}$$

$$C_F^j = T^j(F) \cdot P_{Bb} + (1 + k_{Aa})\, T^j(F) \cdot P_{Aa}, \tag{24}$$

or, in double tariff scheme,

$$C_F^j = T^j(F) \cdot P_{Bb} + 2T^j(F) \cdot P_{Aa}, \tag{25}$$

$$C_F^r = T^r(F) \cdot P_{Aa}. \tag{26}$$

The total capacity tariffs for *B&b* customers and *A&a* customers will be differentiated, respectively:

$$T_{Bb}(F) = T^m(F) + T^j(F), \tag{27}$$

$$T_{Aa}(F) = T^m(F) + (1 + k_{Aa}) \cdot T^j(F) + T^r(F), \tag{28}$$

or, in double tariff scheme,

$$T_{Aa}(F) = T^m(F) + 2T^j(F) + T^r(F). \tag{29}$$

Using the same three assumptions, the equivalent energy tariffs for customers will be derived from the following expressions (30)–(33):

$$C_F^m = t^m(F) \cdot (W_{Aa} + W_{Bb}), \tag{30}$$

$$C_F^j = t^j(F) \cdot W_{Bb} + (1 + k_{Aa}) \cdot t^j(F) \cdot W_{Aa}, \tag{31}$$

or, in double tariff scheme,

$$C_F^j = t^j(F) \cdot W_{Bb} + 2t^j(F) \cdot W_{Aa}, \tag{32}$$

$$C_F^r = t^r(F) \cdot W_{Aa}. \tag{33}$$

The total energy tariffs for the use of MV lines will be differentiated as follows:

$$t_{Bb}(F) = t^m(F) + t^j(F), \tag{34}$$

$$t_{Aa}(F) = t^m(F) + (1 + k_{Aa}) \cdot t^j(F) + t^r(F), \tag{35}$$

or, in double tariff scheme,

$$t_{Aa}(F) = t^m(F) + 2t^j(F) + t^r(F). \tag{36}$$

8 Allocation of Local Transformer Related Costs

Since local transformers perform on LV level, their cost C_{tr} should not be imposed on *A&B* customers. The application of those transformers in categories of "only major supply", "joint supply" and "only reserving supply" as an instance case is determined in Table 4.

Single tariff case

If no tariff differentiation for the use of local transformers is undertaken, *b*-class customers will cross-subsidize *a*-class customers through a single tariff. Then a single capacity tariff will be:

$$T(tr) = \frac{C_{tr}}{P_b + P_a} \tag{37}$$

Table 4 Application of local transformers in the sample grid (Fig. 2)

Local transformer	Major supply loads	Reserving supply loads	Transformer application form
tr(*b*1)	*b*1	–	"m"
tr(*b*2)	*b*2	–	"m"
tr(*b*3, *b*4, *a*1)	*b*3, *b*4, *a*1	–	"m"
tr(*b*5, *a*2(1))	*b*5, *a*2(1)	*a*2(2)	"j"
tr(*a*2(2))	*a*2(2)	*a*2(1)	"j"
tr(*b*6)	*b*6	*a*1	"j"

Note: no one transformer for "only reserve supply" in the sample grid

and its equivalent in energy terms, i.e. single energy tariff:

$$t(tr) = \frac{C_{tr}}{W_b + W_a}. \tag{38}$$

Differentiation of tariffs according to local transformers application

The differentiation will be based on the same rationale as for MV lines. The cost C_{tr} is split to the terms C_{tr}^m, C_{tr}^j, C_{tr}^r (the latter does not exist in the sample grid). Each of them could be determined by the weight of its transformer capacity in the total capacity of local transformers in real grid, e.g.:

$$C_{tr}^m = C_{tr} \frac{S_{tr}^m}{S_{tr}^m + S_{tr}^j + S_{tr}^r}, \tag{39}$$

where $S_{tr}^m, S_{tr}^j, S_{tr}^r$ are installed capacities of local transformers in the categories "*m*", "*j*" and "*r*", respectively.

The cost components will be expressed through the capacity tariff components similarly as in (23)–(26):

$$C_{tr}^m = T^m(tr) \cdot (P_b + P_a), \tag{40}$$

$$C_{tr}^j = T^j(tr) \cdot P_b + (1+k_a)\, T^j(tr) \cdot P_a, \tag{41}$$

where k_a – the weight of *a*-class customers in total capacity of LV customers:

$$k_a = \frac{P_a}{P_b + P_a}. \tag{42}$$

C_{tr}^j also might be based on tariff double-sizing:

$$C_{tr}^j = T^j(tr) \cdot P_b + 2T^j(tr) \cdot P_a, \tag{43}$$

$$C_{tr}^j = T^r(tr) \cdot P_a. \tag{44}$$

The total capacity tariffs for $a\&b$ customers, respectively:

$$T_b(tr) = T^m(tr) + T^j(tr), \tag{45}$$

$$T_a(tr) = T^m(tr) + (1 + k_a) \cdot T^j(tr) + T^r(tr), \tag{46}$$

or, in double tariff scheme,

$$T_a(tr) = T^m(tr) + 2T^j(tr) + T^r(tr). \tag{47}$$

The equivalent cost expressions through energy tariff components would be as follow:

$$C^m_{tr} = t^m(tr) \cdot (W_b + W_a), \tag{48}$$

$$C^j_{tr} = t^j(tr) \cdot W_b + (1 + k_a) \cdot t^j(tr) \cdot W_a, \tag{49}$$

or, in double tariff scheme,

$$C^j_{tr} = t^j(tr) \cdot W_b + \cdot 2t^j(tr) \cdot W_a, \tag{50}$$

$$C^r_{tr} = t^j(tr) \cdot W_a. \tag{51}$$

The total energy tariffs for the use of local transformers will be differentiated as follows:

$$t_b(tr) = t^m(tr) + t^j(tr), \tag{52}$$

$$t_a(tr) = t^m(tr) + (1 + k_a) \cdot t^j(tr) + t^r(tr), \tag{53}$$

or, in double tariff scheme,

$$t_a(tr) = t^m(tr) + 2t^j(tr) + t^r(tr). \tag{54}$$

9 Allocation of LV Lines Related Costs

The application of LV lines (f-type lines) in the categories "m", "j" and "r" in the sample grid is determined in Table 5.

Table 5 Application of LV lines in the sample grid (Fig. 2)

LV line	Major supply loads	Reserving supply loads	Line application form
$f(1,1)$	$b3$, $b4$, $a1$	–	m
$f(1,2)$	–	$a1$	r

Note: no one "joint supply" line in the sample grid

Single tariff case

If there is no tariff differentiation for the use of LV lines, b-class customers will cross-subsidize a-class customers through a single tariff. The single capacity and energy tariffs, respectively, will be:

$$T(f) = \frac{C_f}{P_b + P_a}, \tag{55}$$

$$t(f) = \frac{C_f}{W_b + W_a}. \tag{56}$$

Differentiation of tariffs according to application of LV lines

Repeating the analogical procedure of cost decomposition as described by preceding (22), the expressions of C_f^m, C_f^j, C_f^r are formulated in the same manner as in (40)–(44) for C_{tr}^m, C_{tr}^j, C_{tr}^r and further the capacity tariff components $T^m(f)$, $T^j(f)$ and $T^r(f)$ determined thereof. Consequently, the total capacity tariffs for $a\&b$ customers will be derived as follows:

$$T_b(f) = T^m(f) + T^j(f), \tag{57}$$

$$T_a(f) = T^m(f) + (1 + k_a) \cdot T^j(f) + T^r(f), \tag{58}$$

or, in double tariff scheme,

$$T_a(f) = T^m(f) + 2T^j(f) + T^r(f). \tag{59}$$

As for energy tariffs, their components $t^m(f)$, $t^j(f)$ and $t^r(f)$ should be found from expressions analogical to (48)–(51), and the total energy tariffs for $a\&b$ customers will be:

$$t_b(f) = t^m(f) + t^j(f), \tag{60}$$

$$t_a(f) = t^m(f) + (1 + k_a) \cdot t^j(f) + t^r(f), \tag{61}$$

or, in double tariff scheme,

$$t_a(f) = t^m(f) + 2t^j(f) + t^r(f). \tag{62}$$

10 Differentiated End Tariffs

After the total capacity and energy tariffs having been defined as in Sects. 6–9, the end tariffs for grid-provided distribution service can be derived for each customer class. End tariff is an integrated parameter which covers the respective total tariffs:

it is an additive function where all total tariffs (covering all types of grid equipment) are summed up as independent terms. Since the total tariffs have been differentiated by reliability level, the end tariffs get differentiated, too.

End capacity tariffs for 2nd reliability category customers

The end capacity tariff ET_A for a A-class customer will include the following total tariffs:

- Equations 8 or 15 – for the use of distribution transformers,
- Equations 28 or 29 – for the use of MV lines,

and, consequently, in the instance of capacity-proportional differentiation scheme, (15) and (28) should be summed up:

$$ET_A = T''_{Aa}(TR) + T_{Aa}(F) = (1 + k_{Aa}) \cdot T''(TR) + \\ + T^m(F) + (1 + k_{Aa}) \cdot T^j(F) + T^r(F). \tag{63}$$

The end capacity tariff ET_a for a a-class customer will include the same total tariffs as imposed on a A-class customer plus LV grid related total tariffs, i.e.:

- Equations 46 or 47 – for the use of local transformers,
- Equations 58 or 59 – for the use of LV lines,

and, consequently, in the instance of capacity-proportional differentiation scheme, (46) and (58) should be added to (63):

$$ET_a = ET_A + T_a(tr) + T_a(f) = ET_A + T^m(tr) + (1 + k_a) \cdot T^j(tr) + \\ + T^r(tr) + T^m(f) + (1 + k_{Aa}) \cdot T^j(f) + T^r(f). \tag{64}$$

End capacity tariffs for 3rd reliability category customers

The end capacity tariff ET_B for a B-class customer will include the following total tariffs:

- Equations 7 or 14 – for the use of distribution transformers,
- Equation 27 – for the use of MV lines,

and, similarly, in the instance of capacity-proportional differentiation scheme, (14) and (27) should be summed up:

$$ET_B = T''_{Bb}(TR) + T_{Bb}(F) = T''(TR) + T^m(F) + T^j(F). \tag{65}$$

The end capacity tariff ET_b for a b-class customer will include the same total tariffs as imposed on a B-class customer plus LV grid related total tariffs, i.e.:

- Equation 45 – for the use of local transformers,
- Equation 57 – for the use of LV lines,

and, consequently, in the instance of capacity-proportional differentiation scheme (45) and (57) should be added to (65):

$$ET_b = ET_B + T_b(tr) + T_b(f) = ET_B + T^m(tr) + T^j(tr) + T^m(f) + T^j(f). \quad (66)$$

End energy tariffs for 2nd reliability category customers

Being equivalent to (63), end energy tariff et_A for a A-class customer will consist of total tariffs defined by (11) or (18) and by (35) or (36). In the instance of capacity-proportional differentiation scheme, (18) and (35) should be taken:

$$et_A = t''_{Aa}(TR) + t_{Aa}(F) = (1 + k_{Aa}) \cdot t''(TR) + t^m(F) + (1 + k_{Aa}) \cdot t^j(F) + t^r(F). \quad (67)$$

Being equivalent to (64), end energy tariff et_a for a a-class customer will consist of the end tariff et_A defined by (67) and total tariffs defined by (53) or (54) and (61) or (62). In the instance of capacity-proportional differentiation scheme, namely (53) and (61) are taken:

$$\begin{aligned} et_a = et_A + t_a(tr) + t_a(f) = et_A + t^m(tr) + (1 + k_{Aa}) \cdot t^j(tr) + \\ + t^r(tr) + t^m(f) + (1 + k_a) \cdot t^j(f) + t^r(f). \end{aligned} \quad (68)$$

End energy tariffs for 3rd reliability category customers

Being equivalent to (65), end energy tariff et_B for a B-class customer will consist of the total tariffs defined by (10) or (17) and by (34). In the instance of capacity-proportional differentiation scheme, (17) and (34) should be taken:

$$et_B = t''_{Bb}(TR) + t_{Bb}(F) = t''(TR) + t^m(F) + t^j(F). \quad (69)$$

Being equivalent to (66), end energy tariff et_b for a b-class customer will consist of the end tariff et_B defined by (69) and total tariffs defined by (52) and (60). It is expressed as:

$$et_b = et_B + t_b(tr) + t_b(f) = et_B + t^m(tr) + t^j(tr) + t^m(f) + t^j(f). \quad (70)$$

The major condition for the application of model (2) to (70) by the operators or regulators would be to assess the adequate proportions of grid equipment participating in major, joint and reserving supply in the real grid.

11 Numerical Setup

The numerical setup is presented below to demonstrate the applicability of the model (2) to (70). This setup is a hypothetical case study for the above mentioned sample grid (Fig. 2). The initial data presented in Tables 6 and 7 seem to be fairly realistic for modeling and interpretation of results. For the purpose of simplicity, the energy consumption of each customer was set using the same annual duration time

Table 6 Loads and energy consumption in the sample grid (Fig. 2)

	Customer code	Declared load (capacity), kW	Consumption (energy), MWh/year
1.	$A1(1)$	300	1,350
2.	$A1(2)$	325	1,462.5
3.	$A2$	325	1,462.5
4.	A class	$P_A = 950$	$W_A = 4{,}275$
5.	$a1$	220	990
6.	$a2(1)$	300	1,350
7.	$a2(2)$	300	1,350
8.	a class	$P_a = 820$	$W_a = 3{,}690$
9.	$A\&a$ classes	$P_{Aa} = 1{,}770$	$W_{Aa} = 7{,}965$
10.	$B1$	450	2,025
11.	$B2$	650	2,925
12.	B class	1,100	4,950
13.	$b1$	60	270
14.	$b2$	160	720
15.	$b3$	50	225
16.	$b4$	120	540
17.	$b5$	100	450
18.	$b6$	180	810
19.	b class	$P_b = 670$	$W_b = 3{,}015$
20.	$B\&b$ classes	$P_{Bb} = 1{,}770$	$W_{Bb} = 7{,}965$
21.	All classes	3,540	15,930
22.	Weight k_{Aa} (pursuant to (12))	50.0%	50.0%
23.	Weight k_a (pursuant to (42))	55.0%	55.0%

of declared load (capacity) 4,500 h/year. Therefore the weight k_{Aa} (as well as k_a) has the same value both as load (capacity) ratio and as consumption (energy) ratio (Table 6).

The annual cost as supposed for individual item of grid equipment is aligned in Table 7. It is expressed in national currency units (NCU). The total costs for major equipment types are C_{TR}, C_F, C_{tr}, C_f.

Firstly, the single tariffs have been calculated (Table 8). They serve as reference values to evaluate the extent of differentiation.

Further the total costs were split to components which correspond to major, joint and reserving supply. For instance, the components C_F^m, C_F^j, C_F^r were calculated in accordance with (22) and C_{tr}^m, C_{tr}^j, C_{tr}^r – in accordance with (39). Then the capacity tariff components were derived from the respective components of total costs, e.g.:

- $T^m(F)$, $T^j(F)$, $T^r(F)$ – using expressions (23) to (26),
- $T^m(f)$, $T^j(f)$, $T^r(f)$ – using expressions (40), (41), (43), and (44).

Having the capacity tariff components, the differentiated capacity tariffs were calculated, both for each type of grid equipment and each customer class (see Table 9).

Table 7 Characteristics of equipment in the sample grid (Fig. 2)

	Grid equipment	Parameter	Annual cost, NCU
1.	Distr. transformers	Load: 3,540 kW	$C_{TR} = 49{,}500$
2.	TR1	2,225 kW	24,750
3.	TR2	1,315 kW	24,750
4.	MV lines	Length: 73.0 km	$C_F = 91{,}250$
5.	F(1,1)	5.0 km	6,250
6.	F(1,2)	5.0 km	6,250
7.	F(1,3)	10.5 km	13,125
8.	F(1,4)	12.0 km	15,000
9.	F(2,1)	6.0 km	7,500
10.	F(2,2)	7.5 km	9,375
11.	F(2,3)	8.0 km	10,000
12.	F(2,4)	9.0 km	11,250
13.	F(2,5)	10.0 km	12,500
14.	Local transformers	Capacity: 1,650 kVA	$C_{tr} = 24{,}350$
15.	$tr(b1)$	100 kVA	3,500
16.	$tr(b2)$	180 kVA	3,200
17.	$tr(b3, b4, a1)$	400 kVA	4,500
18.	$tr(b5, a2(1))$	400 kVA	4,500
19.	$tr(a2(2))$	320 kVA	4,850
20.	$tr(b6)$	250 kVA	3,800
21.	LV lines	Length: 8.3 km	$C_f = 9{,}500$
22.	f(1,1)	4.5 km	4,500
23.	f(2,1)	3.8 km	5,000

Table 8 Single (non-differentiated) tariffs for the sample grid (Fig. 2)

	Grid equipment	Capacity tariff, NCU/kW/year	Formula	Energy tariff, NCU/kWh	Formula
1.	Distr. transformers	$T(TR) = 14.0$	(4)	$t(TR) = 0.0031$	(3)
2.	MV lines	$T(F) = 25.8$	(19)	$t(F) = 0.0057$	(20)
3.	End tariffs for MV customers	$ET_{AB} = 39.9$	(4) + (19)	$et_{AB} = 0.0088$	(3) + (20)
4.	Local transformers	$T(tr) = 16.3$	(37)	$t(TR) = 0.0036$	(38)
5.	LV lines	$T(f) = 6.4$	(55)	$t(f) = 0.0014$	(56)
6.	End tariffs for LV customers	$ET_{ab} = 62.5$	(4) + (19) + (37) + (55)	$et_{ab}(tr) = 0.0138$	(3) + (20) + (38) + (56)

Similarly, the alternative energy tariff components were derived from the same components of total costs, e.g.:

- $t^m(F)$, $t^j(F)$, $t^r(F)$ – using expressions (30) to (33),
- $t^m(f)$, $t^j(f)$, $t^r(f)$ – using expressions analogical to (48) to (51).

The differentiated energy tariffs were calculated both for each type of grid equipment and each customer class (see Table 10).

Table 9 Differentiated capacity tariffs for the sample grid (Fig. 2)

Grid equipment	Capacity-proportional tariff, NCU/kW/year, and formula		Double-proportional tariffs, NCU/kW/year, and formula	
	3rd category	2nd category	3rd category	2nd category
1. Distribution transformers	$T''_{Bb}(TR) =$ 11.02 (14)	$T''_{Aa}(TR) =$ 16.8 (15)	$T'_{Bb}(TR) = 9.3$ (7)	$T'_{Aa}(TR) =$ 18.6 (8)
2. MV lines	$T_{Bb}(F) =$ 20.5 (24), (27)	$T_{Aa}(F) =$ 31.0 (24), (28)	$T_{Bb}(F) =$ 18.8 (25), (27)	$T_{Aa}(F) =$ 32.8 (25), (29)
3. End tariffs for MV customers	$ET_B =$ 31.7 (65)	$ET_A =$ 47.8 (63)	$ET_B = 28.1$	$ET_A = 51.4$
4. Local transformers	$T_b(tr) =$ 14.1 (41), (45)	$T_a(tr) =$ 18.2 (41), (46)	$T_b(tr) = 12.9$ (43), (45)	$T_a(tr) = 19.1$ (43), (47)
5. LV lines	$T_b(f) =$ 3.5 (57)	$T_a(f) =$ 8.8 (58)	$T_b(f) = 3.5$ (57)	$T_a(f) = 8.8$ (59)
6. End tariffs for LV customers	$ET_b =$ 49.3 (66)	$ET_a =$ 74.8 (64)	$ET_b = 44.5$	$ET_a = 79.3$

Table 10 Differentiated energy tariffs for the sample grid (Fig. 2)

Grid equipment	Capacity-proportional tariff, NCU/kW/year, and formula		Double-proportional tariffs, NCU/kW/year, and formula	
	3rd category	2nd category	2nd category	2nd category
1. Distribution transformers	$t''_{Bb}(TR) =$ 0.0025 (17)	$t''_{Aa}(TR) =$ 0.0037 (18)	$t''_{Bb}(TR) =$ 0.0021 (10)	$t''_{Aa}(TR) =$ 0.0042 (11)
2. MV lines	$t_{Bb}(F) =$ 0.0046 (31), (34)	$t_{Aa}(F) =$ 0.0069 (31), (35)	$t_{Bb}(F) =$ 0.0042 (32), (34)	$t_{Aa}(F) =$ 0.0074 (32), (36)
3. End tariffs for MV customers	$et_B =$ 0.0071 (69)	$et_A =$ 0.0106	$et_B =$ 0.0063	$et_A =$ 0.0116
4. Local transformers	$t_b(tr) =$ 0.0031 (49), (52)	$t_a(tr) =$ 0.0040 (49), (53)	$t_b(tr) =$ 0.0029 (50), (52)	$t_a(tr) =$ 0.0043 (50), (54)
5. LV lines	$t_b(f) =$ 0.0008 (60)	$t_a(f) =$ 0.0019 (61)	$t_b(f) =$ 0.0008 (60)	$t_a(f) =$ 0.0019 (62)
6. End tariffs for LV customers	$et_b = 0.0110$	$et_a =$ 0.0165 (68)	$et_b =$ 0.0100	$et_a =$ 0.0177

To summarize the exemplary results of tariff differentiation (Tables 9 and 10), it is evident that the model (2) to (70) performs well to differentiate the tariffs for customers with one independent supply point (3rd category) and two points (2nd category). For a sample grid where the distribution transformers share 29% (49,500 NCU) of total annual grid cost, MV lines – 52% (91,250 NCU), local transformers – 14% (24,350 NCU) and LV lines – 5% (9,500 NCU), and the proportions of MV and LV customers constitute a half of total declared loads, the differentiation of single capacity tariff disjoints it to the values less/larger by 20–21% for 3rd/2nd category customers, respectively. Such changes were found for a capacity-proportional differentiation scheme. In a double-tariff differentiation scheme, such values diverge by ±(27–29)% from a single tariff, respectively.

It should be noted that the question *to what extent such differentiation could be justified as a fair allocation* is a problem. First and foremost, such an extent depends on the structure and configuration of real distribution grids in national control areas and sub-areas. The national statistical data about (1) proportions of grid equipment-related cost in categories of distribution transformers, MV lines, local transformers and LV lines, and (2) usage of those categories of grid equipment by various customer classes would provide a basis for fair differentiation. Anyway, it seems preliminarily that differentiation extent close to double end tariff for higher reliability would be a hyper differentiation.

12 Conclusions

Differentiation in power distribution reliability pricing is not yet regarded as research issue aimed at development of effective differentiation methods. Nevertheless, it could be regarded as relevant opportunity for the customers to receive better quality of energy service and for the distribution grid operators to offer additional products to electricity marketplace.

At present European energy regulators do not apply the differentiation of reliability of power supply for peer customers. We suppose that soon grid operators will be challenged to provide the higher reliability categories (levels) for customers who need them. The higher reliability category for a customer could be ensured by connecting it to the second independent supply point (backup bus or feeder) provided by a grid operator. Differentiated reliability prices could be derived from network topology, i.e. from availability of independent supply points.

The rationale of the presented reliability pricing model is to derive the differentiated tariffs from the grid operation, management and amortization cost components for major, joint and reserving supply, according to the participation of a customer in the use of grid equipment. The numerical setup for this pricing model yielded quite reasonable and coherent results.

This model might be interesting as a new viewpoint and practically applicable procedure for energy regulators, grid operators and customers.

References

1. Glachant JM, Lévêque F (2009) Electricity reform in Europe: towards a single energy market. Edward Elgar Publishing, Cheltenham
2. Rau NS, Hegazy Y (1990) Reliability differentiated pricing of electricity service. NRRI series, 90–5. National Regulatory Research Institute, Columbus
3. Barot H, Bhattacharya K (2010) Load service probability differentiated nodal pricing in power systems. IET Gen Transm Distrib 4(3):333–348
4. Hegazy Y, Guldmann JM (1996) Reliability pricing of electric power service: a probabilistic production cost modeling approach. Energy 21(2):87–97

5. Siddiqi SN, Baughman ML (1993) Reliability differentiated real-time pricing of electricity. IEEE Trans Power Syst 8(2):548–554
6. Chung KH, Kim BH, Hur D, Park JK (2005) Transmission reliability cost allocation method based on market participants' reliability contribution factors. Electr Power Syst Res 73(1):31–36
7. Hur D, Parka JK, Leeb WG, Kim BH, Chunb YH (2004) An alternative method for the reliability differentiated transmission pricing. Electr Power Syst Res 68(1):11–17
8. Woo ChK, Horowitz I, Martin J (1998) Reliability differentiation of electricity transmission. J Regul Econ 13(3):277–292
9. Sugihara H, Kita H, Hasegawa J, Nishiya K (1999) An evaluation of reliability differentiated supply system considering local characteristics. Trans Inst Elect Eng Jpn B 119-B(3):354–361, in Japanese
10. Gonzalez-Cabrera N, Gutierrez-Alcaraz G (2010) Pricing reliability service based on end-users choice. In: 11th IEEE international conference on probabilistic methods applied to power systems (PMAPS), Singapore, 14–17 June 2010, pp 24–29
11. Wolf G (2007) Technologies advance transmission system. Transmission and distribution world, 4–10 Nov 2007 (in special supplement *Digital Transmission*)
12. Council of European Energy Regulators (2005) Third benchmarking report on quality of electricity supply. Ref: C05-QOS-01-03, Brussels, 2005
13. CIGRE (International Council on Large Electric Systems) (2004) Power quality indices and objectives. Report 2004, London
14. The European Commission, Directorate-General for Research (2006) European smart grids technology platform: vision and strategy for Europe's electricity networks of the future, Luxembourg 2006
15. Klementavicius A (2008) Customer reliability categories as the tool to improve electricity delivery in Lithuania. In: 8th international conference control of power systems'08, Strbske Pleso, 11–13 June 2008, pp 1–10

Compromise Scheduling of Bilateral Contracts in Electricity Market Environment

Sergey I. Palamarchuk

Abstract The paper formulates an electricity delivery scheduling problem in accordance with the bilateral contracts in the competitive wholesale market. Bilateral forward contracts are generally used in electricity markets to stabilize prices and hedge risks of electricity shortage. A contract party is able to draw electricity from the contract and resell it to the day-ahead wholesale and retail markets. Contract parties schedule electricity deliveries over contract period to get the highest profit. The problem solution results in determination of a contract price acceptable for both parties. Normally the contract parties are interested in different delivery strategies during the contract period. A compromise in determination of the delivery strategy implies obtaining the equality of relative concessions from supplier and buyer.

Conclusion of bilateral contracts bears certain risks due to price and demand uncertainty. Both contract parties forecast price levels in the spot market and electricity demand. Both parties estimate expected profits (or losses) from caring out the contract and participation in the spot market. It is important to adjust the contract in time.

The paper formulates optimization problems for contract scheduling. A numerical example demonstrates the efficiency of the algorithm.

The sequence of actions to be performed for contract correction is considered. The statements of optimization problems are given for decision making on contract correction and cancellation. The statements take into account financial compensation to another party in the case of prescheduled contract correction or cancellation. A numerical example illustrates applicability of the proposed procedures for decision making.

S.I. Palamarchuk (✉)
Energy Systems Institute, Irkutsk, Russia
e-mail: palam@isem.sei.irk.ru

A. Sorokin et al. (eds.), *Handbook of Networks in Power Systems I*,
Energy Systems, DOI 10.1007/978-3-642-23193-3_9,

Keywords Bilateral contracts • Compromise approach • Contract arrangement • Delivery scheduling • Electricity market

1 Introduction

Bilateral contracts (BCs) for physical delivery (forward contracts) are widely applied in the competitive electricity markets. A forward contract determines that some asset will be delivered at a given time in the future at an agreed price and in a defined location [1]. Bilateral contracts stabilize prices and reduce the possibility of market power abuse in the "day ahead" (spot) market [2]. The bilateral contracts secure delivery of a certain amount of electricity at agreed prices. They allow producers to choose solvent customers and to use generating facilities in an optimal way. In many countries bilateral trade covers the main portion of electricity supply.

The spot market and bilateral contract market exhibit synergistic interactions. BC volumes, duration and prices influence trade volumes and prices in the spot market. The spot market risks and price behavior have an impact on the contract arrangement strategy of partners. Electricity received under BC may be resold in the spot market and electricity bought in the spot market may be delivered to fulfill bilateral obligations [3, 4].

The BC parties are an electricity supplier and a buyer. The supplier is a generation company (GC) or an individual power plant. The buyer is an electricity supply company (ESC) that sells electricity to end consumers in the retail market. ESC considered in the paper is supposed to have no electricity sources of its own.

The contract period is divided into time intervals (months, weeks, days). The paper considers medium-term bilateral contracts that cover a time period from several weeks to one year.

Conclusion of bilateral contracts bears certain risks due to demand fluctuations and price uncertainty in the competitive spot market. The electricity price for contract period can deviate considerably from that expected at the time of contract arrangement. Prior to BC conclusion, potential BC parties forecast demand and price levels and schedule the electricity deliveries by time intervals. Both parties aim to gain a maximum profit. They elaborate their strategies of participation in the spot market. After BC has been concluded and come into force, its parties are liable to follow delivery schedule stipulated in it.

Monitoring of economic efficiency of existing BC is an important problem for supplier and buyer. It is often necessary to correct or cancel BC to hedge commercial losses. Contract parties keep track of the market price behavior during the contract period and update price and demand forecasts. Each party estimates a contract profit or financial loss expected in changing market conditions. Each contract party can make a decision on contract correction or cancellation, if the expected profit is not acceptable. The other party has the right to accept or reject a proposed correction. Certainly, decision making on contract correction has to take into account financial compensations to the other partner.

Delivery scheduling is a sophisticated procedure, since at the same prices in the spot market the BC parties pursue different interests. High prices in the spot market are favorable for the electricity supplier to increase sales in this market and reduce deliveries under BC. At the same time, it is profitable for buyer to decrease purchase from the spot market and increase contract deliveries. At low levels of spot prices the situation is opposite.

Two types of contracts introduced in [3, 5] are a possible way to coordinate interests of the BC parties. *Type I*: The buyer determines electricity amount to be delivered under BC at each time interval t. The supplier must guarantee electricity delivery in accordance with the buyer requirement. *Type II*: The supplier determines electricity amount to be sold according to the contract at each time interval t. The buyer accepts the delivered electricity according to the supplier decision.

According to this approach one party determines delivery amounts under BC, the other is required to implement the suggested delivery schedule. In this case the parties appear to be in different conditions. One party is granted with the privilege of delivery scheduling and has a greater opportunity to increase its profit. The other party plays a role of a subordinate partner. The technique of delivery scheduling for two types of contracts was worked out in [3, 5]. However, the schedule obtained by one party is not always acceptable for the other one.

Determination of the contract price is an important goal of BC scheduling. For two types of contracts both parties determine their own acceptable contract price. The price that is acceptable for both sides in BC to be concluded is found as a result of bilateral negotiations [3]. Such a procedure can bring to different benefits for the contracting parties.

Another way to balance the interests of the BC parties is a compromise approach to delivery scheduling, in which both BC parties are regarded as equal partners. This approach offers an opportunity to determine the schedule of electricity deliveries and the contract price that bring equal relative economic benefits for the parties. Deliveries can be scheduled jointly by the contract parties or by the third neutral party involved in coordinating interests of the BC parties.

There is a number of studies dealing with the problem of BC arrangement. The paper [3] discusses price setting in flexible contracts, in which deliveries can be changed at some time intervals, whereas the total contract volume is preserved. The works [6–8] present the analysis of BC with possible interruption of deliveries by either supplier or buyer. In [9] weather forecasts are suggested to be taken into consideration to reduce the risk of purchasing expensive electricity in the spot market. The papers [4, 10] provide a risk-constrained technique to solve the electricity procurement problem faced by a retailer and a large consumer. In [11] an approach to designing the bilateral bargaining between two utilities with incomplete information is introduced. The papers [12, 13] analyze the bilateral contract arrangement in the electricity markets under substantial network constraints.

The papers mentioned above consider different conceptual issues of BC arrangement. At the same time, the problem is so complex and many-sided that further studies are justified. The approach presented in this paper is different from most of others in treatment of contract scheduling and correction of its parameters as separate problems.

The studies [4, 10] take into account a pre-specified risk level on profit volatility during delivery scheduling. The authors consider the possibility for the end consumers to select a rival supplier if the electricity price is not low enough. However, it is difficult to reach validity in risk assessments at the stage of BC scheduling because the information about market conditions is highly uncertain. It is more reasonable to assess such risks during contract period.

The papers [6–8] consider the interruptible contracts that allow for delivery reduction in exchange for reduction in electricity price or for financial compensation at time of reduction. But, delivery curtailment can be caused by the external factors, which are unpredictable at the stage of BC preparation. In this case a supplier will suffer unjustified financial losses.

This paper proposes scheduling deliveries and negotiating a contract price before the BC conclusion on the basis of available forecasted information. Both contract parties assess the risks during contract period and make a decision on contract correction or cancellation using updated current information.

All studies on bilateral contract arrangement are oriented to certain market requirements and rules. For example, in [4, 10] both ESC and end consumers assumed to own some self-generating facilities. Large consumers seek to minimize the expected cost of delivery under BC rather than maximize their profit. These assumptions make distinctions in problem statements, applied methods and computation procedures. The approaches and models proposed in this paper can be useful in some competitive electricity markets.

2 Notation

2.1 A. Variables

x_k^t	Amount of electricity delivered under BC and sold to end consumers at interval t, MWh.
x_s^t	Amount of electricity bought by ESC in the spot market at interval t, MWh.
x_{ss}^t	Amount of electricity received by ESC under BC and sold in the spot market at interval t, MWh.
x^t	Amount of electricity received by ESC under BC at interval t, MWh.
x_c^t	Amount of electricity produced by GC and delivered under BC at interval t, MWh.
x_{gss}^t	Amount of electricity produced by GC and sold in the spot market at interval t, MWh.
x_{gs}^t	Amount of electricity bought by GC in the spot market and delivered under BC at interval t, MWh.
x_g^t	Output of GC at interval t, MWh.
x_d^t	Amount of electricity sold to end consumers at interval t, MWh.

(continued)

R_1, R_2	Revenues earned by contract parties from participation in the spot market, managing bilateral contract and supplying electricity to end customers, \$.
S_1, S_2	Profits earned by contract parties from participation in the spot market, managing bilateral contract and supplying electricity to end customers, \$.
ΔS_1^0, ΔS_2^0	Relative concessions of the contract parties applying the compromise delivery schedule.
J	Contract price equal to the total agreed cost of electricity delivered under BC, \$.

2.1.1 Random Variable

$\tilde{p}_s^t$	Forecasted electricity price in the spot market at interval t, \$/MWh.

2.2 B. Constants

p_d^t	Electricity price for end consumers in the retail market at interval t, \$/MWh.
V	Contract volume, i.e. total amount of electricity received by ESC under the BC during the contract period, MWh.
p^t	Agreed price of electricity delivered under BC at interval t, MWh.
$x_{\min}^t$, $x_{\max}^t$	Limits for contract deliveries at interval t, MWh.
$x_{g\min}^t$, $x_{g\max}^t$	Limits for electricity production of GC at interval t, MWh.
p_{int}^t	Fine on GC in the case of delivery reduction initiated by producer, \$/MWh.
p_{ovd}^t	Fine on ESC in the case of delivery overdrawing initiated by buyer, \$/MWh.
ΔF	Financial compensation to the other party in the case of prescheduled contract cancellation, \$.

2.3 C. Function

$C^t(x_g^t)$	Production cost function of GC at interval t, \$.

2.4 D. Number

N	Number of time intervals in the contract period.

3 Independent Bilateral Contract Scheduling

Bilateral contract scheduling means determination of the electricity amount to be delivered to the contract partner at each time interval t during the contract period. For the sake of simplicity we consider the case, where GC and ESC are going to arrange only one BC and its period is divided into N equal time intervals.

Both contract parties participate in the spot market. Amounts of electricity received from and delivered to the spot market at interval t are shown in Fig. 1.

The supplier has information about its electricity production cost. Traditionally this is known as the cost function $C^t(x_g^t)$ at interval t. The contract parties forecast behavior of prices $\tilde{p}_s$ in the spot market and take them as random variables when scheduling deliveries. Note that the prices forecasted by ESC can differ from those predicted by GC.

The parties can schedule deliveries independently of one another. For simplicity of presentation the power transmission costs are not taken into consideration. The electric network is assumed to be capable to transmit the required amounts of electricity. In real cases the BC parties should take into account the transmission cost according to existing transmission pricing. For the firm point-to-point transmission service the transmission cost may be included as expenses in the objective function. In the case of non-firm transmission access the BC parties should model distribution of electricity in the network at time intervals. Transmission cost consideration may influence the optimal delivery schedule under BC.

Suppose that ESC considers a desirable contract volume V and schedules $x^t,\ t = \overline{1,N}$ independently of GC, in order to gain the maximum of its expected profit S_1

$$\max_{x_s^t, x_{ss}^t, x_k^t} S_1 = E\left\{\sum_{t=1}^{N} \left[p_d^t\left(x_k^t + x_s^t\right) - \tilde{p}_s^t\, x_s^t + \tilde{p}_s^t x_{ss}^t\right] - J\right\}, \tag{1}$$

where E is the mathematical expectation symbol. Note that electricity delivery to end customers at interval t is $x_d^t = x_k^t + x_s^t$ and delivery from GC to ESC under contract x^t is $x_k^t + x_{ss}^t$.

Since the price J for the considered contract is constant, profit maximization is equivalent to maximization of the expected sales revenue R_1 in the spot and retail markets:

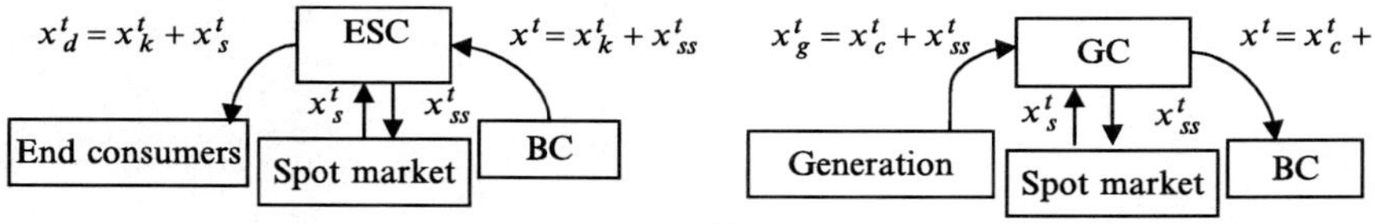

Fig. 1 Interaction of BC with the spot market for ESC and GC

$$\max_{x_k^t, x_s^t, x_{ss}^t} R_1 = E\left\{\sum_{t=1}^{N} \left[p_d^t \left(x_k^t + x_s^t\right) - \tilde{p}_s^t\, x_s^t + \tilde{p}_s^t\, x_{ss}^t)\right]\right\} \tag{2}$$

subject to the following constraints on:

(a) The total contract volume

$$\sum_{t=1}^{N} (x_k^t + x_{ss}^t) = V, \tag{3}$$

(b) The sales to end consumers

$$x_k^t + x_s^t = x_d^t, \quad t = 1, \ldots, N, \tag{4}$$

(c) The amount of deliveries under BC at certain intervals

$$x_{\min}^t \le x_k^t + x_{ss}^t \le x_{max}^t, \quad t = 1, \ldots, N, \tag{5}$$

(d) The non-negativity of variables

$$x_k^t \ge 0, \;\; x_s^t \ge 0, \;\; x_{ss}^t \ge 0, \;\; t = 1, \ldots, N. \tag{6}$$

Denote by $x_{s,}^{t*}\, x_{ss}^{t*},\, x_k^{t*},\, t = 1, \ldots, N$ the solution to problem (2)–(6). Let R_1^* be the maximum revenue of ESC (2) at the point of independent solution.

If GC schedules deliveries $x^t,\;\; t = 1, \ldots, N$ under BC independently of ESC, it considers its own desirable contract volume V and achieves the maximum of its expected profit S_2

$$\max_{x_c^t, x_{gs}^t, x_{gss}^t} S_2 = E\left\{\sum_{t=1}^{N} \left[\tilde{p}_s^t x_{gss}^t - \tilde{p}_s^t x_{gs}^t - C^t\left(x_c^t + x_{gss}^t\right)\right] + J\right\}. \tag{7}$$

Note that GC's output is $x_g^t = x_c^t + x_{gss}^t$ and electricity delivery from GC to ESC under bilateral contract x^t is $x_c^t + x_{gs}^t$.

Since J is constant, expression (7) can be replaced by:

$$\max_{x_c^t, x_{gs}^t, x_{gss}^t} R_2 = E\left[\sum_{t=1}^{N} \left(\tilde{p}_s^t x_{gss}^t - \tilde{p}_s^t x_{gs}^t - C^t(x_c^t + x_{gss}^t)\right)\right] \tag{8}$$

subject to the following constraints on:

(a) The total contract volume

$$\sum_{t=1}^{N} (x_c^t + x_{gs}^t) = V, \tag{9}$$

(b) Electricity generation at each interval

$$x_{g\,min}^t \leq x_c^t + x_{gss}^t \leq x_{g\,max}^t, \quad t = 1, ..., N, \tag{10}$$

(c) The amount of deliveries under BC at certain intervals

$$x_{min}^t \leq x_c^t + x_{gs}^t \leq x_{max}^t, \quad t = 1, ..., N, \tag{11}$$

(d) The non-negativity of variables

$$x_c^t \geq 0, \; x_{gs}^t \geq 0, \; x_{gss}^t \geq 0, \quad t = 1, ..., N. \tag{12}$$

Denote by $x_{gs,}^{t*}$, x_{gss}^{t*}, x_c^{t*}, $t = 1, ..., N$ the solution to problem (8)–(12). Let $|R_2^*|$ be expenses of GC (8) at the point of independent solution.

Other constraints, for example the constraints on sales to and purchases from the spot market, can be taken into account in addition to (3)–(6) and (9)–(12).

Problems (1)–(6) and (7)–(12) use the assumption that amounts of electricity x_s^t and x_{ss}^t do not affect the spot market prices $\tilde{p}_s^t$, $t = 1, ..., N$. The assumption is correct if amounts x_s^t and x_{ss}^t are relatively small as compared to the total amount of electricity in the spot market. If it is not true, functions $\tilde{p}_s^t(x_s^t, x_{ss}^t)$, $t = 1, ..., N$ should be taken into account.

The stochastic dynamic programming technique is developed as a numerical method to solve problems (2–6) and (8–12).

The BC price J can be determined by the no-profit principle [3]. It means that for ESC the contract price J in (1) is chosen equal to the expected trade revenueR_1. If J is below R_1 ESC makes a profit. For GC the BC price J in (7) should equal the expected revenue R_2. If the J is above R_2 GC makes a profit.

The ESC can construct its bid curve $J(V)$ by varying the amount of V and solving the optimization problem (2)–(6). Similarly, the GC can construct its offer curve $J(V)$ by solving problem (8)–(12). The bid and offer curves, Fig. 2, reflect the no-profit conditions and are helpful for the BC arrangement.

Independent solution of problems (2)–(6) and (8)–(12) results in different delivery schedules and BC prices that represent interest of one party only.

The optimization problems (2)–(6) and (8)–(12) have numerous variables because normally number N of time intervals t is large. The stochastic nature of spot prices $\tilde{p}_s^t$ requires implementation of a multistage stochastic programming technique. Constraints (3) and (9) make deliveries at time intervals x^t mutually dependent. Increase in delivery at one interval results in decrease at the others.

A non-trivial numerical procedure has to be implemented to solve (2)–(6) and (8)–(12).

Some studies, e.g. [4, 10, 14], implement algorithms with the scenario tree generation. Consideration of the full-size scenario tree makes problems (2)–(6) and (8)–(12) intractable. The studies based on the stochastic programming techniques pay significant attention to the scenario tree reduction. Thus, the authors of [4, 10] implement the heuristic 'fast forward algorithm' to compose a smaller subset of scenarios. Generalized regression neural network is used in [14] to create a 'more realistic' tree with tractable number of nodes and branches. Both techniques introduce several assumptions to obtain reduced scenario trees.

The dynamic programming approach is used for BC schedule in [15]. It is based on preliminary construction of revenue functions, and treats forecasted stochastic information about prices $\tilde{p}_s^t$ at each interval t independently. It does not require introduction of significant simplifying assumptions. In the case of adequate presentation of random prices $\tilde{p}_s^t$ at each interval the dynamic programming algorithm is capable of providing a decision maker with rather accurate results of delivery scheduling.

4 Compromise Approach to the Contract Scheduling

Normally, cooperation between BC parties implies consent to consider the combinations of J and V that bring profit to both partners simultaneously. Figure 2 presents the bid and offer curves $J(V)$ for ESC (*I*) and GC (*II*). The dashed region between curves *I* and *II*, called a negotiation region, represents a set of points, where both ESC and GC make a profit. Certainly, ESC is interested in getting a contract with price J which belongs to curve *II*, while GC is interested in getting price J at curve *I*. Bounds of the negotiation region provide BC parties with maximum profits. However, BC with the maximum profit for one party is not acceptable for the other.

The Nash point from the Pareto-optimal set provides players with the maximum profits at their relative concessions being equal [16]. The main aim of the compromise approach is to achieve the Nash point. The compromise solution does not provide BC parties with their maximum profits. Both contract parties should

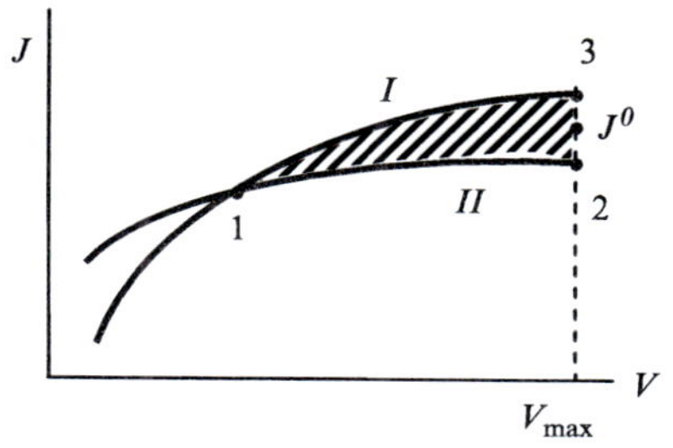

Fig. 2 The bid and offer curves of the BC parties: *I* for ESC; *II* for GC

"sacrifice" part of profit to conclude a contract with mutually acceptable parameters.

In the negotiation region the maximum profit values are

$$S_1^* = S_2^* = R_1^* - |R_2^*| = E\left\{\sum_{t=1}^{N}\left[p_d^t(x_k^{t*} + x_s^{t*}) - \tilde{p}_s^t x_s^{t*} + \tilde{p}_s^t x_{ss}^{t*}\right]\right\}$$
$$- \left|E\left\{\sum_{t=1}^{N}\left[\tilde{p}_s^t x_{gss}^{t*} - \tilde{p}_s^t x_{gs}^{t*} - C^t(x_c^{t*} + x_{gss}^{t*})\right]\right\}\right|. \qquad (13)$$

Let both parties agree to have minimum and equal relative concessions with respect to their maximum possible profits $S_1^*$and S_2^* (13).

Denote by $x_{s,}^{t0}\, x_{ss}^{t0}, x_k^{t0}, x_{gs}^{t0}, x_{gss}^{t0}, x_c^{t0}$, $t = 1, \ldots, N$ the amounts of deliveries corresponding to the compromise variant of BC. The revenues of ESC and GC with compromise deliveries are

$$R_1^0 = E\left\{\sum_{t=1}^{N}\left[p_d^t(x_k^{t0} + x_s^{t0}) - \tilde{p}_s^t x_s^{t0} + \tilde{p}_s^t x_{ss}^{t0}\right]\right\}$$

and

$$R_2^0 = E\left\{\sum_{t=1}^{N}\left[\tilde{p}_s^t x_{gss}^{t0} - \tilde{p}_s^t x_{gs}^{t0} - C^t(x_c^{t0} + x_{gss}^{t0})\right]\right\}.$$

Relative concessions are

$$\Delta S_1^0 = [S_1^* - (R_1^0 - J^0)]/S_1^*$$

and

$$\Delta S_2^0 = [S_2^* - (J^0 - |R_2^0|)]/S_2^*.$$

Denote by S_1^0 and S_2^0 the profits of ESC and GC at the compromise variant of BC. The necessary condition for the Nash point to be achieved is $\max\,(S_1^0 \times S_2^0)$ on the set of possible profits [16]. It is not difficult to prove that the Nash point is achieved when

$$J^0 = (R_1^* + |R_2^*|)/2 \qquad (14)$$

and

$$S_1^0 = S_1^*/2, \quad S_2^0 = S_2^*/2$$

Taking into account (14)

$$\Delta S_1^0 = [S_1^* - R_1^0 + (R_1^* + |R_2^*|)/2]/S_1^*$$

and

$$\Delta S_2^0 = [S_2^* + |R_2^0| - (R_1^* + |R_2^*|)/2]/S_2^*.$$

An important aim of the compromise contract scheduling is to obtain electricity deliveries by time intervals $t = 1, \ldots, N$. The following problem can be solved for this purpose[1]

$$\min_{k, x_k^t, x_s^t, x_{ss}^t, x_c^t, x_{gs}^t, x_{gss}^t} \quad k \tag{15}$$

subject to the constraints on:

(a) The equality of relative profit decreases (relative concessions)

$$\left\{ S_1^* - E\left\{ \sum_{t=1}^{N} \left[p_d^t (x_k^t + x_s^t) - \tilde{p}_s^t x_s^t + \tilde{p}_s^t x_{ss}^t \right] \right\} + (R_1^* + |R_2^*|)/2 \right\} / S_1^* = k, \tag{16}$$

$$\left\{ S_2^* + |E\left\{ \sum_{t=1}^{N} \left[\tilde{p}_s^t x_{gss}^t - \tilde{p}_s^t x_{gs}^t - C^t (x_c^t + x_{gss}^t) \right] \right\}| - (R_1^* + |R_2^*|)/2 \right\} / S_2^* = k, \tag{17}$$

(b) The consideration of points from the set of mutual profits (negotiation region)

$$E\left\{ \sum_{t=1}^{N} \left[p_d^t (x_k^t + x_s^t) - \tilde{p}_s^t x_s^t + \tilde{p}_s^t x_{ss}^t \right] \right\} \geq$$

$$\left| E\left[\sum_{t=1}^{N} \left(\tilde{p}_s^t x_{gss}^t - \tilde{p}_s^t x_{gs}^t - C^t (x_c^t + x_{gss}^t) \right) \right] \right|, \tag{18}$$

(c) The total contract volume

$$\sum_{t=1}^{N} (x_k^t + x_{ss}^t) = V, \tag{19}$$

[1] Superscripts "0" at variables are removed for simplicity of the problem statement.

(d) The sales to end consumers

$$x_k^t + x_s^t = x_d^t, \quad t = 1, \ldots, N, \tag{20}$$

(e) The electricity generation at each interval

$$x_{g\,min}^t \leq x_c^t + x_{gss}^t \leq x_{g\,max}^t, \quad t = 1, \ldots, N, \tag{21}$$

(f) The amount of deliveries under the contract at certain intervals

$$x_k^t + x_{ss}^t = x_c^t + x_{gs}^t, \quad t = 1, \ldots, N, \tag{22}$$

$$x_{min}^t \leq x_k^t + x_{ss}^t \leq x_{max}^t, \quad t = 1, \ldots, N, \tag{23}$$

(g) The non-negativity of variables

$$k \geq 0,\ x_k^t \geq 0,\ x_s^t \geq 0,\ x_{ss}^t \geq 0,\ x_c^t \geq 0,\ x_{gs}^t \geq 0,\ x_{gss}^t \geq 0, \quad t = 1, \ldots, N \tag{24}$$

Solution of problem (15)–(24) makes it possible to schedule electricity deliveries under the BC with equal and minimum decreases of possible profit for its parties. The contract price can be determined using Eq. 14.

5 Numerical Example 1

Let an ESC and a GC be going to conclude a bilateral contract. They have agreed on the duration, the starting point and the total volume of the BC $V = 145$ MWh. Initial data are given in Tables 1 and 2.

Table 1 Initial data for BC scheduling made by ESC

Time intervals	t_1			t_2			t_3		
Spot price forecasts made by ESC									
Spot price scenarios, p_{sj}^t, \$/MWh	10.0	10.4	10.8	11.0	11.2	11.8	11.0	11.4	11.8
Expected probabilities, ξ_j^t	0.1	0.8	0.1	0.3	0.5	0.2	0.2	0.4	0.4
Electricity price for end consumers, p_d^t, \$/MWh	16			16			16		
Constraints for optimization problem									
Electricity consumption by end consumers, MWh	x_d^1 9.8			x_d^2 11.4			x_d^3 14.5		
Limits on electricity deliveries under BC, MWh	$x_{\min}^1$ 8	$x_{\max}^1$ 60		$x_{\min}^2$ 5	$x_{\max}^2$ 68		$x_{\min}^3$ 6	$x_{\max}^3$ 62	

Non-negativity constraints are set for variables x_k^t, x_s^t, x_{ss}^t

Table 2 Initial data for BC scheduling made by GC

Time intervals	t_1			t_2			t_3		
Spot price forecasts made by GC									
Spot price scenarios, p_{sj}^t, \$/MWh	10.8	11.2	11.6	11.0	11.6	12.0	11.0	11.8	12.4
Expected probabilities, ξ_j^t	0.2	0.6	0.2	0.25	0.5	0.25	0.1	0.6	0.3
Production cost functions, $C^t(x_g^t)$, \$	$8.4+1.4x_g^1+0.4(x_g^1)^2$			$10.4+1.52x_g^2+0.44(x_g^2)^2$			$11.2+1.4x_g^3+0.32(x_g^3)^2$		
Constraints for optimization problem									
Electricity generation limits, MWh	$x_{g\,\min}^1$	$x_{g\,\max}^1$		$x_{g\,\min}^2$	$x_{g\,\max}^2$		$x_{g\,\min}^3$	$x_{g\,\max}^3$	
	14	50		15	60		16	65	

Non-negativity constraints are set for variables x_c^t , x_{gs}^t, x_{gss}^t

Table 3 Results of independent BC scheduling made by ESC

Time intervals	x_k^t	x_s^t	x_{ss}^t	Deliveries under BC $x^t = x_k^t + x_{ss}^t$
t_1	9.8	0	5.2	15
t_2	11.4	0	56.6	68
t_3	14.5	0	47.5	62

Table 4 Results of independent BC scheduling made by GC

Time intervals	x_c^t	x_{gs}^t	x_{gss}^t	Deliveries under BC $x^t = x_c^t + x_{gs}^t$
t_1	14	46	0	60
t_2	0	68	15	68
t_3	0	17	16,4	17

A compromise solution is obtained after solving the problems of independent scheduling. The results of solving problem (2)–(6) by ESC for the numerical example are shown in Table 3. With independent scheduling ESC gets a revenue R_1^* (2) equal to \$1,807.9.

Table 4 shows the results of independent scheduling made by GC. The results are obtained by solving problem (8)–(12). In the GC case independent scheduling leads to expenses $|R_2^*|$ (8) in the amount of \$1,493.32.

Based on results of independent scheduling ESC proposes BC deliveries $x^1 = 15$, $x^2 = 68$, $x^3 = 62$ MWh. If GC accepts this schedule it has additional expenses. If ESC accepts schedule $x^1 = 60$, $x^2 = 68$, $x^3 = 17$ MWh proposed by GC it decreases its revenue. They can obtain a compromise solution to get mutually acceptable contract conditions.

If the BC parties agree to consider the contract, which brings profit to both partners, the maximum profits for ESC and GC according to (13) are $S_1^* = S_2^* = \$1,807.9 - \$1,493.32 = \$314.58$. The contract price J^0 in the case of compromise scheduling at the Nash point (14) is \$1,650.61.

To obtain the compromise delivery schedule problem (15)–(24) is solved by a nonlinear programming algorithm. For the numerical example $N = 3$, the total

Table 5 Results of compromise delivery scheduling

Time intervals	Values of variables for ESC				Values of variables for GC			
	x_k^t	x_s^t	x_{ss}^t	Deliveries under BC $x_k^t + x_{ss}^t$	x_c^t	x_{gs}^t	x_{gss}^t	Generation $x_c^t + x_{gss}^t$
t_1	9.8	0	22.9	32.7	9.5	23.2	4.5	14
t_2	11.9	0	56.6	68	15	53	0	15
t_3	14.5	0	29.8	44.3	5.12	39.18	11.25	16.37

amount of electricity V to be delivered under the contract is 145 MWh. Parameters S_1^*, S_2^*, R_1^*, R_2^* are obtained from the independent delivery scheduling (Tables 3, 4). The data for constraints (16)–(23) are given in Tables 1 and 2.

The results of compromise scheduling are summarized in Table 5. ESC gets the revenue R_1^0 equal to $1,788.8. The total expenses of GC $|R_2^0|$ are $1,512.4.

Comparison of the results presented in Tables 5, 3, and 4 shows that the compromise approach enables both contract parties to adjust their strategies of participation in the spot market. For example, ESC changes its sales in the spot market x_{ss}^tat intervals t_1 and t_2. Deliveries under BC are different as well. GC changes the amounts of electricity to be sold and bought in the spot market.

The compromise approach is a way to arrange an acceptable BC. The ESC's revenue decreases from $1,807.9 to $1,788.8. The GC's expenses increase from $1,493.32 to $1,512.4. Both contract parties have the minimum and equal relative concessions ΔS_1^0 and ΔS_2^0. They reduce their profits from the maximum value of $314.58 to a compromise level of $157.29. Profits in the case of compromise scheduling make up 56% of the maximum value of $314.58.

The very simple numerical example considered in this section demonstrates the main ideas of the compromise approach and applicability of proposed model (15)–(24). The author and his colleagues have experience in scheduling lager time frames. For example, BC scheduling for a half year period included 26 week time intervals. The numerical technique based on the stochastic dynamic programming procedure [15] provided results of independent scheduling (1)–(6) for ESC and (7)–(12) for GC. Problem (15)–(24) with 156 variables was successfully solved with acceptable computation time.

6 Sequence of Actions for Contract Correction

The sense of BC monitoring and correction differs from the interruptible contract approach [6–8]. The interruptible forward contracts presume that consumer grants the right to curtail a given unit of load to power supplier in return for a price discount. Consumer agrees with possible curtailment in advance. Supply is normally interrupted for congestion management, rejection of price spikes or maintenance of system security. Decision on contract correction that is considered here is made during the existing BC.

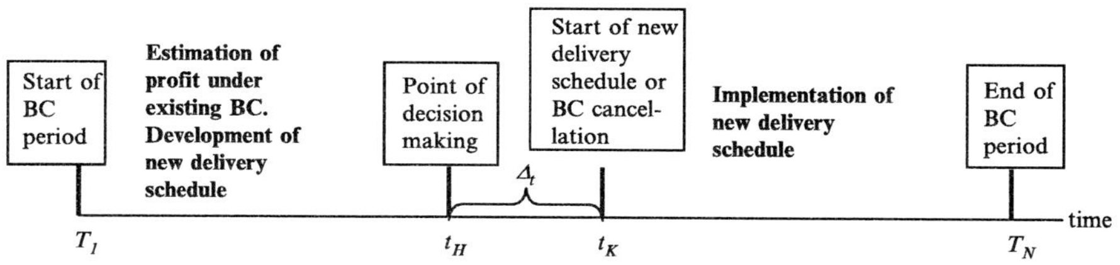

Fig. 3 The sequence of actions for contract correction

Correction of BC consists of several stages. The sequence of actions is shown in Fig. 3.

Stage 1. Starting with the first time interval T_1 each BC party follows the delivery schedule originally stipulated in BC. Each party keeps track of actual demand and price behavior and updates price and demand forecasts up to the end of contract period T_N. Actual prices and new forecasts may deviate significantly from those predicted before the contract conclusion. Both parties estimate their actual and expected profits (or losses), provided the original BC schedule x^t, $t = T_{1,\ldots,}T_N$ is followed strictly. If the estimation shows significant reduction in expected profit or losses one of the BC parties goes to Stage 2.

Stage 2. The party-initiator of correction develops a new delivery schedule and estimates its possible profit or loss in the case of contract correction. The profit is obtained taking into account the fine for changing contract obligations. The party-initiator makes a decision on the contract correction during interval $T_1 - t_H$.

Stage 3. The party-initiator of correction considers possible cancellation of BC during the interval $T_1 - t_H$ and estimates the profit or loss in the case of contract cancellation. The profit is obtained taking into account the financial compensation for contract termination.

Stage 4. The party-initiator of correction chooses the best decision considering the analysis made at stages 1, 2 and 3. It proposes the BC correction or cancellation to the other partner at time t_H.

Stage 5. The party-initiator of correction informs the other partner about its decision during period Δt and proposes a new delivery schedule. The partner estimates consequences of the proposed correction during interval Δt and approves (or rejects) the new delivery schedule. If the contract correction is appropriate for the second party the new delivery schedule should be approved by the System Operator. Both parties follow the original delivery schedule x^t, $t = T_{1,\ldots,}T_N$ during Δt, while the actual situation in the market may differ from that originally predicted. New delivery schedule or contract cancellation starts with interval t_K.

Stage 6. Electricity delivery under BC continues from t_K according to the new delivery schedule. Both BC parties continue to analyze the levels of spot prices and update price and demand forecasts up to the end of the BC period T_N. They continue to estimate expected profits (or losses) using improved forecasts. Decision on the next contract correction can be made by the same or the other contract party.

7 Decision Making on Contract Correction

Let us consider the case where BC is concluded between an electricity supply company (ESC) and a generation company (GC) for the period of time between T_1 and T_N. ESC sells electricity to end consumers and participates in the spot market. ESC receives $x^t,\ t = T_1, ..., T_N$ under the contract.

Stage 1. Let t_C be a current interval where ESC analyzes the economic efficiency of existing BC. Denote by $\tilde{p}^t_{sF}$ and x^t_{dF}, $t = t_C, ..., T_N$ the updated forecasts of spot prices and electricity demand. Forecasts of spot prices are considered as random values. Denote by x^t_{kF} a portion of x^t received under BC and delivered to end consumers, x^t_{ssF} a portion of x^t received under BC and sold in the spot market, and x^t_{sF} the amount of electricity bought in the spot market and delivered to end consumers. So, $x^t_{kF} + x^t_{sF}$ is amount of electricity delivered to end consumers, and $x^t_{kF} + x^t_{ssF}$ is electricity received under BC at interval t.

ESC revises its participation in the spot market and estimates expected profit S_F, provided if it receives electricity under BC according to the fixed delivery schedule $x^t,\ t = T_1, ..., T_N$. ESC solves the problem

$$S_F = \max_{x^t_{kF}, x^t_{ssF}, x^t_{sF}} E\left[\sum_{t=t_C}^{T_N} \left(p^t_d(x^t_{kF} + x^t_{sF}) + \tilde{p}^t_{sF}(x^t_{ssF} - x^t_{sF}) - p^t(x^t_{kF} + x^t_{ssF})\right)\right] \quad (25)$$

subject to the following constraints:

$$x^t_{kF} + x^t_{sF} = x^t_{dF}, \quad t = t_C, ..., T_N \quad (26)$$

$$x^t_{kF} + x^t_{ssF} = x^t, \quad t = t_C, ..., T_N \quad (27)$$

$$x^t_{kF} \geq 0, \quad x^t_{sF} \geq 0, \quad x^t_{ssF} \geq 0, \quad t = t_C, ..., T_N \quad (28)$$

Other constraints can be taken into account in addition to (26)–(28), for example limits on amounts of electricity sold or bought in the spot market, prohibition of sale and purchase in the spot market at interval t. The negative value of S_F means expected losses of ESC.

If S_F significantly differs from the value of profit expected at the time of contract conclusion ESC continues the process of decision making.

Stage 2. ESC develops a new delivery schedule on the basis of updated forecasts $\tilde{p}^t_{sF}$ and x^t_{dF} $t = t_C, ..., T_N$. It estimates the expected profit S_N provided if it implements a new delivery schedule. The new schedule may be implemented starting with interval t_K in the case of BC correction.

Denote by x^t_N the amount of electricity received by ESC under BC according to the new delivery schedule at interval t. Let x^t_{kN} be a portion of x^t_N delivered to end consumers, x^t_{ssN} a portion of x^t_N sold in the spot market, and x^t_{sN} the amount of electricity bought in the spot market and delivered to end consumers. Amounts $x_N^t = x_{kN}^t + x_{ssN}^t,\ t = t_K, ..., T_N$ form a new delivery schedule.

ESC solves the problem[2]

$$S_N = \max_{x_{kN}^t, x_{ssN}^t, x_{sN}^t} E\left[\sum_{t=t_C}^{T_N} (p_d^t(x_{kN}^t + x_{sN}^t) + \tilde{p}_{sF}^t(x_{ssN}^t - x_{sN}^t) - p^t(x_{kN}^t + x_{ssN}^t)) - p_{ovd}^t(\left|x^t - x_{kN}^t - x_{ssN}^t\right|)\right] \quad (29)$$

subject to

$$x^t - x_{kN}^t - x_{ssN}^t = 0, \quad t = t_C, ..., t_K \quad (30)$$

$$x_{kN}^t + x_{sN}^t = x_{dF}^t, \quad t = t_C, ..., T_N \quad (31)$$

$$x_{\min}^t \le x_{kN}^t + x_{ssN}^t \le x_{\max}^t, \quad t = t_K, ..., T_N \quad (32)$$

$$x_{kN}^t \ge 0, \quad x_{sN}^t \ge 0, \quad x_{ssN}^t \ge 0, \quad t = t_C, ..., T_N \quad (33)$$

Stage 3. ESC considers consequences of the contract cancellation. It takes into account financial compensation (fee) ΔF to the other contract party. ESC estimates an expected profit for the case of contract cancellation on the basis of updated forecasts $\tilde{p}_{sF}^t$ and $x_{dF}^t, t = t_C, ..., T_N$. The contract can be cancelled starting with interval t_K.

Denote by x_C^t the amount of electricity received by ESC under BC at interval t before contract cancellation. Let x_{kC}^t be a portion of x_C^t delivered to end consumers, x_{ssC}^t- a portion of x_C^t sold in the spot market, and x_{sC}^t the amount of electricity bought in the spot market and delivered to end consumers.

ESC solves the problem

$$S_C = \max_{x_{kC}^t, x_{ssC}^t, x_{sC}^t} E\left[\sum_{t=t_C}^{T_N} (p_d^t(x_{kC}^t + x_{sC}^t) + \tilde{p}_{sF}^t(x_{ssC}^t - x_{sC}^t) - p^t(x_{kC}^t + x_{ssC}^t)) - \Delta F\right] \quad (34)$$

subject to

$$x^t - x_{kC}^t - x_{ssC}^t = 0, \quad t = t_C, ..., t_K, \quad (35)$$

[2] Objective function (29) is written on the assumption that reduction and increase in delivery under BC is paid at the same price p_{ovd}^t. Real contracts can provide other requirements.

$$x_{kC}^{t} + x_{ssC}^{t} = 0, \quad t = t_K, ..., T_N, \tag{36}$$

$$x_{kC}^{t} + x_{sC}^{t} = x_{dF}^{t}, \quad t = t_C, ..., T_N, \tag{37}$$

$$x_{kC}^{t} \geq 0, \quad x_{sC}^{t} \geq 0, \quad x_{ssC}^{t} \geq 0, \quad t = t_C, ..., T_N. \tag{38}$$

Stage 4. ESC compares the values of S_F, S_N, S_C and chooses the best decision concerning the existing BC. It continues to receive electricity according to the originally stipulated delivery schedule, proposes the BC correction or cancellation to the other partner.

Stage 5. If ESC decides to correct the existing BC it proposes a new delivery schedule to generation company (GC). GC estimates its commercial benefits or losses of the proposed correction.

For interval t denote by x_{gc}^{t} the amount of electricity generated by GC and delivered to ESC under BC, x_{gss}^{t}- the amount of electricity generated by GC and sold in the spot market, x_{gs}^{t}- the amount of electricity bought in the spot market and delivered to ESC under BC. The total generation of GC is $x_{g}^{t} = x_{gc}^{t} + x_{gss}^{t}$. The amount of electricity delivered under BC is $x_{gc}^{t} + x_{gs}^{t}$. Assume that the production cost function $C^{t}(x_{g}^{t})$ is known and obtained by GC based on the optimal unit commitment.

Generation company solves the following problem to estimate its expected profit (or loss) of new delivery schedule implementation[3]

$$S_G = \max_{x_{gc}^{t}, x_{gss}^{t}, x_{gs}^{t}} E\left[\sum_{t=t_C}^{T_N} (\tilde{p}_{sF}^{t}(x_{gss}^{t} - x_{gs}^{t}) - C^{t}(x_{gc}^{t} + x_{gss}^{t}) \right.$$
$$\left. + p^{t}(x_{gc}^{t} + x_{gs}^{t}) + p_{int}^{t}(|x^{t} - x_{gc}^{t} - x_{gs}^{t}|))\right] \tag{39}$$

subject to

$$x_{g\,min}^{t} \leq x_{gc}^{t} + x_{gss}^{t} \leq x_{g\,max}^{t}, \quad t = t_C, ..., T_N, \tag{40}$$

$$x_{gc}^{t} + x_{gs}^{t} = x^{t}, \quad t = t_C, ..., t_K, \tag{41}$$

$$x_{gc}^{t} \geq 0, \quad x_{gs}^{t} \geq 0, \quad x_{gss}^{t} \geq 0, \quad t = t_C, ..., T_N. \tag{42}$$

Analysis of S_G allows GC to make a decision on approval or rejection of a proposed new delivery schedule.

---

[3] It is assumed in objective function (39) that GC uses the price forecast $\tilde{p}_{sF}^{t}$, $t = t_C, ..., T_N$ updated by ESC. In real life it can obtain a new forecast itself.

8 Numerical Example 2

It is assumed in the example that BC is concluded between supply and generation companies. Contract period consists of four time intervals of equal duration.

Both contract parties agree that the total amount of electricity to be delivered under BC is $V = 78$ MWh. The price of electricity delivered under BC is $p^t = 70.51$ \$/MWh for all time intervals $t = 1, ..., 4$. The necessary period Δt for correction approval is one time interval. Electricity price for end consumers is $p_d^t = 100$ \$/MWh for all intervals.

The following parameters are stipulated in the contract:

- Delivery schedule $x^1 = 8.0$, $x^2 = 5.0$, $x^3 = 58.0$ and x^4=7.0 MWh;
- Fine on GC in the case of delivery reduction initiated by producer, p_{int}^t =75 \$/MWh;
- Fine on ESC in the case of delivery overdrawing, initiated by buyer, p_{ovd}^t=72.5 \$/MWh;
- Financial compensation to the other party in the case of contract cancellation $\Delta F = p_{\text{int}}^t \sum_{t=t_K}^{4} \Delta x^t$, where $\sum_{t=t_K}^{4} \Delta x^t$is an amount of undelivered electricity.

The ESC forecasted the following average levels of spot prices p_s^t and demand x_d^t at the contract conclusion, Table 6[4].

The strategy of participation in the spot market for ESC is shown in Table 7. The strategy is developed before the BC conclusion taking into account forecasts, Table 6. The total profit of ESC expected at contract conclusion is \$1,503.75 including \$298.8 at interval 1.

Table 6 Initial price and demand forecasts

	Time intervals, t			
	1	2	3	4
Spot price, \$/MWh	65.00	70.37	71.75	70.00
Demand, MWh	9.8	11.4	14.5	13

Table 7 Distribution of electricity received under BC

Time intervals	x_k^t	x_{ss}^t	x_s^t	x_k^t+x_{ss}^t
1	8.0	0	1.8	8.0
2	5.0	0	6.4	5.0
3	14.5	43.5	0	58.0
4	7.0	0	6.0	7.0

[4] Random character of prices and demand values is not taken into account in the example for simplicity.

Let us assume that ESC has received 8.0 MWh under BC during the first time interval and gained \$298.8 of profit. At the beginning of the second interval ESC updates forecasts for spot prices and demand behavior. According to new expectations the spot prices are $p^2_{sF} = 68.26$, $p^3_{sF} = 69.6$, $p^4_{sF}=74.9$ \$/MWh. Amounts of demand are $x^2_{dF} = 11.4$, $x^3_{dF} = 10.0$, $x^4_{dF} = 11.0$ MWh.

ESC solves problem (25)–(28) at *Stage 1* to revise its participation in the spot market and estimate the expected profit S_F on the assumption that it receives electricity under BC according to the fixed delivery schedule. Intervals $t_C = 2$ and $t_K = 3$ for problem (25)–(28). New amounts of sales and purchases are given in Table 8. According to (25) the expected profit of ESC S_F for intervals 2–4 is \$908.3 including profit of \$350.5 at interval 2. Taking into account the profit in the first interval the total expected profit of ESC is \$1,207.1.

Stage 2 implies development of a new delivery schedule for intervals 3 and 4. ESC solves problem (29)–(33) and obtains new deliveries under the contract, Table 9. According to (29) the expected profit of ESC S_N for intervals 3 and 4 with the new delivery schedule is \$697.0. Taking into account profits at intervals 1 and 2 the total expected profit is \$1,346.3.

Estimation of the profit S_C expected after the BC cancellation needs solution of problem (34)–(38) at *Stage 3*. If the contract is cancelled and delivery under the contract is interrupted from interval 3, ESC will buy electricity in the spot market and $x^3_{sc} = 10.0$, $x^4_{sc} = 11.0$ MWh. GC will get a financial compensation ΔF in amount of 75(58.0 + 7.0) = \$4,875 from ESC. Taking into account profits gained at intervals 1 and 2 the expected total profit of ESC in this case S_C is equal to $-$\$3,645.4. The negative value of S_C means loss for ESC.

Comparison of S_F, S_N, and S_C shows that the best decision for ESC is correction of BC. Correction aims to fulfill new deliveries under BC at intervals 3 and 4, Table 8. After the decision is made ESC proposes new contract parameters to GC. If the new delivery schedule is approved by GC and System Operator it comes into force from interval 3. Otherwise, ESC should continue to receive electricity under BC according to the fixed delivery schedule.

Table 8 Revised participation of ESC in the spot market

Time intervals	x^t_{kF}	x^t_{ssF}	x^t_{sF}	$x^t_{kF}+x^t_{ssF}$
2	5.0	0	6.4	5.0
3	10.0	48.0	0	58.0
4	7.0	0	4.0	7.0

Table 9 Updated deliveries under BC

Time intervals	x^t_{kN}	x^t_{ssN}	x^t_{sN}	$x^t_{kN} + x^t_{ssN}$
3	10.0	48.0	0.0	58.0
4	11.0	54.0	0.0	65.0

9 Conclusion

Bilateral contracts for electricity delivery play an important role in the competitive electricity markets. Bilateral trade covers the main portion of electricity supply in the electricity markets. Potential contract parties schedule electricity delivery over a contract period during contract arrangement to gain the highest profit.

Optimization problems that are solved to schedule a bilateral contract between the electricity supply and generation companies are formulated. The problems are multi-dimensional, embrace many time intervals and use stochastic initial data.

The compromise approach is useful for delivery scheduling at the minimum and equal relative decrease of profit for both partners. The problem statement for the compromise scheduling is presented. The problem allows the contract parties to distribute electricity deliveries by time intervals and elaborate their strategies for participation in the spot and retail markets.

Mutually acceptable contract price corresponds to the Nash point from the Pareto-optimal set and provides players with the maximum possible profits.

A sequence of actions to be performed for contract correction during the BC period is considered. Decision making process consists of several stages. The stages deal with price and demand forecasting, estimation of actual and expected profits (losses), development of a new delivery schedule. The correction may reduce the risk of financial loss and increase the profit of the contract in force.

The statements of optimization problems are developed for decision making on BC correction and cancellation.

Numerical examples illustrate applicability of the suggested techniques to bilateral contract scheduling and correction.

References

1. Hunt S, Shuttleworth G (1996) Competition and choice in electricity. Wiley, Chichester
2. Borenstein S (2001) The trouble with electricity markets (and some solutions). The University of California, the Energy Institute, Berkeley. Paper PWP-081
3. Bjorgan R, Song H, Liu C-C et al (2000) Pricing flexible electricity contracts. IEEE Trans Power Syst 15(2):477–482
4. Carrion M, Conejo AJ, Arroyo JM (2007) Forward contracting and selling price determination for a retailer. IEEE Trans Power Syst 22(4):2105–2114
5. Palamarchuk SI (2003) Forward contracts for electricity and their correlation with spot markets. In: Proceedings of IEEE PowerTech conference, Bologna. Paper BPT03-223
6. Gedra TW, Varaiya P (1993) Markets and pricing for interruptible electric power. IEEE Trans Power Syst 8(1):122–128
7. Kamat R, Oren SS (2000) Exotic options for interruptible electricity supply contracts. Oper Res 50(5):835–850
8. Baldick R, Kolos S, Tompaidis S (2006) Interruptible electricity contracts from an electricity retailer's point of view: valuation and optimal interruption. Oper Res 54(4):627–642

9. Mount TD (2002) Using weather derivatives to improve the efficiency of forward markets for electricity. In: Proceedings of the international conference on system sciences, Hawaii, pp 21–31
10. Carrion M, Philpott AB, Conejo AJ et al (2007) A stochastic programming approach to electric energy procurement for large consumers. IEEE Trans Power Syst 22(2):744–754
11. David AK, Wen FS (2001) Bilateral transaction bargaining between independent utilities under incomplete information. IEE Proc Gener Transm Distrib 148(5):448–454
12. Bompard E, Ma Y (2008) Modeling bilateral electricity markets: a complex network appoach. IEEE Trans Power Syst 23(4):1590–1600
13. Cheng JWM, Galiana FD, McGillis DT (1998) Studies of bilateral contracts with respect to steady-state security in a deregulated environment. IEEE Trans Power Syst 13(3):1020–1025
14. Shrestha GB, Pokharel BK, Lie TT et al (2005) Medium term power planning with bilateral contracts. IEEE Trans Power Syst 20(2):627–633
15. Palamarchuk S (2010) Dynamic programming approach to the bilateral contract scheduling. IET Gener Transm Distrib 4(2):211–220
16. Fudenberg D, Tirole J (1996) Game theory, 5th edn. MIT Press, Cambridge, MA

Equilibrium Predictions in Wholesale Electricity Markets

Talat S. Genc

Abstract We review supply function equilibrium models and their predictions on market outcomes in the wholesale electricity auctions. We discuss how observable market characteristics such as capacity constraints, number of power suppliers, load distribution and auction format affect the behavior of suppliers and performance of the market. We specifically focus on the possible market power exerted by pivotal suppliers and the comparison between discriminatory and uniform-price auctions. We also describe capacity investment behavior of electricity producers in the restructured industry.

Keywords Capacity investment. • Electricity auctions • Electricity markets • Markov perfect equilibrium • Pivotal suppliers • Supply function equilibrium

1 Introduction

This research mainly focuses on the generation side of the electric power industry and reviews the recent findings on bidding behavior and market power issues in wholesale electricity auctions, and examines the role of production capacity constraints and auction institutions (discriminatory and uniform-price auctions) on behavior and equilibrium outcomes. Supply function equilibrium approach is the common tool in explaining the price formation process and bidding behavior in electricity markets. Hence, we mostly review the results predicted by this approach. We also discuss the incentives and behavior of power producers investing in production capacity in oligopolistic electricity markets.

Most of the analysis studied here is based on modeling without transmission network constraints because bidding behavior or investment analysis with these

T.S. Genc (✉)
Department of Economics, University of Guelph, Guelph, ON, N1G2W1, Canada
e-mail: tgenc@uoguelph.ca

A. Sorokin et al. (eds.), *Handbook of Networks in Power Systems I*,
Energy Systems, DOI 10.1007/978-3-642-23193-3_10,

constraints complicates equilibrium predictions and computations. Hence, the supply function equilibrium literature is missing the transmission network effects on equilibrium analysis in wholesale electricity markets. However, the results in this literature are still meaningful as they build insights as to market behavior in the power markets. The organization of this paper is the following. Section 2 of this paper describes the models of supply function equilibrium and explains the recent findings on market equilibrium predictions. Section 3 considers market power issues, in particular, (pivotal) suppliers who have potential market power and may exercise this power during certain market conditions. Section 4 compares the outcomes of two popular auction formats, discriminatory (or known as pay-as-bid) and uniform-price auctions, used in wholesale electricity markets. In Sect. 5, we briefly review the capacity investment behavior of power producers. Section 6 concludes with future research directions.

2 Supply Function Equilibrium Models and Bidding Behavior

There are two approaches for examining market outcomes (prices, outputs, profits, welfare losses, etc.) in electricity markets: Cournot model (the quantity choice model) and the supply function equilibrium approach (price-quantity pairs choice model). It is debatable which model better predicts the realized outcomes; however, it is clear that bidding behavior of generators is best characterized by the supply function equilibrium (SFE) concept [7, 20]).

The supply function equilibrium model has been extensively used to study the bidding behavior and the exercise of market power by sellers in multi-unit auction formats (see, for example, Anderson and Philpott [2], Baldick et al. [7], Holmberg [24, 25], Genc [16], Genc and Reynolds [17]). A 'supply function' is a strategy specifying the quantity that a firm is willing to produce as a function of the market price. SFE types of strategies are common in electricity auctions [5]. In the day-ahead market (in many jurisdictions it is the main market in which most of the electricity is traded a day before market opens, and small portion of it is traded in the real-time market), before the demand (or electric load) is observed, that is a day before the actual auction, each firm submits an offer schedule (non-decreasing supply function) specifying the quantity that they are willing to produce as a function of its price. The offer schedule may be viewed as a continuous approximation of the discrete-unit offer schedules that are submitted in these auctions. The independent system operator (ISO) takes these offers, and clears the market based on the demand and supply forecasts. In most of the electricity markets in the world, a uniform-price auction is employed. The uniform market price is determined by the intersection of aggregate demand and aggregate supply functions. Each firm produces at the price-quantity bundle at which its own supply function intersects with its own residual demand curve.

A supply function, specifying the quantity that a firm is willing to produce as a function of price, may be viewed as a firm's strategy in a game. A supply function equilibrium is a Nash equilibrium in supply function strategies. A model utilizing strategies of this type was first formulated by Grossman [22], and later studied by Hart [23]. However, there are two problems in studying SFE in this environment. First, the number of equilibria supported by supply functions is enormous. Second, with deterministic demand the firm knows its equilibrium residual demand for sure. Hence, by choosing either a fixed price or a fixed quantity, the firm can optimize its objective function. Thus, there is no incentive to implement a supply function strategy. However, it is shown in Klemperer and Meyer [31] that under demand uncertainty firms are willing to choose a supply function strategy rather than choosing simple price or quantity strategies. A supply function strategy affords a firm greater flexibility, and correspondingly greater profits, than fixed price or fixed quantity strategies when demand is uncertain. Under demand uncertainty, for each outcome of the random variable, the firm can find a price and a quantity that optimizes its objective function. Hence, the supply function maps each optimum level of price onto optimum quantity. Therefore, this strategy is better than committing to fixed price (Bertrand type) or fixed quantity (Cournot type) strategies under demand uncertainty (see [31]). Although with deterministic demand there are an enormous number of equilibria in supply functions, in the uncertain environment, the set of equilibria shrinks. Under certain demand and cost assumptions, unique supply function equilibrium can even be obtained for symmetric oligopolies [24]. Klemperer and Meyer [31] (hereafter KM) solve a system of differential equations to characterize symmetric SFE in environments for which product demand is uncertain. For n-firm symmetric model they show that there are multiple equilibria when the range of demand variation is bounded. These equilibria predict equilibrium prices between the Cournot price and the most competitive marginal cost price.

Several papers have utilized the SFE concept to analyze various aspects of electricity auctions. Examples include Green and Newbery [20], Newbery [35], Rudkevich, et al. [41], Green [21], Baldick and Hogan [5], Anderson and Philpott [2], Baldick et al. [7], Holmberg [24, 25], Genc [16], Anderson et al. [3], and Genc and Reynolds [17], among others. These papers consider a variety of extensions and modifications of the KM model, including production capacity constraints, asymmetric firms, potential entry, multi-step cost functions, forward contracting, mixed strategies, and auction format comparisons. Below we review some of the above mentioned SFE papers, and others will be discussed in the following sections. A common feature of these papers is that they do not consider the role of transmission constraints on optimum bidding behavior. There are a few recent papers, which we will also mention, that study equilibrium predictions in oligopoly in a transmission network.

The first SFE application paper is Green and Newbery [20] who have studied competition in the British electricity spot market, which was run as a uniform-price auction until 2001 after which it has been changed to discriminatory auction (or pay-as-bid auction). In their analysis, they follow the Klemperer and Meyer [31] paper set up. Rather than assuming uncertain demand, Green and Newbery assume that demand varies deterministically over time during the course of a market day; deterministic variation in demand over time is mathematically equivalent to KM's model of uncertain demand with bounded variation. They show that at the Nash equilibrium the generators, National Power and PowerGen that bid supply functions to the grid dispatchers who meet the demand at the lowest cost, make so much profit far above marginal costs and cause deadweight losses. Thus, to increase the competition they suggest a number of firms to be increased although entry takes 2–3 years and requires significant capital investment. Wolfram [45] using actual pool outcomes shows that Green and Newbery's model does not describe the market very well, and the pool prices that they predict are much higher than the observed prices. One explanation for high price prediction by Green and Newbery could be that they assume in the symmetric model suppliers should select the symmetric equilibrium that yields the highest profit.

Newbery [35] studies competition, contracts and entry in the electricity spot markets using analytically tractable models. He employs a supply function type of strategy to model the spot market and a Cournot type strategy to model the contract market. He finds that first, if the number of players (competitors) increases, then the maximum price reached in the pool and the average pool prices decrease. Second, if the industry has insufficient capacity and new investment has a lower marginal cost than existing investment, then forward contracts can deter entry (in the sense that entrants could not offer lower priced contracts). Generators covering themselves with forward contracts would yield more competition in the spot market, and hence reduce average pool prices.

Green [21] studies the electricity contract market in England and Wales. He shows that competition in the contract markets would cause generators to sell much of their power in these markets and hence would result in spot prices (at the Pool) close to marginal production costs. He employs supply function type strategies in the two stage spot market, where there are two suppliers and many buyers. He also allows conjectural variations to model different degrees of competition in the contract markets. He finds that with the Bertrand conjecture (taking other's price fixed), generators will set prices equal to marginal cost. This result is similar to Allaz and Villa's [1] competitive market outcome.

Baldick and Hogan [5, 6] study capacity constrained supply function equilibria in electricity spot markets. They also consider stability issues of the equilibria and propose a so-called 'function space iteration' method to solve the equilibria numerically. Baldick and Hogan argue that asymmetries among suppliers are common in electricity markets and that SFE models should take this into account. They state that if the firms are asymmetric in capacities and in cost functions then the differential equation approach of solving supply functions may not be effective, because the resulting supply functions may fail to have the non-decreasing property. Moreover, many of the proposed possible equilibria are unstable due to the capacity constraints. This instability restricts the range of equilibria and eliminates some equilibria that may be observed in the markets. However, they do not consider how the extent of excess capacity affects equilibrium predictions, nor do they consider the role that pivotal suppliers might play.

In a recent paper, Holmberg and Newbery [26] review supply function equilibrium and its policy implications for wholesale electricity auctions. They provide a literature review of supply function models applied to analyze bidding behavior in oligopolistic electricity markets in the presence of price caps, forward contracts, different auction formats/mechanisms, capacity constraints, and some behavior restrictive market rules. They summarize the results of theoretical and empirical papers in the supply function literature applied to the electricity markets. Apart from explaining supply function equilibrium predictions in the literature, they also provide a competition policy recommendation as the number of power producers varies. They deliver detailed explanations of the market power issues and measure the welfare loss in England and Wales market in 1999.

3 Pivotal Electricity Suppliers and Market Power

Suppliers in many markets are able to exercise market power. By withholding some production from the market a firm may be able to raise the price of its output and increase its profit. The Cournot oligopoly model is a well-known and often-used framework for analyzing market power. In that model, the amount of market power that any single firm has depends on factors such as the price elasticity of demand, the number of firms, the nature of costs of production, and on firms' capacity constraints (if applicable).

A number of recent assessments of wholesale electricity market performance have emphasized how a single firm could affect the market price in an auction by withholding some output from production (see Joskow and Kahn [28], Lave and Perekhodtsev [32], Rothkopf [40], Borenstein et al. [8], Perekhodtsev et al. [36], Wolak [44], and Genc and Reynolds [17]). This single firm, so called "pivotal supplier", could exercise market power and set the market price when his rivals are capacity constrained. Precisely, a firm is a pivotal supplier if the total capacity of its rivals is not enough to meet the market demand. A pivotal firm or a group of pivotal firms emerge when the market demand/load is high, and/or market capacity is low relative to the peak demand. Alternatively, a pivotal supplier may be defined as a supplier with positive residual demand, in which residual demand for a supplier is total market demand minus the summation of capacity of other generators and total imported power.

Wolak [44] finds a recent evidence on how pivotal suppliers exercise market power in the New Zealand wholesale electricity market. In Sect. 3 (pp. 82–127) of his paper, Wolak explains how pivotal suppliers emerge and exercise market power. Furthermore, in his Sect. 4 (pp. 127–171), Wolak provides empirical evidence on the ability and incentive to exercise unilateral market power by pivotal suppliers. He writes (p. 163), "...In fact, a number of the market power mitigation mechanisms in United States wholesale markets are based on this supposition. The short-term market operator takes the offers and bids of all market participants and determines whether a supplier is pivotal or a set of suppliers are jointly pivotal. If this is the case, then the offers of this supplier or this set of suppliers are mitigated to some reference offer level that is based on that supplier's marginal cost of production. Our analysis examines whether being pivotal or net pivotal predicts higher offer prices by the supplier after controlling for the opportunity cost of water and input fossil fuel prices." He says, (p. 153) as a result of the empirical study, "...We find that when a supplier is a pivotal its offer prices are higher by economically significant magnitudes." He estimates that a supplier is pivotal more than 50% of the time during the trading periods in the New Zealand market. In a recent empirical study, Philpott et al. [37] investigate production inefficiencies in the New Zealand wholesale electricity market. They argue that their model could be used to identify the sources of extensive exercise of market power indicated by Wolak [44] in the New Zealand market.

A number of studies have examined how production capacity constraints influence the range of equilibrium prices under the SFE concept (e.g., see Green and Newbery [20] and Baldick and Hogan [5]). Yet these studies have not examined the potential role of the extent of excess capacity in the market on equilibrium prices, nor have they shown how the presence of pivotal suppliers affects predicted equilibrium supply functions and prices. These studies point out that production capacity constraints may rule out some supply functions as equilibria because quantities supplied at equilibrium prices violate one or more capacity constraints. What prior SFE studies seem to have missed is that capacity constraints may limit the ability of rival sellers to respond to a low supply/high price deviation by any single firm. A deviation from a proposed SFE can be profitable when demand is high and rivals' ability to increase supply is limited by capacity constraints. Capacity constraints can influence the set of supply function equilibria even when there is excess capacity in the competitive equilibrium.

Genc and Reynolds [17] explore how capacity constraints influence the incentive to deviate from proposed supply function equilibria and thereby limit the set of equilibria. They formulate a simple model of a wholesale electricity auction in which pivotal suppliers dictate the market price. They examine the connection between pivotal suppliers and the set of SFE. They assume that demand varies over time (during the trading period), and is perfectly inelastic. In the symmetric model, they consider the case in which players' marginal cost is fixed up to capacity. In another case they assume suppliers have step marginal costs and total capacity is equally divided among them. In the asymmetric model, they assume firms are different in capacities, and have a common marginal cost for production up to capacity. The market price is bounded by a price cap. By withholding output, a pivotal supplier can move the market price to the maximum price, or price cap for the market. There is a continuum of SFE and the presence of pivotal suppliers along with capacity constraints helps refine these multiple equilibria. In the symmetric and asymmetric versions of the model, they show that when pivotal suppliers are present the set of SFE is reduced relative to when no suppliers are pivotal. When the pivotal suppliers are present some of the most competitive SFE from the set of equilibria are eliminated. These SFE are eliminated even though they do not violate capacity constraints anywhere along the proposed equilibrium path. The extent to which the equilibrium set is reduced depends on observable market characteristics such as the extent of excess capacity, the demand distribution, the number of suppliers, and the base load capacity factor. As the amount of industry excess capacity falls, and/or the load factor rises, and/or the number of suppliers decreases, and/or the low-cost base load capacity falls in which the base load is less than the off-peak load level, the set of SFE becomes smaller; the SFE that are eliminated are the lowest-priced, most competitive equilibria. The firm with the larger share of capacity has an incentive to deviate from a wider range of SFE, and it is the larger firm's deviation incentives that determine which SFE are ruled out as equilibrium.

Another relevant research concerning pivotal suppliers is Perekhodtsev et al. [36], who formulate and analyze a game theoretic model in which symmetric, capacity constrained firms submit offers to supply into a uniform price auction. They assume that demand for electricity is perfectly inelastic. Their aim is to assess the role that pivotal suppliers play in price formation process. They restrict attention to simple bidding strategies in which a firm bids either a "Low" price equal to marginal cost or a "High" price equal to the price cap. Equilibrium bidding involves mixed strategies in which each firm bids either low or high with specific probabilities. The equilibrium probability that the price is high depends on the supply margin, the difference between industry capacity and the fixed demand (load). As the supply margin increases the expected price in equilibrium falls. The presence of a single pivotal supplier is associated with a high price in their model. They also discuss the notion of a pivotal group of firms – a group of firms whose total capacity exceeds the supply margin. They show that market power gradually declines as the number of firms that are jointly pivotal rises. To examine the role of pivotal suppliers, they assess how observed price-cost margins in the California wholesale electricity market during late 2000 vary with the number of pivotal suppliers in the market. They find that price cost margins were higher the fewer the number of pivotal suppliers.[1]

4 Discriminatory Versus Uniform-Price Electricity Auctions

This section reviews market outcome predictions under two popular auction formats; discriminatory and uniform-price auctions. The common auction institution used for day-ahead or balancing electricity markets is the uniform-price auction under which sellers whose bids accepted are paid at the market clearing price. On the other hand sellers are paid only at their bid price under the discriminatory auction. In 2001 the British Regulatory Authority in the England and Wales changed the auction format from uniform-price to discriminatory auction in the hope of lower wholesale electricity prices. Recently the Regulatory Authority for Electricity and Gas of Italy has adopted a discriminatory auction in their day-ahead electricity market. Several research papers have examined market performance of

[1] It should be noted that their theoretical conclusions are based on a very simple model with only two possible bids, and symmetric costs and capacities.

these auctions under various assumptions. Examples include Anwar [4], Federico and Rahman [13], Rassenti et al. [38], Son et al. [42], Fabra et al. [12], Holmberg [25], and Genc [16]. Below we discuss these papers in detail.

Anwar [4] compares the discriminatory and uniform auctions in terms of expected cost to the auctioneer in a procurement auction. He studies equilibria in multi-unit common value auction model that sometimes provides a positive residual market demand to suppliers by means of capacity constraints. His model, similar to Fabra et al. [12], considers discrete step supply offers (i.e., there is a limit on the number of price-quantity pairs offered). The quantity for the auction is uncertain and the demand distribution is a common knowledge. There are multiple firms each with a unit capacity of supply. Each firm has the same constant marginal cost up to capacity, which is a common value. He shows that discriminatory auction provides more competitive outcomes and is more efficient for the auctioneer than the uniform auction, when capacity constraints are present. Moreover he finds that when demand is low, both auction formats lead to competitive pricing. When demand is high and firms face some residual demand, the uniform auction leads to higher prices than the discriminatory auction. This is because his model predicts a unique type of pure strategy equilibrium such that one firm sets its bid at the choke-price (i.e., maximum willingness to pay price) when the rivals do not have enough capacity to meet demand with sure probability. Also in the partially pivotal region (the region in which firms are pivotal for sometime during the trading period), he finds that there is no pure-strategy equilibrium in the uniform auction. These and some of his other findings are very similar to the results in Fabra et al. [12].

Wolfram [46] is in favor of the uniform-price auction in the England Wales Electricity Pool, but she admits that which auction format is better (in terms of prices and efficiency) depends on the market concentration and factors such as winner's curse[2] and infra-marginal capacity[3] that may significantly affect prices. Kahn et al. [29] favor the uniform-price auction, and claim that the discriminatory auction may cause inefficiencies, because generators will no longer bid at their marginal costs, and the tacit collusion that exists within the uniform auction may persist in the discriminatory auction. Although wholesale electricity prices have decreased in England and Wales after switching to discriminatory auction,

[2] In the sale auctions, winner's curse refers to winner's overpayment for a product in the common value with incomplete information auction. In electricity context, which is a common value procurement auction, winner's curse occurs, whether it is a uniform-price or a discriminatory auction, when generators offer low prices for their production units. However, to avoid the curse, savvy generators may tend to overbid in the repeated electricity auctions.

[3] Infra-marginal capacity is the production capacity that is less than the market clearing quantity supplied; whereas marginal capacity is the quantity that helps clear the market. Infra-marginal plants generally supply base-load capacity and marginal plants like thermal generators (petroleum-fired generators) clear the market at higher prices. In the uniform-price auction both infra-marginal and marginal plants are paid at the market clearing price. In the discriminatory auction, they are paid at their own bid prices.

Newbery [34] argues that this decrease is due to other factors such as excess market capacity and increased imports.

Kahn et al. [29] reject the idea of switching to discriminatory auction in the following reasoning. First, discriminatory auction may cause inefficiencies, if the generators do not bid in their marginal costs. Indeed, all of them have incentives to raise their bids so that their fixed and common costs are ensured to be paid. However, we note that, under the discriminatory auction generators' (high-price) bidding strategy concerning recovery of their fixed and common costs associated with the commitment of the generating units is futile in the electricity markets that employ uplift or make-whole payments. In many electricity markets fixed costs like startup costs and no-load costs are covered by system operators through uplift or make-whole payments. Under the uniform-price auction these costs are likely to be recovered due to the difference between market-clearing price and marginal cost. This may give generators incentives to bid at their marginal costs in the uniform-price auction. Under the discriminatory auction, nevertheless, it is likely that more costly generators might be dispatched more often than less costly generators, if they could not predict the clearing prices with accuracy. Furthermore, another source of inefficiency would be the extra payments made for forecasting the market prices. Second, small suppliers might be more disadvantaged under pay-as-bid auction. Collecting information about rival bidders and estimating market outcomes period-by-period are more costly per unit of output for small firms than for larger firms. Besides, under the uniform-price auction, smaller firms can benefit from the high prices stemming from the market power exercised by larger firms. However, under the discriminatory auction, since bidders are paid at their offer prices, high prices resulting from market power do not benefit the smaller firms. To avoid it, smaller firms would tend to bid at the higher prices. That would increase the overall market prices and might cause smaller firms' bids not being accepted if they overestimated the clearing prices. Finally, tacit collusion that is attributed to the uniform price auction would persist for the discriminatory auction, because firms would learn how to collude over time.

Klemperer [30] gives several examples of pitfalls in auction design. His examples mostly focus on sale auctions (demand-side bidding) rather than procurement auctions (supply-side bidding). He notes that uniform-price auctions are very vulnerable to collusion, and very likely to deter the entry, because the repeated interactions among bidders more often enable them use signaling and punishment strategies. Hence, they learn to cooperate; otherwise, deviation from the collusive agreement is unprofitable since higher market-clearing bid would be paid by all bidders. However, in the pay-as-bid sealed-bid-auctions, he notes that, bidders who would require small amounts to trade would be discouraged since their bids rely on the distribution of the rivals' values, which is costly to obtain.

Federico and Rahman [13] compare the two auction formats for perfect competition and monopoly structures. These are benchmark cases and do not reflect the structure of the real wholesale electricity markets. They analyze a model in which each supplier, in a perfectly competitive model, sells one infinitesimal unit of capacity to the auctioneer who meets a uniformly distributed elastic demand.

Each supplier has increasing continuous costs and is risk neutral and strives to maximize its expected profit. When they assume that costs are common knowledge and the demand is fixed and perfectly inelastic, they find that these two auctions result in the same prices and payoffs. However, these results change if demand is inelastic and uncertain. In perfect competition, suppliers' expected profits are lower under the discriminatory auction than under the uniform auction. In the monopoly structure, they find that the comparison of the auction formats, in terms of average prices, consumer surplus and expected profits, leads to mixed results which depend on model parameters.

Rassenti et al. [38] do experiments to rank the market outcomes under the discriminatory and uniform auctions. Players face computer-generated step-wise elastic demand schedules, which vary among off-peak, shoulder and on-peak periods. Each seller has multiple technologies with fixed capacities and submits step function offer schedule to the market. Their first finding is that changing auction format from uniform to discriminatory leads to significant electricity price increases in the off-peak and shoulder periods. However, auction format change has no effect on the on-peak period prices when greater excess capacity exists in the market. Their next finding is that for the same level of demand the price variability from trading period to trading period is lower under the discriminatory auction than under the uniform auction. They state that since in the experimental design there is a greater excess capacity during the peak period, low volatility (fewer price spikes) is predicted. However, they admit that this pattern of excess capacity is a specific feature of their experimental design. Thus, their volatility results cannot be generalized to field environments.

Son et al. [42] compare performance of two strategic players, one is with large capacity the other is with small capacity, under both auction formats in a market game. Players bid energy blocks (with a discrete number of price-quantity pairs) in the auction. They show that expected total revenues of players are higher under uniform pricing than under the pay-as-bid pricing. They discuss the mixed strategy Nash equilibrium attained under the discriminatory auction, and are able to compute it by using an algorithm.

Fabra et al. [12] analyze a game-theoretic model in which firms with asymmetric capacities and costs submit discrete unit offer schedules (step offer functions) to the auctioneer. Most of the their analysis assumes a perfectly inelastic demand with a fixed market reserve (maximum) price, constant marginal cost of production, and production capacity constraints. They compare the Nash equlibria of both auction formats in terms of average prices paid to suppliers and productive efficiency. For the fixed demand case, they find that the uniform-price auction yields higher average prices than the discriminatory auction and their numerical examples suggest that price differences can be substantial depending on the total industry capacity, the extent of asymmetry in capacity levels and the price cap. For the uncertain demand and perfectly symmetric case, they find that expected payments to suppliers are the same for both auctions. They also find that for low demand realizations, equilibrium is both unique and identical; the equilibrium is bidding at the marginal cost of the inefficient supplier for the two auction formats. For the asymmetric duopoly case, in the discriminatory auction they find that there is no pure strategy equilibrium but only in mixed strategies.

Most of analysis of Fabra et al. [12] is based on the assumption that bids are "short-lived" and are discrete step supply offers. However, these assumptions may not hold for some electric power markets. Under uniform-price auction, for some parameter regions in which for some periods of time industry demand is higher than rival firms' total available capacity (partially pivotal region) any single firm is pivotal for part, but not all, of the trading period. For step function bidding, Fabra et al. find that pure strategy equilibrium does not exist for parameters in the partially pivotal region; the equilibrium is in mixed strategies. However, they do not characterize the mixed strategy equilibrium. In the continuous SFE model, however, Genc and Reynolds [17] find that there are multiple pure-strategy equilibria. Genc and Reynolds conjecture that predicted market clearing prices for the step function model may be either higher or lower than SFE market clearing prices depending on parameter values, when parameters are in the partially pivotal region.

Holmberg [25] compares the two auction institutions using inelastic and stochastic demand. He assumes convex marginal costs and derives SFE with the condition that demand exceeds total available industry capacity with positive probability. This is a quite strong assumption. Based on this condition he solves the ordinary differential equations of the optimality conditions. He notes that pure strategy equilibrium may not exist in the discriminatory auction, if demand follows some specific probability distribution, and concludes that average prices are weakly lower in the discriminatory auction. However, he does not characterize any mixed strategy equilibrium under the discriminatory auction.

Genc [16] compares the performance of the two auction formats in the presence of capacity constraints and pivotal suppliers using continuous offer schedules. He assumes time dependent stochastic and perfectly inelastic electric load. Marginal cost of production is common knowledge and constant up to the production capacity. The total industry capacity is greater than or equal to the peak demand (load). He considers both mixed strategy and pure strategy Nash equilibrium in continuous supply function strategies in oligopoly. When capacity constraints are non-binding he finds that in the discriminatory auction optimal equilibrium supply function is unique and suppliers bid competitively. However, in the uniform auction there is a continuum of equilibria as in other SFE models in which equilibrium prices range from marginal cost to the price cap. Therefore, each player's profit in the uniform-price auction is always weakly greater than the profit in the discriminatory auction at any time during the trading period. He also finds that in the single-step marginal cost case, the functional form of the demand is irrelevant of the equilibrium strategies in both auction institutions. When capacity constraints are binding and a pivotal supplier emerges he finds that there is no pure strategy SFE under discriminatory auction. But there exists mixed strategy equilibrium and he characterizes this equilibrium. Firms offer their entire capacity to the market and mix the prices over the equilibrium probability distribution functions. Nevertheless the equilibrium strategies under the uniform-price auction are pure strategies and multiple. As a consequence, expected profit per firm under the uniform-price auction is greater or equal to the expected profit per firm under the discriminatory auction.

5 Capacity Investments in Electricity Markets

One of the essential arguments of electricity industry restructuring is to promote capital investments. The importance of capacity investments in restructured electricity markets has been stressed by Roques et al. [39], Murphy and Smeers [33], Joskow [27], among others. Production capacity investments may help play a key role for mitigating market power, entailing more competitive outcomes and ensuring network system reliability. Since power producers face uncertain demand and their investment costs are largely sunk and they face competition, they have to make right decisions on timing of investment, type of technology to acquire, and an optimal investment behavior before investing. To invest they have to project future profitability through growing demand and more efficient production technologies. What follows is a brief review of the recent literature examining capacity expansion behavior of firms in electricity markets.

Chaton and Doucet [10] study Hydro-Quebec's capacity expansion planning in a stochastic linear programming model. Hydro-Quebec is provincially owned monopoly with hydroelectric capacity close to 90% of the total available capacity in the province. The objective function is minimization of total expected costs subject to market clearing constraint, and transmission and production constraints. The uncertainty stems from fuel costs and demand growth. The aim is to meet the final period demand by capacity additions (with option values) made in earlier periods. They calibrate the model, using the GAMS software, with the data from Hydro-Quebec and neighboring jurisdictions to forecast investment behavior of Hydro-Quebec. They conclude that the market conditions do not justify the expansion plan of Hydro-Quebec.

Murphy and Smeers [33] study generation capacity investments in open-loop and closed-loop Cournot duopolies. Each duopolist has a different technology (one is a base-load plant, the other is a peak-load plant) and makes investment to increase production capacities in the face of growing demand. Demand is price sensitive and varies over time deterministically. They study two types of settings. In the first, the open-loop game, they assume that production capacities are simultaneously built and sold in long-term contracts. In the second, the closed-loop game, they assume that there is a time-to-build constraint and the capacities invested in the first stage will be available to sell in the second stage in a spot market. They find that equilibrium investment levels and production quantities in the closed-loop game are in between the values in the open-loop game and the efficient outcomes.

Bushnell and Ishii [9] examine an equilibrium model of capacity investments in electricity markets in which firms make lumpy investment decisions. The model incorporates short-run spot market Cournot competition and long-run Markov perfect equilibrium of investments, and the results are based on simulations. They find that incentives to invest depend on market positions of the firms. Retail or contractual obligations of the firms also affect the investment decisions of the firms, for example, more retail obligations decrease the market power of the firms, hence less incentives to invest. When demand growth uncertainty increases, they find that firms may delay their investments as the "option-value" of the investment theory suggests.

Garcia and Shen [15] characterize Markov perfect equilibrium capacity expansion plans for oligopoly in which firms face demand uncertainty and investment is not productive immediately (i.e. there is a lag between investment and production). They find, not surprisingly, that Cournot firms underinvest relative to the social optimum.

Garcia and Stacchetti [14] study a finite horizon discrete time dynamic duopoly game. Production is subject to capacity constraints; firms have constant marginal cost of production and meet perfectly inelastic demand that has random demand growth component. They find that in some equilibria total capacity falls short of demand, and hence system security is jeopardized. They also find that increasing price caps does not affect the market excess capacity and decreasing the price cap benefits the consumers.

In a recent paper, Genc and Thille [18] study competition between thermal and hydro electric producers and analyze the choice of capacity by the thermal producer under demand uncertainty and characterize both the Markov perfect and S-adapted open-loop equilibria.[4] They assume a low cost hydro generator with a fixed stock of water (since water is renewable on a yearly basis through the cycle of inflows). They find that investment is higher under Markov perfect information, and this investment may be either higher or lower than the efficient investment depending on model parameters. Optimal investment function is discontinuous in initial capacity under Markov-perfect equilibrium and continuous in initial capacity under the open-loop equilibrium. These results are different than the findings of Murphy and Smeers [33] and Garcia and Stacchetti [14] who mostly assume symmetric technologies with constant cost of production.

6 Conclusions

The supply function equilibrium (SFE) approach has been employed to study bidding behavior of firms and market power issues in power markets as well as in the Treasury bill auctions. In the electricity context there is a growing literature analyzing different aspects of power markets using the SFE concept. This literature has considered various extensions of the original SFE model introduced by

[4] Both Markov perfect equilibrium and (S-adapted) open-loop equilibrium are Nash equilibrium in production and investment strategies. Markov perfect strategies are state-dependent, whereas open-loop strategies are mainly conditioned on calendar-time and the decisions are made at the outset of the game. In a stochastic game, the appropriate equilibrium concept with the features of open-loop information is S-adapted open-loop equilibrium. The key difference between the open-loop equilibrium and S-adapted open-loop equilibrium is that S-adapted equilibrium strategies in open-loop equilibrium allow the decisions to be adapted to the demand shock realization.

Klemperer and Meyer [31]. These extensions include equilibrium characterization with capacity constraints, pivotal suppliers, forward contracts, price caps, asymmetric players, multi-step-costs, mixed strategies, and different auction institutions. In this paper, we review the SFE literature analyzing the effects of above mentioned aspects in the wholesale electricity markets. It appears that new research papers will embed other characteristics of electricity markets into the SFE models. The new research directions may include further refinement of multiplicity of SFE, analysis of bidding behavior under transmission network constraints, and equilibrium characterization in the presence of exports and imports made through neighboring jurisdictions in a network.

We review power generators' bidding behavior in the discriminatory and uniform-price auctions under various assumptions regarding equilibrium bidding function types (discrete or continuous), cost, capacity and the number of firms. We discuss the relevance of the continuous supply offers in bidding as opposed to the discrete offers. Importantly, for empirically relevant parameter region in many electricity markets we argue that to be able to compute equilibrium outcomes it is useful to use continuous supply function bidding rather than step function bidding. The characteristics of equilibrium bidding strategies in both auction formats have been understood. The SFE under the discriminatory auction is unique, but equilibrium is multiple in the uniform-price auction when capacity constraints do not bind. When capacity constraints bind and pivotal suppliers face positive residual demand there is no pure strategy supply function equilibrium in the discriminatory auction. The mixed strategy supply function has the property that suppliers tend to dump all of their capacity into the market and they employ a mixed strategy in which prices are mixed along horizontal supply functions. We argue that offering all of the capacity at a single price is more profitable than using multiple bid prices for capacity tranches. In a recent paper, Anderson et al. [3] extend the results of Fabra et al. [12], Genc [16], and Holmberg [25], and show that in the discriminatory auction mixtures over strictly increasing supply functions are possible in markets with non-pivotal producers, inelastic demand and no price cap. With the pivotal suppliers, they obtain the same result as in Genc [16] that the equilibrium is in horizontal supply function mixtures. We conclude that although discriminatory auction is not easily tractable and gives difficulties to power producers to form their optimal supply functions due to the nature of mixed strategies, consumers would gain and expected electricity prices would be lower than the ones under the uniform-price auction.

We also summarize the recent findings on capacity investments in electricity markets. The investment in production capacity becomes an important issue in the power industry given the growth in electricity demand and the concerns like phasing out environmentally hazardous and economically expensive some old smokestack technologies. Several papers examine capital investment issues in the wholesale electricity markets with or without transmission network constraints. Examples include Chaton and Doucet [10], Murphy and Smeers [33], Bushnell and Ishii (2007), Garcia and Shen [15], Garcia and Stacchetti [14], and Genc and Thille [18]. These papers consider state-controlled monopoly, perfect competition and oligopoly market structures with the open-loop, closed loop, and Markov perfect equilibrium concepts to examine the capacity expansion behavior of power producers in the electricity markets in which transmission constraints do not bind. We emphasize on several game theoretic settings such as investment game among hydro producers, and investment game between thermal and hydro producers under various assumptions. In particular, the degree of overinvestment in thermal capacity and the efficiency of water-use are analyzed. We address hydro and thermal player's output and investment behavior under different equilibrium concepts. We observe that the thermal player has a strategic motive when choosing to invest in production capacity: overinvesting in the Markov perfect equilibrium. However, this investment may not be efficient and the level of investment could be above or below the social welfare maximizing level depending on availability of water in the reservoir.

There are recent papers addressing investment issues in electricity markets in the presence of transmission constraints, production capacity constraints, and time-to-build constraints. For example, Genc and Zaccour [19] study long-run capacity investment dynamics in oligopoly under demand uncertainty. This paper is related to capital accumulation in network industries, and, in particular, in electric power generation industry. The main finding of the paper is that firms invest in capacity incrementally over time and the investment rule is that firms invest as if high demand would unfold in the future. Implications of this finding for the electricity industry are, (a) firms need to see growth in electricity demand to be able to invest in generation capacity; (b) firms' investments are not lumpy and not made just at once, but made incrementally over time; (c) firms have to invest before the realization of the demand, whether it turns out to be high or low, because of the lag between investment and production. Another example of such papers is Dijk et al. [11], where capacity investment and access regulation in electricity markets are examined using the real options approach. They consider entry and investment incentives of firms in the presence of a transmission network in which nodal prices guide the efficient use of transmission system. They find that with nodal pricing and financial transmission rights firms tend to overinvest relative to the efficient investment.

A future research direction would be studying supply function equilibria in the presence of transmission constraints, which could change the biding behavior significantly. Wilson [43] presents first order optimality conditions of a firm submitting supply functions to ISO in a simple transmission network. However, he cannot solve for the optimal bid schedules (supply functions) since they depend on the probability distribution of random demand shocks and transmission capacity, unlike in optimal bidding models with no network constraints. It would be challenging but valuable to know the characteristics of optimum supply functions and price distributions, and whether the market outcomes are less competitive when transmission system limits the power flow in certain directions in the network.

Acknowledgements I thank the corresponding editor Panos Pardalos and the anonymous referees for useful comments and suggestions. This research is supported by the Social Sciences and Humanities Research Council of Canada (SSHRC).

References

1. Allaz B, JL Vila (1993) Cournot competition, futures markets and efficiency. J Econ Theory 59: pp. 1–16
2. Anderson EJ, Philpott AB (2002) Using supply functions for offering generation into an electricity market. Oper Res 50:477–489
3. Anderson EJ, Holmberg P, Philpott AB (2010) Mixed strategies in discriminatory divisible-good auctions. Working paper, August, 2010
4. Anwar AW (1999) The case for a discriminatory pricing rule in competitive electricity pools, Mimeo, Edinburgh University, Edinburgh, UK1999
5. Baldick R, Hogan W (2002) Capacity constrained supply function equilibrium models of electricity markets: stability, non-decreasing constraints, and function space iterations, PWP-089 Working Paper, Harvard University, MA, USA
6. Baldick R, Hogan W (2006) Stability of supply function equilibrium: implications for daily versus hourly bids in a poolco market. J Regul Econ 30(2):119–139
7. Baldick R, Grant R, Kahn E (2004) Theory and application of linear supply function equilibrium in electricity markets. J Regul Econ 25:143–167
8. Borenstein S, Bushnell J, Wolak F (2002) Measuring market inefficiencies in California's restructured wholesale electricity market. Am Econ Rev 92:1376–1405
9. Bushnell J, Ishii J (2006) Equilibrium model of investment in restructured electricity markets, Mimeo, December, University of California Berkeley, CA, USA 2006
10. Chaton C, Doucet J (2003) Uncertainty and investment in electricity generation with an application to the case of hydro-Quebec. Ann Oper Res 120:59–80
11. Dijk J, Petroulos G, Willems B (2009) Generation investment and access regulation in the electricity market: a real option approach, Mimeo, Tilburg University, Tilburg, Netherlands, December 2009
12. Fabra N, von der Fehr N, Harbord D (2006) Designing electricity auctions. Rand J Econ 37(1):23–46
13. Federico G, Rahman D (2003) Bidding in an electricity pay-as-bid auction. J Regul Econ 24(2):175–211
14. Garcia A, Stacchetti E (2011) Investment dynamics in electricity markets, Econ Theory 46(2):149–187
15. Garcia A, Shen J (2010) Equilibrium capacity expansion under stochastic demand growth. Oper Res 58:30–42
16. Genc TS (2009) Discriminatory versus uniform-price auctions with supply function equilibrium. J Optim Theory Appl 140:9–31
17. Genc TS, Reynolds SS (2011) Supply function equilibria with capacity constraints and pivotal suppliers. Int J Indus Organ 29(4):432–442
18. Genc TS, Thille H (2011) Investment in electricity markets with asymmetric technologies. Energy Econ 33(3):379–387
19. Genc TS, Zaccour G (2010) Capacity investment dynamics under demand uncertainty, Working paper, University of Guelph, Ontario, Canada; 2010
20. Green R, Newbery D (1992) Competition in the British electricity spot market. J Polit Econ 100(5):929–953
21. Green R (1999) The electricity contract market in England and Wales. J Ind Econ 47 (1):107–124
22. Grossman S (1981) Nash equilibrium and the industrial organization of markets with large fixed costs. Econometrica 49:1149–1172
23. Hart J (1985) Imperfect competition in general equilibrium: an overview of recent work. In: Arrow K, Honkapohja S (eds) Frontiers of economics. Basil Blackwell, Oxford
24. Holmberg P (2008) Unique supply function equilibrium with capacity constraints. Energy Econ 30(1):148–172

25. Holmberg P (2009) Supply Function Equilibria of pay-as-bid auctions. J Regul Econ 36:154–177
26. Holmberg P, Newbery D (2010) The supply function equilibrium and its policy implications for wholesale electricity auctions. Utilities Policy 18(4): 209–226
27. Joskow P (2007) In: Dieter H (ed) Competitive electricity markets and investment in new generating capacity. The new energy paradigm. Oxford University Press, Oxford
28. Joskow P, Kahn E (2002) A quantitative analysis of pricing behavior in California's wholesale electricity market during 2000. Energy J 23:1–35
29. Kahn AE, Crampton P, Porter R, Tabors R (2001) Uniform pricing or pay-as-bid pricing: a dilemma for California and beyond. Electricity J 14:70–79
30. Klemperer P (2002) What really matters in auction design. J Econ Perspect 16:169–190
31. Klemperer PD, Meyer MA (1989) Supply function equilibria in oligopoly under uncertainty. Econometrica 57(6):1243–1277
32. Lave L, Perekhodtsev D (2001) Capacity withholding equilibrium in wholesale electricity markets, Mimeo, Carnegie Mellon University, Pittsburgh, USA 2001
33. Murphy F, Smeers Y (2005) Generation capacity expansion in imperfectly competitive restructured electricity markets. Oper Res 53(4):646–661
34. Newbery DM (2003) What Europe can learn from British privatisations. Economia Pubblica 33(2):63–76
35. Newbery DM (1998) Competition, contracts and entry in the electricity spot market. Rand J Econ 29:726–749
36. Perekhodtsev D, Lave L, Blumsack S (2002) The pivotal oligopoly model for electricity markets, Mimeo, Carnegie Mellon University, Pittsburgh, 2002
37. Philpott AB, Guan Z, Khazaei J, Zakari G (2010) Production inefficiency of electricity markets with hydro generation. Utilities Policy 18(4):175–185
38. Rassenti S, Smith V, Wilson B (2003) Discriminatory price auctions in electricity markets: low volatility at the expense of high price levels. J Regul Econ 23:109–123
39. Roques F, Newbery D, Nuttall W (2005) Investment incentives and electricity market design: the British experience. Rev Netw Econ 4(3):93–128
40. Rothkopf MH (2001) Control of market power in electricity auctions. Mimeo, Rutgers University, New Brunswick, New Jersey, USA 2001
41. Rudkevich A, Duckworth M, Rosen R (1998) Modeling electricity pricing in a deregulated generation industry: the potential for oligopoly pricing in a Poolco. Energy J 19(3):19–48
42. Son YS, Baldick R, Lee KH, Siddiqi S (2004) Short-term electricity market auction game analysis: uniform and pay-as-bid pricing. IEEE Trans Power Syst 19(4):1990–1998
43. Wilson R (2008) Supply function equilibrium in a constrained transmission system. Oper Res 56(2):369–382
44. Wolak F (2009) An assessment of the performance of the New Zealand wholesale electricity market. Investigation report. http://www.comcom.govt.nz/BusinessCompetition/Publications/Electricityreport/DecisionsList.aspx
45. Wolfram CD (1999) Measuring duopoly power in the British electricity spot market. Am Econ Rev 89:805–826
46. Wolfram C (1999) Electricity markets: should the rest of the world adopt the United Kingdom's reforms? Regulation 22(4):48–53

The Economic Impact of Demand-Response Programs on Power Systems. A Survey of the State of the Art

Adela Conchado and Pedro Linares

Abstract Demand Response (DR) programs, which aim to reduce electricity consumption in times of high energy cost or network constraints by allowing customers to respond to price or quantity signals, are becoming very popular in many electricity systems, frequently associated to smart-grid developments. These programs could entail significant benefits for power systems and the society as a whole. Assessing the magnitude of these benefits is crucial to determine their convenience, especially when there are non negligible costs associated to their implementation (if advanced metering infrastructure or control technologies are needed). Quantifying DR benefits requires first to estimate the changes in demand patterns that can potentially be achieved and then to evaluate the effects of those changes on the complex behavior of power systems, neither of these analyses being trivial. This paper presents a survey of the state of the art of these assessments.

Keywords Benefits • Cost-benefit analysis • Demand response • Power systems • Survey

1 Introduction

In the present energy context, in which growing concerns about environmental sustainability and security of supply need to be tackled as cost-effectively as possible, demand side management can play an important role [1, 2]. In the case of electricity, given that its cost and environmental impact vary over time, consuming more efficiently implies not only consuming less, but also managing consumption in time. Tools aiming to promote a more efficient distribution of electricity

A. Conchado (✉) • P. Linares
Instituto de Investigación Tecnológica, Universidad Pontificia Comillas, Santa Cruz de Marcenado 26, 28015 Madrid, Spain
e-mail: adela.conchado@iit.upcomillas.es; pedro.linares@upcomillas.es

A. Sorokin et al. (eds.), *Handbook of Networks in Power Systems I*,
Energy Systems, DOI 10.1007/978-3-642-23193-3_11,

demand in time constitute what is known as Demand Response (DR). In essence, DR programs consist in providing incentives to consumers to get a reduction in demand in times of high energy cost or network constraints. In most cases, this would imply shifting some consumption from peak hours to off-peak hours.

Although DR is not a new concept, it has been gaining interest recently, as power systems become more congested, smart grids develop, and the penetration of renewable energy increases. While most DR programs in past years have consisted on interruptible or curtailable services from large customers, nowadays the development of smart meters, home automation and advanced communication and control technologies enables more sophisticated forms of DR even at the household level, with domestic customers being able to adapt demand at their discretion in response to time-varying price signals.

The current interest in DR is materialized in numerous research projects (GAD (www.proyectogad.es) in Spain, Smart-A (www.smart-a.org) and ADDRESS (www.addressfp7.org) in Europe, Demand Response Research Center (http://drrc.lbl.gov) in the USA and IEA Demand Side Management Programme (www.ieadsm.org) internationally), trials and initiatives. Faruqui and Sergici [3] presented a survey of the 15 most recent experiments with dynamic pricing at the household level. RRI [4] reviewed the current status of DR in the USA, and Goldfine et al. [5] the major developments in DR programs and initiatives. Some countries and regions have carried out studies to assess the cost-effectiveness or potential for advanced metering and DR (e.g. FERC [6] for the USA, NERA [7] for Australia, Vasconcelos [8] for the European Union, Navigant [9] for Ontario (Canada)), and many countries have started deploying smart meters or have set roll-out targets, which will facilitate the implementation of DR programs and broaden their possibilities. In Europe, the penetration rate of smart meters is about 85% in Italy and 25% in France. UK, Spain, Ireland, the Netherlands, Norway and France have set deployment targets to achieve nearly 100% smart meter installation by 2020 [10].

DR programs can result in significant benefits for power systems (e.g. US DOE [11]), but can also entail non negligible costs, especially if an advanced metering, communication or remote control infrastructure is put in place to facilitate automatic demand response. For this reason, assessing the benefits of DR is a necessary step to determine the interest of DR programs, both from the perspective of regulators and market agents.

However, the assessment of benefits *a priori* is not trivial: it is difficult to estimate how demand patterns would change, and understanding the effects of such changes on intrinsically complex power systems requires a thorough analysis. A range of studies have analyzed these effects both qualitatively and quantitatively, providing valuable insights and constituting a useful starting point for future studies.

The purpose of this paper is to present the state of the art of the analysis and assessment of the economic impacts of DR on power systems. Since the evaluation of implementation costs is not too complex, the focus will be on the evaluation of benefits.

After providing some background on DR in Sect. 2, the potential benefits of DR will be identified and described in Sect. 3, and the methodologies to quantify them that have been proposed in the literature will be reviewed in Sect. 4. Throughout these sections, a special emphasis on providing useful references on the different topics will be made. The paper will finish in Sect. 5 with some conclusions about the current status of this topic and the issues that still need to be addressed.

2 Background: Categorization of Demand Response Programs

Being aware of the broad range of DR programs is important to understand the potential benefits that can be achieved and to put the variety of studies that have analyzed them in context. This section provides some background on the different designs and applications of DR programs.

There are many types of DR programs, which can be classified according to various criteria. Table 1 summarizes some classifications proposed in the literature. As shown in Table 1, DR can have reliability or economic purposes [12]. Depending on the factor that triggers demand response, programs can be emergency-based or price-based [13]. With a similar meaning, but referring to the source of the trigger signal, they can be called system-led and market-led programs respectively [14]. According to the type of signal provided (quantity or price), there are load response and price response programs [12]. According to the method used to motivate DR, incentive-based programs or time-based rates can be distinguished (FERC [6]; US DOE [11]). Finally, there are direct load control programs, in which load reductions are controlled by a system operator, or passive load control programs, in which load reductions are controlled by customers (DTE [15]).

To simplify, the whole range of DR programs may be reduced to two types, which correspond to each of the columns in Table 1. On the one hand, DR aiming to improve system reliability is generally implemented through emergency-based, system-led, load-response, incentive-based, direct-load control programs. On the other hand, DR aiming to reduce system costs is generally implemented through price-based, market-led, price-response (using time-based rates), passive load control programs.

Load response programs include direct load control, curtailable load, interruptible load and scheduled load. Price response programs include time-of-use (TOU)

Table 1 Categorization of DR programs

Classification criteria	Dualities		Source
Purpose	Reliability	Economics	[12]
Trigger factor	Emergency-based	Price-based	[13]
Origin of signal	System-led	Market-led	[14]
Type of signal	Load response	Price response	[12]
Motivation method	Incentive-based	Time-based rates	[6, 11]
Control	Direct load control	Passive load control	[15]

Table 2 Some other differentiating factors of DR programs

Other criteria	Dualities	
System/market structure	Vertically-integrated regulated system	Liberalized market
Promotion and financing	By regulator	By market agents
Targeted customers	High-voltage (industrial and large commercial)	Low-voltage (small commercial and domestic)
Automation of response	Manual response (without enabling technologies)	Automatic response (with AMI and/or other smart devices)

tariffs, dynamic pricing (such as critical-peak pricing (CPP) or real-time pricing (RTP)) and demand bidding [12]. In general, in load response programs demand is remotely controlled upon conditions contracted with customers, while in price response programs, customers respond on their discretion to time-varying prices [16].

Some other factors that would influence the characteristics of DR programs, summarized in Table 2, are the following:

- The incentives to undertake DR and the program design differ significantly between liberalized market environments and centralized regulated environments [14, 17].
- Similarly, it is important to consider if the promotion and financing of DR – or the installation of enabling technologies – is assumed by the regulator or is left to the initiative of market agents [7].
- The targeted segment of customers, from large industries to small commercial or domestic loads, is another relevant factor.
- Finally, the installation of enabling technologies critically determines DR options. For example, direct load control programs require remote control capabilities and real-time pricing requires an advanced metering infrastructure.

3 Potential Benefits of Demand Response

DR has a broad range of potential benefits. The benefits that will materialize in practice will depend on the purpose, design and performance of the DR program implemented, as well as on other factors such as the structure of the market/system and the enabling technologies in place.

DR programs can have impacts on system operation, system expansion and market efficiency (the last only applicable in liberalized market environments). In this section, the potential benefits arising in those three aspects of power systems will be identified and described from a theoretical point of view (and summarized in Table 3). Some further considerations about the distribution of benefits among different agents and about smart metering will be pointed out as well.

3.1 System Operation

DR programs where customers are able to respond to price signals that reflect to a certain extent real operational costs (generation and/or network costs) can achieve savings in system operation.

If prices reflect the cost of generation, part of the demand in times of high generation costs may be avoided or shifted to less expensive periods, resulting in some savings in the production of electricity.

If the cost of environmental impact is conveniently internalized in energy prices, the response of demand will also consider the impact on the environment [18]. However, the change in net emissions will be very dependent on the generation mix. In systems in which marginal electricity in peak hours is produced from technologies emitting less CO_2 than marginal technologies in off-peak hours (e.g. on-peak gas and off-peak coal, as occurs in many power systems), shifting some peak demand to off-peak could imply an increase in CO_2 emissions, at least in the short-term [19]. Nevertheless, if not only shifting but also conservation effects from DR are taken into account, the overall emissions are likely to be reduced [20].

Another positive effect of DR on the operation of generation systems is facilitating the real-time balance of supply and demand, which is especially important when intermittent generation has large shares of production [21]. In fact, DR is considered as a major option to decrease problems caused by the variable and uncertain output of intermittent renewable sources [22].

This contribution of DR to real-time balancing, coupled with the fact that DR can help compensate supply shortages with load reductions in case of generation outages, may entail a reduction in the requirements of operating reserves for a certain level of short-term reliability of supply (or to increase short-term reliability of supply for a certain level of operating reserves) [23].

Regarding network operation, if network-driven DR actions are promoted (either through prices or through other agreed incentives), demand can respond to alleviate network constraints or to avoid outages in case of contingencies [24]. Moreover, DR can contribute to reduce lines losses [25]. DR programs can even provide ancillary services for electricity network system operators, such as voltage support, active/reactive power balance, frequency regulation and power factor correction [26]. All these effects on networks can mean an increase in network reliability and quality of supply.

3.2 System Expansion

As already mentioned, DR can potentially reduce demand peaks, both local peaks in a particular area and system peaks.

At the local level, since networks are dimensioned for the highest expected demand, demand clipping can mean a reduction in the need for network reinforcement for a certain level of reliability (or an increase in long-term network reliability for the same level of investment).

At the system level, leveling the demand pattern reduces the need for installed capacity in peaking units. Moreover, it reduces the need of investment in capacity reserves [27] for a certain level of reliability of supply (or increases long-term reliability of supply for a certain level of capacity reserves).

Another effect of DR on the expansion of generation systems, which can be considered a benefit in countries where renewable energy is encouraged, is that it enables higher penetration of intermittent sources (by facilitating supply and demand balancing).

3.3 Market Efficiency

In liberalized environments, market-driven DR programs, most frequently implemented in the form of time-varying tariffs, can allow an active participation of the demand side in the market and thereby achieve significant improvements in market efficiency.

This gives consumers the opportunity to maximize their utility by adjusting their demand in response to price signals. If price signals are accurate (in the sense that they reflect actual costs), only those consumers for whom consuming electricity at a certain time is worth at least as much as the cost it represents at that time would consume, resulting in a more efficient allocation of resources [28].

On the supply side, increasing the elasticity of demand would mitigate the generators' capacity to exercise market power (IEA [14]: 54; Braithwait et al. [27]), which would also entail a reduction in the magnitude and number of price spikes [17, 29].

Prices would also be moderated by the smoothing of the demand profile [14]. However, it should be noticed that price reductions only represent wealth transfers from generators to consumers and not real savings for the society as a whole [27].

DR may allow generators and retailers to reduce the cost of imbalances [14]. Similarly, DR can also be seen as a way of hedging against price and production volatility [30] and extreme system events difficult to predict [31].

With the implementation of DR programs, retailers may increase their business opportunities and offer contracts to customers better suited to their demand profile. At the same time, consumers can benefit from a greater choice of contracts and save money if their consumption profile is favorable to the system (in the sense that demand is low in times of high cost).

3.4 Summary of Benefits

Table 3 summarizes the potential benefits of DR that have been mentioned, categorized according to the activity of power systems where they originate. Notice that benefits are assigned to the activity where they originate regardless of the activity that finally receives them (the distribution of benefits among agents will be discussed in 3.5). In line with this, benefits included in Table 3 are only those that represent actual savings or gains in efficiency for the society as a whole, and not wealth transfers among agents.

Table 3 Potential benefits of DR

	Operation	Expansion	Market[a]
Transmission and Distribution	Relieve congestion Manage contingencies, avoiding outages Reduce overall losses Facilitate technical operation[b]	Defer investment in network reinforcement or increase long-term network reliability	
Generation	Reduce energy generation in peak times: reduce cost of energy and - possibly-emissions[c]	Avoid investment in peaking units	Reduce risk of imbalances
	Facilitate balance of supply and demand (especially important with intermittent generation)	Reduce capacity reserves requirements or increase long-term reliability of supply	Limit market power
	→ Reduce operating reserves requirements or increase short-term reliability of supply	Allow more penetration of intermittent renewable sources[d]	Reduce price volatility
Retailing[a]			Reduce risk of imbalances Reduce price volatility New products, more consumer choice
Demand	Consumers more aware of cost and consumption, and even environmental impacts Give consumers options to maximize their utility: opportunity to reduce electricity bills or receive payments	Take investment decisions with greater awareness of consumption and cost	Increase demand elasticity

[a]Only applicable in liberalized systems
[b]Keep frequency and voltage levels, balance active and reactive power, control power factor, etc.
[c]Depends on the electricity mix
[d]It can be considered a benefit in systems where renewable generation is encouraged

3.5 *Further Considerations on Potential Benefits from Demand Response*

3.5.1 Distribution of Benefits Among Agents

The benefits arising in generation or network activities will not necessarily be received by generation companies and network operators, respectively. The distribution of benefits among the agents is a key issue that needs to be properly assessed considering the particular regulatory framework in place when performing an economic evaluation of a DR program.

In general terms, under a centralized paradigm, the benefits would be directly transferred to consumers through lower tariffs. In liberalized systems, if there is an incentive-based remuneration scheme, benefits arising in distribution would be earned by distribution companies in the short-term, and would be transferred to customers in the long-term through lower access tariffs. On the contrary, the savings arising in the generation system would be transferred directly to customers through lower energy prices (if markets are efficient), meaning at the same time a reduction in the revenues of generation companies. In any case, a more efficient use of energy due to DR should translate into benefits for consumers [32].

According to IEA [14], the distribution of benefits among agents in liberalized environments entails a dispersion of the incentives to undertake DR in the following way:

- Base-load generators have little incentive and see DR only as a means of hedging to unplanned outages, whereas peaking generators view DR as direct competition.
- System operators may be interested in DR to facilitate supply and demand balance and to improve reliability.
- Network operators can use DR to relieve network congestion, improve local reliability or quality of supply or reduce network investments, but their incentives would depend crucially on their regulated remuneration scheme.
- Retailers can be interested in DR as a means to balance their contracted supply with the demand of their consumers.
- Consumers may use DR to reduce their electricity expenses, their incentives to respond basically depending on the incentives they are offered by retailers or utilities.

3.5.2 Smart Metering and Other Enabling Technologies

The potential benefits of DR can be broadened or amplified with the installation of enabling technologies. Indeed, most of the benefits mentioned in Table 3 can only be realized if an advanced metering and/or control infrastructure is in place. Thus, the implementation of dynamic tariffs requires an Advanced Metering Infrastructure (AMI), including the installation of "smart meters" and communication systems, and managing network contingencies through load interruptions or curtailments requires remote control devices.

In the literature, smart metering and demand response are usually related concepts. Many studies that have evaluated the cost-effectiveness of smart meters include the benefits associated to demand response in their assessment (e.g. [16, 33–35]). In fact, the benefits of DR dominate the societal benefits that have been attributed to smart metering in recent business cases [36]. Other studies that analyze smart metering from a regulatory perspective can also provide interesting clues about demand response (e.g. [28, 32]).

However, it should be noticed that the deployment of smart meters would entail some operational benefits not related to DR, such as savings in meter reading and network fault detection. These should be taken into account when performing a cost-effectiveness analysis of advanced metering infrastructures, but will not be included in this review.

Haney et al. [16] mention the following operational benefits of smart meters:

- Improvement in the efficiency of metering services: avoided cost of meter reading, better outage detection, faster response times to outages, improved quality of supply recording and accurate billing.
- Reduction in customer service costs due to a lower level of customer complaints.
- Non-technical losses reduction.
- Others such as greater level of choice in terms of payment options, improved consumption information or micro-generation facilitation.

Smart meters also enable detailed locational data and more efficient pricing to network users of usage and system charges (IEA [14]: 110). Moreover, the knowledge of demand patterns that can be gained with smart metering may allow more efficient network investment and operation (just because of the value of information, even without considering DR). Finally, smart meters may provide greater scope for innovative tariffs and more competition in retailing (Frontier Economics [35]).

There are other technologies (apart from smart meters) that can contribute to DR, such as smart thermostats (the Smart Thermostat Program is an interesting pilot in California that tested smart thermostats to control air conditioning of 5.000 residential customers [37])., lighting control systems, under-voltage and under-frequency relays or thermal storage systems [38]. Lockheed Martin Aspen [39] examines in detail the current status of enabling technologies for homes and small business for either reliability-based or price-based DR programs, and SCE [40] presents an inventory of emerging demand response technologies.

4 Quantitative Assessment of Benefits of Demand Response

A proper quantitative assessment of DR benefits requires first an estimation of the changes induced in the demand, and secondly a thorough analysis of the effect of those changes on power systems. A review of studies that have approached each of these issues will be presented in the following sections.

4.1 Estimating Changes in Demand

Most of the benefits associated to DR programs are directly dependent on the changes in demand achieved with them. Thus, in order to evaluate DR benefits, the shifting and conservation effects in the load profiles of the participating customers need to be properly assessed. Generally, each customer segment needs to be evaluated separately to take into account differences in load patterns and in sensitivities to prices.

Notice that the level of detail in which demand changes are assessed needs to be consistent with the methodology that would be applied to estimate DR benefits. For some simple studies based on estimates, measuring variations in discrete demand blocks (e.g. on-peak and off-peak) may be sufficient, whereas complex analyses using simulation techniques would frequently require hourly or sub-hourly demand patterns before and after DR.

It is also worth noticing the difficulties in predicting changes in demand. The response of consumers is uncertain and can be influenced by multiple factors, such as climate, tariff design (prices), customer type (available electric devices, incomes, level of consumption, etc.), enabling technologies, the way in which critical prices or system alerts are notified, feedback information about consumption reported to consumers, awareness and education campaigns launched, etc. [41–46].

Next, three different ways to approach the assessment of the effects of DR on consumption patterns will be presented: (1) using estimates from previous studies or experiences (mainly price elasticities of demand), (2) developing an econometric demand model or (3) simulating demand with a bottom–up model. An interesting review of methods to evaluate demand response using a different classification than the one proposed here can be found in Woo and Herter [47].

4.1.1 Price Elasticity Estimates

Using price elasticity estimates based on previous studies or experiences is one possible and simple approach to evaluate changes in demand. The price elasticity of demand expresses the demand increment in percentage terms in response to a one percentage point increase in price.

In the context of DR, three types of elasticity are commonly used: own-price elasticity, cross-price elasticity and elasticity of substitution. Own-price elasticity expresses the demand change in one period for a 1% increment in the price of that period, whereas cross-price elasticity expresses the variation of demand in one period for a 1% increment in the price of other period (generally, between on-peak and off-peak periods). The elasticity of substitution expresses the demand shifted from on-peak periods to off-peak periods given a 1% increment in the relative price on-peak to off-peak [48].

Some authors have compiled price elasticity measures observed in different types of DR programs and different regions, or have presented their own estimates. Next, Table 4 presents some elasticity values for different target customers and time-varying tariffs adapted from US DOE [11].

Table 4 Summary of price elasticity estimates (Adapted from [11])

Target customers	Type of program	Own-price elasticity	Elasticity of substitution	Region
Residential (and small commercial)	TOU		0.07 to 0.21 (0.14 average)	US
	TOU/CPP	−0.1 to −0.8 (−0.3 average)		US - International
	CPP		0.04 to 0.13 (0.09 average)	California
	RTP	−0.05 to −0.12 (average −0.08)		Illinois
Medium or large commercial and industrial	RTP	−0.01 to −0.28		Georgia
		−0.01 to −0.27		UK
		<−0.01 to −0.38		N-S Carolina
			0.10 to 0.27	Southwest US
			0.02 to 0.16 (0.11 average)	New York

A number of studies have applied this kind of price elasticity estimates to evaluate changes in demand when assessing DR benefits. For example, Berg et al. [49] use hourly own-price and cross-price elasticities (both short-run and long-run values), whereas Navigant [9] applies elasticities of substitution to compute the load shifts in the demand profile of different segments of consumers.

4.1.2 Econometric Demand Models

Econometric models try to derive the level of demand from some explicative factors based on microeconomic theory. The usual formulation is the maximization of the utility for consumers of their electricity consumption. The explicative factors most frequently used are the electricity price and the income (or budget) of the consumer. Social and demographic conditions, dwelling characteristics or technological aspects can be taken into account as well. Econometric models are developed from data of real experiences, and then used to evaluate other programs. Price-elasticity estimates are generally obtained from this type of model.

Some early examples of this approach are the studies of Lawrence and Braithwait [50] and Hausman et al. [51], which develop an econometric demand model to analyze the effect of TOU tariffs. In both studies, demand is given by the maximization of the consumer utility function subject to a budget constraint. Demand in different periods is considered as a different product, in such a way that load shifting can be modeled as the substitution between two products.

Similar studies analyzing TOU pricing can be found in Caves et al. [52], Parks and Weitzel [53] and Hill [54]. More recently, the model proposed by Reiss and White [55] enables the evaluation of different tariff designs (not only TOU).

From a different perspective (and based on a real RTP experiment), Allcott [56] estimates hourly residential demand as a utility function depending on household characteristics, daily prices and some load substitution and shifting parameters (based on temperature and dummy variables).

Econometric demand models can provide an accurate representation of the demand if the most relevant factors affecting consumption are included and the parameters expressing how demand changes with respect to those factors are properly assessed. However, since these parameters are adjusted for a given set of data under some particular conditions, econometric demand models may not be valid when the underlying conditions change. Thus, the main limitation of this type of models is that their results may be very dependent on the underlying conditions and difficult to extrapolate.

A final consideration is that econometric models that overlook the heterogeneous nature of electric loads may not be sufficiently accurate to evaluate DR actions. The utility of consumption strongly depends on its final use (the specific electric device used) and on circumstantial aspects (the service provided by the electric device in the time of consumption). Moreover, the capacity to reduce or shift demand also depends strongly on the technical potential of the electric devices in use. Therefore, including technological considerations into the demand model can improve the representation of DR actions.

4.1.3 Bottom–Up Demand Models

Bottom–up demand models, unlike econometric models, aim to capture the specific loads that constitute the demand. The demand profile is obtained by aggregation of elemental loads (that may represent individual customers or the consumption of each appliance).

There are some interesting studies that have modeled electric demand using this approach but without considering any demand response action. It is the case of Cappasso et al. [57] and Boonekamp [58] for domestic demand. Even if these models would not be valid to evaluate DR programs, their contributions may be useful to develop models adapted to DR evaluation.

Other studies have incorporated into their bottom–up models the possibility to shift or reduce demand as a consequence of direct load control. The work of Paatero and Lund [59] is a good example. The authors try to overcome the lack of detailed data about domestic consumption by using statistical data easily available, and incorporate stochastic processes to take into account the random nature of the demand.

Finally, some studies have considered the response of demand to prices explicitly. Conchado and Linares [60] suggest allocating simplified consumption cycles of electric appliances into the demand curve of individual consumers, and shift or reduce the consumption of each appliance according to its technical potential with the objective of minimizing the cost of consumption. In a different way, Lu et al. [61] model thermostatically controlled loads by means of equivalent electric circuits that represent their heat transfer properties, and non-thermostatically controlled loads as tasks in a queue system.

4.2 Assessing Benefits

The assessment of DR benefits can be approached from a range of possible methods, such as those based on avoided costs, resource planning, welfare analysis, value of system reliability, transmission planning and forward capacity auctions identified by Heffner [62]. The suitability of each method will depend on the type of benefit to be assessed, and there is no single approach able to capture accurately the whole range of effects of DR on power systems.

In this section, a survey of the state of the art of the quantification of DR benefits will be presented, distinguishing between two types of studies: those based on estimates and those based on simulation techniques. Both approaches have advantages and disadvantages. Studies based on estimates are simple and transparent but may fail to represent accurately the complexity of power systems. In contrast, analyses with simulation techniques allow for detailed representations of power systems but are more complicated and difficult to track back and compare.

4.2.1 Studies Based on Estimates

In this type of studies, DR benefits are derived analytically from some estimates, necessarily making simplifications about consumer and market behavior to express all relationships in algebraic terms. Generally, only a few periods (such as on-peak and off-peak) are considered for the analysis.

An illustrative example of this type of study is found in Baer et al. [63], who evaluate the GridWise initiative in the USA. Given a set of input data (market penetration and price-elasticity of demand by end-use sector, wholesale peak and off-peak prices in a baseline scenario, projected generating reserve margin in 2025, discount rate, etc.), the system peak-load reduction is calculated. From this estimate, generation capital cost deferrals are computed by multiplying peak-load reduction by the capital cost of peaking units (gas-fired combustion turbines or diesel generators). Similarly, the operating and fuel costs associated to deferring new capacity are calculated by multiplying peak-load reduction by the fuel costs and operation and maintenance costs of the avoided peaking units. T&D capital cost deferrals are directly estimated as a function of generation capacity deferral.

Another example of a study based on estimates is a cost-benefit analysis of advanced metering in France by CapGemini [33]. Three scenarios representing different technology levels are considered, and it is assumed that the level of demand response depends on the technological capabilities implemented. The final estimates of benefits are allocated to generators, distributors and suppliers, which is very useful for regulatory design.

Other similar studies have been presented by Faruqui et al. [10] for the European Union, Ofgem [34] for the UK, Siderius and Dijkstra [64] for the Netherlands, Faruqui and George [65] for the USA, Navigant [9] for Ontario, or ESC [66] for Australia.

The strong points of this approach are its simplicity and transparency. The results can be easily tracked back to the original assumptions, and it facilitates comparing the results of alternative DR designs. The drawback, however, is that it may not represent with enough level of detail the complex behavior of the market and the numerous interactions occurring in power systems. Moreover, since stochasticity is not considered (only point estimates of the parameters are used), the dependence of the results on the assumptions is magnified with respect to other methods [36].

4.2.2 Analyses with Simulation Techniques

Instead of using estimates, DR benefits can be evaluated using models that simulate the behavior of power systems. Since simulation models allow for a detailed representation of the expansion and/or operation of generation systems and networks, or the performance of the market, this approach seems to be the most accurate for the evaluation of DR benefits.

Using simulation models, DR benefits are generally computed as the difference in the results between two simulations, one for a baseline scenario without DR and another for a scenario with DR.

Most of the studies assessing DR benefits with simulation techniques have focused on the impacts on the generation system or the wholesale market, but there are some studies addressing network impacts as well. A review of these studies is presented in this section, showing first those with a focus on the generation system and wholesale market, and next those focusing on a network impacts. At the end of the section, Table 5 provides a brief summary of the purpose, scope and methodology of the studies mentioned.

Impacts on the Generation System and Wholesale Market

When modeling generation systems, whether DR is evaluated in a regulated environment or in a liberalized market affects the simulation framework. The model can be developed from the viewpoint of an utility (e.g. [49]) or from a market perspective, either assuming perfect competition (e.g. [67]) or imperfect competition (e.g. [56]).

DR is sometimes included endogenously into the simulation models, either by considering price-elastic demand-side bids in market equilibrium models (e.g. [76]) or introduced as an available resource under centralized market approaches (e.g. [77]). Some other studies determine the demand exogenously, either assuming certain load reductions (e.g. [69]) or evaluating *ex-ante* the changes in demand (e.g. [70]).

Some authors have analyzed DR benefits considering the stochasticity of future outcomes for key variables, as Andersen et al. [68], who use Monte Carlo

Table 5 Summary of the purpose, scope and methodology of the references mentioned for the quantification of DR benefits with simulation techniques

Reference	Summary
[56]	Following Borenstein [67], evaluates the effects of RTP in the PJM market, using a two-stage model that simulates the entering of new units in the first stage and the day-ahead market in the second stage. DR is incorporated changing the slope of the demand.
[68]	Assess the short-term value of DR in the Nordic Market using the partial equilibrium model Balmorel coupled with Monte Carlo simulations to include extreme situations, and address market power mitigation simulating supply function equilibrium competition in the Danish system.
[49]	Simulate the impact of TOU pricing on generation operation and expansion under a utility perspective, using price elasticities to evaluate changes in demand.
[67]	Evaluates the impact of RTP on long-run efficiency in a competitive electricity market with simplified simulations (representative parameters for US).
[69]	Estimates short-term impacts of day-ahead demand curtailment on locational marginal prices in the PJM market, using a model that simulates the generation dispatch together with the transmission network load flow.
[20]	In their assessment of DR benefits and costs for Spain, the authors analyze the impact of DR in the expansion of distribution networks by means of a model able to quantify the cost of reinforcements in real networks for expected increases in demand.
[70]	Simulate the effect of RTP (at the residential level) in the Spanish electricity market with a detailed generation expansion model. DR effect on the load profile is estimated ex-ante with a bottom–up model of domestic electricity demand.
[71]	Evaluate several time-varying tariffs for large customers in New England using a prospective price formation simulation model able to compute how locational marginal prices are impacted by load changes.
[25]	Provide a quantitative estimate of the possible reduction in losses associated to domestic demand shifting in Great Britain using a spreadsheet model.
[72]	Estimates the value of smart domestic appliances to reduce balancing costs in European countries with high penetration of wind generation by simulating the annual system operation based on simultaneous scheduling of both generation and DR, including DR as part of the standing reserve providers.
[73]	Explore how RTP prices could contribute to diminish wind spillages by increasing the flexibility of demand in response to wind availability or to system constraints that limit wind generation, using a unit-commitment model with DC power-flow of the Texas power system.
[74]	Analyze the potential of DSM to mitigate network congestion, increase the utilization of network assets and avoid wind spillages by means of a multi-period DC-OPF applied to a urban distribution network, considering historical time series data for demand and wind output.
[75]	Evaluate congestion costs in constrained electricity networks (with high penetration of wind generation) by incorporating DSM options (thermal load management and appliance shifting algorithms) into a DC-OPF model.
[76]	Quantify the effect of increased participation of the demand side in the electricity market using a centralized complex-bid market-clearing mechanism that considers the load shifting behavior of consumers who submit price-sensitive bids. Algorithm tried only in a test system.

(continued)

Table 5 (continued)

Reference	Summary
[77]	Use a resource planning approach to assess the value of DR in a 19-year horizon from an utility perspective, considering 100 Monte Carlo scenarios (case study for a region in the USA).
[78]	Perform an economic welfare analysis of DR in the PJM electricity market with a simulation of demand-side bidding, analyzing the tradeoff between the distortions introduced by the subsidies provided to responsive consumers and the social welfare gains.

simulations (following Violette et al. [77]) to evaluate the potential of DR not only in average but also in extreme situations.

In order to consider network congestion, it is possible to include in the generation model a representation of the transmission network and simulate the power flow through lines. DR impacts can then be assessed by computing locational marginal prices (LMP), as done by Neenan et al. [71] or Brattle [69]. Walawalkar et al. [78] also use LMP in their economic welfare analysis, but compute them by means of an econometric model.

The potential of DR to facilitate real-time balancing on supply and demand in systems with large penetration of wind generation has been investigated by Sioshansi and Short [73] and Silva [72]. The former quantify the reduction of wind spillages if demand is elastic to real-time prices that reflect wind availability and network constraints, whereas the latter evaluates the contribution of DR to balancing considering it as a reserve resource that can be scheduled to minimize system costs.

Impact on Networks

The quantification of potential impacts of DR on the network system has not been sufficiently investigated. Only a few studies have been found providing estimates of DR network benefits.

The impact of DR on the investments of distribution networks has been evaluated by Conchado and Linares [20] using a detailed network expansion simulation model and assuming certain reductions in the peak demand of participating customers.

Regarding network operation, the reduction in distribution losses due to domestic load-shifting has been assessed by Shaw et al. [25], evaluating how the network power flow profiles can be changed by load-shifting and assuming that the overall potential for loss reduction is a function of the preexisting losses and the demand patterns.

A method to quantify the value of DR to alleviate network congestion has been proposed by Stanojević and Silva [75]. Using a DR model (that includes thermal load management and appliance shifting at the domestic level) integrated in a

DC-OPF model allows for estimating the reduction of congestion costs due to the modification of the demand patterns.

In a similar way, the study by Stanojević et al. [74] shows how the modification of daily demand patterns can improve the utilization of existing network capacity by reducing network critical loading and congestion in a stressed distribution network. The model used combines the dispatch of the generation units with a multi-period optimal power flow where DR is applied to minimize the re-dispatch.

Table 5 summarizes the purpose, scope and methodology of the references mentioned throughout this section.

5 Conclusions

Demand response programs, usually in the context of smart grid developments, show a large potential and a promising future, particularly if the benefits that they may provide for generation systems, networks, retailers and consumers are realized. As such, they are currently being evaluated in many electricity systems across the world, either by rather simplified estimations, or by complex simulation and econometric approaches. This paper has presented a comprehensive survey of the state of the art of these assessments.

Our survey shows that the major challenges of these analysis lie in the estimation of the changes in demand profiles from these demand-response programs, and also on the assessment of the impact of these changes in demand on generation and network systems. Although many interesting studies have been identified, and given the relative infancy of these programs, much more research is needed in order to provide reliable estimates of the benefits of demand response programs, and therefore to give an rigorous indication of their interest in the short and long term.

A first point of research should be the extension of the assessment of demand changes to many countries which are considering the implementation of demand response programs. The simple extrapolation of estimates will probably fall short of providing a reliable assessment, and therefore more pilot studies to quantify the extent and direction of changes of demand expected from DR programs would be much welcome.

Then, two major issues regarding the impact of these demand changes on the power system require more attention. First, the estimation of the benefits of these changes for network systems, both in terms of operation and investment, but particularly regarding short term emergency actions, quality of service considerations, or black starting capabilities. The assessment of the benefits for network systems is also more complicated given the dependence of these benefits on the specific regulation of the distribution companies. The second issue concerns the interaction of DR programs with intermittent generation: DR may contribute to a large extent to a stronger integration of intermittent energy sources into power systems, and therefore a careful estimation of their interaction is warranted. Again, this is not an easy task, given the non deterministic character of both DR and intermittent generation, and the sequential nature of DR actions to counteract generation deviations.

On the positive side, our survey also shows that this is a burgeoning field, with much activity in many lines of research. As such, we hope that, by identifying and categorizing them, it will provide a good starting point for many of these new inquiries.

Acknowledgements This research has been funded by the GAD Project (www.proyectogad.es) and by the ADDRESS project (funded by the European Community's Seventh Framework Programme (FP7/2007-2013) under grant agreement n° 207643).

The GAD (Active Demand Management) project, which is financed by the Center for Industrial Technological Development (CDTI), Spanish Ministry of Industry, Trade and Tourism, has as objective the research and development of solutions for the optimization of electricity consumption in low and middle voltage consumers. The project consortium is led by Iberdrola Distribución Eléctrica S.A, and features other 14 firms: Red Eléctrica de España, Unión Fenosa Distribución, Unión Fenosa Metra, Iberdrola, Orbis Tecnología Eléctrica, ZIV Media, DIMAT, Siemens, Fagor Electrodomésticos, BSH Electrodomésticos España, Ericsson España, GTD Sistemas de Información, Acceda Mundo Digital and Airzone. There are also other 14 research centers collaborating in the project.

Some parts of the text draw extensively from reports issued within the ADDRESS project.

References

1. IEA (2008) World energy outlook. International Energy Agency, Paris
2. EC (2005) Green paper on energy efficiency or doing more with less. COM (2005) 265 final. Commission of the European Communities, Brussels
3. Faruqui A, Sergici S (2009) Household response to dynamic pricing of electricity – a survey of the experimental evidence. Working paper series, Jan 10 2009
4. RRI (2008) The status of U.S. demand response. Research Reports International, 7th edn
5. Goldfine D, Haldenstein M, Moniz A, Traverso D (2008) Demand response review – a survey of major developments in demand response programs and initiatives. Edison Electric Institute, Washington
6. FERC (2006) Assessment of demand response and advanced metering. Staff report, AD-06-2-000, Federal Energy Regulatory Commission
7. NERA (2008) Cost benefit analysis of smart metering and direct load control. Overview report for consultation. Report for the Ministerial Council on Energy Smart Meter Working Group, NERA Economic Consulting
8. Vasconcelos J (2008) Survey of regulatory and technological developments concerning smart metering in the European Union electricity market. RSCAS Policy Papers 2008/01. Florence School of Regulation
9. Navigant Consulting (2005) Benefits of smart metering for Ontario. Discussion draft, April 5. Presented to Ontario Ministry of Energy
10. Faruqui A, Harris D, Hledik R (2009) Unlocking the €53 billion savings from smart meters in the EU. Discussion paper. The Brattle Group
11. US DOE (2006) Benefits of demand response in electricity markets and recommendations for achieving them. United States Department of Energy, Feb 2006
12. RMI (2006) Demand response: an introduction - Overview of lessons, technologies, and lessons learned. Rocky Mountain Institute, 2006

13. Faruqui A, Hledik R (2007) The state of demand response in California. Prepared for California Energy Commission, final consultant report, CEC-200-2007-003-F, September 2007
14. IEA (2003) The power to choose – demand response in liberalised electricity markets. OECD/International Energy Agency, Paris. ISBN 92-64-10503-4
15. DTE Energy (2007) Demand response overview and pilot concepts, 26 July 2007
16. Haney AB, Jamasb T, Pollit M (2009) Smart metering and electricity demand. Technology economics and international experience. Cambridge working paper in economics 0905, EPRG working paper 0903. Electricity Policy Research Group, University of Cambridge
17. Borenstein S, Jaske M, Rosenfeld A (2002) Dynamic pricing, advanced metering and demand response in electricity markets. CSEM WP 105, University of California Energy Institute
18. Spees K, Lave LB (2007) Demand response and electricity market efficiency. Electricity J 20:69–85
19. Holland SP, Mansur ET (2007) Is real-time pricing green? The environmental impacts of electricity demand variance. Working paper 1350, October, National Bureau of Economic Research. http://www.nber.org/papers/w13508
20. Conchado A, Linares P (2009b) Gestión activa de la demanda eléctrica doméstica: beneficios y costes. IIT Working Paper, Universidad Pontificia Comillas, Madrid, 2009
21. Zibelman A, Krapels EN (2008) Deployment of demand response as a real-time resource in organized markets. Electricity J 21(5):51–56
22. Kärkkäinen S, Ikäheimo J (2009) Integration of demand side management with variable output DG. In: 10th IAEE European conference, Vienna, 7–10 Sep 2009
23. Earle R, Kahn EP, Macan E (2009) Measuring the capacity impacts of demand response. Electricity J 22(6):47–58
24. Affonso CM, da Silva LCP, Freitas W (2006) Demand-side management to improve power security. In: Transmission and distribution conference and exhibition, 2005/2006 IEEE PES, Dallas, May 2006
25. Shaw R, Attree M, Jackson T, Kay M (2009) The value of reducing distribution losses by domestic load-shifting: a network perspective. Energ Policy 37:3159–3167
26. Crossley D (2008) Assessment and development of network-driven demand-side management measures. IEA Demand Side Management Programme, Task XV, Research report No 2. Energy Futures Australia Pty Ltd, Hornsby Heights, NSW
27. Braithwait SD, Hansen DG, Kirsch LD (2006) Incentives and rate designs for efficiency and demand response. Lawrence Berkeley National Laboratory, LBNL-60132
28. EEI (2006) Responding to EPAct 2005: looking at smart meters for electricity, time-based rate structures, and net metering. Edison Electricity Institute, Washington
29. Kirschen D (2003) Demand-side view of electricity markets. IEEE T Power Syst 18(2):520–527
30. PLMA (2002) Demand response: principles for regulatory guidance. Peak Load Management Alliance, Feb 2002
31. Violette D, Freeman R, Neil C (2006a) Valuation and market analyses. Volume I: overview. Prepared for: International Energy Agency, Demand Side Programme, Jan 2006
32. ERGEG (2007) Smart metering with a focus on electricity regulation. European Regulators' Group for Electricity and Gas. E07-RMF-04-03
33. CapGemini (2007) Comparatif international des systèmes de télé-relève ou de télégestion et étude technico-économique visant à évaluer les conditions d'une migration du parc actuel de compteurs. March
34. Ofgem (2006) Domestic metering innovation, 1 Feb 2006
35. Frontier Economics (2006) Current prices, anybody? The costs and benefits of "smart" electricity meters, Feb 2006
36. Neenan B, Hemphill R (2008) Societal benefits of smart metering investments. Electricity J 21 (8):32–45

37. KEMA-XENERGY (2003) Smart thermostat program impact evaluation, Madison, 24 Feb 2004
38. Batlle C, Rodilla P (2008) Electricity demand response tools: current status and outstanding issues. Working Paper IIT-08-006A. Prepared for: Special Issue on Incentives for a low-carbon energy future, European Review of Energy Markets
39. Lockheed Martin Aspen (2006) Demand response enabling technologies for small-medium businesses. Technical report prepared in conjunction with the 2005 California Statewide Pricing Pilot, R.02.06.001
40. SCE (2006) Inventory of emerging demand response technologies. Southern California Edison
41. Kohler DF, Mitchell BM (1984) Response to residential time-of-use electricity rates - How transferable are the findings? J Econometrics 26:141–177
42. Faruqui A, George SS (2005) Quantifying customer response to dynamic pricing. Electricity J 18:53–63
43. Herter K (2007) Residential implementation of critical-peak pricing of electricity. Energ Policy 35:2121–2130
44. EEE (2006) A survey of Time-Of-Use (TOU) pricing and Demand-Response (DR) programs, Energy and Environmental Economics, San Francisco, July 2006
45. Summit Blue Consulting (2006) Evaluation of the 2005 energy-smart pricing plan. Summit final report, Jan 2006
46. Darby S (2006) The effectiveness of feedback on energy consumption: a review for DEFRA of the literature on metering, billing and direct displays. Environmental Change Institute, University of Oxford
47. Woo CK, Herter K (2006) Residential demand response evaluation: a scoping study. LBNL-61090, collaborative report. Demand Response Research Center (DRCC), Ernest Orlando Lawrence Berkeley National Laboratory
48. King CS, Chatterjee S (2003) Predicting California demand response. Public Utilities Fortnightly, July 2003, pp 27–32
49. Berg SV, Capehart BL, Feldman J, LaTour S, Sullivan RL (1983) An interdisciplinary approach to cost/benefit analysis of innovative electric rates. Resour Energ 5:313–330
50. Lawrence A, Braithwait S (1979) The residential demand for electricity with time-of-day pricing. J Econometrics 9:59–77
51. Hausman JA, Kinnuca M, McFadden D (1979) A two-level electricity demand model. Evaluation of the Connecticut time-of-day pricing test. J Econometrics 10:263–289
52. Caves DW, Christensen LR, Schoech PE (1984) A comparison of different methodologies in a case study of residential time-of-use electricity pricing – cost-benefit analysis. J Econometrics 26:17–34
53. Parks RW, Weitzel D (1984) Measuring the consumer welfare effects of time-differentiated electricity prices. J Econometrics 26:35–64
54. Hill LJ (1991) Residential time-of-use pricing as a load management strategy. Effectiveness and applicability. Utilities Policy 1:308–318
55. Reiss PC, White MW (2005) Household electricity demand, revisited. Rev Econ Stud 72:853–883
56. Allcott H (2008) Real time pricing and imperfect competition in electricity markets. Working paper. http://economics.stanford.edu/seminars/real-time-pricing-and-electricity-markets
57. Capasso A, Grattieri W, Lamedica R, Prudenzi A (1994) A bottom-up approach to residential load modeling. IEEE T Power Syst 9(2):957–964
58. Boonekamp PGM (2007) Price elasticities, policy measures and actual developments in household energy consumption – a bottom up analysis for the Netherlands. Energ Econ 29:133–157
59. Paatero JV, Lund PD (2006) A model for generating household electricity load profiles. Preprint version. http://users.tkk.fi/patte/pub

60. Conchado A, Linares P (2009a) Gestión activa de la demanda eléctrica: simulación de la respuesta de los consumidores domésticos a señales horarias de precio. IV Congress AEEE, Sevilla, 2009
61. Lu N, Chassin DP, Widergren SE (2004) Simulating price responsive distributed resource. In: Power systems conference and exposition, IEE PES, New York,, 2004
62. Heffner G (2007) A framework for demand response valuation, Demand Response Research Center
63. Baer W, Fulton D, Mahnowski S (2004) Estimating the benefits of the GridWise initiative. Phase I report, May. Rand Science and Technology
64. Siderius HP, Dijkstra A (2006) Smart metering for households: cost and benefits for the Netherlands. SenterNovem
65. Faruqui A, George SS (2002) The value of dynamic pricing in mass markets. Electricity J 15:45–55
66. ESC (2004) Mandatory rollout of interval meters for electricity consumers. Final decision, July 2004. Essential Services Commission, Australia
67. Borenstein S (2005) The long-run efficiency of real-time electricity pricing. Energy J 26(3):93–116
68. Andersen FM, Jensen SG, Larsen HV, Meibom P, Ravn H, Skytte K, Togeby M (2006) Analyses of demand response in Denmark. Risø National Laboratory, Risø-R-1565(EN), October
69. Brattle Group (2007) Quantifying demand response benefits in PJM. Prepared for PJM Interconnection, LLC and the Mid-Atlantic Distributed Resources Initiative (MADRI) Jan 2007
70. Linares P, Conchado A (2009) Gestión activa de la demanda eléctrica: Evaluación de su impacto en el sistema de generación. IV Congress AEEE, Sevilla
71. Neenan B, Cappers P, Pratt D, Anderson J (2005) Improving linkages between wholesale and retail markets through dynamic retail pricing. Prepared for New England ISO. www.iso-ne.com
72. Silva V (2009) Value of smart appliances in systems balancing. Prepared for EIE project "Smart domestic appliances in sustainable energy systems" (SMART-A), WP 4
73. Sioshansi R, Short W (2009) Evaluating the impacts of real-time pricing on the usage of wind generation. IEEE T Power Syst 24(2):516–524
74. Stanojević V, Strbac G, Silva V, Lang P, Pudjianto D, Macleman D (2009) Application of storage and demand side management to optimise existing network capacity. In: CIRED, 20th international conference on electricity distribution, Prague, 8–11 June 2009
75. Stanojević V, Silva V (2009) Role of demand side management to reduce system constraints and investment. Prepared for EIE project "Smart domestic appliances in sustainable energy systems" (SMART-A), WP 4
76. Su C, Kirschen D (2009) Quantify the effect of demand response on electricity markets. IEEE T Power Syst 24(3):1199–1207
77. Violette D, Freeman R, Neil C (2006b) Valuation and market analyses. Volume II: Assessing the DRR benefits and costs. Prepared for: International Energy Agency, Demand Side Programme, Jan 2006
78. Walawalkar R, Blumsack S, Apt J, Fernands S (2008) An economic welfare analysis of demand response in the PJM electricity market. Energ Policy 36:3692–3702

Investment Timing, Capacity Sizing, and Technology Choice of Power Plants

Ryuta Takashima, Afzal S. Siddiqui, and Shoji Nakada

Abstract Deregulation of electricity industries has created wholesale markets with uncertain prices and offered greater flexibility to investors to make decisions. In this chapter, we consider the problem of a typical investor who has discretion over not only the timing, but also the sizing of a new power plant. The interaction between these two types of managerial flexibility may be addressed analytically using the real options approach. Since an investor may also have discretion over technology choice, we allow for an investment opportunity in two mutually exclusive projects with embedded timing and sizing options. Via numerical examples, we illustrate how an investor may make decisions about timing, sizing, and technology choice. Sensitivity analyses to key parameters also highlight the intuition for how decisions are made.

Keywords Electricity price • natural gas-fired power • nuclear power • optimal investment decision • real options • technology choice

---

R. Takashima (✉)
Department of Risk Science in Finance and Management, Chiba Institute of Technology, Chiba, Japan
e-mail: takashima@sun.it-chiba.ac.jp

A.S. Siddiqui
Department of Statistical Science, University College London, London, UK

Department of Computer and Systems Sciences, Stockholm University/KTH, Stockholm, Sweden
e-mail: afzal@stats.ucl.ac.uk

S. Nakada
Department of Nuclear Engineering and Management, The University of Tokyo, Bunkyo-ku, Tokyo, Japan
e-mail: nakada@n.t.u-tokyo.ac.jp

A. Sorokin et al. (eds.), *Handbook of Networks in Power Systems I*,
Energy Systems, DOI 10.1007/978-3-642-23193-3_12,

1 Introduction

Reform of electricity industries worldwide has been based on the premise that greater economic efficiency may be achieved if participants were allowed to make investment and operational decisions based on price signals. In particular, while the transmission sector still requires regulatory oversight due to its natural monopoly characteristics, there is no reason for the potentially competitive generation and retail sectors to be subject to state regulation (see Wilson [1]). Hence, even though the outcome of these reforms is subject to debate (see Hyman [2]), there is a greater role for markets now.

In tandem with these reforms, a variety of spot and forward markets for electricity have been set up in order to reflect its relative scarcity. Unlike in the vertically integrated paradigm, generators and retailers must account for wholesale prices when making their investment and operational decisions. These prices may be subject to great uncertainty, which makes the timing of investment and operational decisions and the role of managerial flexibility more crucial. Consequently, traditional methods, such as now-or-never net present value (NPV) and internal rate of return (IRR), may be inadequate to cope with dynamic aspects of existing electricity markets, and decision support based on real options (see Dixit and Pindyck [3] and McDonald and Siegel [4]) is one of the alternatives that has been proposed.

The amenability of the real options approach to the energy sector has been widely demonstrated at least as far back as Ekern [5]. Recent work has explored the valuation of generation and transmission assets (Deng et al. [6]), optimal operation of power plants (Tseng and Barz [7] and Deng and Oren [8]), and incentives for construction of nuclear power plants (Rothwell [9]). Modeling decision-making for investment opportunities with embedded options to abandon, expand, or switch projects after adoption is particularly relevant in the electricity industry. Towards this end, real options has been applied to problems of modularity (Gollier et al. [10], Näsäkkälä and Fleten [11], and Siddiqui and Maribu [12]), capacity sizing (Bøckman et al. [13]), and equipment replacement (Takashima et al. [14]). Furthermore, recent advances in real options theory, which address the selection of mutually exclusive projects (Décamps et al. [15]) and the game-theoretic implications of duopolistic competition (Huisman and Kort [16]), have been applied to problems of technology choice (Fleten et al. [17] and Wickart and Madlener [18]) and strategic investment (Takashima et al. [19]).

Here, we take the perspective of a firm that has the perpetual right, but not the obligation, to invest in a power plant. After the plant is constructed, it will operate continuously for a fixed number of years before being decommissioned. During its operating lifetime, the plant will provide the firm with a profit flow related to the stochastic electricity price. We assume that the plant's operating cost is constant, and its construction cost depends on its size. Thus, the firm's problem is to determine not only when to invest, but also its plant's capacity. Furthermore, it may choose between two types of technologies, e.g., nuclear or natural gas. We illustrate how the real options method is able to provide decision support to handle the three types of managerial flexibility (timing, sizing, and technology choice) given a stochastic electricity price, which would not have been possible via traditional valuation tools.

The structure of this chapter is as follows:

- Section 2 provides the assumptions for the model
- Section 3 introduces the basic real options model for investment under uncertainty with fixed capacity for a single technology
- Section 4 formulates and solves the problem with endogenous capacity sizing
- Section 5 extends the model to enable mutually exclusive investment in two types of technologies with discretion over timing and sizing
- Section 6 provides numerical examples to illustrate the concepts
- Section 7 summarizes the results of this chapter and outlines directions for future research in this area

2 Model Setup

Consider a firm that starts operating a power plant of production output Q (in MWh/annum) by incurring investment cost $I(Q)$ (in \$) that is a non-linear function of Q, i.e., $I(Q)=\delta Q^{\varepsilon}$ where $\delta>0$ and $\varepsilon>1$. We assume that the investment cost function is convex in output due to diminishing marginal returns. After the investment decision and the construction lead time, T (in years), instantaneous cash flow, π_t, is generated from the power plant over its operating lifetime, L (in years).

Suppose that the investment decision and the output capacity are dependent on the dynamics of the electricity price. If we assume that the firm is a price taker, i.e., its actions have no influence on the dynamics of the electricity price, then the electricity price, P_t(\$/MWh), may be considered as exogenous. Here, we model it as evolving according to a geometric Brownian motion (GBM) process,

$$dP_t = \mu P_t + \sigma P_t dW_t, \quad P_0 = p, \tag{1}$$

where μ and σ are the risk-adjusted expected growth rate and the volatility of P_t respectively, and W_t is a standard Brownian motion.

Consequently, the profit flow can be represented by the following equation,

$$\pi_t \equiv \pi(P_t, Q) = (P_t - c)Q, \tag{2}$$

where c is the operating cost (in \$/MWh) that is composed of the fuel cost as well as operating and maintenance costs.[1] If the plant has operating flexibility, then the

[1] In this paper, the cost, which we consider, contains no costs for building and operation of electric networks, and also on transport of electric energy to consumers. Thus, this paper doesn't deal necessarily with transmission costs, but more with transmission planning (see Sauma and Oren [22]).

profit flow $\pi(P_t,Q)$ becomes zero when $P_t < c$. For simplicity, however, we assume that the power plant does not have operating flexibiltiy.

3 Basic Model

We begin by describing the model of Gollier et al. [10] that extends the McDonald and Siegel model [4] deriving the investment timing and its option value by introducing fixed construction time and project lifetime. Suppose that the firm can determine the investment timing of a power plant with a fixed output,Q. The value of the investment opportunity is:

$$F(p) \equiv \sup_{\tau \in \mathcal{S}} \mathbb{E}_p \left[\int_{\tau+T}^{\tau+T+L} e^{-rt} \pi(P_t, Q) dt - e^{-r\tau} I(Q) \right], \tag{3}$$

where $\mathbb{E}_p$ is expectation with respect to the probability law of P_t given an initial value p, τ is the investment time, $\mathcal{S}$ is the set of stopping times of the filtration generated by the electricity price process, and $r > 0$ is an arbitrary discount rate. We must have $r > \mu$ in order to ensure that the firm's value is finite for $L \rightarrow \infty$.

Given the investment threshold, P^*, the optimal investment time, τ^*, has the following form:

$$\tau^* = \inf\{t \geq 0 | P_t \geq P^*\}. \tag{4}$$

Prior to determining P^* and $F(p)$, we calculate the now-or-never expected NPV, $V(p,Q)$, of a power plant with fixed output, Q:

$$V(p, Q) = \mathbb{E}_p \left[\int_T^{T+L} e^{-rt} \pi(P_t, Q) dt - I(Q) \right] = K_1 \frac{Qp}{r - \mu} - K_2 \frac{cQ}{r} - \delta Q^{\varepsilon}, \tag{5}$$

where $K_1 = e^{-(r-\mu)T}$ and $K_2 = e^{-rt}(1 - e^{-rL})$. Note that as $T \rightarrow 0$ and $L \rightarrow \infty$, both K_1 and K_2 tend to one, i.e., we have the case of an infinitely lived power plant that is constructed with no lead time.

The value of the investment opportunity is then determined using the standard methodology as in Dixit and Pindyck [3]. In particular, this is an optimal stopping time problem, which is cast as a non-linear maximization problem using the conditional expectation of the stochastic discount factor, $\mathbb{E}_p[e^{-r\tau}]$, and applying the strong Markov property of the GBM process along with the law of iterated expectations:

$$F(p) = \sup_{\tau \in \mathcal{S}} \mathbb{E}_p[e^{-rt} V(P_\tau, Q)] = \max_{P^* \geq p} \left(\frac{p}{P^*} \right)^{\beta_1} V(P^*, Q), \tag{6}$$

where $\beta_1 > 1$ is the positive root of the characteristic equation. $\frac{1}{2}\sigma^2\beta(\beta-1)+\mu\beta-r=0$. The optimal investment threshold after taking the first-order necessary condition is:

$$P^* = \frac{\beta_1}{\beta_1 - 1}\frac{r-\mu}{K_1}\left(\frac{cK_2}{r} + \delta Q^{\varepsilon-1}\right) \tag{7}$$

The investment threshold, P^* is dependent on the degree of the uncertainty, σ. Since $\beta_1 > 1$, and $\frac{\partial\beta_1}{\partial\sigma}<0$, the investment threshold is an increasing function of σ. By contrast, the now-or-never investment threshold price is $P^{NN} = \frac{r-\mu}{K_1}\left(\frac{cK_2}{r} + \delta Q^{\varepsilon-1}\right) < P^*$, i.e., having the deferral option provides a value to waiting, which then increases the opportunity cost of investing.

4 Investment Timing and Sizing

In this section, following Dangl [20],[2] we develop a model for analyzing not only the investment timing, but also the plant sizing. The value of the investment opportunity is now:

$$\begin{aligned} F(p) &\equiv \sup_{\tau\in\mathcal{S},Q\geq 0} \mathbb{E}_p\left[\int_{\tau+T}^{\tau+T+L} e^{-rt}\pi(P_t,Q)dt - e^{-rt}I(Q)\right] \\ &= \sup_{\tau\in\mathcal{S}} \mathbb{E}_p[e^{-rt}V(P_\tau, Q^*(P_\tau))] \\ &= \max_{P^*\geq p}\left(\frac{p}{P^*}\right)^{\beta_1} V(P^*, Q^*(P^*)) \end{aligned} \tag{8}$$

Here, the firm first optimizes the plant's productive capacity, $Q^*(p)$, before deciding on investment timing. Thus, the optimal size of the plant for any p is:

$$Q^*(p) = \arg\max_{Q\geq 0} K_1\frac{Qp}{r-\mu} - K_2\frac{cQ}{r} - \delta Q^\varepsilon = \left[\frac{1}{\delta\varepsilon}\max\left(\frac{K_1 p}{r-\mu} - \frac{K_2 c}{r}, 0\right)\right]^{\frac{1}{\varepsilon-1}} \tag{9}$$

When we substitute the optimal size of the plant for any p, Eq. 9, back into the now-or-never expected NPV, Eq. 5, we obtain the maximized now-or-never expected NPV:

---

[2] Although Dangl [20] considers the investment timing and the plant sizing with operational flexibility, as described previously, in this chapter we assume that the power plant does not have such discretion.

$$V^*(p) \equiv V(p, Q^*(p)) = \left(\frac{1}{\delta\varepsilon}\right)^{\frac{1}{\varepsilon-1}} \left(\frac{\varepsilon-1}{\varepsilon}\right) \left[\max\left(\frac{K_1 p}{r-\mu} - \frac{K_2 c}{r}, 0\right)\right]^{\frac{\varepsilon}{\varepsilon-1}} \tag{10}$$

Inserting Eq. 10 into Eq. 8, we are now able to solve the investment timing problem with endogenous capacity sizing:

$$F(p) = \max_{P^* \geq p} \left(\frac{p}{P^*}\right)^{\beta_1} V^*(P^*) \tag{11}$$

Taking the first-order necessary condition as before and solving for P^* and $Q^* \equiv Q(P^*)$, we obtain the following:

$$P^* = \frac{\beta_1(\varepsilon-1)}{\beta_1(\varepsilon-1)-\varepsilon}(r-\mu)\frac{K_2 c}{K_1 r} \tag{12}$$

$$Q^* = \left(\frac{1}{\delta\varepsilon}\right)^{\frac{1}{\varepsilon-1}} \left[\frac{K_2 c\varepsilon}{r(\beta_1(\varepsilon-1)-\varepsilon)}\right]^{\frac{1}{\varepsilon-1}} \tag{13}$$

We must have $\beta_1(\varepsilon\text{-}1)\text{-}\ \varepsilon > 0$ to ensure that P^* and Q^* are non-negative. However, if $p > P^*$, then it is optimal to invest immediately and to construct a plant of size greater than Q^*. In particular, Eq. 9 would be used to determine the optimal capacity.

The degree of the uncertainty affects not only the investment threshold but also the optimal capacity. Since $\beta_1 > 1$, and $\frac{\partial\beta_1}{\partial\sigma} < 0$, the investment threshold and the optimal capacity are increasing functions of σ.

5 Technology Choice

We now consider the full investment problem in which the firm also has a choice of two technologies for power plants: the first type is capital intensive, but with low operating costs, while the second one has low capital, but high operating, costs. Thus, the first technology may be thought of as nuclear power, while the second one may be based on natural gas combined-cycle combustion. In order to make the tradeoff relevant, we have $c_N < c_G$ and $\delta_N. > \delta_G$ Other aspects of both projects remain identical, i.e., both types of power plants face the same price shocks, construction lead times, and operating lifetimes. We follow the framework of [15] in order to analyze this problem of mutually exclusive investment in two projects. Unlike [15], we also have the issue of endogenous capacity sizing in addition to investment timing and technology choice.

Formally, the value of investment opportunity here is:

$$F(p) \equiv \sup_{\tau \in \mathscr{S}, Q \geq 0} \mathbb{E}_p \left[1_{\{\tau_N \leq \tau_G\}} e^{-r\tau_N} V_N\left(P_{\tau_N}, Q_N^*(P_{\tau_N})\right) \right.$$
$$\left. + 1_{\{\tau_N > \tau_G\}} e^{-r\tau_G} V_G\left(P_{\tau_G}, Q_G^*(P_{\tau_G})\right)\right], \tag{14}$$

where $1_{\{\cdot\}}$ is the indicator function, $\tau_i, i = N, G$, is the investment time of technology i in the case where the firm has two alternative plants, and τ is the investment time of a plant for either technology, i.e.,

$$\tau = \min\{\tau_N, \tau_G\}. \tag{15}$$

In order to solve this problem, we first let $\tilde{p}$ be the indifference price between the two projects, i.e.,$V_N^*(\tilde{p}) = V_G^*(\tilde{p})$. Next, we note that if the nuclear power plant has a higher maximized expected NPV than the one for the natural gas plant for some $p<\tilde{p}$, then the option value for the entire investment opportunity, $F(p)$, may be dichotomous.

The option value for the entire investment opportunity, $F(p)$, can be obtained from the Bellman equation (see Dixit and Pindyck [3]),

$$\frac{1}{2}\sigma^2 p^2 F''(p) + \mu p F'(p) - rF(p) = 0 \tag{16}$$

This ordinary differential equation is an Euler-type one. Therefore, this differential equation has the following general solution:

$$F(p) = D_1 p^{\beta_1} + D_2 p^{\beta_2}, \tag{17}$$

where D_1 and D_2 are positive unknowns, and $\beta_2 < 0$ is the negative root of the characteristic equation $\frac{1}{2}\sigma^2\beta(\beta - 1) + \mu\beta - r = 0$..

The procedure for checking it is as follows:

1. If F_G(p) $>$ F_N(p), then $F(p) = F_G$(p);
2. Else $F(p) = F_N$(p) for $0 \leq p < P_N^*$ and $F(p) = D_1 p^{\beta_1} + D_2 p^{\beta_2}$ for $P_L^* < p < P_R^*$,

where $P_{\{\cdot\}}^*$ denotes an investment threshold, that is, an optimal investment rule for each case. In the first case, it is optimal to skip the nuclear power plant and focus on the gas one. Consequently, the option value is simply the value of the opportunity to invest in the gas power plant. By contrast, in the second case, it may be optimal to invest in the nuclear power plant for $0 \leq p < P_N^*$. Thus, the option value is dichotomous: there is a lower waiting region for the opportunity to invest only in the nuclear power plant should the price increase sufficiently as well as an upper waiting region around the indifference price in which it may be optimal to invest in either technology. This latter waiting region for $P_L^* < p < P_R^*$ has an option value that reflects the opportunity to invest in either gas or nuclear power plants via $D_1 p^{\beta_1}$

and $D_2 p^{\beta 2}$, respectively, where β_2 is the negative root of $\frac{1}{2}\sigma^2\beta(\beta-1)+\mu\beta-r=0$, and D_1 and D_2 are both positive endogenous constants.

The investment threshold prices, P_L^* and P_R^*, along with endogenous constants, D_1 and, are determined via the following value-matching and smooth-pasting conditions:

$$F\left(P_L^*\right) = V_N^*\left(P_L^*\right), \tag{18}$$

$$F^{'}\left(P_L^*\right) = V_N^{'*}\left(P_L^*\right), \tag{19}$$

$$F\left(P_R^*\right) = V_G^*\left(P_R^*\right), \tag{20}$$

$$F^{'}\left(P_R^*\right) = V_G^{'*}\left(P_R^*\right), \tag{21}$$

where the primes denote derivatives, that is, $F' = \frac{dF}{dp}$. Since these four equations are highly non-linear, it is not possible to find an analytical solution to the system. However, numerical solutions may be obtained for specific parameters as we illustrate in the next section. Upon solving for the optimal investment and waiting regions, the capacity size is then scaled accordingly.

6 Numerical Examples

We use the following parameter values for our numerical examples:

- $r = 0.10$
- $\mu = 0.01$
- $P_0 = 25(\$/\mathrm{MWh})$
- $c_N = 20(\$/\mathrm{MWh})$
- $c_G = 40(\$/\mathrm{MWh})$
- $T = 2(\mathrm{years})$
- $L = 40(\mathrm{years})$
- $\varepsilon = 2$
- $\delta_N = 4.17 \times 10^{-5}\left(\$/\mathrm{MWh}^2\right)$
- $\delta_G = 2.09 \times 10^{-5}\left(\$/\mathrm{MWh}^2\right)$
- $Q = 8{,}760 \times 500 \ = 4{,}380{,}000(\mathrm{MWh})$

The δ_N and δ_G parameters are calibrated so that a nuclear plant of 500 MW capacity has an investment cost of \$800 million based on a reported cost of \$1,500/kW in [9]. A gas-fired power plant of the same capacity is assumed to be half as expensive to build. In addition, we assume a base value of $\sigma = 0.20$, but allow it to vary in order to perform sensitivity analyses.

6.1 *Timing Flexibility*

At the initial electricity price, P_0, both the nuclear and gas projects are out of the money if investment is to be undertaken for a fixed capacity of 500 MW. In particular, the expected NPVs are -\$516 million and -\$820 million for the nuclear and gas projects, respectively. On the other hand, the option values are \$98 million and \$74 million, respectively (see Fig. 1). Therefore, it is optimal to wait to invest in either technology.

As uncertainty increases, the value of waiting also goes up. Indeed, a more volatile electricity price increases the opportunity cost of immediate action. Consequently, although the option value of the entire investment opportunity increases, the cost of killing the deferral option also increases, thereby making it optimal to delay investment. For example, if volatility increases to 0.25, then the investment thresholds increase to \$71/MWh and \$85/MWh for nuclear and gas technologies, respectively, from the base levels of \$63/MWh and \$76/MWh (see Fig. 2). By comparison, the now-or-never investment thresholds are \$38/MWh and \$46/MWh, respectively.

6.2 *Timing and Capacity Sizing*

With discretion over capacity, both the maximized expected NPV and option value of each project increases. In particular, the former is now never negative because it is always possible to set capacity size to zero in case of an unprofitable investment.

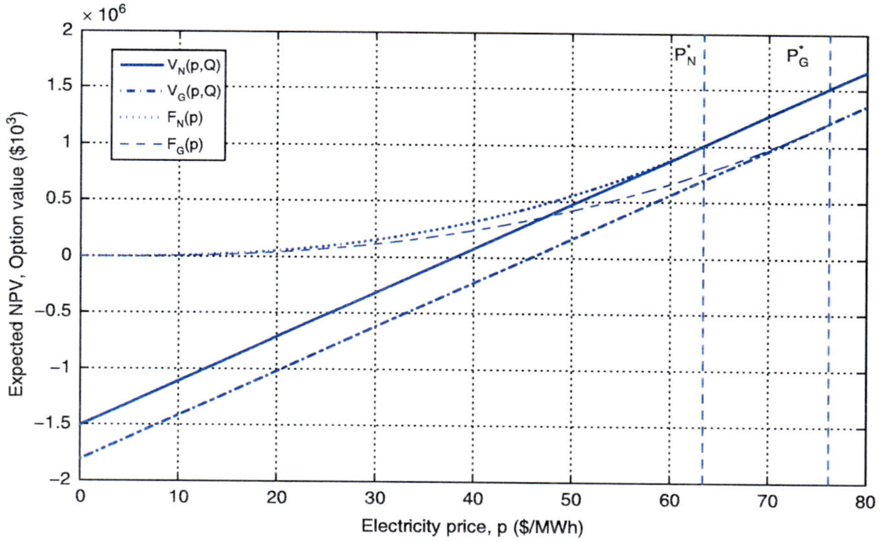

Fig. 1 Expected NPV and option values of nuclear and gas power plants with fixed capacities ($\sigma = 0.20$)

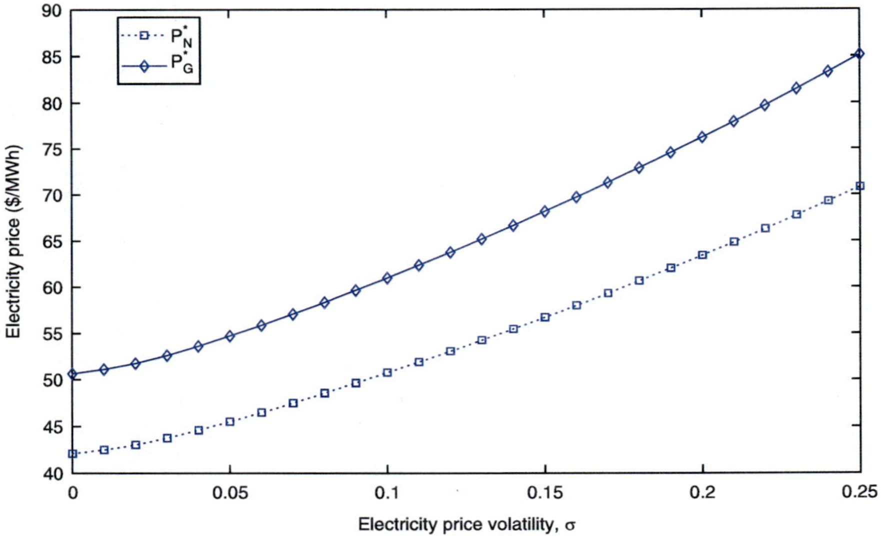

Fig. 2 Investment threshold prices as a function of volatility

Now, atP_0, the expected maximized NPVs, $V_N^*(P_0)$and $V_G^*(P_0)$, are \$25 million and \$0, respectively. In other words, the nuclear project is in the money from a maximized expected NPV perspective, but the gas project is still out of the money. At this initial price, however, it is not optimal to invest in either project because the option values of the two opportunities are \$104 million and \$146 million, respectively (see Fig. 3). Thus, even though the gas plant is out of the money, its option value is more than that of the nuclear one.

By allowing the volatility parameter to vary, we illustrate its effect on the optimal investment threshold prices and capacities. Specifically, the endogeneity of the capacity-sizing decision promotes a further delay in the investment decision. Therefore,compared to the case in Sect. 6.1, the investment thresholds are now \$89/MWh for nuclear and \$178/MWh for gas (see Fig. 4). The delay is seemingly greater for the gas power plant because its lower marginal investment cost can take particular advantage of higher electricity prices. Consequently, the optimal capacity size also increases with volatility (see Fig. 5), i.e., plants of capacities 880 MW and 3520 MW are selected for nuclear and gas, respectively. On the other hand, if the current price were greater than the investment threshold price, then it would be optimal to build immediately and scale the plant's size using Eq. 9. The effect of the initial price on the capacity size is indicated in Fig. 6 for $\sigma = 0.20$.

With a lower electricity price volatility, although the maximized expected NPVs are not affected, the option values decrease for both technologies. In fact, with $\sigma = 0.15$, it is the case that the option value of investing in nuclear (\$62 million) is greater than that for gas (\$60 million) even though the maximized expected NPV of the latter dominates for higher prices (see Fig. 7). Consequently, while investment occurs sooner in each technology ($P_N^* = 52.13$ and $P_G^* = 104.27$), a lower capacity level is

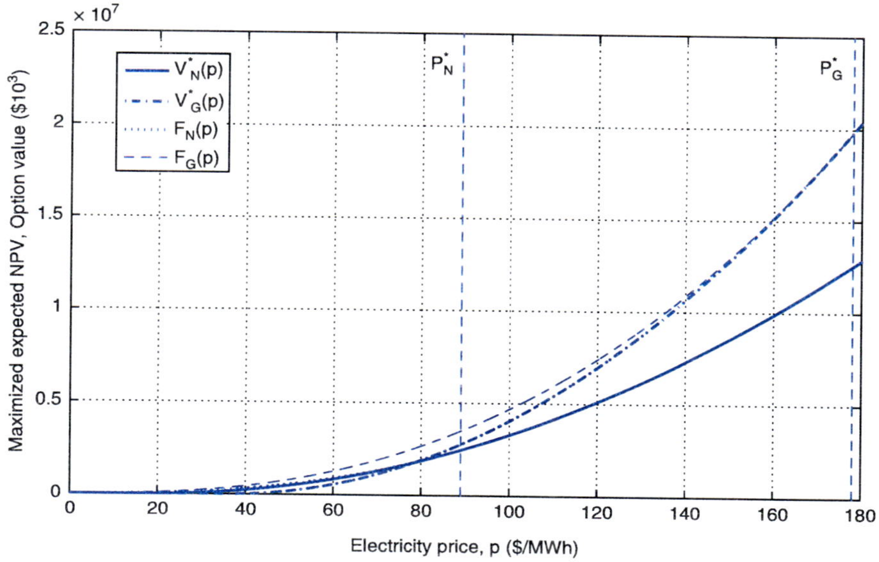

Fig. 3 Maximized expected NPV and option values of nuclear and gas power plants with optimal capacities ($\sigma = 0.20$)

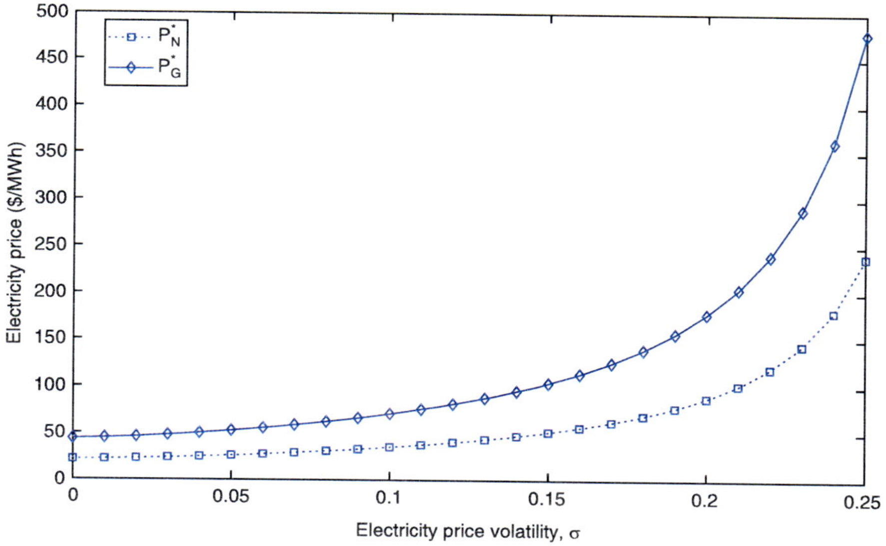

Fig. 4 Investment threshold prices as a function of volatility with endogenous capacity sizing

installed, i.e., 424 MW and 1,700 MW for nuclear and gas, respectively (see Fig. 8). As we will show in Sect. 6.3, this conundrum may lead to a dichotomous option value when the investor considers the two technologies as being mutually exclusive projects.

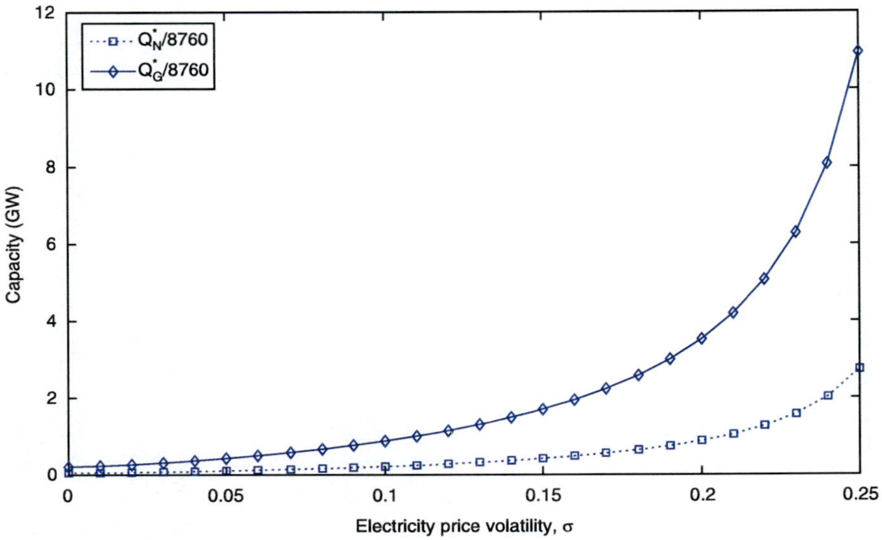

Fig. 5 Optimal capacity size as a function of volatility

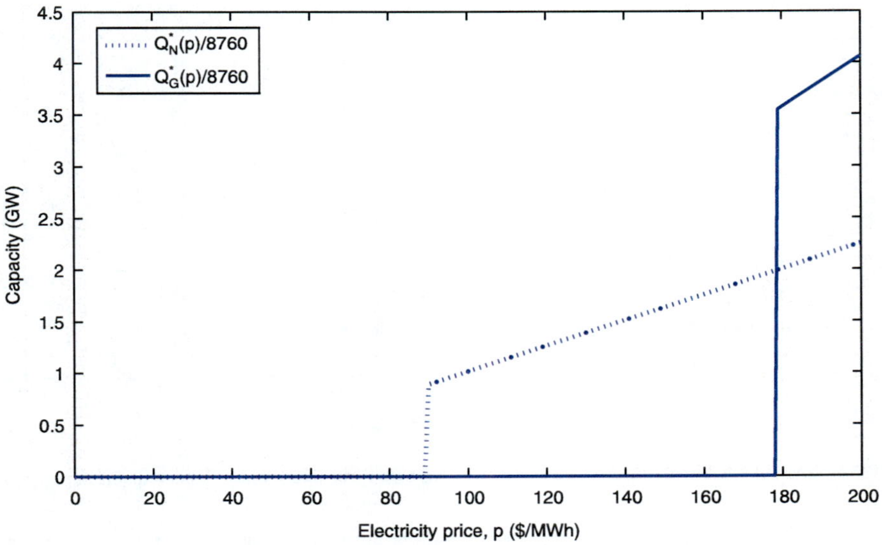

Fig. 6 Optimal capacity size as a function of price ($\sigma = 0.20$)

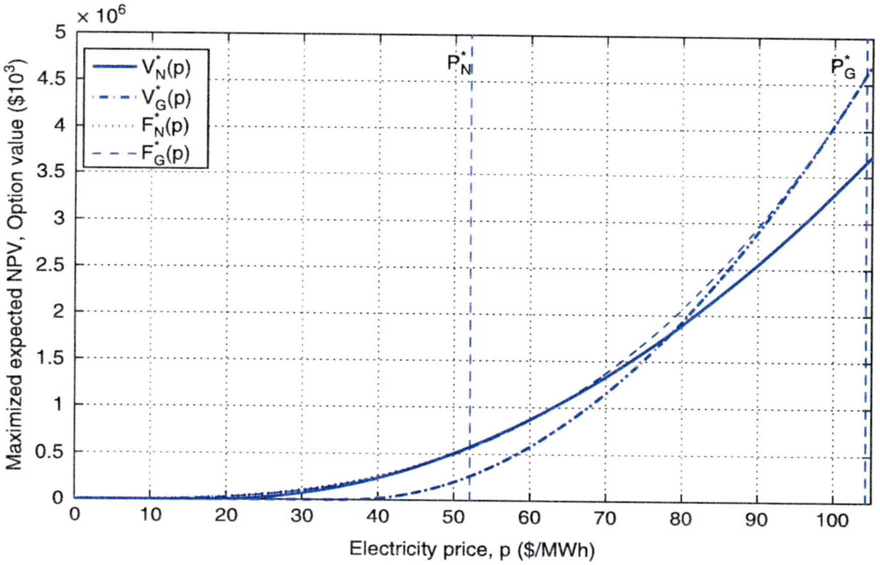

Fig. 7 Maximized expected NPV and option values of nuclear and gas power plants with optimal capacities ($\sigma = 0.15$)

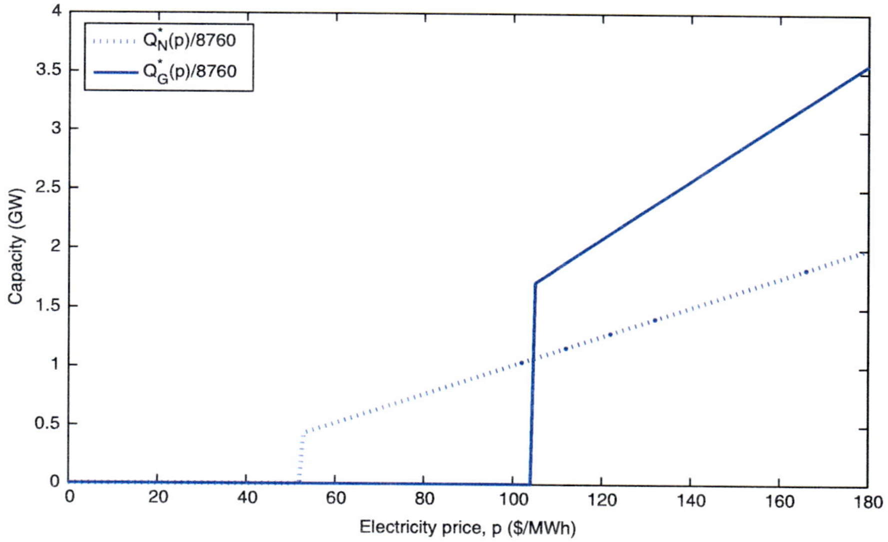

Fig. 8 Optimal capacity size as a function of price ($\sigma = 0.15$)

6.3 Timing, Sizing, and Technology Choice

When a mutually exclusive investment opportunity in the two projects is considered, the value of the option to invest is dichotomous around the indifference price, $\widetilde{P} = 78.60$, at a relatively low level of uncertainty. In this example, it is the case that $F_N(P_0) > F_G(P_0)$ for $\sigma < 0.152$, which leads to the situation in Fig. 9. For an initial electricity price of \$25/MWh, this implies that it is optimal to wait until the electricity price hits a level of $P_N^* = 52.13$ before investing in a nuclear power plant of capacity 424 MW (as determined via Eq. 13). In fact, for electricity prices between \$52.13/MWh and \$56.80/MWh, it is optimal to invest immediately in a nuclear power plant with capacity that is scaled to maximize the NPV (as determined via Eq. 9), i.e., it is optimal to construct an even larger plant. However, for an initial price in the range $P_L^* = \$56.80$/MWh to $P_R^* = \$104.49$, it is optimal to wait: if the electricity price drops (rises) to the lower (upper) threshold, then it is optimal to invest immediately in a nuclear (gas) power plant. In the latter case, a gas-fired power plant of capacity 1,702 MW is constructed, where the optimal size is determined via Eq. (9). Finally, it should be noted that $P_R^* > P_G^* = \$104.27$, i.e., it is optimal to delay investment in the gas-fired power plant longer when the mutually exclusive option to proceed with nuclear is also considered. Intuitively, the inclusion of a second possible project increases the value of the entire investment opportunity, but it also makes investment in any specific technology less

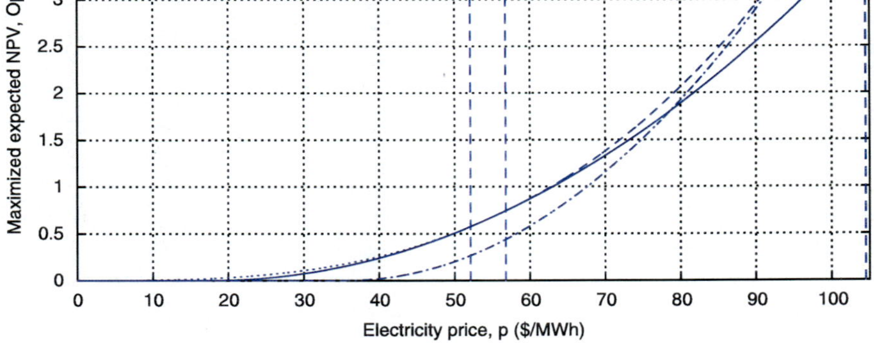

Fig. 9 Maximized expected NPV and option value of a mutually exclusive investment opportunity in nuclear and gas power plants with optimal capacities ($\sigma = 0.15$)

likely. For example, $F(\widetilde{P})$ = \$1.957 billion, which is greater than $F_G(\widetilde{P})$= \$1.952 billion by \$4.88 million, i.e., a difference of 0.25%.

As uncertainty increases, however, the immediate investment region in the nuclear technology shrinks until for $\sigma>0.152$ it disappears completely. In fact, for volatility estimates greater than 0.152, the value of the option to invest in the gas technology dominates the one for the nuclear technology. Thus, it is preferable to skip the nuclear technology and consider only the option to invest in the gas one. Figure 10 illustrates the maximized expected NPV and option value curves for $\sigma = 0.20$. Since $F_N(P_0)$=\$104 million and $F_G(P_0)$=\$146 million, the value of the mutually exclusive option to invest is simply the value of the option to invest in the gas technology, i.e., $F(p) = F_G(p)$. As in Sect. 6.2, we wait until the electricity price is \$178.07/MWh before investing in a gas-fired power plant of capacity 3,520 MW. Of course, if the initial price were higher than this threshold, P_G^*, then we would invest immediately in a larger power plant, which would be scaled using Eq. 9. Finally, Fig. 11 traces the effect of varying the volatility parameter on investment threshold prices and technology choices. The dichotomous waiting region appears only for low levels of uncertainty.

In Figs. 12 and 13, we illustrate the impact of the initial electricity price on the optimal capacity size when the investment opportunity is mutually exclusive for low and high levels of uncertainty. When $\sigma = 0.15$, the waiting region is dichotomous, i.e., it is optimal to invest immediately in the nuclear technology for prices in the region (52.13, 56.80). Using Eq. 9, we can also determine the optimal capacity of the nuclear power plant,which starts off at 424 MW and increases linearly until

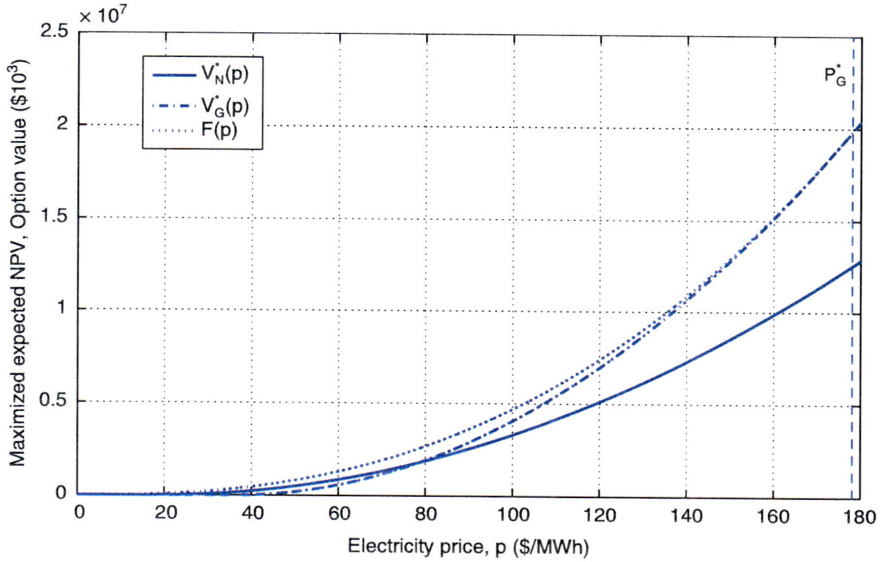

Fig. 10 Maximized expected NPV and option value of a mutually exclusive investment opportunity in nuclear and gas power plants with optimal capacities ($\sigma = 0.20$)

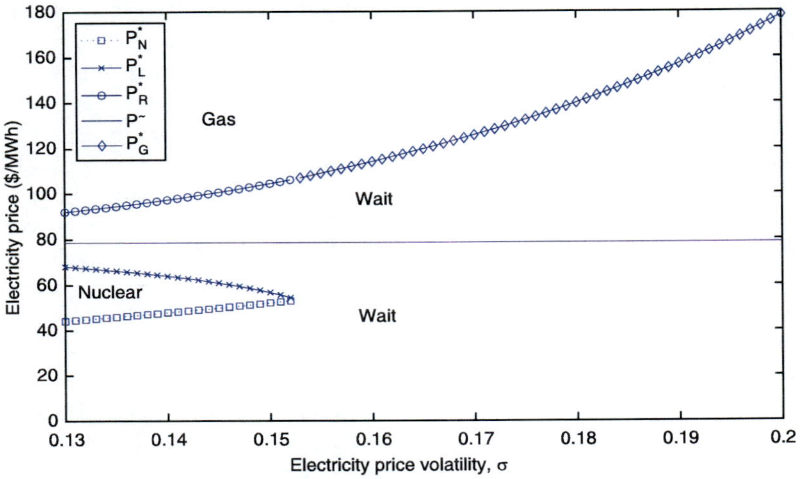

Fig. 11 Investment threshold prices as a function of volatility with endogenous capacity sizing and mutually exclusive investment

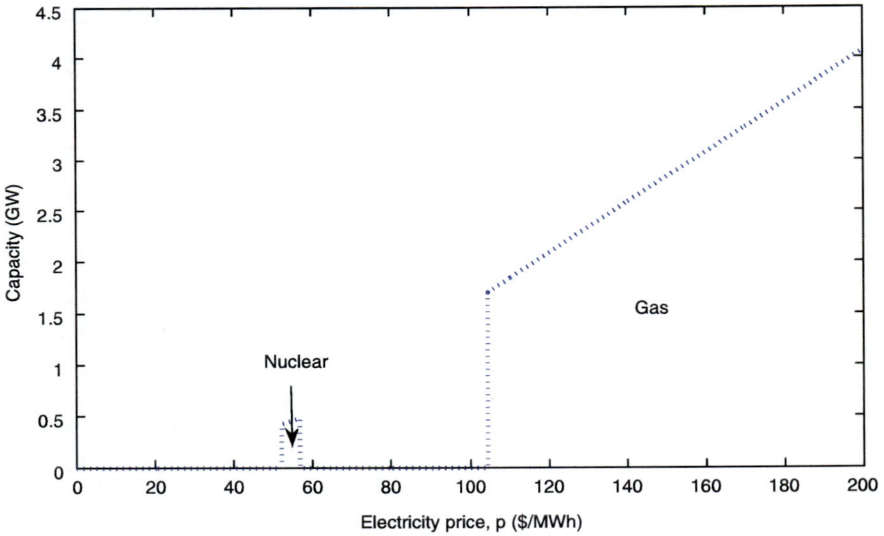

Fig. 12 Optimal capacity size as a function of price with mutually exclusive investment ($\sigma = 0.15$)

the electricity price is high enough to warrant waiting again. However, when the price increases to $104.49/MWh, it is optimal to invest immediately in a gas-fired power plant, and its optimal capacity is agian determined from Eq. 9. By contrast, if $\sigma = 0.20$, then we skip the nuclear technology and wait for the price to increase

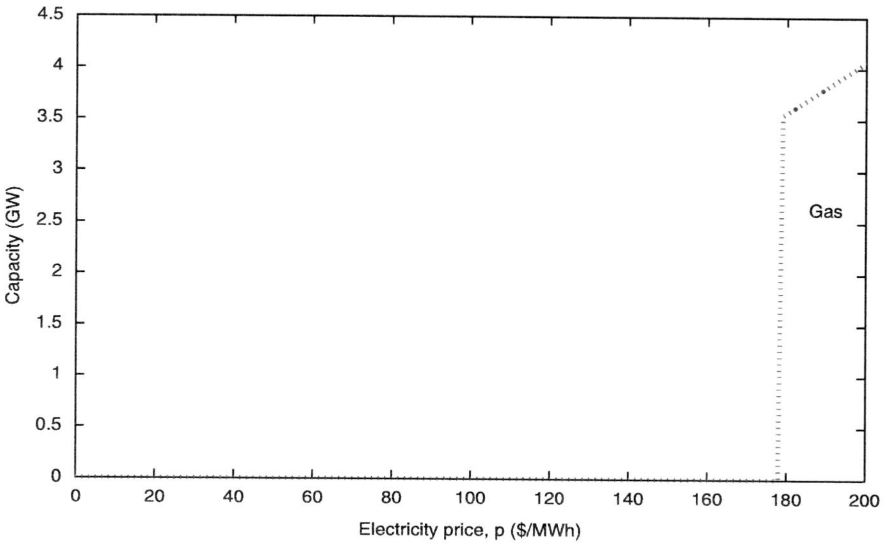

Fig. 13 Optimal capacity size as a function of price with mutually exclusive investment ($\sigma = 0.20$)

sufficiently to warrant immediate investment in the gas technology with the optimized capacity as in Fig. 13.

7 Conclusions

Due to the ongoing deregulation of the electricity industry, investors in generation and transmission assets have more flexibility over their decisions. Such flexibility can encompass timing, sizing, and technology choice, to name a few. Under the regulated paradigm, relatively stable energy prices made such flexibility nearly obsolete as most investment opportunities could be appraised via the standard NPV approach. Uncertain prices, however, provide an incentive to defer decisions in order to receive more information especially when there is additional flexibility in the form of capacity sizing and technology choice.

In this chapter, we have developed an analytical model to value the option to invest in mutually exclusive generation technologies when there exists flexibility over timing and sizing. We note first that due to the value of waiting, investment using the real options approach occurs later than via the now-or-never NPV one. Second, uncertainty affects not only the timing of investment, but also its scale, i.e., investors try to maximize their expected profit by waiting longer and building larger plants. Finally, consideration of mutually exclusive projects increases the option value of the entire investment opportunity while deferring adoption of any particular technology. Indeed, in a situation with a relatively low level of uncertainty, the investor may not be able to rule out the technology that performs better under relatively low long-term electricity prices, e.g., nuclear in our case, and, thus, has the additional incentive to wait longer.

It should be noted, however, that the real options approach does not necessarily imply that investment and operational decisions are deferred under uncertainty. For example, it may be the case with lags in construction stages that it is optimal to invest sooner due to greater uncertainty (see [21]). The intuition for this result is that by proceeding sooner, the investor learns more about the cost of construction and may make more informed decisions about subsequent modules. Therefore, extension ofthis chapter's approach towards modularized capacity addition with uncertain construction or operating costs would be warranted. Other directions for future work in this area include the use of more realistic price processes, e.g., geometric mean-reverting ones, the inclusion of capacity constraints, operational flexibility, i.e., the power plant may be switched on or off, oligopolistic competition by different investors over power plant timing, sizing, and technology choice, and the development for an agent-based simulation model. Additionally, it would be interesting to consider reliability and safety of power plants along with economic risks as mentioned in this paper.

Acknowledgments The authors would like to thank two anonymous referees for providing valuable comments.

References

1. Wilson RB (1998) Architecture of power markets. Econometrica 70:1299–1340
2. Hyman LS (2009) Restructuring electricity policy and financial models. Energy Econ, 32 (4):751–757
3. Dixit AK, Pindyck RS (1994) Investment under uncertainty. Princeton University Press, Princeton
4. McDonald R, Siegel D (1986) The value of waiting to invest. Q J Econ 101:707–727
5. Ekern S (1988) An option pricing approach to evaluating petroleum projects. Energy Econ 10:91–99
6. Deng S-J, Johnson B, Sogomonian A (2001) Exotic electricity options and the valuation of electricity and generation assets. Decis Support Syst 30:383–392
7. Tseng C-L, Barz G (2002) Short-term generation asset valuation: a real options approach. Oper Res 50:297–310
8. Deng S-J, Oren SS (2003) Incorporating operational characteristics and startup costs in option-based valuation of power generation capacity. Probability Eng Informational Sci 17:155–181
9. Rothwell G (2006) A real options approach to evaluating new nuclear power plants. Energy J 27:37–53
10. Gollier C, Proult D, Thais F, Walgenwitz G (2005) Choice of nuclear power investments under price uncertainty: valuing modularity. Energy Econ 27:667–685
11. Näsäkkälä E, Fleten S-E (2005) Flexibility and technology choice in gas fired power plant investments. Rev Financ Econ 14:371–393

12. Siddiqui AS, Maribu K (2009) Investment and upgrade in distributed generation under uncertainty. Energy Econ 31:25–37
13. Bøckman T, Fleten S-E, Juliussen E, Langhammer H, Revdal J (2008) Investment timing and optimal capacity choice for small hydropower projects. Eur J Oper Res 190:255–267
14. Takashima R, Naito Y, Kimura H, Madarame H (2007) Decommissioning and equipment replacement of nuclear power plants under uncertainty. J Nucl Sci Technol 44:1347–1355
15. Décamps J-P, Mariotti T, Villeneuve S (2006) Irreversible investment in alternative projects. Econ Theory 28:425–448
16. Huisman KJM, Kort PM (2003) Strategic investment in technological innovations. Eur J Oper Res 144:209–233
17. Fleten S-E, Maribu KM, Wangensteen I (2007) Optimal investment strategies in decentralized renewable power generation under uncertainty. Energy 32:803–815
18. Wickart M, Madlener R (2007) Optimal technology choice and investment timing: a stochastic model of industrial cogeneration vs. heat-only production. Energy Econ 29:934–952
19. Takashima R, Goto M, Kimura H, Madarame H (2008) Entry into the electricity market: uncertainty, competition, and mothballing options. Energy Econ 30:1809–1830
20. Dangl T (1999) Investment and capacity choice under uncertain demand. Eur J Oper Res 117:415–428
21. Bar-Ilan A, Strange WC (1996) Investment lags. Am Econ Rev 86(3):610–622 A real-options approach. Rev Financ Stud 16:1239–1272
22. Sauma EE, Oren SS (2006) Proactive planning and valuation of transmission investments in restructured electricity markets. J Regul Econ 30:261–290

Real Options Approach as a Decision-Making Tool for Project Investments: The Case of Wind Power Generation

José I. Muñoz, Javier Contreras, Javier Caamaño, and Pedro F. Correia

Abstract This chapter develops a decision-making tool to invest in renewable power plants using a real options approach. The model is validated for a wind energy plant. To build a base for the investment model, market prices and wind regimes are obtained from Geometric Brownian Motion (GBM) and Weibull models, respectively. Then, considering these and other values, such as investment, maintenance and operation costs, the Net Present Value (NPV) is obtained. As a result, an NPV curve is drawn by shifting the initial time of investment. From the NPV curve obtained, a trinomial lattice is built and applied to a real options valuation method. From this model, it is possible to estimate the probabilities of investing right now, deferring, or not investing at all. This decision tool allows wind energy investors to decide whether to invest or not in different scenarios. Several realistic case studies are presented to illustrate the decision-making method.

Keywords Investments assessment • net present value curves • real options • risk management • stochastic modelling • trinomial trees

J.I. Muñoz (✉) • J. Contreras
E.T.S. de Ingenieros Industriales, University of Castilla – La Mancha, Ciudad Real, Spain
e-mail: joseignacio.munoz@uclm.es; javier.contreras@uclm.es

J. Caamaño
E.T.S. de Ingenieros Industriales, University of País Vasco, Bilbao, Spain
e-mail: eppcaerj@bi.ehu.es

P.F. Correia
Instituto Superior Técnico, Technical University of Lisboa, Lisbon, Portugal
e-mail: pfcorreia@ist.utl.pt

A. Sorokin et al. (eds.), *Handbook of Networks in Power Systems I*,
Energy Systems, DOI 10.1007/978-3-642-23193-3_13,

1 Introduction

Environmental protection and clean sources of energy are increasingly demanded in our world. That is not by chance, but due to economic reasons. High fuel prices and instability in production countries has pushed other countries to reduce dependence on fossil fuel-based energy. Producers should take into account future medium/long-term uncertainties, even more with an increasing worldwide demand.

Different developing countries have been supporting renewable energy sources. Among them, wind energy seems to be the most successful in terms of performance and market penetration. One of main problems in wind farm operation comes from the intermittency of the power source (wind). It has encouraged the use of stochastic models in future investment modelling.

Traditional methods, such as Net present Value (NPV) and its associated cash flow, are not suitable for the new investment models because of the lifetime of this type of plants and the uncertainties involved.

One method to manage uncertainty is a real options model. It permits to develop flexible investment strategies where the investor has three alternatives at any time: execute, wait, or abandon the construction project of a power plant.

1.1 State of the Art

Social welfare is usually a measurement of the impact of renewable energy from an overall system benefits perspective [1]. The decision whether to build or not a new power plant, and when to do it, depends on multiple factors. All these factors should to be taken into account by the investors to maximize their own profits. A deep evaluation of the financial issues involved is necessary [2] as well as having good analytical decision tools.

Real options are decision tools widely used to analyze situations in several fields [3, 4]. They are also applied in power systems (investment and operation), taking into account not only the assessment of the investment buy also the best time to execute the project. In particular, real option methods have been applied to invest in nuclear and CCGT power plants [5, 6], and they have been taken into consideration in investment decisions by utilities.

The operation of a power plant can be studied from a generation asset perspective, where fuel and electricity prices are modelled using mean reverting processes [7, 8] considering seasonal patterns [9]. Financial option theory is used to assess a plant using the concept of Value at Risk. Sophisticated models include operational constraints, as in [10], or risk aversion and neutral risk-adjustment [11] or even define multi-stage stochastic models of the operational decisions, with two-factor lattices describing fuel and electricity prices, such as in [12]. Some authors use a multi-stage model to describe investments under uncertainty [5].

In [13], spark spread and emission costs are used to value a gas-fired plant. The same authors study the transformation of a base-load plant into a peak-load plant [6]. The evaluation of an investment for two inter-related plant projects is described in [14], where a quadrinomial tree represents the market value of the units. Some authors apply real options methods to assess investments in renewable generation. They use price volatility to decide the best time for investment [15]. In addition, the NPV break-even price is determined under price uncertainty.

1.2 *Objectives*

This chapter develops and applies a decision making tool based on a real options methodology to assess the possibility to build a wind power plant under uncertainty.

This tool allows the investor to estimate and use different future investment scenarios in which can make the decision whether to invest and build the power plant next.

In short, the presented method should provide the investor with the capability to evaluate the price of the option to invest in a project.

2 Proposed Investment Model

The following flowcharts, part 1 (Fig. 1) and part 2 (Fig. 2), show the structure of the model consisting of several sequential steps.

Part 1 ends at the NPV calculation and content the classical steps used in investment assessment. Part 2 starts at the last block of part 1, NPV calculation, and follows obtaining the NPV curves and their parameters, constructing trinomial trees and estimating the value of the option and the probability of each alternative.

Briefly, the main steps that are necessary to obtain the final value of the real option related to the investment in each scenario are:

- Wind speed estimation using scale and shape parameters of a Weibull function. These parameters depend on the location of the power plant. The model generates hourly wind speeds.
- Calculation of the production curve of the wind turbine using the hourly wind speed data transformed into monthly data.
- Calculation of the electricity prices to remunerate the wind producer using a Geometric Brownian Motion with Mean Reversion (GBM-MR) model that takes into account parameters (volatility, strength of reversion and trend) obtained from the Spanish market historical data. It is assumed that wind generation is paid at the spot prices resulting from the day-ahead electricity market, but any incentive scheme can be easily added to the model.

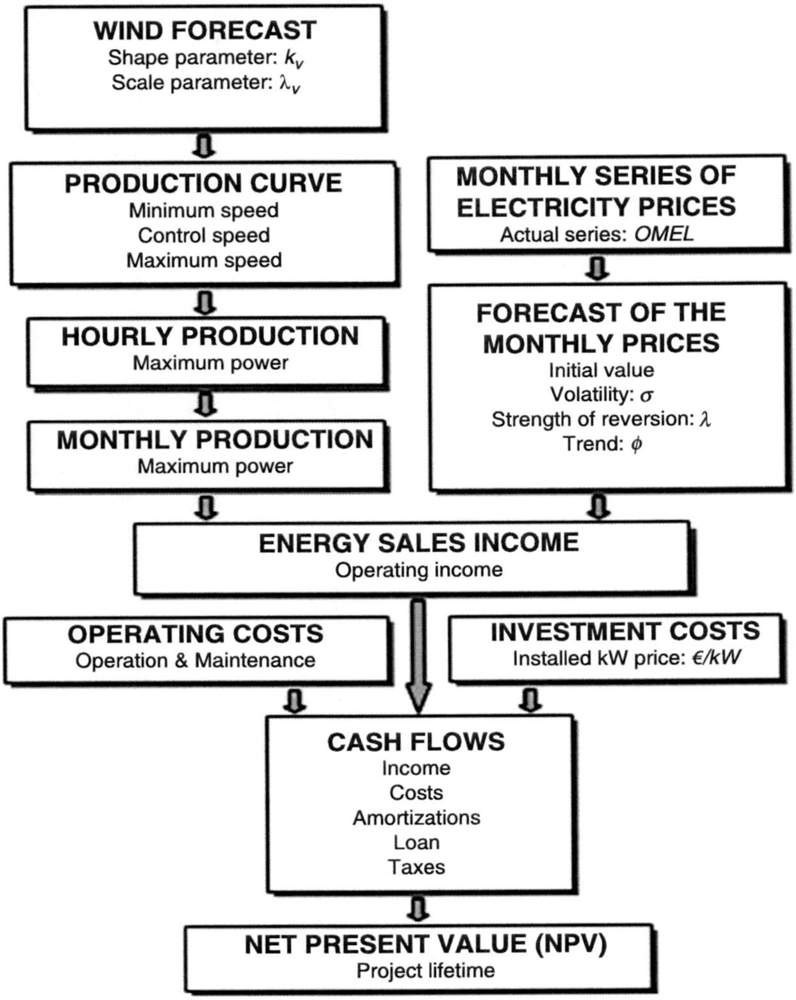

Fig. 1 Model's flowchart (part 1)

- With all this, addition of the investment cost estimation from national standards, the stochastic cash flows and NPVs are obtained.
- Next, estimation of the NPV curves by shifting the initial time for investment. Monte Carlo simulation is used in order to obtain enough data points to achieve high accuracy in the results. The NPV curves represent the evolution of the NPVs that would result if the investment were done at different times.
- A real options methodology is applied from these NPV curves and their GBM-MR parameters.

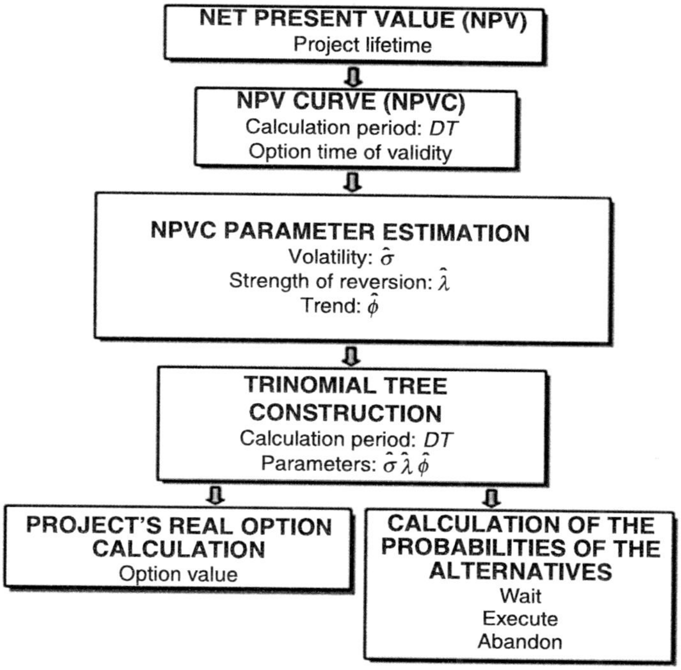

Fig. 2 Model's flowchart (part 2)

- A trinomial investment decision tree is built, where the values of the parameters are only valid for specific time intervals, since a unique GBM-MR model only holds for prices (not NPVs) during the entire lifetime of the project.
- To evaluate the attractiveness of the investment, an American option of the project (that allows investing before the expiration of the option takes place) is formulated, as well as the probabilities of three possible alternatives: execute, wait, or abandon.

The main steps of the overall model that obtains the final value of the real option of the investment are described in the following sections.

2.1 *Stochastic Modelling of Wind Speed and Electricity Price*

It is assumed that the probability function that most closely resembles the wind speed regime is the Weibull distribution function. Its density function is given by:

$$f(x,k,\lambda) = \frac{k}{\lambda}\left(\frac{x}{\lambda}\right)^{k-1} e^{\left(\frac{x}{\lambda}\right)^k}, \tag{1}$$

where λ and k are the scale and shape factors, respectively. The λ parameter refers to the maximum speed and the k parameter indicates the degree of dispersion of the samples.

To estimate the wind energy production it is necessary to evaluate the amount of power that goes through the rotor with the formula:

$$P = N_t \cdot 0.5 \cdot C_P \cdot \rho \cdot V_c^3 \cdot \pi \cdot r^2, \tag{2}$$

where the power produced depends on the number of turbines, N_t, the Betz coefficient, C_p, the air density, ρ, the wind speed, V_c, and the blade radius, r. The hourly energy production is obtained by using the stochastic wind speed values as inputs in (2). Figure 3 shows the hourly production curve of a wind turbine whose maximum power is 10MW, for a 1-month period (720 hours).

Having a 20 years horizon, which is the estimated lifetime of a wind plant, and having 40 years of cash flows, the hourly curve in Fig. 3 must be transformed into a monthly curve in order to make the calculation easier, as shown in Fig. 4.

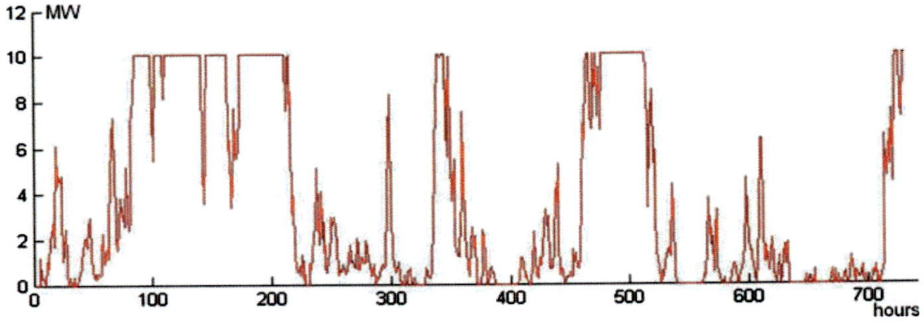

Fig. 3 10 MW wind plant production curve over 1 month

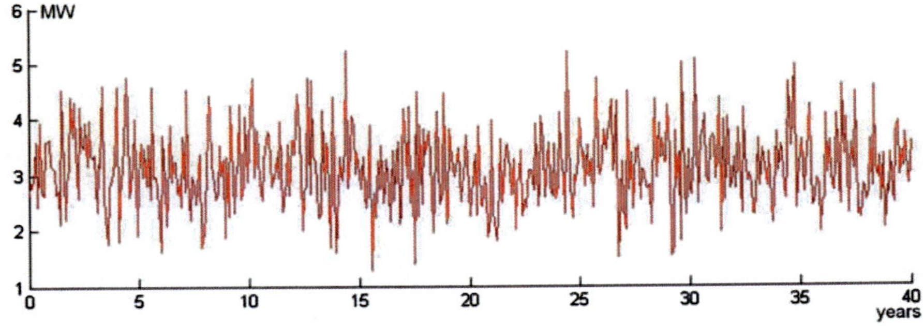

Fig. 4 Estimated monthly wind power production curve over 40 years

2.2 Electricity Price Model and Parameter Estimation

Due to the inherent uncertainty of the parameters involved in the estimation of the NPV of a project, it is advisable to reproduce its behaviour by means of stochastic processes. Electricity prices have been modeled using a geometric Brownian motion with mean reversion, as opposed to a pure short-term geometric Brownian motion, in order to account for the long-term trend, which is characterized by the long-term price and the strength of reversion.

To model electricity prices one of the most commonly used models is the Geometric Brownian Motion with Mean Reversion (GBM-MR) model:

$$dx/x = \lambda[\phi - \ln(x)]dt + \sigma dz, \tag{3}$$

where the price x(t) follows a stochastic process, λ corresponds to the speed of adjustment of the reversion, ϕ is the average long-term value, σ is the standard deviation of the process, and z defines a Wiener process.

The estimation of the parameters of the GBM-MR process is explained in Dixit's book [3], and consists in a simple least squares method.

To facilitate the calculation of the value of the parameters of the GBM-MR process, a transformation $y = \ln(x)$ is needed to apply Itō's lemma:

$$dy = \lambda[\Omega - y]dt + \sigma dz, \tag{4}$$

where:

$$\Omega = \phi - \frac{\sigma^2}{2\lambda}. \tag{5}$$

This is equivalent to the Ornstein-Uhlenbeck process whose mean and variance values are given by the expressions:

$$E[y(t)] \equiv \mu_y(t) = y(t_0)e^{-\lambda(t-t_0)} + \Omega[1 - e^{-\lambda(t-t_0)}], \tag{6}$$

$$Var[y(t)] \equiv \sigma_y^2(t) = \frac{\sigma^2}{2\lambda}[1 - e^{-2\lambda(t-t_0)}]. \tag{7}$$

Using the transformed values, and knowing that $x = e^y$, the mean and variance of the original function are obtained:

$$E[x(t)] \equiv \mu_x(t) = e^{\left[\mu_y(t)+\frac{\sigma_y^2(t)}{2}\right]}, \tag{8}$$

$$Var[x(t)] \equiv \sigma_x^2(t) = e^{[2\mu_y(t)+\sigma_y^2(t)]}\left(e^{\sigma_y^2(t)} - 1\right). \tag{9}$$

Next, the transformed function in (8) is discretized taking discrete time steps δt as follows:

$$y_k = y_{k-1}e^{-\lambda\delta t} + \Omega\left(1 - e^{-\lambda\delta t}\right) + \varepsilon_k, \tag{10}$$

where the uncorrelated independent residual error term, ε_k, has a standard distribution of the form:

$$\varepsilon \approx N\left(0, \frac{\sigma^2}{2\lambda}\left(1 - e^{-2\lambda\delta t}\right)\right), \tag{11}$$

or, equivalently,

$$\mu_\varepsilon = 0 \text{ and } \sigma_\varepsilon^2 = \frac{\sigma^2}{2\lambda}\left(1 - e^{-2\lambda\delta t}\right). \tag{12}$$

Taking differences of adjacent elements of the process in (10), the following expression is obtained:

$$y_k - y_{k-1} = \left(e^{-\lambda\delta t} - 1\right)y_{k-1} + \Omega\left(1 - e^{-\lambda\delta t}\right) + \varepsilon_k. \tag{13}$$

This expression can be assimilated to a first-order autoregressive process AR(1) defined as:

$$y_k - y_{k-1} = \beta_0 + \beta_1 y_{k-1} + \varepsilon_k. \tag{14}$$

Equalizing (13) and (14):

$$\beta_0 = \Omega\left(1 - e^{-\lambda\delta t}\right), \tag{15}$$

$$\beta_1 = e^{-\lambda\delta t} - 1. \tag{16}$$

The Cholesky decomposition expressed as a product of matrixes is used to estimate the values of β_0 and β_1:

$$\hat{\beta} = \begin{bmatrix} \hat{\beta}_0 \\ \hat{\beta}_1 \end{bmatrix} = \left(A' \cdot A\right)^{-1} \cdot A' \cdot Z. \tag{17}$$

The matrixes used in (17) are composed of the transformed (ln) values of the GBM-MR process:

$$A = \begin{bmatrix} 1 & y_2 \\ 1 & y_2 \\ \dots & \dots \\ \dots & \dots \\ 1 & y_n \end{bmatrix}; \; Z = \begin{bmatrix} y_2 - y_1 \\ y_3 - y_2 \\ \dots \\ \dots \\ y_n - y_{n-1} \end{bmatrix} \tag{18}$$

Clearing λ in (16), the expression for the first estimated parameter is attained:

$$\hat{\lambda} = -\frac{\ln\left(1 + \hat{\beta}_1\right)}{\delta t}. \tag{19}$$

On the other hand, from (15), the estimated value of Ω is:

$$\hat{\Omega} = -\frac{\hat{\beta}_0}{\hat{\beta}_1}, \tag{20}$$

and from (5), the estimated value of ϕ is:

$$\hat{\phi} = \frac{\hat{\sigma}^2}{2\hat{\lambda}} - \frac{\hat{\beta}_0}{\hat{\beta}_1}. \tag{21}$$

To calculate the volatility it is necessary to have residual terms resulting from the difference between the values of the matrix Z and the ones in the estimated matrix $\hat{Z}$:

$$\hat{Z} = \hat{\beta} \cdot Y = \hat{\beta}_0 + \hat{\beta}_1 \cdot Y, \tag{22}$$

where,

$$Y = \begin{bmatrix} y_1 \\ y_2 \\ \dots \\ \dots \\ y_{n-1} \end{bmatrix} \text{ and } \varepsilon = \hat{Z} - Z \tag{23}$$

The standard deviation of the residual terms, $\hat{\sigma}_\varepsilon$, is calculated and, from (12), the expression of the estimated variance is attained:

$$\hat{\sigma}^2 = \frac{2\hat{\lambda}\,\hat{\sigma}_\varepsilon^2}{1 - e^{-2\hat{\lambda}\delta t}}. \tag{24}$$

From (25) and (19), the value of the estimated volatility is obtained:

$$\hat{\sigma} = \hat{\sigma}_{\varepsilon} \sqrt{\frac{2 \ln\left(1 + \hat{\beta}_1\right)}{\delta t \left[\left(1 + \hat{\beta}_1\right)^2 - 1\right]}}. \tag{25}$$

2.3 NPV Curve Calculation

The cash flows resulting from the production are the result of two stochastic processes: the energy produced, as shown in Fig. 4, and the price of the electricity sold in the day-ahead market, represented by a GBM-MR model, as in (3). Since the NPV calculation requires long periods, 20 years of lifetime, the estimation of future prices is made for 40 years. This allows shifting the initial point of investment 20 years, which is the length of the NPV curves. The model could also be enriched by using long-term contract values, but this is outside the scope of this work.

The yearly cash flows resulting from the income model, subtracting the investment, maintenance costs, depreciation and interests, and applying corporate taxes, can be used to estimate the NPV of the wind investment. This is the standard calculation of the value of an investment used in Economics and Finance textbooks. However, a more interesting approach is to study the evolution of the NPV for long periods of time. The NPV curve represents the construction of the trajectory of the resulting NPVs when the investment takes place at successive times, updating the investment costs and cash flow values to the corresponding reference periods. In this paper, the NPV curve is calculated on a monthly basis and the annual value corresponds to the average value of the 12 months. Figure 5

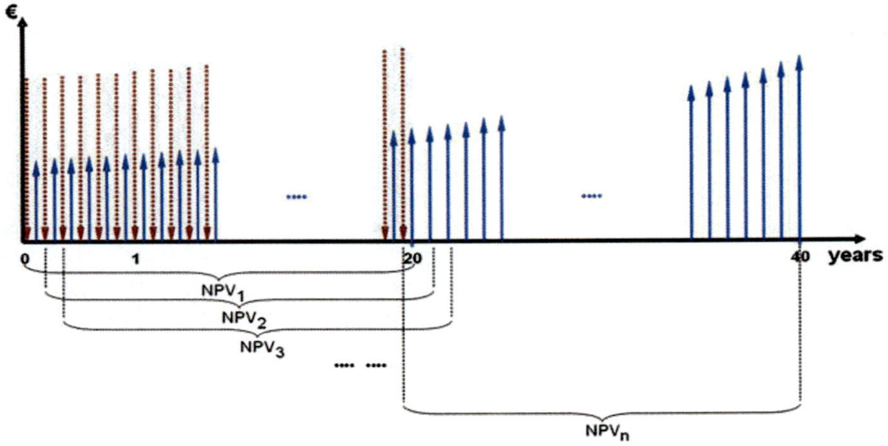

Fig. 5 NPVs calculation for a period of 20 years

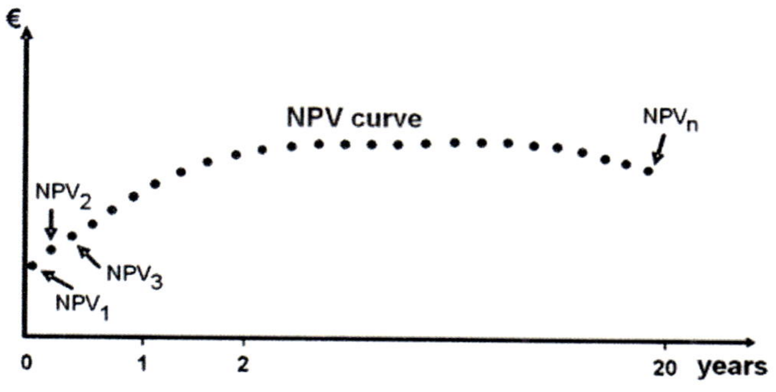

Fig. 6 Example of construction of an NPV curve for a 20-year period

presents an example of how to obtain the different NPVs, and Fig. 6 shows the resulting NPV curve.

2.3.1 NPV Curve Piecewise Parameter Estimation

It is not possible to assume that the NPV curve follows a pure GBM-MR process with constant parameters. The NPV curve stochastic parameters change with time without following the exact price pattern shown in (3) during the whole curve. This is due to the fact that the NPV comes from the combination of several stochastic processed with different characteristics. As said, the values of the NPV parameters change, in particular the trend, ϕ, and the strength of reversion, λ. On the other hand, volatility, σ, remains approximately constant showing a slightly increasing trend.

To solve this problem and to achieve a better fit, the estimation of the parameters is done piecewise. After a Monte Carlo simulation is done in order to obtain an estimation as accurate as possible, the NPV curves are split into several pieces. In this way, the matrixes calculated in Sect. 2.2 (18) are created with the information of each of the pieces so that the parameters, ϕ and λ, are optimally adjusted to the actual trajectory as shown in (26).

$$A_1 = \begin{bmatrix} 1 & y_2 \\ \cdots & \cdots \\ \cdots & \cdots \\ \cdots & \cdots \\ 1 & y_{\left(\frac{n}{t_n}\right)t_1} \end{bmatrix} \cdots A_n = \begin{bmatrix} 1 & y_{\left(\frac{n}{t_n}\right)t_{n-1}} \\ \cdots & \cdots \\ \cdots & \cdots \\ \cdots & \cdots \\ 1 & y_n \end{bmatrix},$$

$$Z_1 = \begin{bmatrix} y_2 - y_1 \\ \cdots \\ \cdots \\ \cdots \\ y_{\left(\frac{n}{t_n}\right)t_1} - y_{\left(\frac{n}{t_n}\right)t_1 - 1} \end{bmatrix} \cdots Z_1 = \begin{bmatrix} y_{\left(\frac{n}{t_n}\right)t_{n-1}} - y_{\left(\frac{n}{t_n}\right)t_{n-1}-1} \\ \cdots \\ \cdots \\ \cdots \\ y_n - y_{\left(\frac{n}{t_n}\right)t_{n-1}} \end{bmatrix}, \tag{26}$$

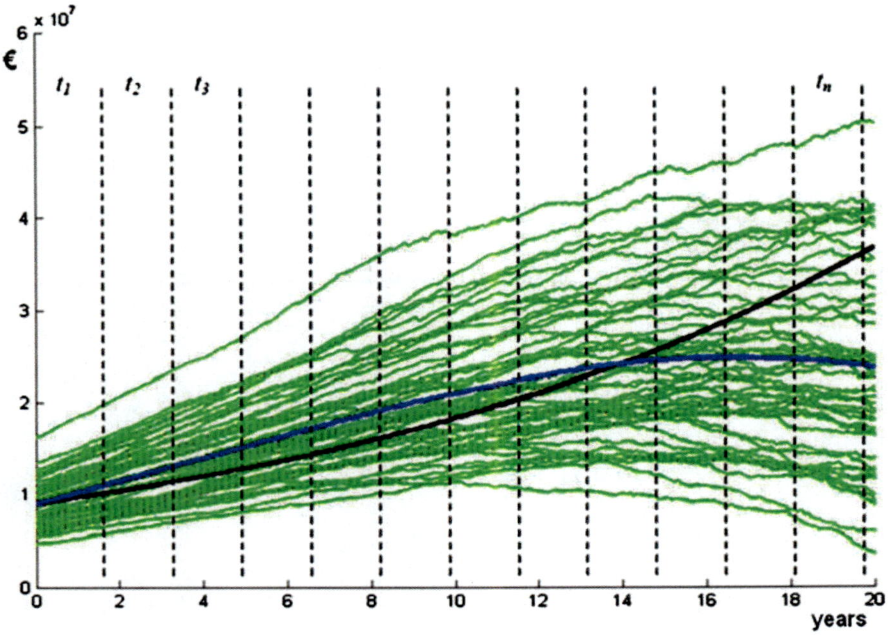

Fig. 7 Piecewise estimation of the NPV curve parameters

Thus, the parameter triads are estimated as follows for each block:

$$\begin{bmatrix} \hat{\lambda}_1 & \hat{\sigma}_1 & \hat{\phi}_1 \\ \hat{\lambda}_2 & \hat{\sigma}_2 & \hat{\phi}_2 \\ \cdots & \cdots & \cdots \\ \cdots & \cdots & \cdots \\ \hat{\lambda}_n & \hat{\sigma}_n & \hat{\phi}_n \end{bmatrix} \tag{27}$$

The more different from a pure GBM-MR the curves are, the more the number of pieces that should be used. The number of pieces ranges between 8 and 16, depending on the shape of the curves.

Figure 7 shows the piecewise estimation of the parameters of the NPV curves. In this case, 12 pieces have been used. Green curves represent each NPV curves from a Monte Carlo Simulation, the blue one represents the average of the curves, and the black one represents the initial investment.

2.4 Calculation of the Real Option Value and Probabilities

To build trinomial trees it is assumed that for each node of the dominion of the solution there are three possible paths to follow: go up, maintain the value, or go down. Therefore three transition probabilities must be defined as: p_u, p_m and p_d, respectively. Moreover, if a mean reversion process is modelled, then, three alternative branching structures are generated.

In [16] a general procedure to build binomial decision trees is proposed to represent processes with a stochastic factor with mean reversion of the Ornstein-Uhlenbeck type.

Once the investment costs are discounted from the NPV in the last period, the real option tree is built backwards up to the initial node. Mathematically this can be modelled by dynamic programming.

The possibility to exercise an American option is available at any time at each node, unlike the European option, that only depends on the values of previous nodes and their associated probabilities. The value of the option is the maximum of three possible choices: invest now (a), wait (b), or abandon (c), as shown in Fig. 8.

To calculate the value of an option from a tree (binomial, trinomial, etc.) that is already built for a specific process, it is necessary to subtract the project investment cost from the NPV of the project once the last period is reached as shown in Fig. 9

In parallel to the construction of the real options decision tree there must be an estimation of the probability of each possible investment situation. Due to the probability distribution of the nodes, the central zone of the tree has the highest weight in the estimation.

Figure 10 shows graphically the three different alternatives: exercise the option, wait, and abandon. It also depicts the probability distribution of each alternative throughout the life of the real option. The probability to exercise the option is marked in red, the waiting option in green, and the abandoning option in black. The visual check corresponding to the sum of the three probabilities is marked in blue.

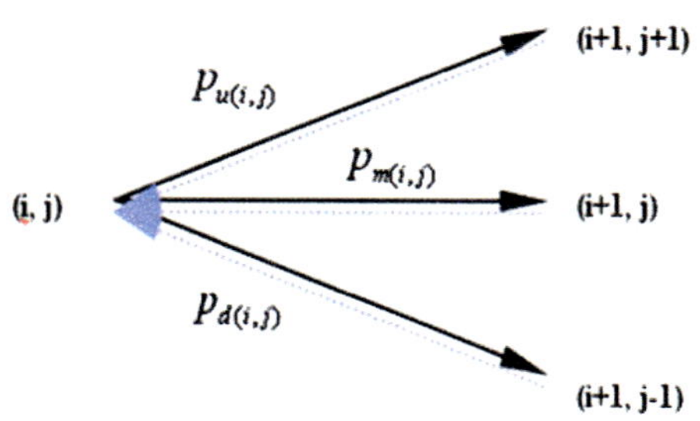

$$Vnode_{(i,j)} = \max(a;b;c)$$

a) $NPV_{(i,j)} - Investment_{(i)}$

b) $NPV_{(i+1,j+1)} \cdot p_{u(i,j)} + NPV_{(i+1,j)} \cdot p_{m(i,j)} + NPV_{(i+1,j-1)} \cdot p_{d(i,j)}$

c) 0

Fig. 8 Valuation of an American option to invest

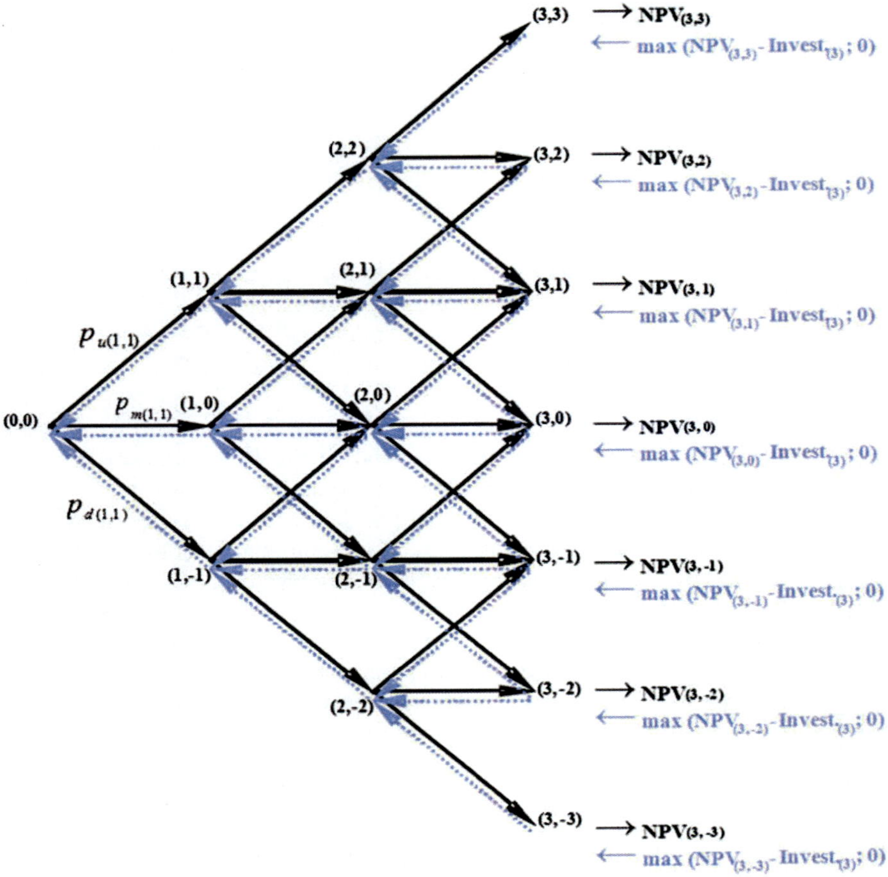

Fig. 9 Calculation of the real option values of the trinomial decision tree

3 Risk Profile Characterization

The model's valuation is carried out in a risk-neutral world, since it is difficult to know which discount rate to apply in the real world. Therefore, the process in (3) has to be risk-adjusted [11]. The drift for the real process is $\alpha = \lambda[\phi - \ln(x)]$, denoting the expected capital gain. If the total expected return is denoted by η, then the dividend-like income stream (or convenience yield) for the plant holder is $\delta = \eta - \alpha$. Consequently, the risk-adjusted drift is given by $\tau - \delta = \tau - \eta + \alpha$($\tau$ denoting the risk-free rate) and the risk-adjusted process is represented as:

$$\frac{dx_r}{x_r} = \lambda[\phi_r - \ln(x_r)]dt + \sigma dz, \tag{28}$$

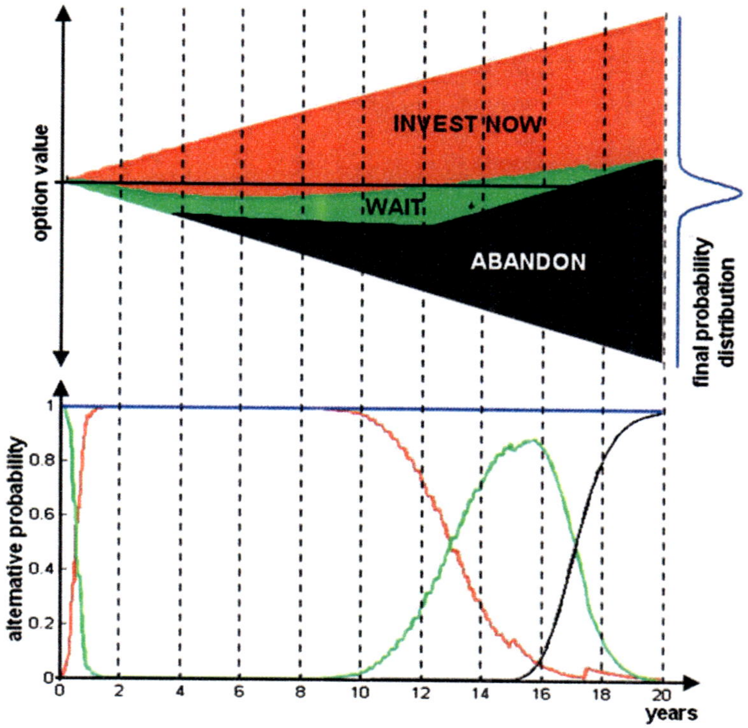

Fig. 10 Graphical depiction of the alternative probabilities: to invest, to wait, or to abandon

where

$$\phi_r = \phi - \left(\frac{\eta - \tau}{\lambda}\right). \tag{29}$$

Note that risk-neutrality valuation assumes that the position in the plant value and the project remains riskless by constant re-balancing. By doing this, we implicitly assume that the licence to build the plant is a tradable derivative [17].

4 Case Study: Parameterization of the Project Financing Percentage

This section contains one case study where the investment model described before is applied. The case study has been coded in MATLAB© [18] and it can be obtained from the authors upon request. A sensitivity analysis is applied with respect to key parameters of the model. The base case assumes a wind farm of

Table 1 Model's Parameters

Type	Parameters	Value
General	NPV calculation period	20 years
General	Estimation period for cash flows	40 years
General	NPV discount rate	7%
General	Monte Carlo simulations	200
Investment	Installed kW Price	€1,000
Investment	Price increase (annual)	7%
Wind	Shape factor	1.8
Wind	Scale factor	5.3
Production	Maximum power	10 MW
Production	Min., control, and max. rotor speed	4, 11, 25 m/s
Electr. price	Initial price	€60 /MWh
Electr. price	Volatility	0.005
Electr. price	Strength of reversion	0.005
Electr. price	Long term trend	16.95
Electr. price	Annual increase of trend	7%
Risk	Estimated expected yield	10%
Risk	Risk-free asset return	5%

10 MW, composed of 5 generators of 2 MW each. The parameters are as follows in Table 1:

This case study studies the effect produced in the model by changing the external financing% of the required investment. The base value of the financing parameter is 80% to mimic what is already in use in Spain, see [19]. The parameter values oscillate between 0% and 100% of external financing, applying a 10% increase for each of the 11 scenarios analyzed. To solve this case study there are four steps to follow:

Step 1: Construction of the NPV curves: Fig. 11 shows the evolution of the NPV curves for each of the scenarios for a period of 20 years. It can be observed that the curves that represent the evolution of the shareholders' equity increase linearly.

Step 2: Estimation of the GBM-MR parameters: From the different NPV curves the parameters of their piecewise GBM-MR associated parameters are obtained. According to the speed of change of the NPV curves, the optimal number of blocks whose parameters remain constant is found. See Fig. 7.

Step 3: Construction of the trinomial trees for the real options: Once tested that the parameters fit with the original data the different values of the American option to invest are calculated. Figure 12 shows their values for a specific NPV curve: wait, execute or abandon. In the figure, the first scenarios, in which the percentage of external financing is high, the "invest now" alternative is present throughout the entire lifetime (20 years). When the external financing decreases, there is more need of equity and the "wait" alternative has more weight. Finally, the "abandon" alternative only shows up in scenarios with more equity and in the latest moments of the lifetime.

Step 4: Calculation of the option to invest: As a result of the above calculations, the value of the real option is attained. As observed in Fig. 13, more equity implies

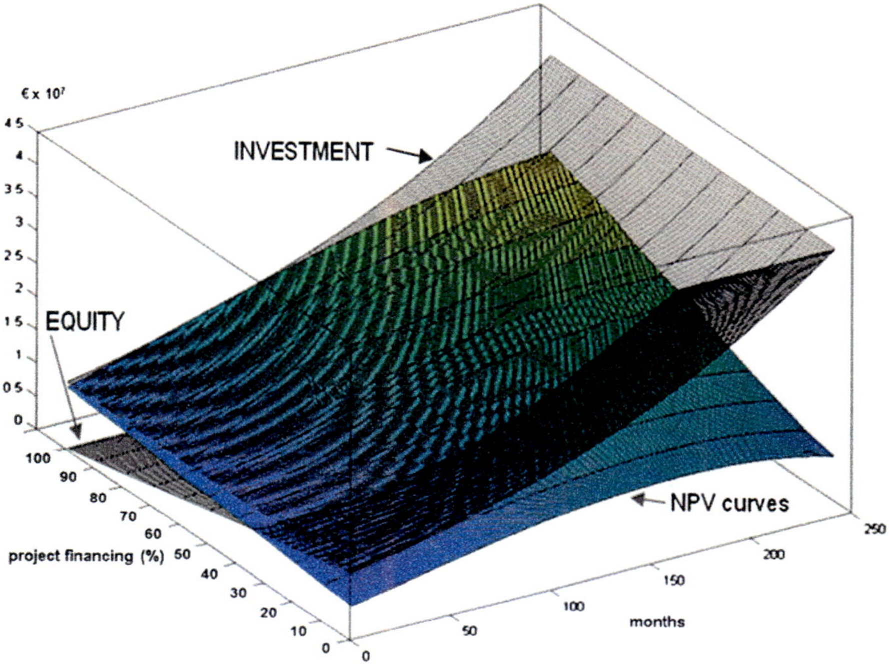

Fig. 11 NPV curves for each scenario

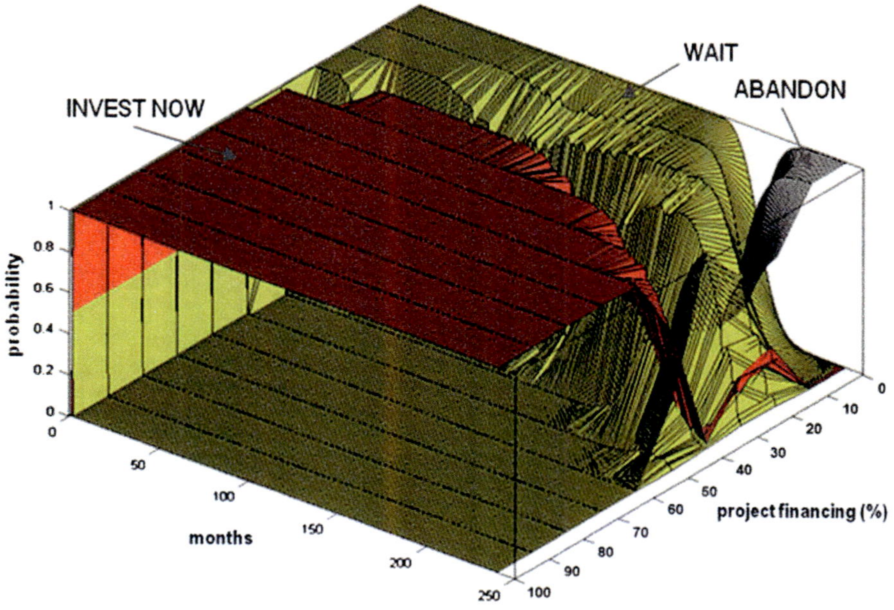

Fig. 12 Probabilities of the alternatives for each scenario

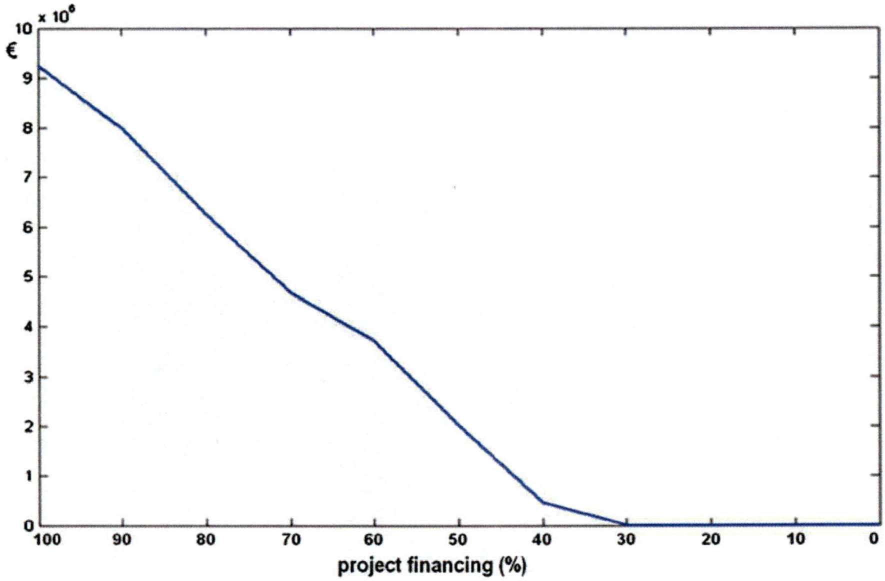

Fig. 13 Real option values to invest for each scenario

that the "invest now" alternative loses value, up to scenarios where the value of the option is 0, therefore, an investment is not advised.

These results can be easily compared to the ones resulting from a more traditional analysis. In particular, a classic investment valuation method would estimate the value of the NPV at the beginning of the lifetime (year 0) and depending on its value would decide to invest or not. Figure 14 shows how these NPV values correspond to the first point of the NPV curves. For this case, it is observed that in the initial scenarios, up to the seventh, the NPV is above the equity and, little by little, the difference is diminishing until it becomes negative in scenario 8.

As a conclusion for this case, the "invest now" alternative is present from the beginning of the option in all scenarios except the last ones.

Finally, Table 2 presents the descriptions of all the cases developed in this work and Table 3 shows the numerical results.

According with this results table, it is necessary to have at least a 40% project financing. Below this value, the project lacks interest, i.e., profitability.

As long as the minimum project financing requirement is fulfilled, the project will be possible if the price of the kW installed is less than €1,100. On the other hand, the variation in the option price is almost linear with respect to the risk aversion. For high risk aversion values (risk-free parameter between 0% and 3%) the "abandon" alternative is shifted to the middle of the option lifetime (10 years). For low risk aversion values (risk-free parameter between 7% and 10%) the "abandon" alternative does not take place until the end of the option lifetime.

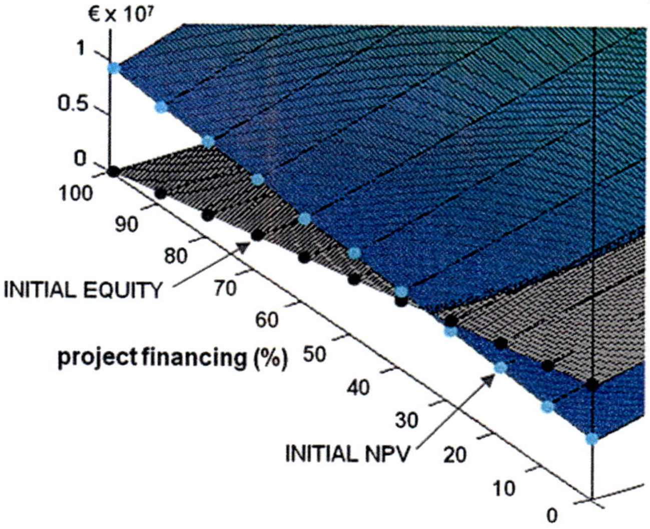

Fig. 14 Initial NPV versus equity

Table 2 Case studies description

Case	Parameters	Range (step)
1	Project financing	100–0% (10)
2	Investment cost (100% of equity)	€750–€1,250 /kW (50)
3	Investment cost (50% of equity)	€750–€1,250 /kW (50)
4	Annual investment cost rise	5–10% (0.5)
5	Risk aversion	0–10% (1)
6	Spot price volatility	0–0.01 (0.001)
7	Spot price trend (annual increase)	5–10% (0.5)
8	Spot prices' strength of reversion	0.025–0.075 (0.005)
9	Wind speed's shape parameter	1.5–2.5 (0.1)
10	Wind speed's scale parameter	5–6 (0.1)

Table 3 Case studies results

	Scenarios														
Case	1	2	3	4	5	6	7	8	9	10	11	Max	Min	Med	S.Dev
1	9.15	8.02	6.21	4.85	3.92	2.08	0.35	0	0	0	0	9.15	0.00	3.14	3.48
2	0.005	0	0	0	0	0	0	0	0	0	0	0.00	0.00	0.00	0.00
3	5.16	4.81	4	3.79	3.02	2.63	1.51	0.97	0.12	0.06	0	5.16	0.00	2.37	1.94
4	2.03	2.05	2.07	2.13	2.18	2.22	2.32	2.41	2.79	3.12	3.54	3.54	2.03	2.44	0.50
5	0.69	0.97	1.24	1.43	1.5	1.71	1.81	1.94	2.08	2.21	2.33	2.33	0.69	1.63	0.52
6	2.33	2.25	2.32	2.19	2.51	2.48	2.11	2.6	1.99	2.81	2.03	2.81	1.99	2.33	0.25
7	0	0	0.24	1.03	2.58	3.18	4.57	6.01	8.02	9.07	10.12	10.12	0.00	4.07	3.76
8	0	0	0	0	1.07	2.31	3.97	4.42	6.05	6.79	8.24	8.24	0.00	2.99	3.07
9	3.64	3.31	2.5	2.02	1.52	1.48	0.63	0.34	0	0	0	3.64	0.00	1.40	1.34
10	0.08	0.51	0.95	1.32	2.3	2.78	3.81	4.09	5.08	6.49	6.89	6.89	0.08	3.12	2.36

In relation to the spot price, its volatility does not affect the results significantly, due to the aforementioned integration effect. However, an increase in the spot price trend or the strength of reversion has considerable effects reaching the option value higher than M€8. On the contrary, due to the great sensibility of the model to these two parameters, for increases in the spot price trend below 6% or strength of reversion below 0.04, the project is not profitable.

Finally, referring to the wind speed, the shape and scale parameters have opposite effects. When the shape parameter increases, the dispersion of the wind speed decreases, and the value of the option decreases up to zero for parameter values higher than 2.2. If the scale parameter increases, the profitability of the project increases. In spite of this, due to the extreme sensitivity to this parameter, for parameter values below the base value, 5.3, the profitability is greatly reduced.

5 Conclusions

This study has presented a model to evaluate investments on wind energy plants based on two main concepts: (1) a stochastic approach to calculate the project NPV and its associated curve and parameters, and (2) a real options model built upon a trinomial tree that evaluates numerically the probabilities of the alternatives of investing now, wait or abandon the project. This trinomial lattice is calculated piecewise in order to make the model more precise.

The NPV curve is a new concept which is made up from the successive points corresponding to the NPV values at each period (month). This concept is introduced and compared with the traditional NPV method.

As regards to risk valuation, the model adapts the results as a function of the risk aversion profile of the investor instead of producing a specific risk associated with the investment.

This model permits an estimation of the best time within the project's lifetime to execute the investment seeking the maximum profit and also an estimation of the probability that a specific future scenario takes place.

In order to demonstrate the usefulness of the method, several case studies are analyzed having a variable range of the most important model's parameters.

References

1. El-Khattam W, Bhattacharya K, Hegazy Y, Salama MMA (2004) Optimal investment planning for distributed generation in a competitive electricity market. IEEE T Power Syst 19(3):1674–1684
2. Khatib H (2003) Financial and economic evaluation of projects in the electricity supply industry, vol 23, IEE Power series. The Institution of Engineering and Technology, Stevenage
3. Dixit A, Pindyck RS (1994) Investment under uncertainty. Princeton University Press, New Jersey

4. Trigeorgis L (1996) Real options: managerial flexibility and strategy in resource allocation. MIT Press, Boston
5. Correia PF, Carvalho PMS, Ferreira LAFM, Guedes J, Sousa J (2008) Power plant multistage investment under market uncertainty. IET Gener Transm Dis 2:149–157
6. Näsäkkälä E, Fleten S-E (2005) Flexibility and technology choice in gas fired power plant investments. Rev Financ Econ 14(3–4):371–393. Special issue on Real Options
7. Tseng C-L, Barz G (2002) Short-term generation asset valuation: a real options approach. Oper Res 50(2):297–310
8. Tseng C-L (2000) Exercising real unit operational options under price uncertainty. In: Proceedings of the 2000 IEEE PES winter meeting, Singapore, 23–27 Jan 2000, pp 436–440
9. Lucía JJ, Schwartz ES (2002) Electricity prices and power derivatives: evidence from the Nordic Power Exchange. Rev Derivatives Res 5:5–50
10. Deng S-J, Oren SS (2003) Incorporating operational characteristics and startup costs in option-based valuation of power generation capacity. Probab Eng Information Sci 17(2):155–182
11. Niemeyer V (2000) Forecasting long-term price volatility for valuation of real power options. In: Proceedings of 33rd Hawaii international conference on system science, HICSS'00. IEEE Computer Society Washington, DC, 2000
12. Tseng C-L, Lin KY (2007) A framework using two-factor price lattices for generation asset valuation. Oper Res 55(2):234–251
13. Fleten S-E, Näsäkkälä E (2003) Gas-fired power plants: investment timing, operating flexibility and abandonment. In: Proceedings of the 7th annual international conference on real options. University Library of Munich, Germany
14. Wang C-H, Min KJ (2006) Electric power generation planning for interrelated projects: a real options approach. IEEE T ENG Manage 53(2):312–322
15. Fleten S-E, Maribu KM, Wangensteen I (2007) Optimal investment strategies in decentralized renewable power generation under uncertainty. Energy 32:803–815
16. Hull JC, White A (1994) Numerical procedures for implementing term structure models I: single-factor models. J Derivatives 2(1):7–16
17. Hull JC (2003) Options, futures, and other derivatives, 5th edn, Finance series. Prentice-Hall – Pearson Education, Inc, Upper Saddle River
18. MATLAB, The Mathworks, Inc. http://www.mathworks.com
19. Spanish Royal Decree (2007) RD 661/2007, on 25th May 2007, for regulating the activity of energy production in a special regime. http://www.boe.es/boe/dias/2007/05/26/pdfs/A22846-22886.pdf
20. Operador del Mercado Ibérico de Energía – Polo Español, S.A. (2008). http://www.omel.es. Accessed 18 May 2009

Electric Interconnections in the Andes Community: Threats and Opportunities

Enzo Sauma, Samuel Jerardino, Carlos Barria, Rodrigo Marambio, Alberto Brugman, and José Mejía

Abstract The increasing costs of electricity and the difficulties to expand the power generation capacity, as well as the need for increasing the energy security levels, have enhanced the potential benefits of the electric interconnection among countries and the formation of sub-regional energy markets.

In this context, this paper identifies some sustainable and technically feasible alternatives for electric exchange through interconnections among the electric systems of Bolivia, Chile, Colombia, Ecuador and Peru. In particular, we assess such interconnections from an economic perspective and identify the main barriers for their development. The analysis is carried out at the pre-feasibility level from both private and social point of views, based on the assessment of different investment alternatives in the transmission systems among the aforementioned countries. The modeling of the different future economic operation conditions for each one of the considered electric systems, and for each one of the assessed scenarios, is a central element of the analysis.

The work reported in this paper was partially funded by the United Nations Development Program (UNDP) through a grand associated to the project "*Estudio para Análisis de Prefactibilidad Técnico Económica de Interconexión Eléctrica entre Bolivia, Chile, Colombia, Ecuador y Perú* (Study for the Technical-Economic Pre-feasibility Analysis of Electric Interconnection among Bolivia, Chile, Colombia, Ecuador, and Peru)"

E. Sauma (✉)
Industrial and Systems Engineering Department, Pontificia Universidad Católica de Chile, Santiago, Chile
e-mail: esauma@ing.puc.cl

S. Jerardino • C. Barria • R. Marambio
Kas Ingeniería S.A, Santiago, Chile
e-mail: samuel.jerardino@kasing.cl; carlos.barria@kasing.cl; rodrigo.marambio@kasing.cl

A. Brugman • J. Mejía
Estudios Energéticos Limitada, Bogotá, Colombia
e-mail: alberto.brugman@gmail.com; josemmejia@gmail.com

A. Sorokin et al. (eds.), *Handbook of Networks in Power Systems I*,
Energy Systems, DOI 10.1007/978-3-642-23193-3_14,

Keywords Interconnected power systems • Power system economics • Power transmission economics • Stochastic dynamic programming • Transmission lines

1 Introduction

The increasing costs of electricity and the difficulties to expand the power generation capacity, as well as the need for increasing the energy security levels, have enhanced the potential benefits of the electric interconnection among countries and the formation of sub-regional energy markets. The electric interconnection among countries allows reducing the need for capacity reserves and, at the same time, providing a higher security level of supply, given the different production and consumption cycles in the countries, which reduces energy prices in the long run.

There are successful examples of electric interconnection all around the world and an increasing interest in its expansion. In the Andes Community, its members – Bolivia, Colombia, Chile, Ecuador, Peru y Venezuela – approved Decision N. 536 in 2002, under which they agreed the electricity integration of the countries of the Andes Community and established the objectives and rules for the operation of the regional interconnections [1, 2].

Colombia and Ecuador are electrically interconnected since more than 4 years, which have helped to partially optimize the joint energy resources, taking advantage of the market conditions that benefit them at every period of the coordinated dispatch. In 2004, Peru and Ecuador built a transmission line (line Zorritos-Machala, in 220 kV), which is still inoperative due to normative differences between both countries. These examples show there is a significant potential for the development of electric interconnection, but also some barriers that limit the use of this potential. An important advantage and benefit of electric interconnections is the potential contribution to reduce greenhouse gas emissions, given the better use of the resources of each country.

There are some few works in the literature about electric-system integration in Latin America. Hammons et al. [3] analyze the energy market integration in South America by studying the electric power, the oil, and the natural gas industries of Argentina, Bolivia, Brazil, Chile, Paraguay, and Uruguay from an institutional-structure perspective. They highlight some legal, institutional, and regulatory issues that must be addressed to facilitate the energy regional integration. They also analyze the potential interactions between gas and electricity markets.

In both [4] and [5], the authors analyze the impacts of the deregulation of Latin-American electric systems over the power-generation sector and its effect on the incentives to accelerate the regional integration of the energy sectors. More recently, Barroso et al. [6] study the integration in South America of the liquefied natural gas (LNG) from the perspectives of the markets, the energy prices, and the security of energy supply.

In this context, this paper identifies some sustainable and technically feasible alternatives for electric exchange through interconnections among the electric systems of Bolivia, Chile, Colombia, Ecuador and Peru, thus allowing optimizing the use of the energy resources in an efficient way. In particular, we assess such interconnections from an economic perspective and identify the main barriers for their development. The analysis is carried out at the pre-feasibility level from both private and social point of views, based on the assessment of different investment alternatives in the transmission systems among the aforementioned countries. For doing this analysis, we study four interconnection scenarios (including the base scenario) in the time horizon 2010–2022, in addition to a sensitivity analysis for each scenario considering the opportunity prices of natural gas for both the exporting and importing countries.

From both the private and the social point of view, the analysis focus on quantifying economic costs and benefits through a methodology that considers different aggregation levels depending on the type of agent considered. The modeling of the different future economic operation conditions for each one of the considered electric systems, and for each one of the assessed scenarios, is a central element of the analysis.

2 Scenarios Analyzed

A base scenario was developed, which incorporates the representation of the main electric systems of the countries, considering the current characteristics of the generation supply, the electricity demand and the main transmission systems, as well as the future evolution of such features according to the planning studies performed by the energy authorities of each one of the countries involved. The base scenario corresponds to the representation of the current state of the electric interconnections among the five countries considered here. That is, it comprises only the current interconnection lines between Colombia and Ecuador (since, despite the fact that there is an electric interconnection between Ecuador and Peru, it does not operate due to the lack of commercial agreements, in addition to the fact that it cannot work in a synchronic manner).

It is important to underscore that in reference to Chile, we define the non-existence of a future interconnection between the main interconnected systems (the *Sistema Interconectado Central*, SIC, and the *Sistema Interconectado del Norte Grande*, SING). Consequently, the international electric integration with Chile is assumed at all times the integration with the SING, which represents relative proximity to both Bolivia and Peru with respect to the location of the SIC. The forecast of the electricity demand for each country is shown in Table 1. Such information was used in all scenarios.

The forecasted average electric energy growths for the 2010–2022 period are as follows: Bolivia: 6.3%; Chile-SING: 5.1%; Colombia: 3.5%; Ecuador: 5.5%; and Peru: 6.7%.

The reference fuel prices for the thermal power plants in each one of the countries considered here were standardized in similar units in order to be able to

Table 1 Forecast of the annual electric-energy demand for each country (GWh)

Year	Bolivia	Chile-SING	Colombia	Ecuador	Peru
2010	5,883	14,320	55,913	18,278	29,956
2011	6,356	15,034	57,849	21,355	31,637
2012	6,717	15,784	59,885	22,417	34,095
2013	7,131	16,573	62,160	23,716	37,665
2014	7,571	17,401	64,547	25,059	43,477
2015	8,041	18,302	66,906	26,504	47,661
2016	8,543	19,249	69,321	28,332	49,478
2017	9,078	20,246	71,821	30,146	51,427
2018	9,649	21,294	74,476	31,712	53,811
2019	10,253	22,396	77,245	33,181	56,320
2020	10,895	23,556	79,734	34,604	58,963
2021	11,578	24,778	81,818	36,294	61,748
2022	12,303	26,067	84,443	38,096	64,684

compare the forecasted profiles. The information and data bases of the electric systems involved were obtained based on the generation and transmission expansion plans that were in force at each country in May of 2009. The associated documents are as follows:

- **Bolivia**: *Plan de Expansión del Sistema Interconectado Boliviano, Periodo 2009–2018* (Expansion Plan for the Bolivian Interconnected System, 2009–2018 period). *Comité Nacional de Despacho de Carga* (National Load Dispatch Committee), November, 2008.
- **Chile**: *Fijación de Precios de Nudo Abril 2009 – Informe Técnico Definitivo, Sistema Interconectado del Norte Grande (SING)* (Nodal Price Setting April 2009 – Final Technical Report, SING). *Comisión Nacional de Energía* (National Energy Commission), April, 2009.
- **Colombia**: *Plan de Expansión de Referencia Generación – Transmisión 2009–2023* (2009–2023 Generation-Transmission Reference Expansion Plan). *Unidad de Planeación Minero Energética, Ministerio de Minas y Energía* (Mining-and-Energy Planning Unit, Ministry of Mines and Energy), March 2009.
- **Ecuador**: *Plan Maestro de Electrificación, Periodo 2009–2020* (Electrification Master Plan, 2009–2020 Period). *Consejo Nacional de Electricidad* (National Electric Council), March 2009.
- **Peru**: *Elaboración del Plan Referencial de Electricidad 2008–2017* (Design of the 2008–2017 Reference Electricity Plan). *Ministerio de Energía y Minas* (Ministry of Energy and Mines), April 2009.

For the different scenarios, the simulations of the operation of the electric systems were carried out taking into consideration the local prices of the generation fuels. However, a sensitivity analysis was performed by using the opportunity prices for natural gas, with an average price of 6 US$/MBTU during the 2009–2022 period.

The generation expansion plans of the countries were not altered in the considered scenarios, excepting for expansion in such plans that do not represent the power systems until 2022. Such extensions were proposed by taking into account the information contained in the aforementioned documents. Table 2 shows a general comparison of the installed generation capacity of each country contained in their respective generation expansion plans.

It is worth to mention that there are countries whose installed capacity up to 2022 is between 2 and 2.5 times the 2009 capacity, in comparison to countries that increase the installed capacity between 30% and 50% during the same period. These differences in the increase of the installed capacity with respect to 2009 levels conditions aspects such as the use of the interconnections towards the end of the horizon, the export levels of electric energy towards countries with smaller capacity growths, the fluctuations in both operational and marginal costs, and the levels of non-served energy.

The interconnection scenarios analyzed in this paper (other than the base scenario) are the following:

(a) **Scenario 1**: considers the reinforcement of the currently available interconnections between Colombia and Ecuador and the construction of a new interconnection between Peru and Ecuador. The rest of the conditions are similar as those in the base scenario.
(b) **Scenario 2:** adds an electric interconnection between Peru and Chile (SING) to the conditions stated for Scenario 1.
(c) **Scenario 3:** adds an electric interconnection between Bolivia and Chile (SING) to the conditions stated for Scenario 2.

Table 3 summarizes these interconnection scenarios for the Andes region, indicating the corresponding start-up dates of the interconnections.

Table 2 Installed generation capacity (MW) of each country, in the period 2009–2022

Type	Bolivia	SING-Chile	Colombia	Ecuador	Peru
Hydro	127	–	3,820	3,724	5,796
Thermal	1,130	1,860	1,159	1,044	3,176
Renewable (no hydro)	97	260	–	–	–
Total	1,354	2,120	4,979	4,768	8,972

Table 3 Interconnection scenarios and start-up dates of the interconnections

Scenario I start-up date	Co-Ec	Ec-Pe	Pe-Ch	Bo-Ch
Base scenario	Current			
Scenario 1	2014	2015		
Scenario 2	2014	2015	2016	
Scenario 3	2014	2015	2016	2017

3 Methodology

The pre-feasibility analysis for the development of electric links is, in general, an iterative process that requires considering not only the particular situations of each country, but also those technologies that are feasible to be used in implementing the interconnection projects.

In the planning process, any change in conditions, assumptions or restrictions that condition the starting point, or in the types of technologies used and their technical-economic characteristics, will result on obtaining a different interconnection alternative. Accordingly, the results presented in this paper are valid within the framework of the assumptions, considerations and restrictions described here. Thus, a greater level of demand, a different relation of the fuel prices used by the thermal power plants, or different investment values for the projects studied would lead the electric systems to another equilibrium point, altering the results (and even the conclusions in the case that such changes are significant).

In general methodological terms, first we performed a long-term analysis of the interconnections in order to establish a maximum potential for complementarities among the different countries. In particular, we compared the current state of the interconnections (where there is only international links between Colombia and Ecuador) with one in which there are interconnections among all the considered countries through links with no capacity restrictions and with different start-up dates (according to an expected feasibility of implementation). This long-term analysis was carried out using the Stochastic Dual Dynamic Programming (SDDP) model, based on a one-node representation of the electric system of each country.

Secondly, the interconnections were designed considering links with sizes that fit the most realistic possibilities in order to be implemented within the horizon considered here. Particularly, we performed a mid-term analysis that considers the interconnection of the main transmission systems of the different countries in a joint and coordinated operation. We use the *Ose2000* model (developed by KAS Ingenieria S.A.), which is similar to the SDDP model, breaking-down the demand in each system from both the temporary and the geographic point of view. Then, using these results, we calculated the benefits associated to the different interconnection scenarios. In addition, we consider that the development of interconnections is carried out by using technologies at efficient voltage levels. Figure 1 shows the general layout used to analyze the different interconnection scenarios.

As it can be observed from Fig. 1, the proposed methodology allows determining the supply-side effects of the interconnection scenarios through quantifying the operational margin[1] of each one of the generators participating in the long-term

[1] The operational margin of a generator corresponds to the difference between the revenues due to the sale of electricity (valued at the respective injection nodes) and the operation costs of each generation power plant.

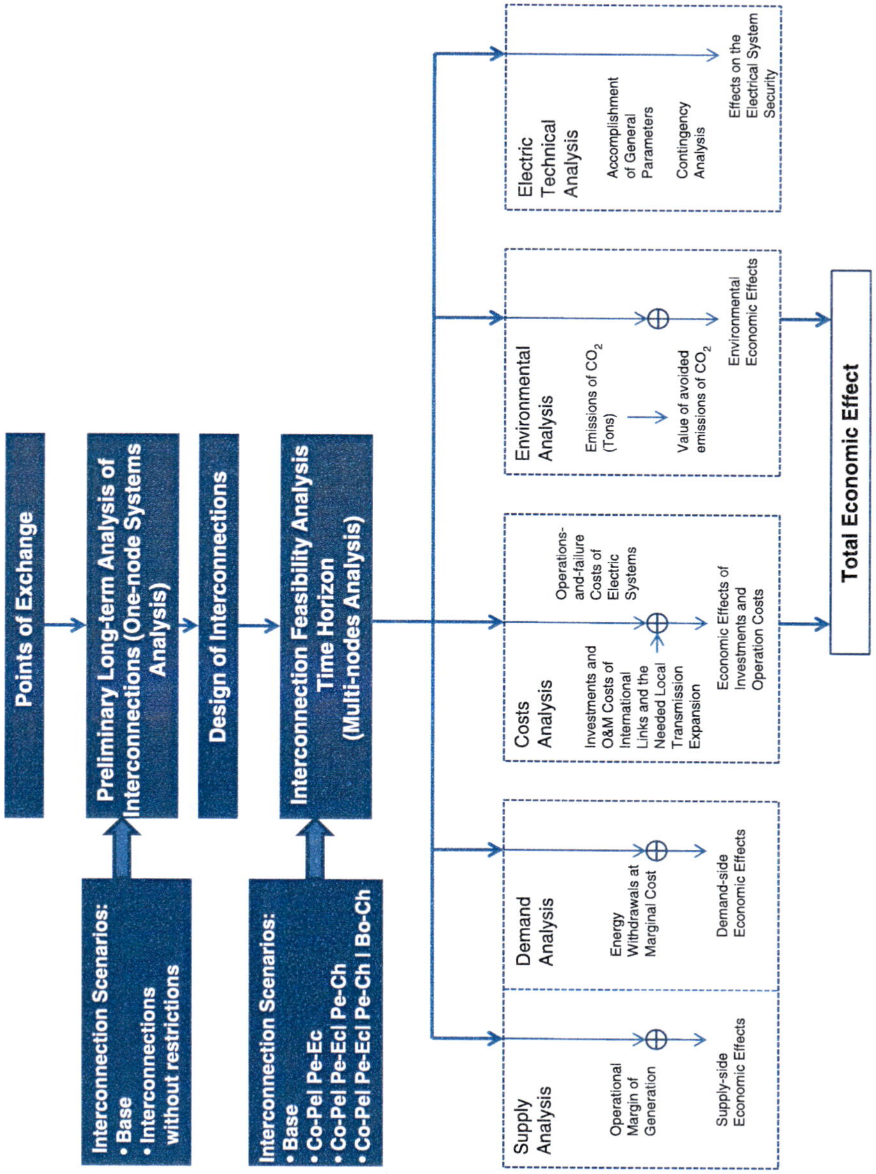

Fig. 1 General layout of the technical-economic analysis of the interconnection scenarios

operation of the systems, in each one of the scenarios analyzed. Similarly, it is possible to determine the economic effects of the interconnection scenarios from the point of view of the electric-energy demand, through quantifying the value of withdrawing energy at every demand node of the electric systems.

The proposed methodology allows quantifying the economic effects of the interconnection scenarios over the system costs. In doing this, we use a cost function that considers the operation and the failure costs in each power system, along with the international-link transmission costs and the costs of the local transmission-system reinforcements necessary for the pre-feasibility of the electric interconnections.

In addition, the proposed methodology allows determining the environmental effect of the interconnection scenarios through the valuation of the avoid emissions of CO_2 on each interconnection scenario as compared to the base scenario.

Finally, the safety and security effects of the electric systems are reviewed through static and dynamic simulations in each one of the proposed scenarios. In every scenario, we use the electric technical analysis model *DigSilent*, which allows evaluating both the transient and the permanent behaviors of the interconnection links under different contingencies.

4 Results

4.1 Interconnection Projects Proposed

Based on the results of the long-term analysis of the interconnections (where we compare the base scenario with a scenario in which there are interconnections among all the considered countries, modeled as a one-node representation, through links with no capacity restrictions),[2] we design interconnections with a high degree of feasibility in terms of their size and start-up date. In this context, we consider earlier the expansion of the current interconnections (i.e. the Colombia-Ecuador and Peru-Ecuador links) and later other expansion projects whose construction time periods may be more uncertain due to aspects such as future negotiations among the countries involved and environmental-dealing issues. Accordingly, we consider transmission investments associated to feasible interconnections in the Andes region, which produce different benefits for each country in consistency with the expected complementarities among the countries.

The links proposed in the different scenarios under an interconnected operation of the electric systems of the five countries are described in Table 4 and shown in Fig. 2.

[2] A detailed description of these long-term analysis results is available in [7].

Table 4 Characteristics of the proposed interconnections in the Andes region

Link	Interconnection point	Long (km)	Main characteristics	.Start-up date	Investment cost (thousands of US$)
Colombia–Ecuador	San Marcos – Jamondino 500 kV (Colombia) – Pifo 500 kV (Ecuador)	551	1,500 MW – 500 kV, AC 60 Hz	Apr-14	210,942
Ecuador–Peru	Yaguachi 500 (Ecuador) – Trujillo 500 kV (Peru)	638	1,000 MW – 500 kV, AC 60 Hz	Jan-15	174,427
Peru–Chile	Montalvo 500 (Peru) – Crucero 500 kV (Chile)	645	1,500 MW – 500 kV, HVDC	Jan-16	401,646
Bolivia–Chile	Chuquicamata 220 kV (Chile) – Chilcobija – Tarija 230 kV (Bolivia)	489	340 MW – 230 kV, AC 50 Hz	Jan-17	163,735

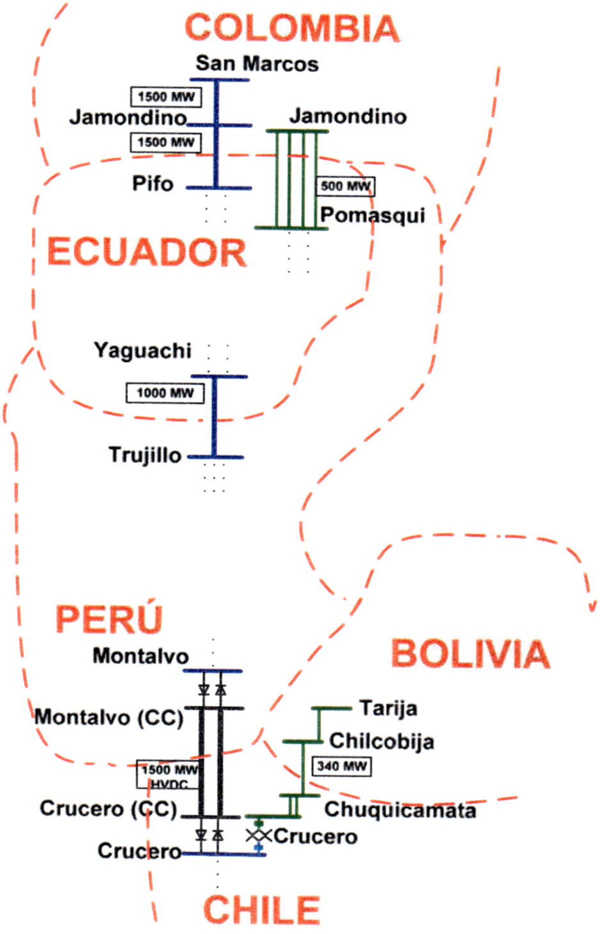

Fig. 2 Proposed interconnections in the Andes region

Table 5 Energy transferences through the interconnections during the 2014–2022 period (GWh)

From	To	Scenario			
		Base	1	2	3
Colombia	Ecuador	14,567	23,435	29,458	29,564
Ecuador	Colombia	10,505	15,031	11,211	11,314
Peru	Ecuador		13,882	4,138	4,237
Ecuador	Peru		12,896	23,321	22,677
Peru	Chile SING			81,663	80,925
Chile SING	Peru			–	–
Chile SING	Bolivia				–
Bolivia	Chile SING				15,503

4.2 Energy Exchange Levels Through the Proposed Interconnections

Using the proposed interconnections and the operation conditions in each scenario, we performed the mid-term analysis in the *OSE2000* model. From the simulation results, it is possible to compute the expected use of every link. The results obtained show that there are important energy blocks feasible to be transferred from one electric system to the other. This is due to multiple reasons, among which it is the possibility of taking advantage of arbitrage (e.g., price differences), thus generating business opportunities for the electricity-market agents in each country.

Table 5 shows a summary of the energy transferences through the interconnections during the 2014–2022 period.

Going from Scenario 1 to Scenario 2, we observe an important increase in the energy transfers from Ecuador to Peru due to the incorporation of the interconnection between Peru and Chile-SING. Similarly, the transfers between these two countries decrease in the opposite direction because the Peruvian energy generation is mainly aimed to supply loads in the north of Chile under this scenario. Something similar occurs with the generation production of Bolivia in Scenario 3.

It is worth to recall that the design of the interconnections considered here is based on a capacity sizing that takes into account the feasibility of implementation within the time horizon, applying conservative criteria regarding the matter. Analyzing ex-post the use level of the interconnections obtained in the scenarios studied, we verify that the interconnections operate in an expected manner at full capacity during most hours in the study period, thus indicating the existence of a greater energy exchange potential than the one established here.

4.3 Effect of the Interconnections on the Systems Marginal Costs

One of the most important effects of the energy transferences between countries is the effect over the local systems marginal costs. Next, we describe the impact of the energy exchange on the marginal costs of each country according to each interconnection scenario.

4.3.1 Base Scenario

This scenario considers that natural gas prices are the prices regulated locally by each country, excepting for Chile that corresponds to the price of the liquid natural gas placed in the future Mejillones re-gasification terminal. This explains why the profile of the system marginal costs for Chile turns out to be much greater than those of the other countries (see Fig. 3). This price condition remains in scenarios 1, 2 and 3.

4.3.2 Scenario 1

In this scenario, due to the expansion of the interconnections between Colombia and Ecuador (in 2014) and an interconnection between Peru and Ecuador (in 2015), the marginal costs of these three countries tend to be similar in the long term (fluctuating from US$30/MWh to US$31/MWh). However, the net effect is an increase in the system marginal costs of Colombia and Peru in reference to the base scenario due to the energy exported towards Ecuador.

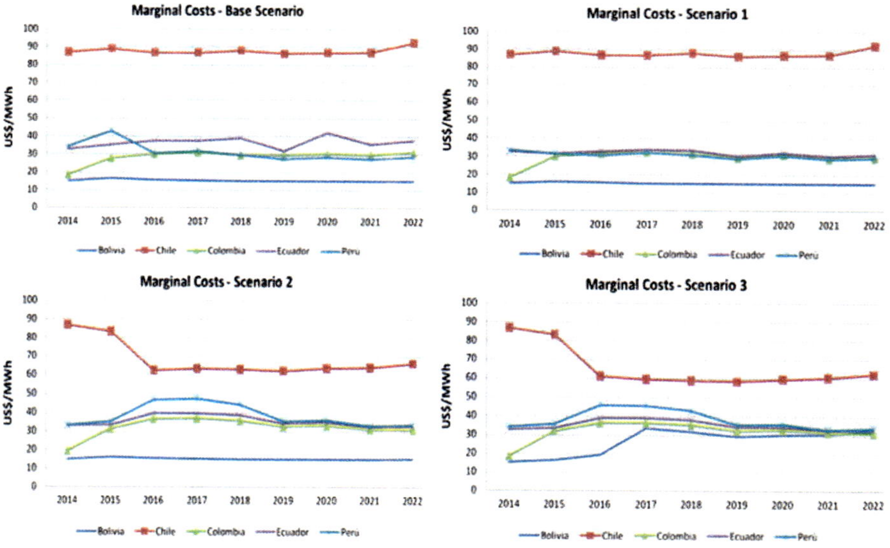

Fig. 3 Annual average marginal costs of the systems under different interconnection scenarios

4.3.3 Scenario 2

In this scenario, the interconnection between Peru and Chile is added. This produces an important increase in the marginal costs during the first years in Colombia, Ecuador and Peru, in reference to the prior scenarios, due to the energy exported towards Chile (see Fig. 3). On reaching the end of the time horizon, the marginal costs in these countries also tend to decrease, approaching values of around US$30/MWh, similar to the stabilization levels provided in Scenario 1. This occurs because, at the end of the time horizon, there is a stronger relation between the installed capacity and demand, including the one coming from the interconnections. On the contrary, the marginal costs in Chile-SING significantly decrease, reaching values close to US$60/MWh. This level of the marginal cost in Chile-SING remains in almost all the rest of the horizon, with a small increase at the end of the horizon, which would indicate that the links are operating at their maximum capacity and it is not possible to inject more low-cost energy to this system.

4.3.4 Scenario 3

In this scenario, the interconnection between the Bolivian electric system and the SING in Chile is added. As observed in Fig. 3, starting from the entrance into operation of this link, the marginal costs in Bolivia tend to approach those of Colombia, Ecuador and Peru in the long-term (reaching values around US$34/MWh), while, in the aforementioned countries, the marginal costs remains in similar levels as in Scenario 2. This indicates an important energy transfer from Bolivia to Chile-SING during this period.

The marginal costs in Chile-SING tend to have values slightly lower than US $60/MWh, in almost all the rest of the horizon, where the final increase in marginal cost is lower than that in Scenario 2 due to the energy imports from Bolivia.

4.3.5 Natural Gas Sensitivity Scenario

This scenario corresponds to a sensitivity analysis of the proposed interconnection scenarios, which considers the opportunity costs for the natural gas prices in the different countries (from both the export and the import viewpoint).

Figure 4 shows the annual average marginal costs with the aforementioned sensitivity analysis over the natural gas prices, when considering the interconnections included in Scenario 3. Under these conditions, the marginal costs of all countries, particularly in Bolivia, increase in comparison to the original Scenario 3. A relative matching is produced in the associated marginal costs to Colombia, Ecuador and Peru in the long-term (at levels between US$40/MWh and US$45/MWh), while marginal costs in Bolivia and Chile-SING tend to be matched (at levels

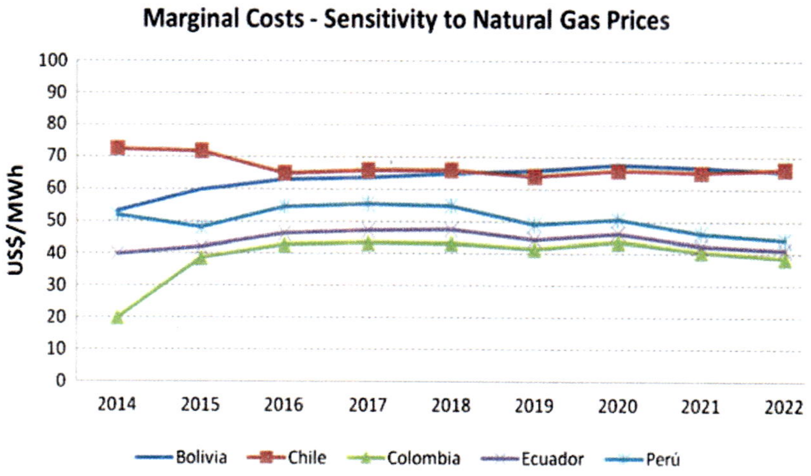

Fig. 4 Annual average marginal costs of the systems under natural gas sensitivity scenario

around US$65/MWh). This would suggest the existence of more equivalent energy exchanges between these last two countries within the planning horizon considered, with possible bi-directional energy flows. This bi-directionality during some periods of the year opens up the possibility of establishing energy-exchange mechanisms based on a better equilibrium of exported and imported energy sales prices.

4.4 *Benefit Analysis*

The benefit analysis is carried out from a whole-horizon behavior perspective of electricity markets. In particular, we quantify the benefits and/or costs on both the electricity supply and the electricity demand sides. We also quantify the total cost savings achieved in every electric system due to the proposed interconnections and the investment costs of both international links and those modifications made to the local transmission systems (and needed to support the international energy exchange). In addition, we quantify the environmental benefits of the proposed interconnections in terms of the avoided emissions of CO_2.

4.4.1 Supply-Side Economic Benefits

The private economic benefits for the generation sector were analyzed considering the sector as a single market agent. Such benefits are calculated based on the "electric-sector operational margin", which is defined as the difference between the valuation (at the monthly marginal cost) of the energy injections coming from all the system's power plants and the operation costs of all the thermoelectric power

Table 6 Present value of the electric-sector operational margin during the 2014–2022 period (millions of US$ discounted to April 2014)

Scenario	Bolivia	Chile	Colombia	Ecuador	Peru	Total	Difference with respect to the base scenario
Base	372	3,139	11,098	5,421	6,903	26,933	0
1	372	3,139	11,546	4,903	6,789	26,748	−185
2	372	1,498	12,623	5,490	9,488	29,470	2,537
3	1,107	1,366	12,548	5,463	9,367	29,852	2,919

plants.[3] Throughout the entire article, a 10% discount rate is used. Table 6 shows the present value of the electric-sector operational margin for every electric system during the time horizon, in each interconnection scenario.

From a private viewpoint, Scenario 1 does not produce a positive margin for the whole set of countries (with respect to the base scenario), although the Colombian generation sector gets higher margins in this scenario than in the base scenario due to its exports to Ecuador (which implies that the Ecuadorian generation sector decreases its margins). On incorporating the interconnection between Peru and Chile in Scenario 2, the operational margins in Colombia, Peru and Ecuador significantly increase, while the operational margin in Chile-SING decreases to less than half (due to the energy imports coming from Colombia, Peru and Ecuador). Something similar occurs with Bolivia, on incorporating the interconnection between this country and Chile-SING in Scenario 3, keeping the other-countries' margins in similar magnitudes as those of Scenario 2. Therefore, one of the most remarkable results of this work is that there are only private economic benefits when the interconnections with Chile are taken into consideration (i.e., scenarios 2 and 3).

It is important to remark that the private benefits obtained by the generation sector of a specific country (due to exporting or importing electricity in some scenario) are computed assuming there are no commercial barriers or restrictions limiting the energy exchanges. In this sense, our results should be understood as referential values, which could vary in practice according to the commercial mechanisms being finally implemented for the regional electric integration.

The natural-gas prices sensitivity analysis allows isolating, up to certain extent, the "price" variable in reference to the natural gas prices, establishing a "floor" for the electricity exchanges. In this case, the private decisions about energy exchanges depend only on the complementarities among the electric systems (they do not depend on arbitrages or price structure fluctuations between countries). Table 7 shows the present value of the electric-sector operational margin for every electric

[3] As it is usual in energy economics, we assume hydroelectric power plants operate at zero marginal cost.

Table 7 Present value of the electric-sector operational margin during the 2014–2022 period, with NG prices sensitivity (millions of US$ discounted to April 2014)

Scenario	Bolivia	Chile	Colombia	Ecuador	Peru	Total	Difference with respect to the base scenario
Base NGs	1,884	1,854	12,685	6,380	9,732	32,535	0
1 NGs	1,884	1,854	14,519	6,257	9,117	33,631	1,096
2 NGs	1,884	1,237	15,360	6,766	11,290	36,537	4,002
3 NGs	1,860	1,255	15,446	6,853	11,533	36,948	4,413

system during the time horizon, when considering interconnections scenarios that use the opportunity costs of the natural gas.[4]

The results of the natural-gas prices sensitivity analysis show that, even under these highly restrictive conditions, there are significant complementarities in the resources use. As previously mentioned, they also show that the proposed interconnections have a high capacity use level, with degrees of bi-directionality higher than in the original scenarios.

4.4.2 Demand-Side Economic Benefits

The effect of the interconnections on the demand side was analyzed considering that there is a single buyer in every electric system who consumes electric energy. The demand-side economic benefits are calculated, for every scenario, by valuating the energy withdrawals of the single buyer, considering he/she pays the monthly average marginal cost of the system (assuming that this monthly average marginal cost should be the electricity price in the long run in order to having generators covering their average production costs and their no-covered investment costs).

Table 8 shows the present value of the valued energy withdrawals, considering the single buyer pays the monthly average marginal cost of the system. We use negative values to symbolize that these are costs (instead of benefits) for the buyers.

From Table 8, it becomes evident that, from a demand-side viewpoint, Scenario 1 is the most favorable scenario and Scenario 3 is the most adverse situation to buyers.

Table 9 shows the present value of the valued energy withdrawals for every electric system during the time horizon, when considering interconnections scenarios that use the opportunity costs of the natural gas.[5] From Table 9, we conclude that considering the opportunity costs of the natural gas leads, in general,

[4] The scenarios labeled "base NGs", "1 NGs", "2 NGs" and "3 NGs" correspond to the equivalent scenarios base, 1, 2 and 3, but incorporating the natural gas prices sensitivity (i.e., using the opportunity costs of natural gas as the natural gas prices).

[5] The scenarios labeled "base NGs", "1 NGs", "2 NGs" and "3 NGs" correspond to the equivalent scenarios base, 1, 2 and 3, but incorporating the natural gas prices sensitivity (i.e., using the opportunity costs of natural gas as the natural gas prices).

Table 8 Present value of the energy withdrawals, valuated at the monthly average marginal cost of the system (millions of US$ discounted to April 2014)

Scenario	Bolivia	Chile	Colombia	Ecuador	Peru	Total	Difference with respect to the base scenario
Base	−909	−11,182	−12,575	−6,840	−10,257	−41,763	0
1	−909	−11,182	−13,032	−6,076	−10,067	−41,266	497
2	−909	−8,803	−14,113	−6,686	−12,552	−43,063	−1,300
3	−1,541	−8,483	−14,031	−6,660	−12,456	−43,171	−1,408

Table 9 Present value of the energy withdrawals, valuated at the monthly average marginal cost of the system, with NG prices sensitivity (millions of US$ discounted to April 2014)

Scenario	Bolivia	Chile	Colombia	Ecuador	Peru	Total	Difference with respect to the base scenario
Base NGs	−3,741	−9,519	−14,313	−8,082	−14,844	−50,499	0
1 NGs	−3,741	−9,519	−16,098	−7,683	−14,122	−51,162	−663
2 NGs	−3,741	−8,512	−16,980	−8,216	−16,183	−53,631	−3,132
3 NGs	−3,708	−8,548	−17,194	−8,297	−16,451	−54,198	−3,699

to higher costs for buyers, which is a direct consequence of the higher production costs faced by the natural gas power plants and the fact that we are considering a pricing scheme based on the marginal costs.

4.4.3 System Cost Savings

The system cost savings due to the proposed interconnections are quantified by comparing the savings in the system operation costs due to the interconnections and the investment costs of such links. The transmission investment costs involve both the costs associated to the construction of the international links and the costs associated to the modifications made to the local transmission systems in order to support the international energy exchange.

Tables 10 and 11 show the present value of the operations-and-failure costs, in the analyzed scenarios, for the cases of without (Table 10) and with (Table 11) sensitivity analysis for the natural gas prices. The operations-and-failure costs include the variable production costs (in terms of both fuels and non-fuels) incurred by the different power plants and the failure cost, which is calculated by valuating the non-supplied energy. Thus, the difference between the total operations-and-failure costs in each scenario and such costs in the base scenario corresponds to the operations-and-failure cost savings due to the proposed interconnections in each one of such scenarios.

From Tables 10 to 11, it is observed that the more interconnections exist, the lower the operations-and-failure costs. The reason of this is the replacement of expensive generation by cheaper energy available in other country with lower

Table 10 Present value of the operations-and-failure costs (millions of US$ discounted to April 2014)

Scenario	Bolivia	Chile	Colombia	Ecuador	Peru	Total	Difference with respect to the base scenario
Base	492	7,482	1,061	910	2,873	12,819	0
1	492	7,482	1,043	670	2,957	12,643	−176
2	492	4,087	1,261	951	4,004	10,793	−2,025
3	660	3,579	1,245	929	4,006	10,420	−2,399

Table 11 Present value of the operations-and-failure costs, with NG prices sensitivity (millions of US$ discounted to April 2014)

Scenario	Bolivia	Chile	Colombia	Ecuador	Peru	Total	Difference with respect to the base scenario
Base NGs	1,668	7,188	1,221	1,051	4,416	15,544	0
1 NGs	1,668	7,188	1,334	910	4,076	15,176	−368
2 NGs	1,668	4,323	1,532	1,099	5,578	14,200	−1,343
3 NGs	1,622	4,373	1,643	1,097	5,564	14,299	−1,244

production costs, which is now available due to the interconnections. In the case of the natural-gas prices sensitivity analysis, the benefits in terms of costs savings decrease, as it is evident in Table 11. This is because the differences among the costs of each system (and, particularly, among systems with a high penetration of natural gas power plants) decrease when considering the opportunity costs of the natural gas.

On the other hand, we calculated the present value of the transmission investments, as well as the costs of operations and maintenance of the international links and the investments needed in the local networks to support the international energy exchanges. All this information can be used for calculating a total cost function, which allows minimizing the total expenditures in the systems in the same way a centralized transmission investment planner will do it. Such total cost function is quantified by means of the present value of the total costs (PVTC), which corresponds to the sum of the present values of the operations-and-failure system costs (O&FC), the investments, operations and maintenance costs of both the international links and the necessary extensions carried out in the local transmission systems (I_{TX} + $O\&MC_{TX}$), and the residual value of the transmission investments ($Residual_{TX}$). We consider a service life of 30 years for the transmission lines, which explains why we consider a residual value of the transmission investments when doing the economic evaluation in the horizon 2014–2022. As mentioned before, we assume 10% as the discount rate. Table 12 shows the present value of the total cost function and its components.

The difference between the total cost function in each scenario and the total cost function in the base scenario corresponds to the system cost savings due to the proposed interconnections in each one of such scenarios. These values are shown in

Table 12 Present value of the total cost function (millions of US$ discounted to April 2014)

Scenario	I_{TX} + O&MC_{TX}	O&FC	$Residual_{TX}$	PVTC[a]	Difference with respect to the base scenario
Base	0	−12,819	0	−12,819	0
1	−240	−12,643	170	−12,713	106
2	−405	−10,793	328	−10,870	1,948
3	−459	−10,420	387	−10,492	2,327

[a]PVTC $= I_{TX} +$ O&$MC_{TX} +$ O&FC $- -$ $Residual_{TX}$

Table 13 Present value of the total cost function, with NG prices sensitivity (millions of US$ discounted to April 2014)

Scenario	ITX + O&MCTX	O&FC	ResidualTX	PVTC	Difference with respect to the base scenario
Base NGs	0	−15,544	0	−15,544	0
1 NGs	−240	−15,176	170	−15,245	299
2 NGs	−405	−14,200	328	−14,277	1,266
3 NGs	−459	−14,299	387	−14,371	1,172

the last column of Table 12. From Table 12, we conclude that the most favorable scenario, from the viewpoint of a centralized transmission planning, is Scenario 3 (with a net benefit of 2,327 millions of dollars in reference to the base scenario). On the other hand, Scenario 1 is the interconnection scenario that shows the least net benefit from this point of view.

Table 13 shows the present value of the total cost function and its components, when considering the opportunity costs of the natural gas.

From Table 13, we observe that the scenarios having opportunity prices for natural gas show a decrease in the net benefit (from the viewpoint of a centralized transmission planning) in comparison to the prior scenarios, with the only exception of Scenario 1. This is a result of the impact of the increase in the prices of natural gas in the system costs for those electric systems having an important participation of such fuel in the generation power plants.

4.4.4 Environmental Benefits

Electric integration allows decreasing the emissions of greenhouse gases. We quantify the avoided CO_2 emissions in a very approximate way, given the pre-feasibility nature of this work. However, although the CO_2 emissions were estimated in a very approximate manner, we found significant emission reductions due to the interconnections, which grows as the interconnection scenarios become more integrating. There is no doubt that this is an element that must be incorporated in future negotiations for the exchange of energy (taking into consideration the

Table 14 Per-technology emission factors used to estimate CO_2 emissions

Type of power plant	Emission factor for CO_2	Unit
Coal	2.64	$TonCO_2/Ton$
Natural gas	0.06	$TonCO_2/MMBtu$
Diesel	3.18	$TonCO_2/Ton$

Table 15 CO_2 emissions in each interconnection scenario, during the 2014–2022 period (thousands of tons of CO_2)

Scenario	Bolivia	Chile	Colombia	Ecuador	Peru	Total	Difference with respect to the base scenario
Base	40,154	190,996	61,910	21,370	75,001	389,431	0
1	40,154	190,996	64,756	16,599	77,356	389,860	429
2	40,154	113,541	71,627	23,255	107,498	356,075	−33,356
3	53,572	96,307	70,974	22,773	107,575	351,202	−38,229

positive or negative impacts on reducing the greenhouse gas emissions that can be produced by a determined electric interconnection).

In order to estimate CO_2 emissions in each scenario, we use per-technology nominal emission factors for the conventional thermal power plants operating in every system. These factors were obtained from the last Intergovernmental Panel on Climate Change (IPCC) report. For those power plants that operate with a mixture of fossil fuels, we consider the emission factor of the main fuel. Table 14 shows the emission factors used.

Table 15 shows the total CO_2 emissions in each interconnection scenario during the planning horizon considered.

The most favorable situation (in reference to avoiding greenhouse gas emissions) is Scenario 3 due to the fact that, in such scenario, an important amount of electric generation coming from coal and diesel power plants located in the SING-Chile is replaced (with respect to the base scenario) by cleaner generation imported from the rest of the countries. This is, although the sale of relatively cheap energy to the SING-Chile market would produce – in expected terms – an increase in the CO_2 emissions in the other countries (situation that should be considered at the time of allocating costs under an interconnection scenario), such energy transactions would allow an overall decrease in the total CO_2 emissions of the Andes region.

Table 16 shows the total CO_2 emissions in each interconnection scenario during the planning horizon, when considering the opportunity prices of the natural gas.

From Table 16, we observe that, in the case of the sensitivity analysis for the natural gas prices, Scenario 3 still represents the most favorable situation in reference to avoiding greenhouse gas emissions. However, in this case, the CO_2 emissions in such scenario are very close to those of Scenario 2 (i.e., going from Scenario 1 to Scenario 2 is where the largest decrease in CO_2 emissions is produced).

Table 16 CO_2 emissions in each interconnection scenario, with NG prices sensitivity, during the 2014–2022 period (thousands of tons of CO_2)

Scenario	Bolivia	Chile	Colombia	Ecuador	Peru	Total	Difference with respect to the base scenario
Base NGs	41,446	180,269	64,250	22,080	74,901	382,946	0
1 NGs	41,446	180,269	74,458	21,017	69,003	386,193	3,247
2 NGs	41,446	117,969	78,575	24,848	97,528	360,365	−22,580
3 NGs	40,489	119,711	78,094	24,520	97,330	360,145	−22,800

Table 17 Present value of the valuation of the avoided CO_2 emissions in each interconnection scenario, during the 2014–2022 period (millions of US$ discounted to April 2014)

Scenario	Bolivia	Chile	Colombia	Ecuador	Peru	Total
Base	0	0	0	0	0	0
1	0	0	0	61	0	61
2	0	884	0	0	0	884
3	0	1,073	0	0	0	1,073

Table 18 Present value of the valuation of the avoided CO_2 emissions in each interconnection scenario, with NG prices sensitivity, during the 2014–2022 period (millions of US$ discounted to April 2014)

Scenario	Bolivia	Chile	Colombia	Ecuador	Peru	Total
Base	0	0	0	0	0	0
1	0	0	0	13	77	89
2	0	699	0	0	0	699
3	10	682	0	0	0	692

For the valuation of the CO_2 emissions, we use the price of the carbon bonds as a proxy for the value of reducing CO_2 emissions. We use the forecast of the carbon bond price made in October of 2009 by the "European Climate Exchange" for December 2012, which is equal to €13.32/ton (i.e., US$19.9/ton approx.).

Tables 17 and 18 show the present value of the valuation of the avoided CO_2 emissions due to the corresponding interconnection projects, for the cases of without (Table 17) and with (Table 18) sensitivity analysis for the natural gas prices. Cases showing an increase of the CO_2 emissions are considered with a null removal of CO_2.

From Tables 17 and 18, we observe that, for both cases (with and without natural gas prices sensitivity analysis), the main benefits for avoiding CO_2 emissions are derived from the emissions reductions occurring in Chile.

5 Conclusions

We have analyzed the electric integration among five countries in the Andes region: Bolivia, Chile, Colombia, Ecuador, and Peru. We have determined technically-feasible interconnection points that allow establishing the approximate operation of interconnected electric markets. We found significant economic benefits of such interconnections from the supply-side, the demand-side, the costs-savings, and the environmental viewpoint.

It is important to remark that the results presented in this paper are valid within the framework of the assumptions, considerations and restrictions described here. Thus, a greater level of demand, a different relation of the fuel prices used by the thermal power plants, or different investment values for the projects studied would lead the electric systems to another equilibrium point, altering the results.

Acknowledgements The authors thanks the valuable comments of Carlos Ferruz, Juan Liu, Andy Garcia, Pia Bravo, Manuel Maiguashca and the comments received during the four meetings organized by the United Nations Development Program (UNDP) in the context of the UNDP project "*Estudio para Análisis de Prefactibilidad Técnico Económica de Interconexión Eléctrica entre Bolivia, Chile, Colombia, Ecuador y Perú*" [7].

References

1. Comunidad Andina de Naciones (CAN) (2002) Decisión del Acuerdo de Cartagena 536: Marco General para la interconexión subregional de sistemas eléctricos e intercambio intracomunitario de electricidad (Cartagena Agreement Decision N. 536: General Framework for the Sub-regional Interconnection of Electric Systems and Intra-communities Electricity Exchange). Year XIX, N.º 878, 19 of December of 2002. http://www.comunidadandina.org/normativa/dec/D536.htm
2. Comunidad Andina de Naciones (CAN) (1969) Acuerdo de Integración Subregional Andino del 26 de mayo de 1969 – Acuerdo de Cartagena (Andes Sub-regional Integration Agreement of May, 26th of 1969 – Cartagena Agreement). http://www.comunidadandina.org/normativa.htm
3. Hammons T, De Franco N, Sbertoli L, Khelil C, Rudnick H, Clerici A, Longhi A (1997) Energy market integration in South America. IEEE Power Eng Rev 17(8):6–14
4. Rudnick H (1997) South American experience in deregulation of the electricity energy industry. IEEE/IEE Future of the Energy Business International Conference, Toronto, Canada, 17–18 Nov 1997
5. Hammons T, Corredor P, Fonseca A, Melo A, Rudnick H, Calmet M, Guerra J (1999) Competitive generation agreements in Latin American systems with significant hydro generation. IEEE Power Eng Rev 19(9):4–21
6. Barroso L, Rudnick H, Mocarquer S, Kelman R, Becerra B (2008) LNG in South America: the markets, the prices and the security of supply. IEEE Power Engineering Society 2008 General Meeting, Pittsburg, PA, July 2008
7. United Nations Development Program (UNDP) (2010) "Estudio para Análisis de Prefactibilidad Técnico Económica de Interconexión Eléctrica entre Bolivia, Chile, Colombia, Ecuador y Perú (Study for the Technical-Economic Pre-feasibility Analysis of Electric Interconnection among Bolivia, Chile, Colombia, Ecuador, and Peru)". January, 2010. Available upon request

Planning Long-Term Network Expansion in Electric Energy Systems in Multi-area Settings

José A. Aguado, Sebastián de la Torre, Javier Contreras, and Álvaro Martínez

Abstract We present a multi-year and multi-area dynamic transmission expansion planning model. We define a set of metrics to rate the effect of the expansion among generators and demands. Additionally, we use congestion and saturation indexes, measuring changes in nodal prices and line saturations, respectively. We set the problem in a multi-area framework, assuming different entities in charge of the expansion simultaneously. The proposed formulation results in a mixed-integer linear optimization problem that can be solved using commercially available software. As a result, our model determines the best overall transmission expansion, reflecting both investment and operation costs as well as long-term financial parameters. The described approach is used to identify the most adequate expansion for a generic multi-area system based upon the IEEE 24-bus RTS. We compare the results obtained with individual expansions vs. multi-area expansions in terms of technical and economic indexes.

Keywords Average nodal price index • Dynamic transmission planning • Multi-area network • Network congestion index • Priority list • Scenario analysis • Welfare metrics

J.A. Aguado (✉) • S. de la Torre • Á. Martínez
Electrical Engineering Department, University of Málaga, c/Doctor Ortiz Ramos s/n, Málaga, Spain
e-mail: jaguado@uma.es; storre@uma.es; martinz78@hotmail.com

J. Contreras
University of Castilla – La Mancha, E.T.S. de Ingenieros Industriales, Ciudad Real, Spain
e-mail: javier.contreras@uclm.es

A. Sorokin et al. (eds.), *Handbook of Networks in Power Systems I*,
Energy Systems, DOI 10.1007/978-3-642-23193-3_15,

1 Nomenclature

The mathematical symbols used throughout this paper are classified below as:

1.1 Constants

b_{srk}	Susceptance of line k in corridor (s, r)
K_{srk}	Construction cost of line k in corridor (s, r)
M	Large enough positive constant
N_d	Number of blocks of the dth demand in all scenarios
N_g	Number of blocks of the gth generation unit in all scenarios
$\bar{p}_{D_{dh}}^{c(y)}$	Size of the hth block of the dth demand in scenario c and year y
$P_{G_g}^{\max(y)}$	Upper bound of the power output of the gth generating unit in year y
$\bar{p}_{G_{gb}}^{y}$	Size of the bth block of the gth generating unit in all scenarios of year y
$p_{srk}^{\max}$	Maximum capacity of line k in corridor (s, r)
$W^{c(y)}$	Weight of scenario c in year y
$\delta_{sr}^{\max}$	Upper bound of the angle difference between nodes s and r
$\lambda_{D_{dh}}^{c(y)}$	Price bid by the hth block of the dth demand in scenario c and year y
$\lambda_{G_{gb}}^{c(y)}$	Price offered by the bth block of the gth generating unit in scenario c and year y
σ	Weighting factor to make investment and operational costs comparable
ε	Economic coefficient
i	Discount rate
$I_{\max}$	Maximum investment in new lines
IRP_{sr}	Investment return period of lines in corridor (s, r)
y_0	Base year
y_n	Total number of years of the period of study
y_a	Total number of years of amortization
τ	Annual amortization rate of the new lines[1]
η	Number of years at the beginning of the time horizon during which no new lines can be built

1.2 Variables

$f_{srk}^{c(y)}$	Lossless power flow in line k of corridor (s, r) in scenario c and year y
$p_{D_d}^{c(y)}$	Total power consumed by the dth demand in scenario c and year y
$p_{D_{dh}}^{c(y)}$	Power consumed by the hth block of the dth demand in scenario c and year y

(continued)

[1] The amortization rate per year is equal to $\frac{i(1+i)^{y_a}}{(1+i)^{y_a}-1}$.

$p_{G_g}^{c(y)}$	Total power produced by the gth generating unit in scenario c and year y
$p_{G_{gb}}^{c(y)}$	Power produced by the bth block of the gth generating unit in scenario c and year y
$p_s^{c(y)}$	Power injection at bus s in scenario c and year y
$p_{srk}^{c(y)}$	Power injection in line k of corridor (s, r) computed at bus s in scenario c and year y
$\delta_s^{c(y)}$	Angle at bus s in scenario c and year y
w_{srk}^y	Binary variable that is equal to 1 if line k from corridor (s, r) is functional during year y, and is equal to 0 otherwise
h_{srk}^y	Binary variable that is equal to 1 if line k from corridor (s, r) was built in a previous year but its investment return period (IRP_{sr}) is not completed, and 0 otherwise

1.3 Sets

Ψ_D^s	Set of all demands located at bus s
Ψ_G^s	Set of all generators located at bus s
Ψ_L^s	Set of all lines connected to bus s
$\Omega_{C(y)}$	Set of all scenarios of year y
Ω_d	Set of indexes of the blocks of the dth demand
Ω_D	Set of indexes of the demands
Ω_g	Set of indexes of the blocks of the gth generating unit
Ω_G	Set of indexes of the generating units
Ω_L	Set of all possible transmission lines, prospective and existing
Ω_{L^+}	Set of all prospective transmission lines
Ω_N	Set of all network buses
Ω_Y	Set of all years of the period of study

1.4 Metrics

NCI	Network congestion index
ANPDI	Average nodal price deviation index
$\mu_1, \ldots, \mu_4$	Metrics to assess the impact of new transmission lines in the network

2 Introduction

Traditional transmission planning methods are based on centralized generation planning and the uncertain growth of demand. With the reform of electricity markets, power plants have broken away from transmission systems and become independent corporations. It is only fairly recently that transmission has demanded more attention from industry participants, regulators and customers. It is generally agreed that traditional transmission expansion planning is no longer viable in an unbundled system. In a deregulated electricity market, transmission planning is faced with more uncertainties. The optimal planning scheme obtained in that environment needs large compensation investments due to uncertainties. Due to that, new methods and tools for transmission planning are needed in competitive electricity markets.

The transition of the transmission sector is not nearly as far advanced as in the generation sector. The broad role of transmission in the new industry structure is reasonably well understood, but the details are not. The integrated and interdependent nature of the transmission system means that the provision of transmission capacity is largely viewed as a natural monopoly, a neutral enabler of the competitive generation sector.

Transmission expansion is a complex multi-period and multi-objective optimization problem that has been studied since the 1970's [18].

Centralized models solve the problem assuming cost minimization of the network. Some of the most important methods [12] used to solve the problem are based on: linear programming [11, 24], mixed-integer linear programming [1], Benders decomposition [2, 23], Bi-level programming [10], heuristic methods [14], genetic algorithms [9, 13], simulated annealing [19], and game theory models [6, 7, 25].

Due to deregulation, new transmission expansion models have been proposed. As a result, cost minimization has been replaced by social welfare maximization produced by the matching of the offers and bids supplied by generators and demands, respectively. The level of network congestion, the bidding structure, and the problem of uncertainty are seen now as the main drivers of long-term transmission expansion planning [22]. Not only deterministic expansions, but also probabilistic models can be used to analyze long-term transmission expansion planning considering risk in a deregulated environment [4].

Long-term effects of bidding in electricity markets can be calculated and quantified by metrics that are useful for generators, consumers and the market regulator. Due to that, economic metrics have been proposed to weigh the appropriateness of the cost of expansion according to generators, demands and the system as a whole [8]. Based on these metrics, economic incentives in valuable investments are not blocked are proposed under a game theory scheme for the allocation of transmission expansion costs [19]. Other incentive schemes are based on the value added to the social welfare through each asset investment in a cooperative game framework, where the Shapley value is used to reward investors according to the added value that they create [5].

One critical aspect of setting incentive schemes is the existence of side payments to distribute the gains among the agents. In [21], a methodology is proposed to evaluate the economic impacts and strategic responses that affect transmission investments. The model considers the optimal expansion under conflicting objectives: social welfare maximization, local market power minimization and the maximization of the agents' surpluses.

An even more complex issue is the interrelation between generation and transmission expansion. In [20], a proactive approach considers the social welfare implication of transmission investment that considers equilibrium models of generation investments affected by the proactive transmission decisions. This problem has also been addressed in [16] in an iterative joint model where auction models of energy and transmission are developed.

In this work, we present a multi-area multi-period formulation for the transmission expansion problem in pool-based electric energy markets. We model the network topology, generator offers and demand bids, and we set a long-term timeframe. Different scenarios are considered to account for different levels of demand. Financial parameters are added to solve the problem in a realistic way. To reduce computational times, a systematic procedure resulting in a priority list is developed to determine the lines that bring the highest increase in social welfare in an orderly fashion.

We also define a set of metrics to account for the improvement brought by the investment both in nodal prices and line congestions. A network congestion index (*NCI*) relating the power flow circulating across the network for the maximum demand scenario and the maximum possible power flow is obtained. An average nodal price deviation index (*ANPDI*) is calculated as the scenario-weighted standard deviation of nodal prices with respect to the average system price.

The remaining sections of this work are organized as follows: Sect. 3 shows the mathematical formulation of the market-based transmission expansion problem in competitive markets. Section 4 introduces the set of metrics used to analyze the technical and economic effects of transmission expansion. Section 5 illustrates the model with multi-area examples based on the IEEE 24-bus RTS, and Sect. 6 gives concluding remarks.

3 Proposed Model for Market-Based Transmission Expansion Planning

In this section we provide the formulation in order to solve the transmission expansion planning problem. As previously stated, this is a compact multiperiod formulation that takes into account both technical and financial constraints in the problem. In this section, first we present the features and assumptions of our model and then we present the complete model formulation.

The formulation is based on the paper by [8], however, the model presented in this paper deals with a multiperiod setting, providing a more accurate description of the expansion process by explicitly considering the inter-temporal links between all the time periods of the planning horizon. The model used to represent the transmission system is a lossless DC model, assuming that the long-term effect of losses is negligible. Note that the introduction of system losses in the problem would increase the amount of variables significantly making realistic problems unsolvable.

Nevertheless, the formulation of a transmission expansion planning problem with losses is relatively easy, as shown in [8]. Many aspects of the formulation in [8] have been changed and improved for this work, specifically: (1) the introduction of a compact multiperiod formulation, and (2) the explicit consideration of the financial constraints directly related to transmission expansion: budget constraints, cost annualization and discount rates.

In this work, a competitive energy market with perfect competition is considered; it is characterized by buyers and sellers making their offers in an electricity wholesale pool. We assume that no market power is exerted and that sellers offer their energy in the market according to their true cost functions and buyers bid in the market according to their true utility functions.

Given the fact that we are considering a long-term problem, it is necessary to take into account the uncertain behavior of many technical and financial parameters. To solve this issue we have considered a set of different scenarios that characterize in a simple way the uncertain nature of the problem. In this chapter, we only consider uncertainties associated with demands. This is taken into account through scenarios modeling different levels of demand. Other uncertainties associated with supplies and prices are not explicitly considered.

The formulation used in this paper (1)–(18) is presented next, providing some details about the role of each constraint in the problem.

The objective function (1) represents the maximization of the social welfare weighted by demand-dependent scenarios over the whole planning horizon. It is comprised of two main elements: firstly, market social welfare and, secondly, expansion costs, *i.e.*, the cost of building new lines. The annual amortization rate, τ, multiplied by the line cost, K_{srk}, expresses the annual amortization cost per line. There are several parameters in the second term of Eq. 1: (a) a weighting factor, σ, that takes into account that the operation costs are measured in €/MWh and the investment costs in M€, and whose value is 1,000/876, and (b) an economic coefficient, ε, whose default value is 1. Other values of this coefficient can be used to implement an algorithm that helps to establish a priority list: starting from a high value of ε, and gradually decreasing it, this method is able to select new lines in an orderly fashion. In fact, there is a value of ε small enough so that only one line is built, which is the one that renders the highest profits and which would be built if the planners were limited to the construction of just one line.

Equations 2–9, 11, 13 and 15 are the multiyear versions of the equations appearing in [8], the rest are the new equations introduced in this paper regarding the new features implemented.

Equation 2 enforces power balance at each node.

Equation 3 enforces flow limits for each line.

Equation 4 implements the DC power flow for the all the actual lines in the system.

Equations 5 and 8 state that the amount of power allocated to each offer/bid block must have a value between zero and the maximum value established for that block.

Equations 7 and 9 state that the total energy generated by one generator or consumed by one consumer must equal the summation of its blocks.

Equation 6 states that each generator must produce less that its maximum generating capacity.

Note that, in order to model the uncertainty in the market, Eqs. 2–9 and 11–12 have to be included, one per scenario, as the formulation presented shows.

Equation 10 limits the total amount of money that can be invested in the expansion plan. This constraint is useful to model a situation under which the expansion plan has a limited budget, and no more money can be spent even if more expenditure implied greater social welfare.

Equation 11 sets the reference node for the system.

Equation 12 is needed to model the technical constraint that states that the angle difference between the two ends of a line cannot exceed a certain amount. This constraint is related to the stability of the operation of the system.

Equation 13 states that only existing lines are functional in year y.

Equation 14 states that during the first η years of the time horizon no new lines can be built. Hence, it simulates the fact that it takes a certain amount of time to install a new line.

Equation 16 expresses the fact that, once built, a line will remain active during the whole planning horizon.

Equation 17 defines the binary variables associated with the payment of new lines.

Equation 18 enforces that the binary variable, h^y_{srk}, to take a value which can only be 0 once a number of years equal to IRP_{sr} have passed since the construction of the line. More specifically, Eq. 18 considers the possibility that the line is built in a year so that there are still more than IRP_{sr} years left in the time horizon. Equation 18 also considers the case when a line is built with less than IRP_{sr} years left in the time horizon. This equation takes into account the aforementioned comment that no lines can be built in the first years.

Maximize

$$\sum_{\forall y \in \Omega_Y} \sum_{\forall c(y) \in \Omega_{C(y)}} W^{c(y)} \left(\sum_{\forall d \in \Omega_D} \sum_{\forall h \in \Omega_d} \frac{\lambda^{c(y)}_{D_{dh}} p^{c(y)}_{D_{dh}}}{(1+i)^{y-y_0}} - \sum_{\forall g \in \Omega_G} \sum_{\forall b \in \Omega_g} \frac{\lambda^{c(y)}_{G_{gb}} p^{c(y)}_{G_{gb}}}{(1+i)^{y-y_0}} \right)$$

$$- \tau \varepsilon \sigma \sum_{\forall y \in \Omega_Y} \sum_{\forall (s,r,k) \in \Omega_{L+}} \frac{K_{srk} h^y_{srk}}{(1+i)^{y-y_0}} \tag{1}$$

subject to:

$$\sum_{g \in \Psi^s_G} p^{c(y)}_{G_g} - \sum_{\forall (r,k) \in \Psi^s_L} f^{c(y)}_{srk} = \sum_{d \in \Psi^s_D} p^{c(y)}_{D_d}; \forall s \in \Omega_N, \forall c(y) \in \Omega_{C(y)}, \forall y \in \Omega_Y : \lambda^{c(y)}_s \tag{2}$$

$$- w^y_{srk} p^{\max}_{srk} \le f^{c(y)}_{srk} \le w^y_{srk} p^{\max}_{srk}; \forall (s,r,k) \in \Omega_L, \forall c(y) \in \Omega_{C(y)}, \forall y \in \Omega_Y \tag{3}$$

$$-(1-w_{srk}^{y})M \leq \frac{f_{srk}^{c(y)}}{b_{srk}} + (\delta_{s}^{c(y)} - \delta_{r}^{c(y)}) \leq (1-w_{srk}^{c(y)})M; \quad \forall (s,r,k) \in \Omega_L, \forall c(y) \in \Omega_{C(y)}, \ \forall y \in \Omega_Y \tag{4}$$

$$0 \leq p_{G_{gb}}^{c(y)} \leq \bar{p}_{G_{gb}}^{y}; \ \forall b \in \Omega_g; \forall g \in \Omega_G, \forall c(y) \in \Omega_{C(y)}, \ \forall y \in \Omega_Y \tag{5}$$

$$0 \leq p_{G_g}^{c(y)} \leq p_{G_g}^{\max(y)}; \ \forall g \in \Omega_G, \forall c(y) \in \Omega_{C(y)}, \ \forall y \in \Omega_Y \tag{6}$$

$$\sum_{\forall b \in \Omega_g} p_{G_{gb}}^{c(y)} = p_{G_g}^{c(y)}; \ \forall g \in \Omega_G, \forall c(y) \in \Omega_{C(y)}, \ \forall y \in \Omega_Y \tag{7}$$

$$0 \leq p_{D_{dh}}^{c(y)} \leq \bar{p}_{D_{dh}}^{c(y)}; \ \forall d \in \Omega_D, \forall h \in \Omega_d, \forall c(y) \in \Omega_{C(y)}, \ \forall y \in \Omega_Y \tag{8}$$

$$\sum_{\forall h \in \Omega_d} p_{D_{dh}}^{c(y)} = p_{D_d}^{c(y)}; \ \forall d \in \Omega_D, \forall c(y) \in \Omega_{C(y)}, \ \forall y \in \Omega_Y \tag{9}$$

$$\tau \sum_{\forall y \in \Omega_Y} \sum_{\forall (s,r,k) \in \Omega_{L^+}} \frac{K_{srk}\,(w_{srk}^{y} - h_{srk}^{y})}{(1+i)^{y-y_0}} \leq I_{\max} \tag{10}$$

$$\delta_{s}^{c(y)} = 0; \ s\text{: reference node,} \quad \forall c(y) \in \Omega_{C(y)}, \ \forall y \in \Omega_Y \tag{11}$$

$$-\delta_{sr}^{\max} - (1-w_{srk}^{y})M \leq \delta_{s}^{c(y)} - \delta_{r}^{c(y)} \leq \delta_{sr}^{\max} + (1-w_{srk}^{y})M; \quad \forall (s,r,k) \in \Omega_L, \forall c(y) \in \Omega_{C(y)}, \ \forall y \in \Omega_Y \tag{12}$$

$$w_{srk}^{y} = 1; \ \forall (s,r,k) \in \Omega_L \backslash \Omega_{L^+}, \ \forall y \in \Omega_Y \tag{13}$$

$$\sum_{y \in (y_1, y_2, \ldots, y_\eta)} \sum_{\forall (s,r,k) \in \Omega_{L^+}} w_{srk}^{y} = 0; \tag{14}$$

$$w_{srk}^{y} \in \{0,1\}; \ \forall (s,r,k) \in \Omega_L, \ \forall y \in \Omega_Y \tag{15}$$

$$h_{srk}^{y} \in \{0,1\}; \ \forall (s,r,k) \in \Omega_{L^+}, \ \forall y \in \Omega_Y \tag{16}$$

$$w_{srk}^{y-1} \leq w_{srk}^{y}; \ \forall (s,r,k) \in \Omega_{L^+}, \ \forall y \in \Omega_Y \tag{17}$$

$$\sum_{m=y}^{\min(y+IRP_{sr}-1, y_n)} \left[h_{srk}^{m} - \left(w_{srk}^{y} - w_{srk}^{y-1} \right) \right] \geq 0 \quad y\text{: } (\eta+1), \ldots, y_n, \ \ \forall (s,r,k) \in \Omega_{L^+} \tag{18}$$

This formulation results in a mixed-integer linear optimization problem that can be solved using commercially available software.

4 Metrics

In this section we present the indexes that have been proposed to help assessing the value of a given transmission expansion plan. We present both the mathematical expressions and the justification for the indexes that have been implemented. Firstly, it must be acknowledged that we do not try to propose new standard indexes, we simply have come up with some that we consider can provide a good insight of the problem solution, also we have tried to obtain indexes that are easy to understand and also easy to compute. Note that some indexes that have been used in the literature, like the HHI and Lerner indexes, are not suitable for our model because they are basically used for market power assessment. Rather, our indexes reflect the effect of new lines in the overall economic and technical performance of the transmission system.

This paper proposes two new indexes (besides the four indexes described later) which were presented in an earlier work by the same authors [8]. The two new indexes take into consideration economic and technical aspects of the system. The first one is the 'Average Nodal Price Deviation Index' (*ANPDI*). This index is defined by the following expression:

$$ANPDI = \frac{\sum_{s \in \Omega_N} \left| \bar{\lambda}_s - \bar{\lambda} \right|}{N\bar{\lambda}} \tag{19}$$

where

$$\bar{\lambda}_s = \frac{\sum_{y \in Y} \left(\sum_{\forall c(y) \in \Omega_{c(y)}} \lambda_s^{c(y)} W^{c(y)} \right)}{y_n} \quad \text{and } \bar{\lambda} = \frac{\sum_{\forall s \in \Omega_N} \bar{\lambda}_s}{N}$$

Note that N in (19) represents the total number of nodes in the system and that y_n is the total number of years in the time horizon. The interpretation for this *ANPDI* metric is as follows: the *ANPDI* is a measure of the nodal price deviation in the expanded system. This deviation is calculated considering all scenarios in each year. A small value of the *ANPDI* index means that prices are, in general, close to the system average price; on the other hand, a high value can be interpreted as a system throughout which nodal prices are quite different.

The second of the new metrics proposed is specifically aimed at assessing the saturation of the lines in the system. Thus, this new metric is a measure of the efficacy of the expansion plan. The new metric is the 'Network Congestion Index' (*NCI*), and is defined as follows:

$$NCI = \frac{\sum\limits_{y \in \Omega_Y} \sum\limits_{\forall(s,r,k) \in \Omega_L} \left| f_{srk}^{c_{\max}(y)} \right|}{\sum\limits_{y \in \Omega_Y} \sum\limits_{\forall(s,r,k) \in \Omega_L} p_{srk}^{\max} w_{srk}^{y}} \tag{20}$$

This metric can be interpreted in a fairly simple, yet powerful, fashion. It assesses the state of the system by measuring the amount of line capacity that is used, dividing it by the total available line capacity in the system. In other words, the *NCI* measures the transmission system usage, hence, a value close to 0 means the system is not used very much, and a value close to 1 means that the system is approaching its capacity limit.

To illustrate these newly proposed metrics, two simple examples are presented next. The importance of each metric will be clarified in the remaining sections of this document.

Case A: Consider a system for which all the lines are at 25% of their capacity, except for one that is congested because it connects one large and inexpensive generator to the rest of the system. For this system, the *NCI* has a low value, because, on average, there is ample capacity in the system. However, the system needs to be expanded. This is shown by the high value obtained for the *ANPDI*, because some nodes have much higher prices than others due to congestion.

Case B: Consider a hypothetical system for which all lines are at exactly 98% of their capacity. This system is clearly in need of expansion and the *NCI* value shows a high value. However, given that no line is at full capacity, there are no significant price differences along the system nodes. As a result, the *ANPDI* value is not particularly high.

Conclusion: Note each of these two indexes provides different, yet valuable, information.

The remaining four indexes used in this paper are the multiperiod version of the set of indexes presented in [8]. A brief description is provided next.

The first index, μ_1, is defined by the following expression:

$$\mu_1 = \frac{SW^* - SW^0}{\sum\limits_{y \in \Omega_Y} \sum\limits_{\forall(s,r,k) \in \Omega_{L+}} \frac{K_{srk} h_{srk}^{y}}{(1+i)^{y-y_0}}} \tag{21}$$

The interpretation of the index is as follows. μ_1 computes the total increase in social welfare obtained for each monetary unit that has been invested in building new lines. In principle, only values bigger than 1 would justify an expansion plan. Note that the numerator results from the difference between the final (optimal) total social welfare and the social welfare in the system before expansion. Note that the calculation of SW^0 is really simple: one can simply run the transmission expansion planning problem without adding possible new lines.

Similarly, to the index presented above, three more indexes μ_2, μ_3 and μ_4 are defined. Only the increase in welfare obtained by part of the participants in the market is taken into consideration, instead of the total social welfare. Index μ_2 is defined considering the increase in social welfare attained by the generators only. Index μ_3 is defined considering the increase in social welfare attained by the demands only. Finally, index μ_4 is defined considering the increase in social welfare attained by the transportation system only, i.e., the merchandising surplus. The mathematical expressions for these three indexes are similar to (21), where the denominators are all equal to the denominator in (21) and the numerators are the social welfare portions assigned to each participant. Also, note that the summation of these three indexes must be equal to μ_1.

The rest of the paper makes an extensive use of the six metrics presented in this section. It is important to state that the last four metrics assess the system in terms of social welfare improvement, whereas the first two metrics consider the technical and economical behavior of the transmission system. We consider that each metric provides unique information. By no means, our metrics are interchangeable or redundant.

5 Case Studies

We illustrate our methodology with three case studies. The first corresponds to the expansion of the IEEE 24-bus RTS [15] and the second and third ones correspond to the expansion of three interconnected systems obtained from the IEEE 24-bus RTS. In the second case study, we show the results of three independent expansions and, in the third, the overall system expansion of the three interconnected systems. All pertinent data in terms of generation, demand and network features are provided in the Appendix. In all cases there are four scenarios to take into account demand fluctuations per year, namely: low (L), low-medium (LM), medium-high (MH), and high demand (H) scenarios. The time frame for our study is 15 years.

To solve the multiarea case, the total running time has been of 1.8 h of CPU using GAMS [3] in a linux system with two processors quadcore at 3.0 GHz and 16 MB of RAM.

5.1 IEEE 24-Bus RTS Expansion

We start from the generation and demand data provided in the Appendix to analyze the possible expansion of the system. To do that, we run an optimal power flow to determine the number of lines that are congested throughout the planning horizon. Figure 1 shows the number of congested lines per scenario for each year, where a red line indicates the average weighted demand per scenario per year.

In Fig. 1, the congestion level of the system increases throughout the years due to generation and demand growths. In addition, the network congestion index (NCI) has a value of 0.73. Figure 2 represents the nodal prices weighted by scenario,

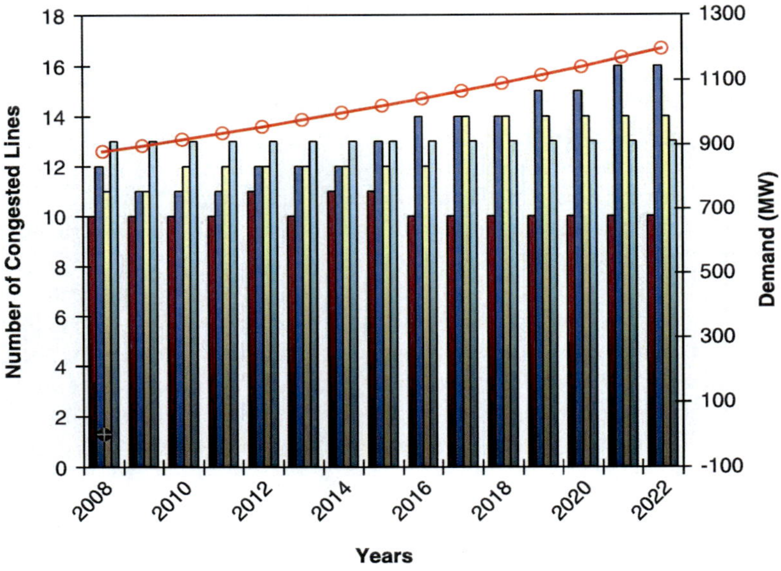

Fig. 1 Congested lines and demand in the IEEE 24-bus RTS before expansion. Demand '⊖'. Congested lines in scenarios L (□), LM (□), MH (□) and H (□)

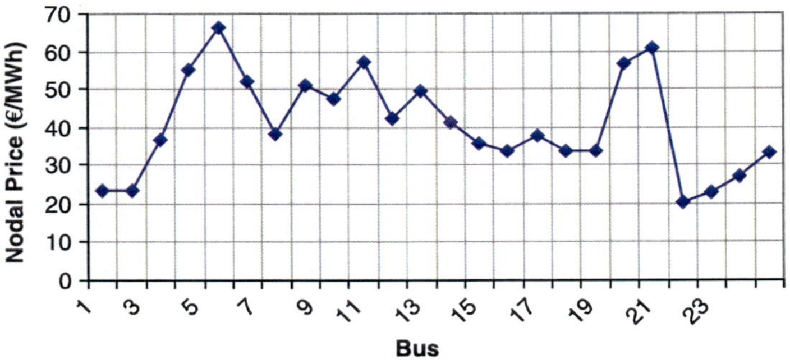

Fig. 2 Nodal prices weighted by scenario in the IEEE 24-bus RTS before expansion

where a wide dispersion of nodal prices occurs, leading to market inefficiency. To quantify it, the average nodal price deviation index (ANPDI) has a value of 0.26, justifying the expansion due to its relatively high value. The pre-expansion economic results are presented in Table 1.

In the expansion, we assume that new lines, up to a maximum of three, will be constructed in parallel to the existing lines in the same corridors. Therefore, there are 68 candidate lines for expansion. It is common practice among engineers in charge of developing transmission expansion plans to consider a set of candidate

Table 1 Economic results in M€ before the expansion

Generators	Costs: 1722.28 Income: 2340.36	Profit: 618.08
Demands	Payments: 2880.85 Utility: 3850.79	Profit: 969.93
Transmission operator	–	Profit: 540.49
Net social welfare	–	Profit: 2128.50

Table 2 Priority list for expansion of the IEEE 24-bus RTS

Order	From	To	Line	E	Δ profit (%)
1	1	5	2	1.70	1.02
2	20	23	2	1.60	2.73
3	1	5	3	1.50	3.48
4	2	4	2	1.20	3.82
5	2	4	3	1.00	3.84
6	2	6	2	0.85	3.69
7	20	23	3	0.80	3.21
8	2	6	3	0.60	2.77
9	18	21	2	0.36	1.09
10	16	17	2	0.32	−4.29
11	16	19	2	0.32	−4.29
12	17	22	2	0.32	−4.29
13	3	9	2	0.25	−5.26
14	17	22	3	0.23	−7.30

transmission lines to reduce the search space of problem (1)–(18). This is usually done based on previous experience or some sort of heuristic.

In this chapter, we propose a systematic procedure for generating a priority list as a set of candidate transmission lines. It can be obtained by adjusting parameter ε and using a single period formulation of the problem defined in (1)–(18) with average values of the demands. As discussed in [8], and from an economic point of view, parameter ε can be viewed as a scaling factor of transmission line construction cost. This parameter is used to generate a set of candidate lines. Initially, ε is set to a certain value, say $\varepsilon=2$, then it is smoothly decreased generating an ordered list of prospective lines according to their economic profitability. In this case, for $\varepsilon = 0.05$ the candidate transmission lines is reduced to a subset of 14. The resulting priority list for this problem is shown in Table 2.

With the 14 candidate lines from the priority list, we obtain the following results for the dynamic expansion, as shown in Fig. 3. Overall, there are nine new lines after expansion with a total investment cost of 124.93 M€. The new lines and its year to enter into service are depicted in Table 3.

It can be verified from these results that the expansion connects low-priced generation buses 1, 2 and 23 with high-priced demand buses 4, 5, 6, and 20. In addition, the expansion is more significant in the lower part of the network, since its voltage level is lower and the construction of lines is more economical there.

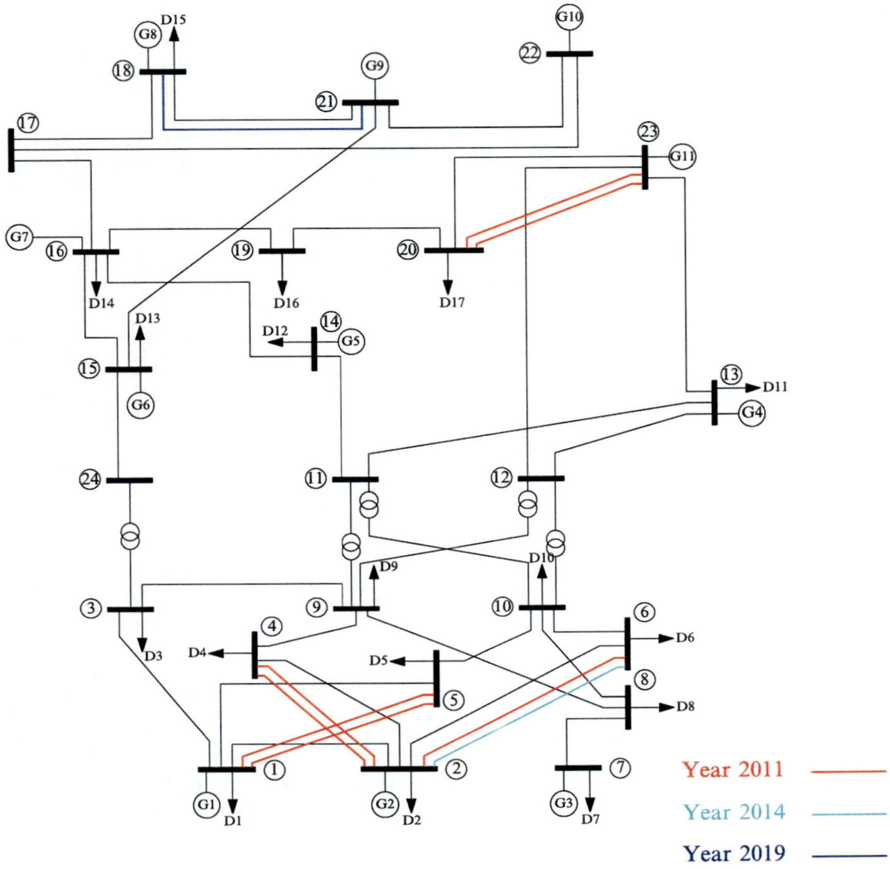

Fig. 3 Expansion results of the IEEE 24-bus RTS before expansion

Table 3 Lines to build in the expansion of the IEEE 24-bus RTS

From	To	Line	Year to enter into service
1	5	2	2011
1	5	3	2011
2	4	2	2011
2	4	3	2011
2	6	2	2011
20	23	2	2011
20	23	3	2011
2	6	3	2014
18	21	2	2019

Table 4 presents the economic results of the expanded system, where the last column shows the increase in profit shown by each of the agents: generators, demands, and transmission operator. Note that the increase in net social welfare is lower than the increase in individual profits by the agents, since the costs of expansion are only considered for the system as a whole. Therefore, the expansion results in an actual profit increase of about 7%.

On the other hand, Table 5 depicts the value of different investment evaluation parameters. It is observed that the expansion plan is profitable, but people who benefit most are the consumers. With regard to the new demand met by generation, Fig. 4 shows that there is an 18% increase, where in the first 3 years there is no

Table 4 Economic results in M€ after the expansion

Generators	Costs: 1983.55 Incomes: 2677.28	Profit: 693.73	Δ Profit: 12.2%
Demands	Payments: 3286.89 Utility: 4380.82	Profit: 1093.93	Δ Profit: 12.8%
Transmission operator		Profit: 609.61	Δ Profit: 12.8%
Net social welfare		2272.33	Δ Profit: 6.8%

Table 5 Evaluation parameters after the expansion of the IEEE 24-bus RTS

Parameter	μ_1	μ_2	μ_3	μ_4
Value	2.15	0.61	0.99	0.55

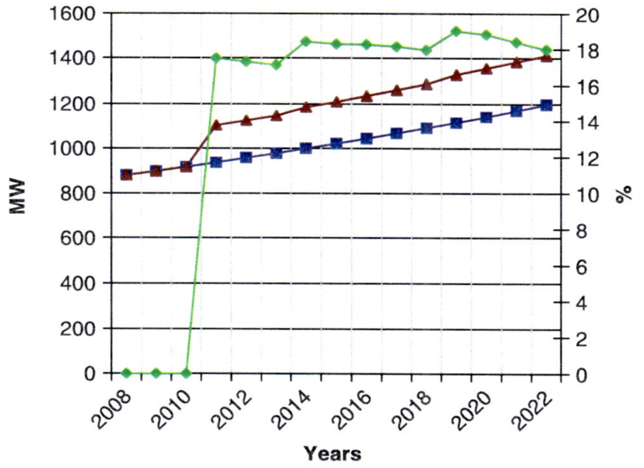

Fig. 4 Demand weighted by scenarios for the IEEE 24-bus RTS. (Before expansion '□', After expansion '▲', Demand increment '×')

change in demand, but in the year 2011 there are seven new lines going into service, and the years 2014 and 2019 have smaller increases in demand.

Finally, the evolution of the expansion indexes is shown. The new NCI and ANPDI are 0.77 and 0.22, respectively. The NCI is slightly higher, but the ANPDI is lower, meaning that the nodal price differences are reduced.

From a computational point of view, it is worth noticing that the results obtained from problem (1)–(18) with a priority list compared against those obtained considering all possible transmission lines as candidate lines are the same results. However, the one with a reduced candidate transmission line set is more than 30 times faster than the other.

5.2 System Based on the IEEE 24-Bus RTS: Three Individual Expansions

In this example, we consider a system composed of three areas, each containing the IEEE 24-bus RTS, but with different data for generation and demand. System 1 corresponds to the original data of the IEEE 24-bus RTS, and systems 2 and 3 share the same topology, but their generation and demand are increased and decreased by 30%, respectively. An independent expansion of the three systems produces the results shown in Fig. 5 and Table 6. Please note that although all systems have a network with similar features, they are subject to different generation and demand patterns. The main expansion results are presented in Table 7.

It can be observed in Table 7 that system 2, being subject to a higher demand, shows a more ambitious expansion plan, having a higher evaluation parameter, which is logical since it is the most congested system and, therefore the one who obtains the highest profit from the overall system expansion. On the one hand, system 2 shows the highest social welfare, as a consequence of having the highest value of traded energy. On the other hand, system 3 is the one with the lowest demand, and, consequently, its expansion plan has the smallest magnitude of the three systems. It is noteworthy that this system shows a negative value of the μ_4, parameter, associated with the system's congestion rents. This is due to the fact that once the system is expanded; it generates lower congestion rents, despite a higher amount of energy traded as a result of network expansion.

5.3 System Based on the IEEE 24-Bus RTS: One Interconnected Expansion

If the three IEEE 24-bus RTS systems are not separated, but interconnected, as shown in Fig. A.1, the new expansion plans are very different from the individual expansions. The real expansion of the interconnected systems is provided in Fig. 6

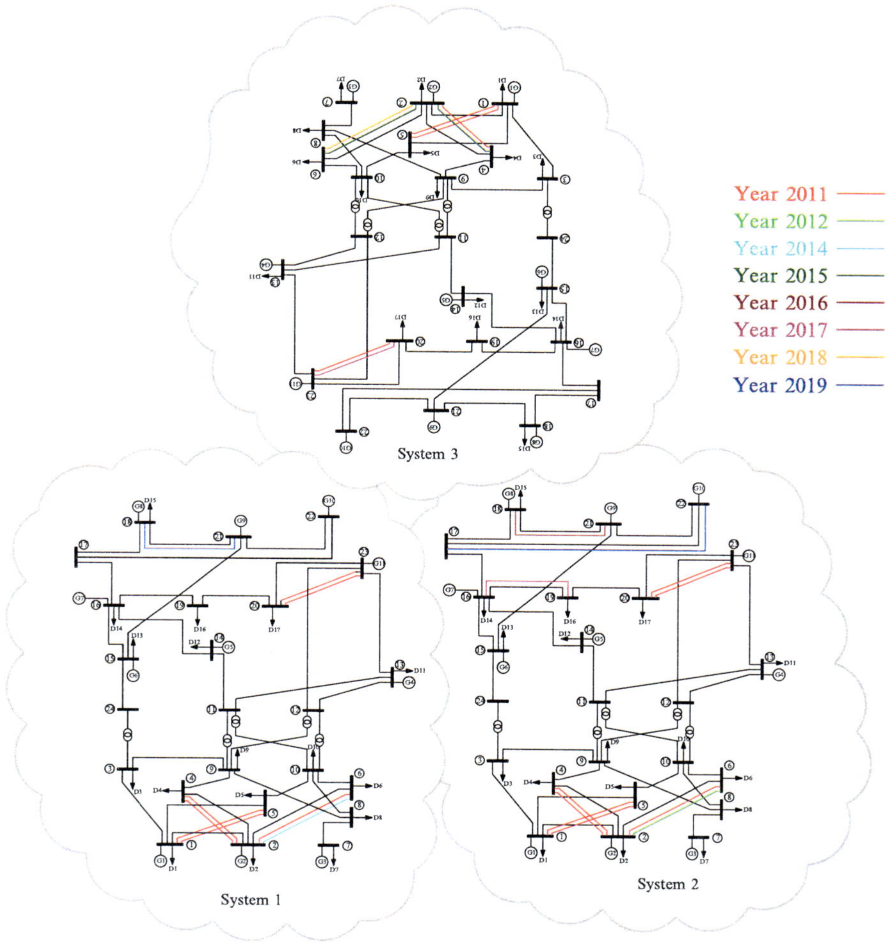

Fig. 5 Independent expansions of the multiarea IEEE 24-bus RTS

and Table 8, where the new lines are shown. The main economic results of the expansion are shown in Table 9. The result of the optimal expansion will not include any of the possible new tie lines.

As it can be seen in Fig. 6 and Tables 8 and 9, the expansion plans per area are quite different from the ones shown in Sect. 2. In particular, the same number of new lines is proposed in system 1, but its temporal distribution is changed, and line 2 of corridor 18–21 is replaced by line 2 of corridor 17–22. This is due to the fact that bus 18 of system 1 is now connected to system 3, which allows for an additional injection of power to this bus, disregarding a possible connection with corridor 18–21 to meet the demand. In addition, the new temporal distribution gives rise to a 4% increase in investment as compared to the individual system expansion. On the

Table 6 New lines after individual expansions of the multiarea IEEE 24-bus RTS

System	From	To	Line	Year
1	1	5	2	2011
1	1	5	3	2011
1	2	4	2	2011
1	2	4	3	2011
1	2	6	2	2011
1	20	23	2	2011
1	20	23	3	2011
1	2	6	3	2014
1	18	21	2	2019
2	20	23	2	2011
2	20	23	3	2011
2	1	5	2	2011
2	1	5	3	2011
2	2	4	2	2011
2	2	4	3	2011
2	2	6	2	2011
2	2	6	3	2012
2	18	21	2	2016
2	16	19	2	2017
2	17	22	2	2019
3	1	5	2	2011
3	1	5	3	2011
3	2	4	2	2011
3	20	23	2	2011
3	2	6	2	2015
3	2	4	3	2015
3	20	23	3	2017
3	2	6	3	2018

Table 7 Economic results after individual expansions of the multiarea IEEE 24-bus RTS

	System 1	System 2	System 3
Number of lines built	9	11	8
Expansion investment (M€)	124.93	147.77	87.80
Net social welfare (M€)	2272.33	2785.55	1712.23
μ_1	2.15	2.31	1.87
μ_2	0.61	0.69	0.78
μ_3	0.99	0.34	1.13
μ_4	0.55	1.28	−0.03
ANPDI evolution	0.26 → 0.22	0.29 → 0.25	0.23 → 0.18
NCI evolution	0.73 → 0.77	0.74 → 0.78	0.71 → 0.75

other hand, the model proposes the construction of two additional lines in system 2 and a new temporal redistribution of the other new lines, producing a 5.5% increase in investment with respect to the individual system expansion. Finally,

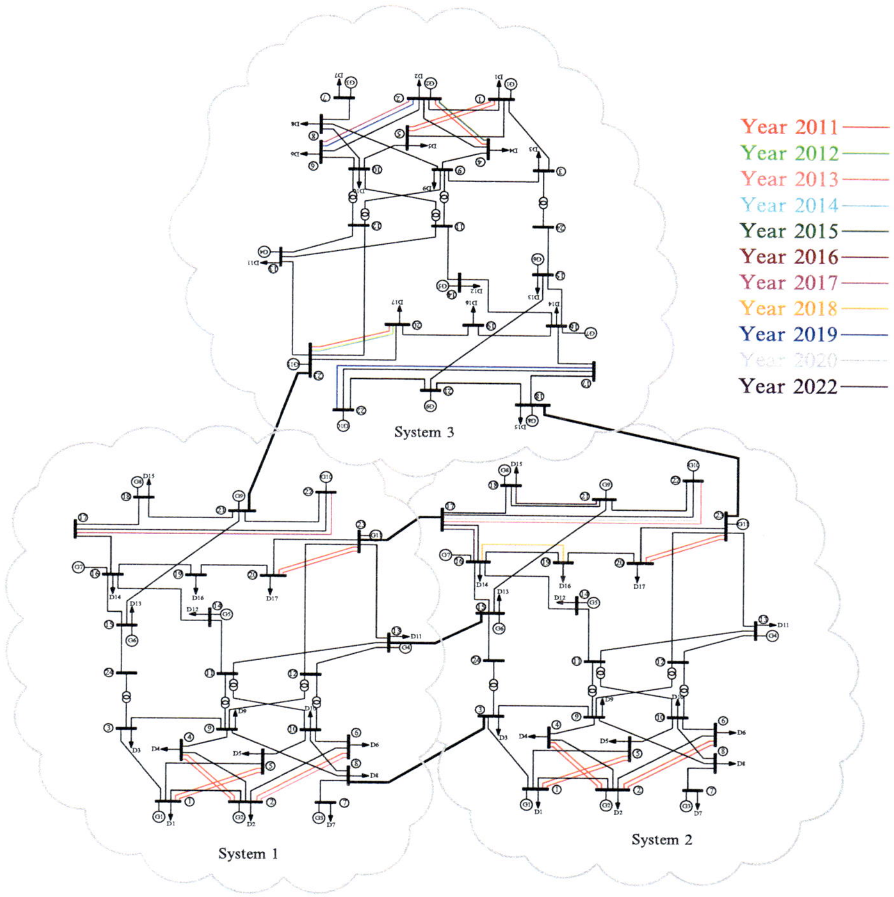

Fig. 6 Interconnected expansion of the multiarea IEEE 24-bus RTS

system 3 experiences a remarkable 21% increase in investment. As a result of all this, the net social welfare remains almost unchanged compared to individual expansions, but the tie lines and the new lines produce a better redistribution of active power flow and a considerable improvement of the global NCI, 0.7, see Table 2. This value is 90% of the global NCI resulting from individual expansions, 0.77 for system 1 (see Table 7). We can conclude by saying that the interconnection of systems generates expansion plans that are well balanced, with better electric and economic indexes, keeping the same values of net social welfare with respect to their individual expansions counterparts.

Table 8 New lines after an interconnected expansion of the multiarea IEEE 24-bus RTS

System	From	To	Line	Year
1	1	5	2	2011
1	1	5	3	2011
1	2	4	2	2011
1	2	4	3	2011
1	2	6	2	2011
1	20	23	2	2011
1	20	23	3	2011
1	2	6	3	2013
1	17	22	2	2017
2	1	5	2	2011
2	1	5	3	2011
2	2	4	2	2011
2	2	4	3	2011
2	2	6	2	2011
2	2	6	3	2011
2	20	23	2	2011
2	20	23	3	2011
2	17	22	2	2013
2	16	19	2	2018
2	17	22	3	2020
2	18	21	2	2022
2	16	17	2	2022
3	1	5	2	2011
3	1	5	3	2011
3	2	4	2	2011
3	20	23	2	2011
3	20	23	3	2012
3	2	6	2	2014
3	2	4	3	2015
3	2	6	3	2019
3	17	22	2	2019

Table 9 Economic results an interconnected expansion of the multiarea IEEE 24-bus RTS

-	System 1	System 2	System 3	Global
# of built lines	9	13	9	31
Expansion investment (M€)	129.87	155.82	106.13	391.82
Net social welfare (M€)	2284.51	2683.52	1802.80	6770.82
μ_1	2.54	2.25	1.62	2.17
μ_2	0.55	0.50	0.73	0.58
μ_3	0.95	0.30	1.09	0.73
μ_4	1.04	1.44	−0.20	0.86
ANPDI evolution	0.26 → 0.21	0.27 → 0.24	0.25 → 0.18	0.26 → 0.21
NCI evolution	0.59 → 0.69	0.66 → 0.75	0.60 → 0.66	0.62 → 0.70

6 Conclusions

We present a mixed-integer model of the transmission expansion planning problem suited for electricity markets. The model is applicable to multi-period and multi-area settings, making it a valuable tool to analyze dynamic investments across different areas. Long-term financial investment constraints are considered through an efficient compact formulation and a methodology is proposed for the selection of an efficient set of candidate transmission lines. We also use a set of physical and economic indexes to measure how nodal prices and the use of the lines can change with different transmission expansions.

We illustrate our methodology for multi-area systems based upon the IEEE 24-bus RTS. As illustrated by the numerical results, the technical and economic indexes properly quantify the quality of the expansion results, helping transmission planners across different areas to coordinate or decide individual expansions.

Acknowledgments This work was supported in part by the Spanish Ministry of Education grants ENE2011-27495, ENE2009-09541 and Junta de Andalucía grant 2008-TEP-4210, and the Junta de Comunidades de Castilla – La Mancha grant PII2I09-0154-7984.

Appendix

This Appendix shows the main features of the IEEE 24-bus RTS. Data refer to the network topology and the generation and demand patterns.

A.1 Single-Line Diagram of the IEEE 24-Bus RTS

Figure A.1 shows the single-line diagram and the topology of the IEEE 24-bus RTS. It is composed of 17 loads and 11 generators connected through two voltage levels: 138 kV and 230 kV. The network is composed of 34 lines connecting 24 buses. Note that the original lines presented in [15] are not the ones contemplated in this work.

A.2 Transmission Data of the IEEE 24-Bus RTS

Table A.1 shows the corresponding line data of the system. The first two columns present the starting and ending buses of the lines. The third and fourth columns show the resistance and the reactance per unit (pu), respectively. The fifth column shows the line capacity in pu, and the sixth column depicts the construction cost in

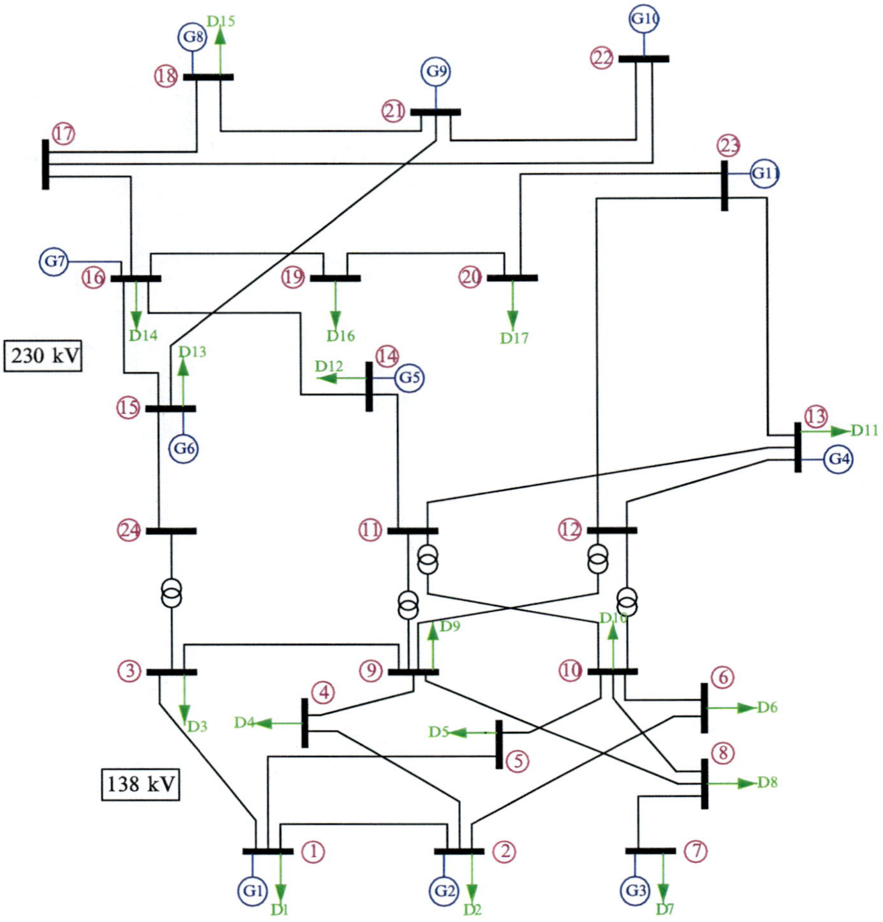

Fig. A.1 Single-line diagram IEEE 24-bus RTS

millions of €. From the table, we see that the estimated cost is 20 M€ for the lines at 138 kV, 40 M€ for the 230 kV lines, and 400 M€ for the lines connecting buses at different voltage levels. The latter cost is very high since lines connecting buses at different voltages must incorporate transformer substations.

A.3 Generation and Demand Data of the IEEE 24-Bus RTS

Tables A.2–A.4 depict detailed information of generation and demand in the system. Table A.2 provides the total demand energy bid as well as the price of each bid block (all the blocks are of equal size).

Table A.1 Transmission line data. IEEE 24-bus RTS

From	To	R [pu]	X [pu]	P_{max} [pu]	Cost [M€]
1	2	0.003	0.014	0.175	20
1	3	0.055	0.211	0.175	20
1	5	0.022	0.085	0.175	20
2	4	0.033	0.127	0.175	20
2	6	0.050	0.192	0.175	20
3	9	0.031	0.119	0.175	20
3	24	0.002	0.084	0.400	400
4	9	0.027	0.104	0.175	20
5	10	0.033	0.088	0.175	20
6	10	0.014	0.061	0.175	20
7	8	0.016	0.061	0.175	20
8	9	0.043	0.165	0.175	20
8	10	0.043	0.165	0.175	20
9	11	0.002	0.084	0.400	400
9	12	0.002	0.084	0.400	400
10	11	0.002	0.084	0.400	400
10	12	0.002	0.084	0.400	400
11	13	0.006	0.048	0.500	40
11	14	0.005	0.042	0.500	40
12	13	0.006	0.048	0.500	40
12	23	0.012	0.097	0.250	40
13	23	0.011	0.087	0.500	40
14	16	0.005	0.059	0.500	40
15	16	0.002	0.017	0.500	40
15	21	0.006	0.049	0.250	40
15	24	0.007	0.052	0.500	40
16	17	0.003	0.026	0.500	40
16	19	0.003	0.023	0.500	40
17	18	0.002	0.014	0.500	40
17	22	0.014	0.105	0.500	40
18	21	0.003	0.026	0.500	40
19	20	0.005	0.040	0.500	40
20	23	0.003	0.022	0.500	40
21	22	0.009	0.068	0.500	40

We consider four scenarios to describe the behaviour of the demand: low, low-medium, medium-high, and high demand. Table A.3 presents the relative weights (% of annual hours with respect to the total) for the four scenarios.

From the combination of Tables A.2 and A.3 we obtain the demand bids per scenario (utility functions).

On the other hand, Table A.4 provides the total energy offered by the generation units, as well as the price of each of the energy blocks that make up the offer (all the energy blocks are equally sized, as in the demand model).

Table A.2 Demand economic data. IEEE 24-bus RTS

Demand	Power (MW)	Bid block 1 (€/MWh)	Bid block 2 (€/MWh)	Bid block 3 (€/MWh)
D1	62	39.0	36.4	33.8
D2	70	44.2	41.6	39.0
D3	48	26.0	20.8	18.2
D4	125	39.0	36.4	33.8
D5	187	44.2	41.6	39.0
D6	62	39.0	36.4	33.8
D7	187	44.2	41.6	39.0
D8	187	26.0	20.8	18.2
D9	125	39.0	36.4	33.8
D10	187	44.2	41.6	39.0
D11	62	39.0	36.4	33.8
D12	70	44.2	41.6	39.0
D13	31	26.0	20.8	18.2
D14	125	39.0	36.4	33.8
D15	148	44.2	41.6	39.0
D16	62	39.0	36.4	33.8
D17	226	44.2	41.6	39.0

Table A.3 Scenario characterization

Scenario	Relative weight	Demand coefficient
1	0.4120	0.47
2	0.3297	0.85
3	0.1592	1.20
4	0.0991	1.70

Table A.4 Generators' economic data. IEEE 24-bus RTS

Bus	Generator	Capacity (MW)	Offer block 1 (€/MWh)	Offer block 2 (€/MWh)	Offer block 3 (€/MWh)	Offer block 4 (€/MWh)
1	G1	250	15.0	18.8	22.5	26.3
2	G2	250	13.0	16.3	19.5	22.8
7	G3	220	15.0	18.8	22.5	26.3
13	G4	100	15.0	18.8	22.5	26.3
14	G5	100	14.0	17.5	21.0	24.5
15	G6	100	16.0	20.0	24.0	28.0
16	G7	100	15.0	18.8	22.5	26.3
18	G8	100	13.0	16.3	19.5	22.8
21	G9	100	14.0	17.5	21.0	24.5
22	G10	300	15.0	18.8	22.5	26.3
23	G11	200	15.0	18.8	22.5	26.3

A.4 Multiarea IEEE 24-Bus RTS

This section presents a system composed of three areas of 24 buses each, constructed from the original IEEE 24-bus RTS. The three systems are interconnected by means of five tie lines. The single-line diagram is shown in Fig. A.2.

The overall system is composed of 107 lines, 33 generators and 51 demands.

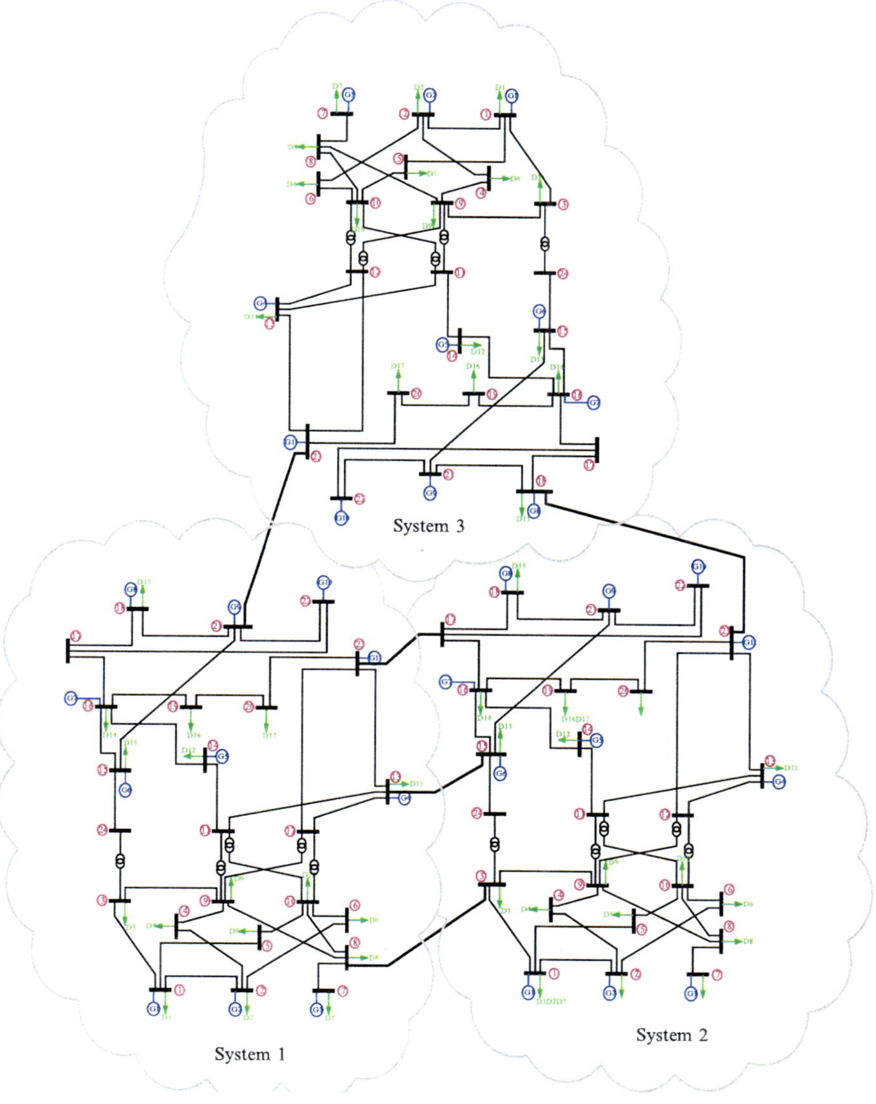

Fig. A.2 Single-line diagram. Multi-area IEEE 24-bus RTS

Table A.5 Tie lines technical data

From bus (System)	To bus (System)	R [pu]	X [pu]	P_{max} [pu]	Cost [M€]	Installed
8 (S1)	3 (S2)	0.043	0.165	0.175	20	1
13 (S1)	15 (S2)	0.011	0.087	0.500	40	1
23 (S1)	17 (S2)	0.003	0.022	0.500	40	1
21 (S1)	23 (S3)	0.009	0.068	0.500	40	1
23 (S2)	18 (S3)	0.003	0.022	0.500	40	1

Transmission data are the same as those in Table A.1, which is the base system, whilst tie lines data are shown in Table A.5.

Regarding generation and demand data, all areas share the same topology and offer prices, but area 1 has the same data shown in Sect. A.3, area 2 has increased its generation and demand by 30%, and area 3 has shrunk its generation and demand by 30%.

References

1. Alguacil N, Motto AL, Conejo AJ (2003) Transmission expansion planning: a mixed-integer LP approach. IEEE Trans Power Syst 18:1070–1076
2. Binato S, Pereira MVF, Granville S (2001) A new benders decomposition approach to solve power transmission network design problems. IEEE Trans Power Syst 16:235–240
3. Brooke A, Kendrick D, Meeraus A, Raman R (2003) GAMS/CPLEX 9.0. User notes. GAMS Development Corp., Washington, DC
4. Buygi MO, Balzer G, Shanechi HM, Shahidehpour M (2004) Market-based transmission expansion planning. IEEE Trans Power Syst 19:2060–2067
5. Contreras J, Gross G, Arroyo JM, Muñoz JI (2009) An incentive-based mechanism for transmission investment. Decis Support Syst 47:22–31
6. Contreras J, Wu FF (1999) Coalition formation in transmission expansion planning. IEEE Trans Power Syst 14:1144–1152
7. Contreras J, Wu FF (2000) A kernel-oriented coalition formation algorithm for transmission expansion planning. IEEE Trans Power Syst 15:919–925
8. de la Torre S, Conejo AJ, Contreras J (2008) Transmission expansion planning in electricity markets. IEEE Trans Power Syst 23:238–248
9. Gallego RA, Monticelli A, Romero R (1998) Transmission system expansion planning by extended genetic algorithm. IEE Proc-Gener Transm Distrib 145:329–335
10. Garces LP, Conejo AJ, Garcia-Bertrand R, Romero R (2009) A bilevel approach to transmission expansion planning within a market environment. IEEE Trans Power Syst 4:1513–1522
11. Garver LL (1970) Transmission network estimation using linear programming. IEEE Trans Power Apparato Syst 89:1688–1697
12. Latorre G, Cruz RD, Areiza JM, Villegas A (2003) Classification of publications and models on transmission expansion planning. IEEE Trans Power Syst 18:938–946
13. Maghouli P, Hosseini SH, Buygi MO, Shahidehpour M (2009) A multi-objective framework for transmission expansion planning in deregulated environments. IEEE Trans Power Syst 24:1051–1061
14. Oliveira GC, Costa APC, Binato S (1995) Large scale transmission network planning using optimization and heuristic techniques. IEEE Trans Power Syst 10:1828–1834
15. Grigg C, Wong, P, Albrecht P, Allan R, Bhavaraju M, Billinton R, Chen Q, Fong C, Haddad S, Kuruganty S, Li W, Mukerji R, Patton D, Rau N, Reppen D, Schneider A, Shahidehpour M,

Singh C (1996) Reliability test system taskforce: the IEEE reliability test system-1996. IEEE Trans Power Syst 14:1010–1020
16. Roh JH, Shahidehpour M, Fu Y (2007) Market-based coordination of transmission and generation capacity planning. IEEE Trans Power Syst 22:1406–1419
17. Romero R, Gallego RA, Monticelli A (1996) Transmission system expansion planning by simulated annealing. IEEE Trans Power Syst 11:364–369
18. Rosellón J (2003) Different approaches towards electricity transmission expansion. Rev Netw Econ 2:238–269
19. Ruiz PA, Contreras J (2007) An effective transmission network expansion cost allocation based on game theory. IEEE Trans Power Syst 22:136–144
20. Sauma EE, Oren SS (2006) Proactive planning and valuating of transmission investments in restructured electricity markets. J Regul Econ 30:358–387
21. Sauma EE, Oren SS (2007) Economic criteria for planning transmission investment in restructured electricity markets. IEEE Trans Power Syst 22:1394–1405
22. Shrestha GB, Fonseka PAJ (2004) Congestion-driven transmission expansion in competitive power markets. IEEE Trans Power Syst 19:1658–1665
23. Tsamasphyrou P, Renaud A, Carpentier P (1999) Transmission network planning: an efficient benders decomposition scheme. In: Proceedings of the 13th power systems computation conference, Trondheim, Norway, pp 487–494
24. Villasana R, Garver LL, Salon SJ (1985) Transmission network planning using linear programming. IEEE Trans Power Apparat Syst 104:349–356
25. Zolezzi JM, Rudnick H (2002) Transmission cost allocation by cooperative games and coalition formation. IEEE Trans Power Syst 17:1008–1015

Algorithms and Models for Transmission Expansion Planning

Alexey Sorokin, Joseph Portela, and Panos M. Pardalos

Abstract This chapter presents an overview of algorithms and optimization models used for solving Transmission Expansion Planning (TEP) problem. Being a very complex problem TEP attracts much attention from both researchers and practitioners. A great number of publications in the technical literature address this problem by providing various optimization models and applying different algorithms to solve the TEP problem. Besides literature review and brief classification of the proposed algorithms and models this survey covers examples for most of the methods.

Keywords Heuristic algorithms • mathematical programming • network expansion • survey • transmission planning

1 Introduction

The transmission network planning or expansion model refers to the use of transmission-related inputs, options, and constraints to obtain an optimal and feasible transmission expansion plan. In general case, the objective of this problem is to minimize construction cost of new lines and operational costs while meeting the forecasted demand. The papers presented in the technical literature vary in their considerations and degrees of complexity. This survey is meant as a starting point for researchers new to transmission planning problems, as well as it reveals new approaches to the problem for people familiar with it. Besides this survey, Latorre et al. [42] provide a very good bibliographical review of publications on TEP problem. There are many publications in the literature addressing this problem and this chapter describes most recent and promising papers to give an overview of

A. Sorokin (✉) • J. Portela • P.M. Pardalos
University of Florida, Gainesville, FL, USA
e-mail: sorokin@ufl.edu; jportela@ufl.edu; pardalos@ufl.edu

A. Sorokin et al. (eds.), *Handbook of Networks in Power Systems I*,
Energy Systems, DOI 10.1007/978-3-642-23193-3_16,

techniques and algorithms currently being used to solve the problem. The paper is structured as follows. There are two main sections of this paper. Section 2 describes heuristic approaches for TEP problem; among them are algorithms based on linear programing relaxations, genetic algorithms, differential evolution, ant colony optimization, greedy randomized adaptive search procedure (GRASP), tabu search, and others. Section 3 focuses primarily on exact techniques and TEP optimization model formulations such as mixed integer disjunctive formulation, models and algorithms for TEP in deregulated market environment, considerations of financial transmission rights, stochastic programming, and models addressing reliability issues in TEP. Finally, Section 4 concludes the chapter.

2 Heuristic Approaches

The problem of transmission expansion planning is a mixed integer program by its nature. The solution of such problems is a very difficult and computationally challenging process for large scale systems. That is the reason why there is a lot of work has been done to develop heuristic algorithms for this problem, which require much less computational time and produce a solution close to optimal in contrast to exact optimization techniques.

Romero et al. [54] applied simulated annealing for TEP problem. The idea and the name of this metaheuristic come from metallurgical industry. During heating, atoms are becoming free from their initials locations and slow cooling allows them to find an alternative configuration with lower energy. By analogy to the physical process a solution can be substituted by a random solution from a neighborhood with probability that depends on a global "temperature" parameter. The two main parameters of the simulated annealing are transition mechanism and cooling scheme. The transition mechanism is the way of switching from current solution to an alternative one. In the paper mentioned above the author defined the transition mechanism as adding a new circuit, swapping two circuits, and removing a circuit. The cooling scheme was defined by initial temperature, final temperature, number of transitions at a given temperature, and temperature rate of change. For Garver's 6-bus system [53] and for Southern Brazilian network [4] the optimal solutions have been found.

Bustamante-Cedeno and Arora [6] developed a constructive heuristic algorithm for TEP problem. The algorithm fixes binary variables to particular values and then solves the resulting linear program. De Oliveira et al. [47] propose a heuristic approach for TEP which represents integer variables by continuous Sigmoid functions and modifies the DC power flow equations of the Optimal Power Flow problem to accommodate these changes considering electrical losses. Da Silva et al. [62] compared the performance of several metaheuristics including particle swamp and ant colony optimizations, differential evolution, artificial immune systems, and tabu search for TEP problem with consideration of Ohmic losses. The solution quality was measured by the quality index and the extensive results are presented in the paper. Wang and Cheng [69] applied plant growth simulation algorithm to TEP problem.

2.1 Linear Programming Relaxations

Many different algorithms have been developed in the past to address this problem using LP relaxation of the original integer problem during the solution process. One example of such algorithms is described in the paper of Hashimoto et al. [34]. In this paper the authors propose a transformation of the original LP relaxation to the problems with only one equality constraint (the power flow balance equation) at a time. This allows to reduce a number of variables in the equality constraint and provides computational advantages using a dual simplex algorithm for solving the modified problem. The authors acknowledge that one of the disadvantages of the proposed algorithm is the requirement of having the inverse of susceptance matrix in explicit form. The other two examples of using linear programming for transmission network planning can be found in the papers of Garver [31] and Villasana et al. [68].

2.2 Genetic Algorithms

Genetic algorithm (GA) is a methodology for solving large scale linear or non-linear problems without necessarily having a good initial approximation. The GA was initially proposed by Holland [36]. It is based on the occurrence in nature of offspring inheriting the genetic traits of parents. Stronger individuals are more likely to survive and reproduce. The GA generally manipulates binary strings that make up chromosomes (solutions) of the problem. Versions of this algorithm are applied by Jalilzadeh et al. [37, 38] and Escobar et al. [19, 20]. We refer the interested reader to the papers [18, 28, 57, 59, 60] for further applications and modifications of genetic algorithms for TEP problem not discussed in this chapter.

Jalilzadeh et al. used the decimal codification genetic algorithm (DCGA) to solve TEP problem considering the voltage level of the transmission lines and network losses, and later also considering the bundle lines (the lines with multiple conductors). DCGA uses a decimal system for easier coding of the transmission lines and to prevent the production of offspring completely different from their parents. The primary difference between the two papers is in the size of the chromosomes. The more recent work includes a longer string to account for the number of bundle lines. A general static (or one stage) transmission network expansion planning formulation presented by the authors minimizes expansion cost of constructing new lines, expansion of substations, and cost of annual losses:

$$min\ C_T = \sum_{i,j\in\Omega} CL_{ij} n_{ij} + \sum_{k\in\Psi} CS_k + \sum_{i=1}^{NY} C_{loss_i} \tag{1}$$

$$C_{loss} = loss \times C_{MWh} \times k_{loss} \times 8760 \tag{2}$$

$$loss = \sum_{i,j \in \Omega} R_{ij} I_{ij}^2 \quad (3)$$

s.t.

$$Sf + g - d = 0 \quad (4)$$

$$f_{ij} - \gamma_{ij}(n_{ij}^0 + n_{ij})(\theta_i - \theta_j) = 0 \quad (5)$$

$$|f_{ij}| \leq (n_{ij}^0 + n_{ij}) f_{ij} \quad (6)$$

$$0 \leq n_{ij} \leq n_{ij} \quad (7)$$

$$0 \leq g \leq \bar{g} \quad (8)$$

$$\text{Line\_Loading} \leq \text{LL}_{\max} \quad (9)$$

where:

C_r	total expansion cost of the network;
CL_{ij}	construction cost of each line in branch *i–j* (different costs for 230 and 400 KV lines);
CS_k	expansion cost of *Kth* substation;
C_{loss}	annual losses cost of network;
$loss$	total losses of network;
C_{MWh}	cost of one MWh;
k_{loss}	losses coefficient;
8,760	number of hours in a year;
R_{ij}	resistance of branch *i–j*;
I_{ij}	flow current of branch *i–j*;
Ω	set of all corridors;
Ψ	set of all substations;
NY	expanded network adequacy (in 1 year);
S	branch-node incidence matrix;
d	demand vector;
Ω	set of all corridors;
θ	phase angle of each bus;
γ_{ij}	total susceptance of circuits in corridor *i–j*;
n_{ij}	number of new circuits in corridor;
$i–jn^0{}_{ij}$	number of initial circuits in corridor *i–j*;

n_{ij}	maximum number of constructible circuits in corridor i–j;
g	generation vector;
$\bar{g}$	generator power limit in generator buses;
f	active power matrix in each corridor;
$\overline{f_{ij}}$	maximum of transmissible active power through corridor $i - j$ (different rates for 230 and 400 KV lines);
Line_Loading	loading of lines at planning horizon year and start of operation time;
LL_{max}	maximum loading of lines at planning horizon year.

The DCGA starts with creation of a chromosome which represents a possible solution for the TEP. The paper of Jalilzadeh et al. [37] describes a two-part chromosome. The first part includes genes which indicate the number of transmission circuits in a given corridor. The genes in part two correspond to the genes in part one and give the voltage of each circuit. The second paper of Jalilzadeh et al. [38] adds a third part to this string that corresponds to the number of bundle lines in each string.

An initial population of chromosomes is randomly created. As stated previously, a good initial solution is not necessary. A selection operator must choose pairs of parent strings from the population to allow a crossover operator to create an offspring later. Probability of the selection is higher for those chromosomes with a higher fitness level; in this case the chromosomes with a higher objective function. Once an even number of chromosomes is chosen, each pair is subject to crossover. This involves choosing a position in each pair of chromosomes and swapping their genes after that position. Finally, a mutation operator ensures the population variety by randomly increasing or decreasing a gene value by one (as long as the change is within its limits). For binary system, this would just be changing the gene to 1 or 0, depending on its current state. The authors chose a mutation probability of .01 per bit, which is large when compared to other applications of GA. The algorithm stops when a certain fitness or maximum number of generations is reached, and the best chromosome is selected as the optimal solution.

Garver's system [53] was used in [37] to test the proposed method. Results showed that networks with mostly 230 kV lines had lower initial expansion costs than those networks with mostly 400 kV lines. The second network, however, was more cost-effective alternative for a long term expansion due to loss of load considerations. The Azerbaijan regional electric system was used in [38] to test the proposed algorithm. Results regarding 230 and 400 kV lines were similar, as the systems with mostly 230 kV lines got overloaded more quickly and were more costly in the long run. Also, bundle conductors were shown to be effective in increasing network adequacy and decreasing loss of load.

Escobar used the GA proposed by Chu and Beasley (CBGA) [14] to solve the STEP considering multiple generation scenarios [19], and uncertainty in generation and demand [20]. The CBGA used by Escobar in each paper has the following characteristics:

1. A *fitness* function to evaluate the objective function of a tested solution, as well as an *unfitness* function that quantifies infeasibility.
2. Substitution of only one individual in the population in each iteration.
3. Efficient means of local improvement for each test individual.

In CBGA, all saved solutions are different, which prevents premature convergence. Infeasible individuals are initially not discarded, but rather are used in the population until there is a point where all individuals are feasible and feasible solutions are added. Infeasible solutions are saved when there is an insufficient number of feasible solutions. Topologies are only discarded when better offsprings are created, which is more efficient than the traditional GA that would eliminate topologies based on some random decisions.

In [19] the authors address deregulation and its effects on the network expansion. Network congestion increases costs for consumers, but may not affect system reliability. Investments that do not affect reliability, but are made anyway, are done so for the sake of the social welfare. The investment is made if the savings from reduced congestion are greater than the required investment cost. The congestion can be measured by the re-dispatch cost, which is defined by the authors as the difference between total generation cost with and without transmission constraints. The expansion should ideally eliminate congestion for all generation scenarios, but this results in extremely high investment costs.

The authors look only at extreme and feasible scenarios to test all feasible scenarios. The infinite set of all feasible scenarios is adequately tested using this method because the constraints for non-extreme scenarios are more relaxed. In the extreme scenarios, some generators function at their upper limit ($\bar{g}_k$), while others function at their lower limit ($\underline{g_k}$) bounds the generator output $\underline{g_k} \leq g_k \leq \bar{g}_k$. Any given test scenario must satisfy the following constraints:

$$d_r \geq \sum_{i \in \Omega_1} g_i + \sum_{i \in \Omega_2} \underline{g_i} + \underline{g_k}, \tag{10}$$

$$d_r \leq \sum_{i \in \Omega_1} g_i + \sum_{i \in \Omega_2} \underline{g_i} + g_k, \tag{11}$$

where:

d_r	total demand of the system;
$\bar{g}_i, \underline{g_i}$	upper and lower limit of generator i, respectively;
$g_k, \underline{g_k}$	upper and lower limit of the free generator, respectively;
Ω_1, Ω_2	set of generators in the upper and lower limit, respectively.

The IEEE 24-bus system [22] is used to evaluate the methodology presented. Results were obtained for the model containing multiple generation scenarios, and then were compared to a centralized planning model. The first model resulted in significantly higher investment costs due to extreme scenarios evaluated. A lower-cost alternative could be obtained if a generator operating range is narrowed.

The role of bundle lines in TEP problem was also investigated by Mahdavi et al. in [45]. In this paper the authors also consider an expansion of substations together with lines and voltage levels. A genetic algorithm has been used for solving the resulting problem.

In [20] Escobar et al. consider uncertainty in demand and generation. Long-term prediction of demand is extremely important in expansion planning and determining capacity requirements. The demand and generation in this paper are not deterministic, but they can assume any deterministic value on a narrow interval. These values are associated with a probability of occurrence that is taken into account when the problem is solved. Two models are solved to show the importance of uncertainty considerations: one with deterministic demand and uncertain generation, and the other with both uncertain demand and uncertain generation. The main difference between these models and the basic STEP is that the upper limits of uncertain component are considered variables of this problem.

The IEEE 24-bus system [22] was also used to evaluate these models. The results were compared to the results from the multiple generation scenarios model from [19]. The first model results in the planning cost that is 83% of the cost from the corresponding model in [19], while the second model results in 74% of the planning cost.

2.3 *Differential Evolution*

Differential Evolution (DE) is a technique very similar to GA. It includes initialization, mutation, crossover, and selection steps to find an optimal solution. The key difference between DE and GA is the mutation phase. Georgilakis [32] applies an improved differential algorithm (IDE) to a market-based transmission expansion planning problem. The objective is to select new transmission lines to be added to an existing system while minimizing the annual generation and transmission investment cost. The problem is constrained by Kirchhoff's laws, transmission line capacities, generator output limits, and contingency constraints to satisfy potential outages.

The initialization procedure consists of generating a uniformly distributed population with NP individuals that fall between the boundary conditions. The DE algorithm keeps a population of vectors $x_i^G, i = 1, 2, \ldots, NP$, where i is the index of that vector and G is the generation index.

The author points out that one of the main differences between GA and DE is in the mutation phase. The following mutant vector is generated for each parent vector, x_i^G:

$$v_i^{G+1} = x_{n1}^G + F * (x_{n2}^G - x_{n3}^G). \quad (12)$$

Where $n1$, $n2$, and $n3$ are random indexes that are different from the current index i. The scaling factor F is constant in the initial DE [63], but the author applies a modification to it for the IDE, where F becomes a random variable within a specified range and is different for each generation. The crossover step determines whether or not this mutant vector will be chosen over the current vector as the trial vector, while the selection phase determines the population for the next generation. The trial vector is discarded if it is worse than the target vector.

The IDE makes use of an auxiliary set to enhance the search process. This set is also of size NP and uses the same initialization process described above. An rejected or discarded trial vector is compared to each member of the auxiliary set with the current index, and replaces that member if it provides a better solution. The best solutions from the auxiliary set replace the worst solutions in the main set after a certain number of generations.

The proposed IDE was tested on IEEE 30-bus, 57-bus, and 118-bus systems [50]. The results were improved when compared to a simple DE and GA. IDE produced the lowest cost solution 85% of the time for a modified version of the 30-bus test system, while others algorithms were unable to converge to the same solution. IDE and DE both performed faster than GA in the test.

An alternative DE algorithm for the transmission expansion problem has been proposed by Sum-Im et al. in [64]. This algorithm works well for networks of medium size and the computational results were demonstrated on Garver's 6-bus and IEEE 25-bus networks. This work has been expanded in [65] where the authors expanded the algorithm for a multistage transmission network expansion problem.

2.4 Ant Colony Optimization

Ant Colony Optimization is inspired by the way ants search for and find nearby food sources. Every ant leaves a trail of pheromone when returning to the nest after finding food. This is detected by others ants in search of food, who base their search decisions on the level of pheromone on each path. Paths with high pheromone intensity are more likely to be chosen, while those with lower intensity will be selected less often. Short paths will be traversed faster by the ants, and will therefore build up pheromone more quickly and continue to be chosen by more ants than longer paths. This results in efficient discovery of the closest food sources to the nest.

Use of the algorithm for TEP based on this occurrence in the nature is discussed by Paulun in [48]. Each potential solution to the network expansion planning problem is represented by a path consisting of all relevant variables, such as the expansion project timeline and types of new lines, transformers, or substations to be built. Unlike GA, new solutions are not created from parent solutions, but created keeping in mind the knowledge base, which consists of n uncertain events by m possible expansion steps. This knowledge base is the only element that needs to be initialized to start the algorithm. Initializing all elements to the same value will be helpful in preventing certain combinations of uncertain events and expansion steps, and in preventing convergence to local optima.

The objective of the optimization problem in [48] is to minimize a cost during the consideration period. Usually this cost is uncertain , that is why the expected value μ should be considered. Besides the expected value, standard deviation σ can be considered as well. Additionally, objective may include some risk measures, such as Value at Risk (VaR) or Conditional Value at Risk (CVaR). The complete objective is given by the following formula, where the multipliers can be changed by a user:

$$\min \alpha \cdot \mu + \beta \cdot \sigma + \gamma \cdot VaR + \phi \cdot CVaR \tag{13}$$

Each solution is evaluated for boundary conditions violation. Repair algorithms are used to correct violations and bring the solution into the valid solution space for economic evaluation (these repair algorithms can also be used in GAs). Net present value of each solution costs is then evaluated to update the knowledge base.

The authors point out the benefits of ant colony optimization when compared to GA: it is better suited for network expansion planning, while GA is better used in determining target networks for use in the future development of a network. The target networks are used as inputs and future goals for the network expansion planning problem. When a particular solution is out of bounds in network expansion planning, repairing the solution involves both adding of components (e.g. lines or transformers) and determining when the components are to be added. This decision can change a current state of the network as well as any other state after the addition. For determining long-term target networks, there are not the same considerations. A standard GA characteristic is that solutions which were evaluated in previous iterations are no longer available to improve the efficiency of repair algorithms.

2.5 *Greedy Randomized Adaptive Search Procedure*

Greedy Randomized Adaptive Search Procedure (GRASP) [23] is a metaheuristic method which is used widely for solving combinatorial problems [23]. GRASP consists of two phases: construction phase and local search. The construction phase produces a feasible solution by drawing one variable at a time from a candidate list. The candidate list is formed from the most promising variables ranked by a greedy function. The construction phase stops when feasibility is obtained. The solutions from the construction phase may be not optimal in a given neighborhood, that is why after the construction phase the local search procedure is applied. The search stops when there exists no better solution in the given neighborhood. Local search may require exponential time to finish, but in most practical cases it performs much faster. It should be noted that local search is heavily depends on a neighborhood definition of a solution. Binato et al. [4] applied GRASP for the transmission expansion planning. The problem is stated as minimization of investment cost and load not supplied:

$$min\ c^t xe + \alpha^t r \tag{14}$$

s.t.

$$Sf + g + r = d \tag{15}$$

$$f - (\gamma^0 + x\gamma^1)\Theta = 0 \tag{16}$$

$$0 \le g \le \overline{g} \tag{17}$$

$$|f| - x\gamma^1\overline{\phi} \le \gamma^0\overline{\phi} \tag{18}$$

$$x \in \{\ 0, 1\} \tag{19}$$

where:

c investment unit cost vector;
x investment decision diagonal matrix;
α load not supplied unit cost vector;
r load not supplied vector;
e vector of ones;
S branch-node incidence matrix;
f flow branch vector;
g generation bus vector;
$\overline{g}$ generation limit bus vector;
d forecasted load bus vector;
Θ difference voltage angle vector of all branches;
γ^0 susceptance diagonal matrix of existing branches;
γ^1 susceptance diagonal matrix of candidate branches;
$\overline{\phi}$ difference voltage angle limit vector of all branches.

At the construction phase GRASP selects one variable at a time by random manner from the candidate list. The authors used a cost of energy not supplied as a greedy function to form the candidate list. The following linear programming model should be solved to evaluate the greedy function:

$$min\ \alpha^t r \tag{20}$$

s.t.

$$Sf + g + r = d \tag{21}$$

$$f - \gamma^0\Theta = 0 \tag{22}$$

$$0 \le g \le \overline{g} \tag{23}$$

$$-\overline{\phi} \le \Theta \le \overline{\phi} \tag{24}$$

The Lagrangian multipliers associated with susceptances are used as indices for the greedy function because susceptance is the characteristic of the prospective transmission line. Denoting π^d the Lagrangian multipliers associated with the first set of constraints and π^Y_{kl} the Lagrangian multiplier associated with the branch susceptance connecting the bus k and l, it was shown that

$$\pi^Y_{kl} = (\pi^d_k - \pi^d_l)(\theta_k - \theta_l). \tag{25}$$

The candidate list can be formed by taking n best indices π^Y or by taking β percent of the best indices π^Y The construction phase returns a feasible solution and local search tries to improve it within a given neighborhood. The neighborhood of a solution is defined as an addition, removal or replacement of one arc in the solution. The authors implemented the algorithm in FORTRAN and used three networks for the case study: 6-bus network, the Brazilian Southern Network, and the reduced Brazilian Southeastern Network, which are described in [4]. In most of the cases GRASP was able to obtain a solution which was known to be optimal so far and in the Brazilian Southeastern case study it was able to obtain a solution which is better than the solutions found by other methods before.

GRASP is a very useful technique for many kinds of decision problems but the construction phase may need some improvements since it is responsible for about 80% of computational time for this problem.

Another application of GRASP for static TEP problem can be found in the paper of Faria et al. [25], where the authors used GRASP with path relinking. After obtaining two good solutions path relinking procedure constructs several paths that can connect these two solutions trying to find a better solution on its way. The selection of path movements is randomized allowing a better exploration of the search space.

2.6 *Tabu Search*

Tabu search is a metaheuristic search which differs from other search procedures because of the use of short-term memory; a user specified parameter how long solutions are to remain on a tabu list, which is the list of movements that are currently forbidden. Da Silva et al. [58] used an implementation of tabu search to solve the TEP. The search presented in this paper includes Expansion, Intensification, and Diversification Phases. The authors begin with a static TEP problem. The objective is to minimize the cost of building new circuits, while also trying to minimize loss of load due to the lack of transmission capacity. It can be formulated as the following mixed-integer non-linear optimization problem:

$$min \sum C_{ij}n_{ij} + \sum \alpha_i r \tag{26}$$

s.t.

$$\beta(x + \gamma^0)\theta + g + r = d \tag{27}$$

$$f_{ij} - (\gamma_{ij}^0 + x_{ij})(\theta_i - \theta_j) = 0 \tag{28}$$

$$|f_{ij}| - x_{ij}\phi_{ij}^{max} \leq \gamma_{ij}^0\phi_{ij}^{max}; \phi_{ij}^{max} = f_{ij}^{max}/\gamma_{ij}^0 \tag{29}$$

$$0 \leq g \leq g^{max}, 0 \leq r \leq d \tag{30}$$

$$x_{ij} = n_{ij}\gamma_{ij}, 0 \leq n_{ij} \leq n_{ij}^{max} \tag{31}$$

$$\forall(i,j) \in \Omega \tag{32}$$

where:

C_{ij}	cost of building a new circuit in branch *i–j*;
n_{ij}	number of circuit additions to branch *i–j*;
α	penalty parameter associated with loss of load caused by lack of transmission capacity;
r	array of load curtailments;
B(.)	susceptance matrix;
γ_{ij}^0	initial susceptance in branch *i–j*;
x_{ij}	total new circuit susceptance added to branch *i–j*;
g	array of bus active powers;
d	array of predicted bus loads;
f_{ij}	active power flow through branch *i–j*;
θ_I, θ_j	voltage angles at buses *i* and *j*;
γ_{ij}	circuit susceptance;
f_{ij}^{max}	flow limit in branch *i–j*;
g^{max}	maximum bus generation capacity array;
n_{ij}^{max}	maximum number of new circuits in branch *i–j*;
Ω	set of all candidate circuits.

The authors then linearized the problem by assuming that the candidate list has already been prepared and that the vector of new susceptances x^k is known. After that the problem is to connect the network with the objective of minimizing loss of load due to lack of transmission capacity:

$$\min \sum \alpha_i r \tag{33}$$

s.t.

$$\beta(x^k + \gamma^0)\theta + g + r = d \quad (34)$$

$$f_{ij} - (\gamma_{ij}^0 + x_{ij}^k)(\theta_i - \theta_j) = 0 \quad (35)$$

$$|f_{ij}| - x_{ij}^k \phi_{ij}^{max} \leq \gamma_{ij}^0 \phi_{ij}^{max} \quad (36)$$

$$0 \leq g \leq g^{max} \quad (37)$$

$$0 \leq r \leq d \quad (38)$$

To begin the tabu search, a current solution is used as the starting point, and a neighborhood and potential movements must be set. Neighborhood for this solution is defined as all solutions obtained by adding or removing a single candidate circuit. Three-part sensitivity index is then proposed to choose which movement to make. The first part consists of the Lagrangian multiplier of the linearized problem with respect to the candidate's susceptance.

The second and the third parts of this index are cost and susceptance of each candidate respectively. The best n candidates are then ranked in each category from 1 to n, with n being the best. The three category rankings are then summed, and a non-tabu candidate with the highest sum is chosen.

The tabu list is formulated in an attempt to avoid cycles and addition of a high volume of parallel circuits. A new circuit is made tabu after it is built, and remains tabu for the next J iterations. This means that it cannot be removed and its parallel circuits cannot be added unless an Aspiration criterion, the set of rules that overrides the tabu list, is satisfied. In this case, a move can be made when the tabu addition leads to improvements. In general, tabu search allows movements to be made which deteriorates the solution with the goal of finding a better one later. The procedure of finding a solution using the sensitivity index is called the Expansion Phase (EP). The EP can be run multiple times. At the end of the EP the local search attempts removing useless additions by trying to remove one circuit at a time, in the reverse order of the investment cost, and then checking feasibility of the solution. The end result is a feasible transmission network expansion plan, which will be further improved in subsequent steps.

Next, the two-part Intensification Phase (IP_1 and IP_2) is used to improve the solution. IP_1 involves removing a previously built candidate circuit and replacing it with another one. This solution is accepted if feasibility is not affected and the total cost is lowered. IP_2 is performed using the IP_1 solution as its starting point. IP_2 allows a movement to higher cost solutions; this is the only difference from IP_1. A check for useless circuits is completed after each IP (same as the end of the EP).

A Diversification Phase (DP) is implemented in an attempt to find solutions in unexplored regions. It adds the most frequently used candidate circuits from the last iteration to the tabu list for the next EP. This results in different trial solutions during the EP and IP. The stopping criterion of this tabu search algorithm is a number of DPs performed.

The authors conclude with the results of two Brazilian networks as case studies. The first is based on the 1980 Brazilian Southern System, which includes 46 busses and 62 circuits. The best solution was found during the first iteration. The expansion phase yielded a result that was feasible, but expensive. The cost was lowered in each of the two IPs, and the IP_2 solution was best. Subsequent iterations provided several other acceptable solutions. The optimal solution matches the result of the tabu search. The second study is based on the 2000 Southeastern System, which includes 79 busses and 155 circuits. The best solution was again found during the first iteration, but in this time it was obtained during IP_1. The results show that tabu search is an effective procedure to solve the TEP. The DP did not improve the solution in either case, but it is still valuable for its ability to explore unexplored regions and provide other acceptable solutions. The key feature of this algorithm is its ability to avoid local optimum solutions.

Gallego et al. [29] proposed another tabu search algorithm for the transmission network expansion planning which employs both genetic algorithms and simulated annealing. For small and medium networks the algorithm has found optimal solutions faster than other methods and for large scale networks the algorithm has found many different alternative configurations within 1% of the required investments. Da Silva et al. [61] proposed a new tabu search algorithm for the transmission expansion problem considering electrical losses and interruption cost. The algorithm employs GRASP for solving the problem considering also dynamic nature of the model, i.e. the expansion of the lines in different time moments.

3 Optimization Models and Exact Algorithms

3.1 *Mixed Integer Disjunctive Model*

A mixed integer disjunctive model for transmission network expansion is an alternative formulation for the classical nonlinear mixed integer model. The nonlinearity appears in the power flow constraints, where bus voltage angles variables are multiplied by circuit investment decision variables. Bahiense et al. [3] analyzed a mixed integer disjunctive formulation, where the nonlinear constraints were avoided by using disjunctive model to which they were equivalent. The following notation will be used through this section:

n number of nodes (busses);
m number of candidate circuits (branches);
Ω_i^0 set of existing circuits connected to bus $i = 1, \ldots, n$;
Ω_i^+ set of candidate circuits connected to bus $i = 1, \ldots, n$;
f vector of circuit flows (existing and candidates);
f^{max} vector of circuit capacities (existing and candidates);
g vector of bus generations;
g^{max} vector of bus generation capacities;

d	vector of bus active loads;
θ	vector of bus voltage angles (in radians);
r	vector of bus load curtailment;
x	binary vector of decision investment on candidate circuits;
c	vector of candidate circuit unit cost (in million dollars);
γ	vector of circuit susceptances;
M	penalty factor for candidate circuits.

One of the typical mixed integer non-linear formulations for the transmission network expansion problem minimizes investment cost subject to operational constraints:

$$min_{\{x,f,\theta,g\}}\ c \cdot x \tag{39}$$

s.t.

$$\sum_{k=(i,j)j\in\Omega_i} f_k + g_i = d_i,\ i = 1,\ldots,n \tag{40}$$

$$f_k - \gamma_k(\theta_i - \theta_j) = 0,\ k = (i,j),\ j \in \Omega_i^0,\ i = 1,\ldots,n \tag{41}$$

$$f_k - x_k\gamma_k(\theta_i - \theta_j) = 0,\ k = (i,j),\ j \in \Omega_i^+,\ i = 1,\ldots,n \tag{42}$$

$$-f_k^{max} \le f_k \le f_k^{max},\ k = (i,j),\ j \in \Omega_i^0,\ i = 1,\ldots,n \tag{43}$$

$$-f_k^{max}x_k \le f_k \le f_k^{max}x_k,\ k = (i,j),\ j \in \Omega_i^+,\ i = 1,\ldots,n \tag{44}$$

$$0 \le g_i \le g_i^{max},\ i = 1,\ldots,n \tag{45}$$

$$x \in \{0,1\}^m \tag{46}$$

$$\theta_{ref} = 0 \tag{47}$$

Constraints (40) represent the first Kirchhoff's law. The second Kirchhoff's law is enforced by the constraints (41) for existing lines and (42) for candidate lines. Line limits are represented by constraints (43) for existing lines and (44) for proposed lines. Generators' limits are imposed by constraints (45) and the reference voltage angle is fixed by the constraint (47). Note that the nonlinearity appears in the second Kirchhoff's law for candidate circuits in the equation (42) due to multiplication of decision variable x and voltage angle θ. The disjunctive mixed integer formulation helps to avoid the nonlinear constraints by using an equivalent disjunctive formulation. This model was independently proposed by Pereira et al. [49] and Villanasa [67]. In this formulation the constraint related to the second Kirchhoff's law for candidate networks is replaced by the following linear constraints:

$$- M_k(1 - x_k) \le f_k - \gamma_k(\theta_i - \theta_j) \le M_k(1 - x_k),\ k = (i,j), j \in \Omega_i^+,\ i = 1, \ldots, n. \tag{48}$$

The objective function and other constraints are the same. When the decision variable x_k is set to zero, the corresponding disjunctive constraint is not binding, and when it is set to one the second Kirchhoff's law is valid. Bahiense et al. [3] proposed an alternative disjunctive formulation with better conditioning properties. Here the second Kirchhoff's law was represented by two inequalities, each one is related to the possible flow direction for each candidate circuit:(second Kirchhoff's law for candidate circuit $k = (i,j)$ upper bound)

$$f_k^+ - \gamma_k \Delta\theta_k^+ \le 0,\ k = (i,j),\ j \in \Omega_i^+,\ i = 1, \ldots, n \tag{49}$$

$$f_k^- - \gamma_k \Delta\theta_k^- \le 0,\ k = (i,j),\ j \in \Omega_i^+,\ i = 1, \ldots, n \tag{50}$$

(second Kirchhoff's law for candidate circuit $k = (i,j)$ lower bound)

$$f_k^+ - \gamma_k \Delta\theta_k^+ \ge -M_k(1 - x_k),\ k = (i,j),\ j \in \Omega_i^+,\ i = 1, \ldots, n \tag{51}$$

$$f_k^- - \gamma_k \Delta\theta_k^- \ge -M_k(1 - x_k),\ k = (i,j),\ j \in \Omega_i^+,\ i = 1, \ldots, n \tag{52}$$

The authors expressed the flow and difference of angle in each candidate circuit as difference of two non-negative variables f_k^+, f_k^- for the flow and θ_k^+, θ_k^- for the voltage angle:

$$f_k = f_k^+ - f_k^-,\ k = (i,j),\ j \in \Omega_i^+,\ i = 1, \ldots, n \tag{53}$$

$$\Delta\theta_k = \theta_i - \theta_j = \Delta\theta_k^+ - \Delta\theta_k^-,\ k = (i,j),\ j \in \Omega_i^+,\ i = 1, \ldots, n \tag{54}$$

The candidate circuit flow variables can be expressed as two inequalities for upper and lower bound:

$$f_k^+ - f_k^{max} x_k \le 0,\ k = (i,j), j \in \Omega_i^+,\ i = 1, \ldots, n \tag{55}$$

$$f_k^- - f_k^{max} x_k \le 0,\ k = (i,j), j \in \Omega_i^+,\ i = 1, \ldots, n \tag{56}$$

This alternative formulation gives a tighter upper bound because there is no penalty term in the RHS. The authors used IEEE 46-bus network as a benchmark for both disjunctive mixed integer formulation and alternative formulation; the case study showed that the alternative formulation requires significantly less computational effort.

3.2 Branch and Bound Algorithm

Branch and bound (B&B) algorithm uses relaxation and separation strategies for solving complex problems. The branching step of B&B results in a tree of subproblems based on the original problem, but with some additional constraints. Bounding occurs when the minimum and the maximum of a particular node are calculated, allowing other nodes to be discarded if a better solution cannot be possibly obtained. The B&B algorithm presented by Rider et al. [51] solves the static TEP using the DC model. The proposed method solves a mixed integer nonlinear problem (MINLP) by solving a nonlinear problem (NLP) in each node on the B&B tree. The MINLP is formulated as minimizing:

$$min\ \upsilon = c_n^T n + c_p^T p \tag{57}$$

$$-\frac{1}{2}|S^T|(N+N^0)[S\theta]\Psi S\theta - S^T(N+N^0)YS\theta + p = d \tag{58}$$

$$(N+N^0)(\frac{1}{2}[S\theta]\Psi S\theta + Y|S\theta| - \bar{f}) \leq 0 \tag{59}$$

$$0 \leq p \leq \bar{p}, 0 \leq n \leq \bar{n} \tag{60}$$

$$n \text{ is integer, } \theta \text{ is unbounded} \tag{61}$$

where:

υ	investment;
c_n	cost vector of circuits that can be added;
c_p	cost vector of energy produced by the generators;
n	vector of circuits added;
$\bar{f}$	maximum power flow vector;
N	diagonal matrix with n_is as diagonal elements;
N^0	diagonal matrix with the circuits in the base case as diagonal elements;
Ψ	diagonal matrix with the conductance of the circuits as diagonal elements;
Y	diagonal matrix with the susceptance of the circuits as diagonal elements;
S	branch-node incidence matrix from the electric system;
θ	angle phase vector;
p	generation vector with maximum value p;
$\bar{n}$	maximum number of vector circuits that can be added;
d	demand vector;
$[x]$	diagonal matrix with x_is as the diagonal elements.

The NLP to be minimized at each node of the B&B tree has the following format:

$$min\ \upsilon = c_n^T n + c_p^T p + \alpha_r^T r \quad (62)$$

s.t.

$$-\frac{1}{2}|S^T|(N+N^0)[S\theta]\Psi S\theta - S^T(N+N^0)YS\theta + p + r = d \quad (63)$$

$$(N+N^0)(\frac{1}{2}[S\theta]\Psi S\theta + Y|S\theta| - \bar{f}) \leq 0 \quad (64)$$

$$0 \leq p \leq \bar{p}, 0 \leq r \leq d, 0 \leq n^{inf} \leq n \leq n^{sup} \leq \bar{n} \quad (65)$$

$$\theta \text{ unbounded} \quad (66)$$

where:

r	artificial generator vector;
α	transformation vector;
n^{inf}, n^{sup}	lower and higher bounds in the n variables that identify the nodes in the B&B tree.

To initialize B&B, set a number of subproblems generated (k) equal to zero, and define a starting incumbent solution ($v^*=\infty$). Also determine an initial list of candidate problems (composed for the solution of the corresponding NLP). A convergence test is the next step. If there are no candidates on the list, the incumbent solution is the optimal solution. If this is not the case, a candidate from this list that is not fathomed must be selected. Solve the candidate's NLP using an interior-point method (detailed in [51]) and store optimal solution as a lower bound for its descendants ($\upsilon_{lower}=\upsilon^*_{NLP}$).

The fathoming test is the next step. A candidate currently under consideration is fathomed if the problem is (a) unfeasible, (b) $\upsilon_{lower} > \upsilon^* + \varepsilon$ (where ε is the safety factor that deals with finding local optimum), or (c) the optimal solution of the relaxed problem is a feasible solution of the original problem. If (c) is true and the optimal solution is less than incumbent (υ^*), then υ_{lower} is a new incumbent. Apply test (b) for all candidates that are not yet fathomed using this new incumbent. If a candidate under consideration has been fathomed, return to the convergence test step. Otherwise, branching is the next step in B&B algorithm. Choose an integer for separation among integer variables that still assume non-integer variables. If n_{ij} is the selected variable (with current value n^*_{ij}), two new problems are generated by including either of these constraints:

$$\text{first descendent} : p+1 \Rightarrow n_{ij} \leq \lfloor n^*_{ij} \rfloor \quad (67)$$

$$\text{second descendent} : p+2 \Rightarrow n_{ij} \geq \lfloor n^*_{ij} \rfloor + 1 \quad (68)$$

where $\lfloor n_{ij}^{*} \rfloor$ is the nearest lower integer of n_{ij}^{*}. Make $k = k + 2$ and go back to choosing the candidate step (after the convergence step).

Garver's 6-bus [53], the IEEE 24-bus [22], the South-Brazilian 46-bus [4], and the Colombian 93-bus [18] systems were used for computational tests. Results were identical to trustworthy metaheuristics when neglecting electrical losses, but involved less computational effort. B&B algorithm found the best solution using Garver's system when electrical losses were considered. The authors indicate that the algorithm is consistent, and intend to implement it to solve the DC model with security and multi-stage planning considerations. Other versions of B&B algorithm applied to transmission network expansion problem are presented in the papers of Haffner et al. [33] and Choi et al. [12].

3.3 *Deregulated Electricity Markets*

In many countries electric power industries are deregulated. Electricity generation and transmission are performed by independent organizations and transmission expansion investments must be planned according to the impact of such investments to electricity market prices and demand response. An interested reader may look at the survey of Buygi et al. [7] which classifies approaches used for solving the TEP problem in deregulated market environments.

Contreras and Wu [15] studied the decentralized coalition formation and cost allocation for TEP problem. Later the authors applied the kernel solution concept described in [17] for the resulting problem of cost allocation in [16]. Cooperative game theory was also used by Zolezzi and Rudnick [72] where the authors proposed a new method for cost allocation.

Fang and Hill [22] simulated future market driven by power flow patterns and reformulated TEP problem in order to take into account these patterns. For the reformulated problem the authors used a decision scheme for minimizing the risk of selected expansion plan.

Maghouli et al. [44] considered a multi-objective TEP problem in deregulated environment. For the objectives the authors considered investment cost, reliability and congestion cost. In order to solve the resulting multi-objective optimization problem the author proposed a genetic algorithm based on fuzzy logic with the linear membership function. The efficiency of the proposed methodology has been demonstrated using IEEE 24-bus system and on the northeastern part of the Iranian transmission grid.

Fan et al. [21] proposed a bi-level programming model for TEP problem. Upper level program considers the maximization of transmission profit for a company in the long run, whereas the lower level problem maximizes social welfare in the short term. For this model the authors impose N-1 reliability constraints and use genetic algorithm along with primal-dual interior point method for solving the problem. The computational results are shown for 18-bus and 77-bus systems.

Garces et al. [30] consider a bi-level TEP problem where both producers and consumers can trade electricity through the market. The upper level problem represents a supplier who wants to minimize investment cost and maximize the social welfare; the lower level problems are market clearing conditions. By using duality theory the authors converted the original bi-level problem to a mixed-integer linear problem and solved it with CPLEX optimization software. The authors conclude that the proposed methodology generates solutions with higher social welfare but also with a higher investment cost.

3.3.1 Social Welfare Impact

Many models for calculating the economic impact of transmission expansion are based on the social impact of the investments in competitive markets and locational marginal pricing. Sauma and Oren [56] formulated an equilibrium-based model to evaluate social-welfare implications of transmission expansion characterizing the competitive behavior of generators companies. This model helps to determine the social-welfare implications of transmission investments by solving a simultaneous Nash-Cournot game that characterizes the market equilibrium with respect to production quantities and prices. The authors present three different behaviors for transmission investment and compare the economic impact of the investments:

1. A proactive network planner (a network planner who anticipates both generation investment and spot market operation);
2. An integrated-resources planner (who co-optimizes generation and transmission expansion);
3. A reactive network planner (who assumes that the generators capacities are given and ignores a correlation between the transmission and generation investments and determines the social-welfare impact of transmission investments based only on changes they induce in the spot market equilibrium).

The authors proposed three-period model to study how generation firms' local market power affects both investments in generation capacity and the valuation of different transmission expansion projects. The model consist of three periods:

1. A network planner evaluates different transmission expansion projects;
2. Each firm invests in new generation capacity, which decreases marginal cost of electricity production;
3. Energy market operation.

It is assumed that at each period the companies can observe actions happened in the previous period and can make decisions based on the expectation of others players behavior. The model allows both the construction of new transmission lines and the upgrades of existing lines. Also the authors assume that the generation cost functions are convex and increasing with respect to the amount of electricity produced and decreasing and convex with respect to the generation capacity. The following notation will be used through this section:

Sets:

N	set of all nodes;
L	set of transmission lines;
C	set of all states of contingencies;
N_G	set of generation nodes controlled by a generation firm G;
Ψ	set of all generation firms.

Decision variables:

q_i^c	quantity generated at node i in state; c
r_i^c	adjustment quantity into/from node i by the system operator in state c;
g_i	expected generation capacity available at node i after implementing the decisions made in period 2;
f_l	expected thermal capacity limit of line l after implementing the decision made in period 1.

Parameters:

g_i^0	expected generation capacity available at node i before period 2;
f_l^0	expected thermal capacity limit of line l before period 1;
g_i^c	generation capacity available at node i in state c, given g_i;
f_l^c	thermal capacity limit of line l in state c, given f_l;
$P_i^c(\cdot)$	inverse demand function at node i in state c;
$Cp_i^c(q_i^c, g_i^c)$	production cost function at node i in state c;
$CIG_i(g_i, g_i^0)$	cost of investment in generation capacity at node i to bring expected generation capacity to g_i;
$CI_l(f_l, f_l^0)$	investment cost in line l to bring expected transmission capacity to f_l;
$\phi_{l,i}^c(L)$	power transfer distribution factor on line l with respect to a unit injection/withdrawal at node i, in state c, when the network properties (network structure and electric characteristics of all lines) are given by the set L.

In the first stage of period 3 the state of the world is being determined. In the second stage generation firm G solves the following profit-maximization problem for a given state c:

$$max_{\{q_i^c, i\in N_G\}}\ \pi_G^c = \sum_{i\in N_G} \{P_i^c(q_i^c + r_i^c)\cdot q_i^c - CP_i^c(q_i^c, g_i^c)\} \tag{69}$$

s.t.

$$q_i^c \geq 0 \tag{70}$$

$$i \in N_G \tag{71}$$

The system operator simultaneously solves the following welfare maximizing re-dispatch problem for a given state c:

$$max_{\{r_i^c, i \in N\}} \Delta W^c = \sum_{i \in N} \left(\int_0^{r_i^c} P_i^c(q_i^c + x_i)\, dx_i \right) \tag{72}$$

s.t.

$$\sum_{i \in N} r_i^c = 0 \tag{73}$$

$$-f_l^c \leq \sum_{i \in N} \phi_{l,i}^c(L) \cdot r_i^c \leq f_l^c, \; \forall l \in L \tag{74}$$

$$q_i^c + r_i^c \geq 0, \; \forall i \in N \tag{75}$$

Both of the problems (69) and (72) are concave, i.e. the KKT conditions are sufficient. The 3rd period model can be solved by applying KKT conditions for (69) for all firms and for (72). The KKT conditions for problem (69) are the following:

$$P_i^c(q_i^c + r_i^c) + P_i^{c\,\prime}(q_i^c + r_i^c) q_i^c - \frac{\partial CP_i^c(q_i^c, g_i^c)}{\partial q_i^c} + \gamma_i^c = 0, \; \forall i \in N_G, \, G \in \Psi, \, c \in C \tag{76}$$

$$\gamma_i^c \cdot q_i^c = 0, \; \forall i \in N_G, \; G \in \Psi, \; c \in C \tag{77}$$

$$q_i^c \geq 0, \; \forall i \in N_G, \; G \in \Psi, \; c \in C \tag{78}$$

$$\gamma_i^c \geq 0, \; \forall i \in N_G, \; G \in \Psi, \; c \in C \tag{79}$$

where γ_i^c is Lagrangian multiplier associated with the non-negativity constraints in (69). The KKT conditions for the problem defined in (72) are:

$$P_i^c(q_i^c + r_i^c) + \alpha^c + \sum_{l \in L} (\lambda_l^c - \lambda_{l+}^c) \cdot \phi_{l,i}^c(L) + \beta_i^c = 0, \; \forall i \in N, \; c \in C \tag{80}$$

$$\sum_{i \in N} r_i^c = 0, \; \forall c \in C \tag{81}$$

$$-f_l^c \leq \sum_{i \in N} \phi_{l,i}^c(L) \cdot r_i^c \leq f_l^c, \; \forall l \in L, \; c \in C \tag{82}$$

$$q_i^c + r_i^c \geq 0, \; \forall i \in N, \; c \in C \tag{83}$$

$$\lambda_{l-}^{c} \cdot \left(f_l^c + \sum_{i \in N} \phi_{l,i}^{c}(L) \cdot r_i^c \right) = 0, \ \forall l \in L, \ c \in C \tag{84}$$

$$\lambda_{l+}^{c} \cdot \left(f_l^c - \sum_{i \in N} \phi_{l,i}^{c}(L) \cdot r_i^c \right) = 0, \ \forall l \in L, \ c \in C \tag{85}$$

$$\beta_i^c \cdot (q_i^c + r_i^c) = 0, \ \forall i \in N, \ c \in C \tag{86}$$

$$\lambda_{l-}^{c} \geq 0, \ \forall l \in L, \ c \in C \tag{87}$$

$$\lambda_{l+}^{c} \geq 0, \ \forall l \in L, \ c \in C \tag{88}$$

$$\beta_i^c \geq 0, \forall i \in N, c \in C \tag{89}$$

where:

α^c Lagrangian multiplier associated with the adjustment-quantities balance constraint;

λ_{l-}^{c} and λ_{l+}^{c} Lagrangian multipliers associated with the transmission capacity constraints;

β_i^c Lagrangian multipliers associated with the nonnegativity constraints.

During the period two, each firm decides how much it should invest in the generation capacity to maximize the expected return of the investments. Thus, it solves the following problem:

$$max_{\{g_i, i \in N_G\}} \ E_c|\pi_G^c| - \sum_{i \in N_G} \{CIG_i(g_i, g_i^0)\} \tag{90}$$

s.t. (76)–(89)

This problem was solved for a 30-bus network using a sequential programming algorithm. During the first period, the network planner decides what lines should be built and upgraded. In order to do this, the proactive network planner solves the following optimization problem:

$$max_{l,f_l} \sum_{i \in N} \left[E_c \left(\int_0^{q_i^c + r_i^c} P_i^c(q) dq - CP_i^c(q_i^c, g_i^c) \right) - CIG_i(g_i, g_i^0) \right] - CI_l(f_l, f_l^0) \tag{91}$$

s.t.(76)–(89) and all optimality conditions of period-2 problem.

Integrated-resources planner solves the following problem during the first period:

$$max_{\{g_i\} l, f_l} \sum_{i \in N} \left[E_c \left(\int_0^{q_i^c + r_i^c} P_i^c(q) dq - CP_i^c(q_i^c, g_i^c) \right) - CIG_i(g_i, g_i^0) \right] - CI_l(f_l, f_l^0)$$

s.t.(76)–(89).

The model for the reactive network planner is the following:

$$max_{l,f_l} \sum_{i \in N} \left[E_c \left(\int_0^{q_i^c + r_i^c} P_i^c(q)dq - CP_i^c(q_i^c, g_i^c) \right) - CIG_i(g_i, g_i^0) \right] - CI_l(f_l, f_l^0) \quad (93)$$

s.t. (76)–(89) and $g_i = g_i^0, \forall_i \in N$

All these problems were used to prove the following theoretical results:

- The optimal solution for integrated-resource planner model is never smaller than the optimal expected social welfare from proactive network planner model.
- The optimal solution for integrated-resource planner model is never smaller than the optimal expected social welfare from reactive network planner model.

The integrated resource planning paradigm performs better than the proactive network planning in terms of social welfare and both of them are better than the reactive network planning, but the integrated resource planning is no longer valid in a system with privately owned generators where investments in generations are decentralized. The proactive network planning paradigm can be implemented for a system operators to estimate the investments.

3.3.2 Mixed-Integer Linear Programming Approaches for Deregulated Planning

Torre et al. [66] presented an approach that maximizes social welfare using mixed-integer LP approach. Also several metrics were created to track an individual benefit to different parties involved in transmission planning investment. Another contribution of the paper is accounting for different scenarios, with the main difference between scenarios being a demand level. The mixed-integer LP formulation includes the maximization of scenario-weighted social welfare:

$$\sigma \left[\sum_{\forall c \in \Omega_C} w^c \left(\sum_{\forall d \in \Omega_D} \sum_{\forall h \in \Omega_d} \lambda_{D_{dh}}^c p_{D_{dh}}^c - \sum_{\forall i \in \Omega_G} \sum_{\forall i \in \Omega_i} \lambda_{G_{ib}}^c p_{G_{ib}}^c \right) \right] - \sum_{\forall (s,r,k) \in \Omega_{L+}} K_{srk} w_{srk} \quad (94)$$

where:

σ weighting factor to make investment and operational costs comparable;
w_c weight of scenario c;
$\lambda_{D_{dh}}^c$ price bid by the h-th block of the d-th demand in scenario c;
$\lambda_{G_{ib}}^c$ price offer by the b-th block of the i-th generating unit in scenario c;
$p_{D_{dh}}^c$ power consumed by the h-th block of the d-th demand in scenario c;

$p^c_{G_{ib}}$ power produced by the b-th block of the i-th generating unit in scenario c;
k_{srk} investment cost of constructing line k in corridor (s, r);
w_{srk} binary variable: $w_{srk} = 1$ if line k from corridor (s, r) is built in the study period; $w_{srk} = 0$ if not;
Ω_c set of all scenarios of the period of study;
Ω_d set of indices of the blocks of the d-th demand;
Ω_D set of indices of the demand;
Ω_G set of indices of the generating units;
Ω_I set of indices of the blocks of the i-th generating unit;
Ω_{L+} set of all prospective transmission lines.

The objective function is expressed as aggregated demand utility bid function minus aggregated generator offer function, minus investment costs in new lines. w_c is used to correctly consider each scenario's relevance. Scenarios are assigned weights based on how often they occur. A scenario which occurs everyday for 3 h would be given a weight of $1/8$ (3 h divided by 24 h); as previously stated, the demand level is the main difference between scenarios. The problem is subject to power flow balance, line limits, operating, and other constraints detailed in [66].

The authors defined four metrics to show how different parties involved are affected by investments in transmission expansion. The metric representing a change in total welfare resulting from the addition of new lines with respect to the investment cost is defined as:

$$\mu_1 = \frac{SW^* - SW^0}{\sum_{\forall(s,r,k)\in\Omega_{L+}} K_{srk} w_{srk}} \tag{95}$$

where SW^* is the optimal aggregate social welfare and SW^0 is the aggregate social welfare without any expansion. μ_1 greater than one justifies an investment, but may not please all parties. Individual metrics were formulated for the generator, demand, and merchandizing surplus. The metric representing a change in the generator surplus with respect to the investment cost is defined as:

$$\mu_2 = \frac{GS^* - GS^0}{\sum_{\forall(s,r,k)\in\Omega_{L+}} K_{srk} w_{srk}} \tag{96}$$

where GS^* is the total generator surplus (total revenue minus total cost of generators) and GS^0 is the total generator surplus without any expansion. μ_2 should be greater than the portion of cost to be paid by generators ($\mu_2 = 5$ is acceptable as long as the generators are paying for less than 50% of the expansion). Similar to the generators, the metric representing the change in the demand surplus with respect to investment cost is defined as:

$$\mu_3 = \frac{CS^* - CS^0}{\sum_{\forall(s,r,k)\in\Omega_{L+}} K_{srk} w_{srk}} \tag{97}$$

where CS^* is the total demand surplus (aggregate demand utility function minus total payment of the demands) and CS^0 is the total demand surplus without any expansion. The final metric represents the change in merchandizing surplus with respect to the investment cost and is defined as:

$$\mu_4 = \frac{MS^* - MS^0}{\sum_{\forall(s,r,k)\in\Omega_{L+}} K_{srk} w_{srk}}, \tag{98}$$

where MS^* is the total merchandizing surplus and MS^0 is the total merchandizing surplus without any expansion.

The authors applied this methodology to the Garver's 6-bus system [53] and the system based on the IEEE RTS 24-bus system. Different scenarios and weighting factors result in an appropriate economic impact for generators, demands, and network planners. In other words, this methodology properly shows the impact network expansion investment has on each of the parties involved.

Alguacil et al. [1] proposed a linearization method for TEP problem with electrical losses and nonlinear constraints. In this paper the authors approximated the nonlinear loss function by a piece-wise linear one and linearized the AC power flow nonlinear constraints. The computational results showed accuracy and efficiency of the proposed model.

Zhao et al. [71] formulated TEP problem as a linear mixed-integer program and addressed market uncertainties by generating a set of possible scenarios. For every scenario an optimal expansion plan is computed using differential evolution algorithm. The authors used expected energy not supplied and stability measures to estimate reliability of a plan. After that the expansion plan is evaluated against others scenarios and an adaptation cost is being computed, which can be considered as a flexibility cost. The plan which has minimum flexibility cost is considered to be the optimal one. The proposed model has been tested on IEEE 14-bus system and computational results are reported in the paper.

3.3.3 Probabilistic Locational Marginal Prices

Buygi et al. [8] presented a new approach for the transmission expansion planning in deregulated environments that utilizes a tool for computing the probabilistic density functions (pdfs) of nodal prices to show performance of an electric market. As mentioned earlier, in the deregulated environment, there are other concerns beyond the usual cost and loss of load minimization. Fair competition and availability of electricity to consumers are considerations addressed in this market-based approach.

Locational Marginal Price (LMP) is a cost of supplying the next MW of load at a specific location, considering generation marginal cost, cost of transmission congestion, and losses. Formulation of these LMPs is shown in the appendix of [8], and consists of the Lagrangian multipliers of the DC power flow constraints.

Monte Carlo simulation is used to compute the pdfs of LMPs. The first task is to compute pdfs for unavailability of each transmission facility (line, generator, load, etc.), as well as for all random inputs. Next, random numbers are generated from each pdf in the previous step to obtain a large sample of network configurations. The optimal load flow is run for each configuration, and a pdf is fit to the samples of each output, including LMPs.

Another consideration in the deregulated environment is competitiveness of the electric market. Perfect competition within this market occurs when there is (1) nothing prevents consumers from buying electricity from their preferred producer and (2) all producers offer their electricity at the same price. The first step is to reduce the congestion. The congestion cost is defined as the opportunity cost of transmitting power through it:

$$cc_i = (lmp_{i_2} - lmp_{i_2})P_{i_1 i_2}, \tag{99}$$

where:

cc_i	congestion cost of line i in \$/h;
lmp_{i_2}, lmp_{i_2}	LMPs of end buses of line i in \$/MWh;
$P_{i_1 i_2}$	power of line i from bus i_1 to bus i_2 in MW.

The total congestion cost of the network is then just a sum of congestion costs of all the lines in the network. LMP differences among busses increase as congestion increases. This makes congestion cost an appropriate measure of a price discrimination and customer constraints, as well as a criterion for measuring a competition within the electric market [9].

The flatness of the price profile is used by the authors to show how equal competitor's prices are. A flatter profile means LMPs have smaller differences. The flatness of the price profile is measured by using the standard deviation of the means of LMPs. The deviation is weighted by a sum of generation power and load at each bus. This results in transmission planning's attempt to equalize LMPs at all load and generator buses.

Two additional criteria (based on congestion and flatness of price) are presented to show how the investment cost can be justified in the deregulated and competitive environment. The first of these is a decrease in the annual congestion cost divided by the annual investment cost. This shows how much congestion is alleviated and therefore competition encouraged per unit of the investment cost. The second criterion is the decrease in LMP weighted standard deviation divided by annual investment cost. This shows how much price profile is flattened and therefore the competition encouraged per unit of the investment.

This approach to the transmission planning involves all the components outlined so far in this section (probabilistic locational marginal prices, congestion cost, flatness of price profile, and investment cost criteria). The model also considers a parameter referred to as specified value (SV). A new line is suggested as an expansion candidate if LMP difference between two buses is greater than a chosen SV. The procedure for an expansion begins with determining a strategy for modeling nonrandom uncertainties and assigning a degree of importance to each one. The next step is to compute pdfs of LMPs for an existing network in each scenario using the steps outlined earlier in this section, then select a SV. The next step is to determine a set of transmission planning candidates. Add the candidates with highest possible capacity to the network and compute pdfs of LMPs for all scenarios. The plan with minimax regret is chosen using a market-based criteria computed for each scenario. Decreasing the SV will result in additional candidates if the price profile flatness is not improved by current candidates. This approach was effective when applied to IEEE 30-bus test system. The values of the congestion cost and the standard deviation of LMP in this system indicate the need for an expansion (additional lines). The planning approach was repeated under different criterion, first under the assumption that there are no nonrandom uncertainties, and then under the assumption that there are nonrandom uncertainties. The average congestion cost proved to be the most effective criterion under the first assumption (it caused the largest decrease in LMP standard deviation). This standard deviation of LMP weighted by the sum of generation power and load, as well as weighted by only generation power, both proved to be acceptable criteria, too. With nonrandom uncertainty, the average congestion cost is most effective in single scenarios, while the standard deviation of LMP weighted by the sum of generation power and load provided a flatter price profile than congestion cost for multi-scenario cases.

3.3.4 Financial Transmission Rights

Long-Term Financial Transmission Rights (LTFTRs) are a way to protect the market participants from nodal price fluctuations. Kristiansen and Rosellón [40] used auctioning of these rights to encourage investments in the long-term expansion of transmission networks. The authors present a set of criteria for auctioning incremental LTFTRs when an expansion is relatively small compared to the market size. The first criterion is a feasibility rule that states that LTFTR increment must be feasible and result in a feasible network. The next criterion is that the increment remains simultaneously feasible under the condition that currently unallocated rights (or proxy awards) must remain unallocated. The third criterion states that the investors should seek to maximize their objective function. The final criterion states that the awarding process should apply for both increases and decreases in grid capacity.

Pay-off from FTRs is determined using the following equation:

$$FTR = (P_j - P_i)Q_{ij}, \tag{100}$$

where:

P_j price at location j;
P_i price at location i;
Q_{ij} directed quantity injected at point i and withdrawn at point j.

Let T be current allocation of LTFTRs. Feasibility can be assumed in this scenario, therefore $K(T,u) \leq 0$, where $K(T, u)$ is a vector of power flows in the lines and u is a control variable that takes into account all others parameters. Due to the first criterion, any incrementally awarded FTR must result in a feasible grid. Due to the second criterion, this expanded grid must also remain feasible when unallocated rights are preserved.

The authors introduced an auction model to represent the third criterion. This model first uses a proxy rule that maximizes a value of the proxy awards, and then maximizes an investor's preference. The proxy awards are maximized in order to ensure revenue adequacy of the transmission system as a whole. Detailed formulation of this auction model can be found in [40].

The authors conclude that LTFTRs are efficient under non-lumpy expansions of transmission network. The authors comment that the regulation plays an important role in expansions that are large when compared to the market size.

Rosellon [55] provides an extensive study of the long-term financial transmission rights, incentive-regulation, and market-power hypotheses. Kristiansen et al. [41] provide a European case study for the merchant transmission expansion considering Germany, France, and Belgium. In this paper the authors model an allocation of FTR in Europe and provide a welfare analysis. Hogan et al. [35] discuss a regulatory mechanism for the transmission network expansion planning which combines both the merchant approach based on long term financial transmission rights, and the regulatory approach.

3.4 Stochastic Programming

Stochastic programming is a way to deal with randomness or imperfect information that is presented in a problem. This uncertainty is accounted for using probability distributions. Risk and uncertainty are inherent parts of the transmission expansion planning, and stochastic programming are used by López et al. [43] to address these uncertainties. The authors present four different models to show the effect of ignoring risk and uncertainty. The models presented are: deterministic, stochastic, stochastic with risk, and stochastic with risk and probabilistic constraints. The authors combine the transmission expansion planning with the generation expansion planning. The objective of the resulting MINLP becomes whether to invest in new generation, new transmission, or some combination of the two to minimize cost and prevent loss of load. The following notation will be used in this section:

Et	index for existing generating technologies;
h,p, i ,j	indices for nodes in the system;
i–j	index for node pairs in the system;
k	ndex for foreseeable scenarios;
m	index for operating nodes;
nt	index for new generating technologies;

K set of all foreseeable scenarios;
c_{i-j} annualized cost per circuit added in the right-of-way between i and j; in \$/year;
cf_m capacity factor determined by m; dimensionless;
pr_k probability of occurrence of k; dimensionless;
q_{et} annualized variable generating cost of et; in \$/MW-year;
q_{nt} annualized variable generating cost of nt; in \$/MW-year;
r_{nt} annualized fixed cost of nt; in \$/MW-year;
x_{nt} typical capacity size of nt; in MW;
θ_r risk factor; dimensionless;
$g_{m,et,h}$ capacity of et effectively used in each m at h; in MW;
$g_{m,nt,p}$ capacity of nt effectively used in each m at p; in MW;
$g_{k,m,nt,p}$ capacity of et effectively used in each m at h for all k; in MW;
$g_{k,m,nt,p}$ capacity of nt effectively used in each m at p for all k; in MW;
n_{i-j} integer variable; number of circuits added in the right-of-way between i and j; dimensionless;
$n_{nt,p}$ integer variable; number of new generators of nt at p; dimensionless.

The deterministic objective function to be minimized is:

$$\sum_{i-j} c_{i-j} n_{i-j} + \sum_{nt,p} r_{nt} x_{nt} n_{nt,p} + \sum_{m} cf_m \left(\sum_{et,h} q_{et} g_{m,et,h} + \sum_{nt,p} q_{nt} g_{m,nt,p} \right), \tag{101}$$

subject to budget, capacity, nodal balance, power flow, transmission line capacity, stability, and upper and lower limit constraints. Formulation of each of the constraints can be found in [43].

The authors formulated a two-stage stochastic program, with the first stage decisions being an investment in a new transmission and a new generation, and the second stage decision being an annual estimate of the generation in order to minimize the annualized generation cost of the existing and potential new generating plants based on the expansion made.

Let z be defined as

$$z = \sum_{m} cf_m \left(\sum_{et,h} q_{et} g_{m,et,h} + \sum_{nt,p} q_{nt} g_{m,nt,p} \right). \tag{102}$$

The expected value of z can be expressed as

$$E\{z\} = \sum_{k} pr_k z_k = \sum_{k,m} pr_k cf_m \left(\sum_{et,h} q_{et} g_{k,m,et,h} + \sum_{nt,p} q_{nt} g_{k,m,nt,p} \right). \tag{103}$$

The new objective function to be minimized becomes

$$\sum_{i-j} c_{i-j} n_{i-j} + \sum_{nt,p} r_{nt} x_{nt} n_{nt,p}$$
$$+ \sum_{k,m} pr_k cf_m \left(\sum_{et,h} q_{et} g_{k,m,et,h} + \sum_{nt,p} q_{nt} g_{k,m,nt,p} \right), \qquad (104)$$

subject to the constraints stated for the deterministic model, but modified to take into account all possible scenarios of demand $k \in K$.

The next model presented is a stochastic model with the risk parameter, which comes from mean-variance Markowitz theory, and results in a minimization of the addition of the first-stage decision, expected value of the second-stage decision, and the standard deviation of the second-stage decision:

$$f(x) + E\{f(y)\} + \theta_r \sqrt{var\{f(y)\}}, \qquad (105)$$

where θ_r is set by a decision maker and determines the importance of standard deviation.

The variance of z is

$$\sigma_z^2 = E\{\, z^2\} \; - E^2\{\, z\} \qquad (106)$$
$$= \sum_k pr_k (\sum_m cf_m (\sum_{et,h} q_{et} g_{k,m,et,h} \qquad (107)$$
$$+ \sum_{nt,p} q_{nt} g_{k,m,nt,p}))^2 \qquad (108)$$
$$- (\sum_{k,m} pr_k cf_m (\sum_{et,h} q_{et} g_{k,m,et,h} + \sum_{nt,p} q_{nt} g_{k,m,nt,p}))^2 \qquad (109)$$

The objective function is therefore the minimization of

$$\sum_{i-j} c_{i-j} n_{i-j} + \sum_{nt,p} r_{nt} x_{nt} n_{nt,p} \qquad (110)$$
$$+ \sum_{k,m} pr_k cf_m (\sum_{et,h} q_{et} g_{k,m,et,h} + \sum_{nt,p} q_{nt} g_{k,m,nt,p}) \qquad (111)$$
$$+ \theta_r \{\sum_k pr_k (\sum_m cf_m (\sum_{et,h} q_{et} g_{k,m,et,h} + \sum_{nt,p} q_{nt} g_{k,m,nt,p}))^2 \qquad (112)$$
$$- (\sum_{k,m} pr_k cf_m (\sum_{et,h} q_{et} g_{k,m,et,h} + \sum_{nt,p} q_{nt} g_{k,m,nt,p}))^2\} \, \frac{1}{2} \qquad (113)$$

subject to the same constraints as the stochastic model without risk. The final model presented by the authors is the same stochastic model with risk, but also includes probabilistic constraints. The derated capacity constraints are modified to represent a random availability of the generating plants. A random availability factor accounts for this in the probabilistic constraint. The transmission line capacity constraints are modified in the similar fashion.

The expected value of perfect information (EVPI) is defined as a difference between stochastic solution and wait and see solution (WSS), with WSS being an average of deterministic solutions for each expected level of demand. EVPI is used by the authors to show an importance of uncertainty considerations in the models. The results presented show that the stochastic solution differed noticeably from the WSS in the 21-bus example network used.

Blanco et al. [5] study a stochastic TEP problem with uncertain fuel prices and demand growth. In this paper the authors also consider uncertainty of wind presence as well as peak and base load scenarios. Further, they provide a financial evaluation of the transmission investment decisions as well as study the utilization of Flexible Alternating Current Transmission System devices.

3.5 *Reliability Considerations*

Most of the traditional transmission network expansion models minimize costs and losses of load. Although unlikely, failures or intentional attacks on a system could cause widespread damage and blackouts. Expansion is costly, thus vulnerability is increased due to operation being common at or near the limit of what current systems can handle.

Choi et al. [11] proposed a method for the TEP problem considering probabilistic reliability criteria. In this paper the authors formulate the TEP problem by minimizing the investments subject to Loss of Load Expectation for a transmission system and Loss of Load Expectation for a bus reliability criteria. The probabilistic branch and bound algorithm was used for solving the resulting problem. The proposed methodology was tested on the 21-bus system, which is a part of Korean power grid and extensive computational results are presented in the paper.

Yu et al. [70] proposed a chance constrained TEP problem considering uncertainties associated with loads and wind turbines power generation. The authors formulated the TEP problem by bounding probabilities of the transmission lines being overloaded. In order to solve this complicated problem the authors employed a two-step genetic algorithm.

Furusawa et al. [27] studied TEP problem with Expected Congestion Cost (ECC) and Expected Outage Cost (EOC) as economic and reliability indices. ECC was calculated as an increase of fuel cost for electricity generation with consideration of N–1 criterion and unexpected generator failure. EOC was calculated as expected energy unserved times unit outage cost. For the case study the authors used a 8-bus system with 11 branches.

Ron et al. [52] considered the TEP problem in competitive electricity market environment with uncertainties caused by failure of generators and transmission lines, as well, as errors in a long-term demand forecasting. In order to model market participants interactions the authors used Bender's decomposition and Lagrangian relaxation methods. Scenario reduction method has beed proposed for reducing the complexity of the problem.

A study of Kazerooni and Mutale [39] describes the TEP problem with consideration of CO_2 emission trading scheme. Monte Carlo simulation was used for simulating CO_2 emission price volatility. The authors also considered in the model the N–1 reliability criterion and a linear model for electrical losses. The CO_2 emission was modeled by two different schemes: in the first one, generators can sell any CO_2 emission surplus and buy a shortage allowance from the market. The second approach disregards any benefits of having a surplus of CO_2 emission allowances and allowing only buying the allowances from the market. The problem was formulated as a mixed-integer optimization problem and solved by Xpress optimization software for the IEEE 24-bus system.

3.5.1 TEP Under Deliberate Outages

Alguacil et al. [2] presented an approach that accounts for deliberate outages, and the trade-off between cost minimization and accounting for such vulnerability.

A set of scenarios Ω represents the randomness of attack plans. Each scenario w represents a plausible attack plan with a level of damage to the transmission lines, with damage being measured in terms of total load shed. Ω is made up of vectors $v(w)$ of 0s and 1s as follows:

$$v(\omega) = \{v_1(\omega), \ldots, v_{n_L}(\omega)\},\ \omega = 0, \ldots, n_\Omega, \tag{114}$$

where:

n_L number of lines in the original transmission network;
n_Ω number of attack plans considered as scenarios;
$v_l(\omega)$ constant equal to 0 if line ℓ is destroyed in scenario ω (1 otherwise).

Scenario generation approach is based on the terrorist threat problem adopted from [46]. A disruptive agent with limited resources acts first in an attempt to cause maximum damage, then a system operator reacts to the disruptive action in an attempt to minimize the resulting damage. Each selected attack scenario is given a weight representing its relative likelihood of occurring. The weights are based on load shed and the number of destroyed lines. The weight of scenario ω is formulated as:

$$\pi(\omega) = \frac{\frac{\Delta P_n^D(\omega)}{I(\omega)}}{\sum_{\omega\prime=1}^{n_\Omega} \frac{\Delta P_T^D(\omega\prime)}{I(\omega\prime)}},\ \omega = 1, \ldots, n_\Omega, \tag{115}$$

where $\Delta P_T^D(\omega)$ is the system load shed in the original network associated with scenario ω and $I(\omega)$ is the number of destroyed lines in scenario ω. The authors use scenarios generated in the following formulation:

$$min \sum_{\omega=1}^{n_\Omega} [\sum_{n \in N} \Delta P_T^D(\omega)] + \beta \sum_{\ell \in L^C} C_\ell^L S_\ell \tag{116}$$

s.t.

$$\sum_{\ell \in L^C} C_\ell^L S_\ell \le C_T^L \tag{117}$$

$$\sum_{g \in G_n} P_g^G(\omega) - \sum_{\ell | O(\ell)=n} P_\ell^L(\omega) + \sum_{\ell | R(\ell)=n} P_\ell^L(\omega) = P_n^D - \Delta P_n^D(\omega), \quad \omega = 0, \ldots, n_\Omega, \ \forall n \in N \tag{118}$$

$$P_\ell^L(\omega) = \frac{1}{X_\ell}[\delta_{O(\ell)}(\omega) - \delta_{R(\ell)}(\omega)] v_\ell(\omega), \ \omega = 0, \ldots, n_\Omega, \ \forall \ell \in L^O \tag{119}$$

$$P_\ell^L(\omega) = \frac{1}{x_\ell}[\delta_{O(\ell)}(\omega) - \delta_{R(\ell)}(\omega)] S_\ell, \ \omega = 0, \ldots, n_\Omega, \ \forall \ell \in L^C \tag{120}$$

$$-\bar{P}_\ell^L \le P_\ell^L(\omega) \le \bar{P}_\ell^L, \ \omega = 0, \ldots, n_\Omega, \ \forall \ell \in \{ L^O \cup L^C \} \tag{121}$$

$$0 \le P_g^G(\omega) \le \bar{P}_g^G; \ \omega = 0, \ldots, n_\Omega, \ \forall g \in G \tag{122}$$

$$\underline{\delta} \le \delta_n(\omega) \le \bar{\delta}, \ \omega = 0, \ldots, n_\Omega, \ldots \forall n \in N \tag{123}$$

$$\Delta P_n^D(\omega) = 0, \ \omega = 0, \ldots, n_\Omega, \ \forall n \in N \tag{124}$$

$$0 \le \Delta P_n^D(\omega) \le P_n^D, \ \omega = 0, \ldots, n_\Omega, \ \forall n \in N \tag{125}$$

$$S_\ell \in \{0, 1\}, \ \forall \ell \in L^C \tag{126}$$

where:

N	set of node indices;
$\Delta P_n^D(\omega)$	load shed in node n and scenario ω;
β	weighting factor for the investment cost;
L^c	set of indices of candidate lines;
C_ℓ^L	investment cost of candidate line *ell*;
S_ℓ	binary variable that is equal to 1 if candidate line ℓ is built (0 otherwise);

C_T^L	expansion planning budget;
G_n	set of indices of generators connected to node n;
$P_g^G(\omega)$	power output of generator g in scenario ω;;
$P_\ell^L(\omega)$	power flow in line ℓ and scenarioω;;
$O(\ell), R(\ell)$	sending and receiving nodes of line ℓ, respectively;
P_n^D	demand in node n;
$P_\ell^L(\omega)$	power flow in line ℓ and scenario ω;;
x_ℓ	reactance of line ℓ;
$\delta_n(\omega)$	phase angle in node n and scenario, ω;;
L^0	set of indices of lines in the original transmission network;
$\bar{P}_\ell^L$	power flow capacity of line ℓ;
$\bar{P}_g^G$	capacity of generator g;
G	set of generator indices;
$\underline{\delta}, \bar{\delta}$	lower and upper bounds for the nodal phase angles, respectively.

The first term of the objective function represents vulnerability of the system, while the second term represents the investment cost. β is a weighting parameter chosen by the network planner that determines importance of the investment cost relative to vulnerability. The problem above is a mixed-integer nonlinear programming problem. The multiplication of s_ℓ and $\delta_n(\omega)$ are subsequently transformed into linear expressions using methods found in [26] for the product of binary and continuous variables.

Garver's 6-bus system [53] and a system based on IEEE 24-bus system were used to test the proposed method. Different values of β were chosen to show options a decision-maker has, along with a range of expansion budgets. Results showed a decrease in vulnerability as investment cost increases.

An alternative method of scenario generation can be found in [10] and another interesting discussion about security and vulnerability related criteria for the transmission network expansion planning can be found in [13].

4 Conclusion

In this chapter we provided a short review of methods for the transmission expansion planning problem appeared in recent literature. Due to space limitations not all of the existing methods have been covered in this survey. The purpose of this chapter was to provide an introduction to the current common methods used for the TEP and to give brief examples for most of the approaches.

The deregulation of electricity industries in many countries brings completely new problem setup and many recent publications address this problem from different perspectives. Also there are many recent publications about reliability issues and robust transmission expansion planning with respect to network component failures and uncertainty in future demand. Unfortunately, the great majority of publications describe only one-time investment models (static problems) and do not consider when the circuits should be constructed in the case of additional expansion in future years.

The transmission expansion planning is a complex problem for the real world networks and as we showed in this chapter there are many models for TEP problem viewing it in different settings. We hope that this chapter will provide a good introduction to people who did not worked on TEP problem before as well as describe alternative models and algorithms for the researchers working in this area.

References

1. Alguacil N, Motto AL, Conejo AJ (2003) Transmission expansion planning: a mixed-integer LP approach. IEEE Trans Power Syst 18:1070–1077
2. Alguacil N, Carrion M, Arroyo JM (2009) Transmission network expansion planning under deliberate outages. Int J Electr Power Energy Syst 31:553–561
3. Bahiense L, Oliveira GC, Pereira M, Granville S (2001) A mixed integer disjunctive model for transmission network expansion. IEEE Trans Power Syst 16:560–565
4. Binato S, Oliveira GC, Araujo JL (2001) A greedy randomized adaptive search procedure for transmission expansion planning. IEEE Trans Power Syst 16:247–253
5. Blanco G, Waniek D, Olsina F, Garces F, Rehtanz C (2011) Flexible investment decisions in the European interconnected transmission system. Electr Power Syst Res 81:984–994
6. Bustamante-Cedeno E, Arora S (2009) Multi-step simultaneous changes constructive heuristic algorithm for transmission network expansion planning. Electr Power Syst Res 79:586–594
7. Buygi MO, Shanechi HM, Balzer G, Shahidehpour M (2003) Transmission planning approaches in restructured power systems, *Power Tech Conference Proceedings, 2003 IEEE Bologna*, vol. 2, pp. 7, 23–26, doi: 10.1109/PTC.2003.1304666 URL: http://ieeexplore.ieee.org/stamp/stamp.jsp?tp=&arnumber=1304666&isnumber=28975
8. Buygi MO, Balzer G, Shanechi HM, Shahidehpour M (2004) Market-based transmission expansion planning. IEEE Trans Power Syst 19:2060–2067
9. Buygi MO, Balzer G, Shanechi HM, Shahidehpour M (2004) Market based transmission expansion planning: fuzzy risk assessment. *Electric Utility Deregulation, Restructuring and Power Technologies, 2004. (DRPT 2004). Proceedings of the 2004 IEEE International Conference on*, vol 2, pp. 5–8, 427–432, doi: 10.1109/DRPT.2004.1337997 URL: http://ieeexplore.ieee.org/stamp/stamp.jsp?tp=&arnumber=1337997&isnumber=29506
10. Carrion M, Arroyo JM, Alguacil N (2007) Vulnerability-constrained transmission expansion planning: a stochastic programming approach. IEEE Trans Power Syst 22:1436–1445
11. Choi J, Tran T, El-Keib AA, Thomas R, Oh H, Billinton R (2005) Method for transmission system expansion planning considering probabilistic reliability criteria. IEEE Trans Power Syst 3:1606–1615
12. Choi J, El-Keib AA, Tran T (2005) A fuzzy branch and bound-based transmission system expansion planning for the highest satisfaction level of the decision maker. IEEE Trans Power Syst 20:476–484
13. Jaeseok Choi, Mount T, Thomas R (2006) Transmission System Expansion Plans in View Point of Deterministic, Probabilistic and Security Reliability Criteria. *System Sciences, 2006. HICSS '06. Proceedings of the 39th Annual Hawaii International Conference on*, vol. 10, pp. 4–7, 247b, doi: 10.1109/HICSS.2006.510 URL: http://ieeexplore.ieee.org/stamp/stamp.jsp?tp=&arnumber=1579806&isnumber=33370

14. Chu PC, Beasley JE (1997) A genetic algorithm for the generalized assignment problem. Comput Oper Res 24:17–23
15. Contreras J, Wu FF (1999) Coalition formation in transmission expansion planning. IEEE Trans Power Syst 14:1144–1152
16. Contreras J, Wu FF (2000) A kernel-oriented algorithm for transmission expansion planning. IEEE Trans Power Syst 4:1434–1440
17. Davis R, Maschler D (1965) The kernel of a cooperative game. Nav Res Logist Quart 12: 223–259
18. Escobar AH, Gallego RA, Romero R (2004) Multistage and coordinated planning of the expansion of transmission systems. IEEE Trans Power Syst 19:735–744
19. Escobar AH, Romero RA, Gallego RA (2008) Transmission network expansion planning considering multiple generation scenarios. In: IEEE/PES transmission and distribution conference and exposition: Latin America, pp 1–6, 2008
20. Escobar AH, Romero RA, Gallego RA (2008) Transmission network expansion planning considering uncertainty in generation and demand. In: IEEE/PES transmission and distribution conference and exposition, Latin America, pp 1–6 2008
21. Fan H, Cheng H, Yao L (2009) A bi-level programming model for multistage transmission network expansion planning in competitive electricity market. In: Power and energy engineering conference, APPEEC 2009, Asia-Pacific, pp 1–6, 2009
22. Fang R, Hill DJ (2003) A new strategy for transmission expansion in competitive electricity markets. IEEE Trans Power Syst 18:374–380
23. Feo T, Resende M (1995) Greedy randomized adaptive search procedures. J Global Optimiz 6: 108–133
24. Festa P, Resende M (2009) An annotated bibliography of GRASP–Part II: applications. Int Trans Oper Res 16:131–172
25. Faria H Jr, Binato S, Resende MGC, Falcao DM (2005) Power transmission network design by greedy randomized adaptive path relinking. IEEE Trans Power Syst 20:43–49
26. Floudas CA (1995) Nonlinear and mixed-integer optimization: fundamentals and applications. Oxford University Press, New York, p 480
27. Furusawa K, Okada K, Asano H (2009) A method of evaluating transmission network expansion plan considering security constraints and supply reliability index, *Power Systems Conference and Exposition, 2009. PSCE '09. IEEE/PES*, vol., pp.1–6, 15–18, doi: 10.1109/PSCE.2009.4839925 URL: http://ieeexplore.ieee.org/stamp/stamp.jsp?tp=&arnumber=4839925&isnumber=4839920
28. Gallego RA, Montecelli A, Romero R (1998) Transmision system expansion planning by an extended genetic algorithm. IEE Proc Gener Transm Distrib 145:329–335
29. Gallego RA, Romero R, Monticelli AJ (2000) Tabu search algorithm for network synthesis. IEEE Trans Power Syst 15:490–495
30. Garces LP, Conejo AJ, Garcia-Bertrand R, Romero R (2009) A Bilevel approach to transmission expansion planning within a market environment. IEEE Trans Power Syst 24:1513–1522
31. Garver LL (1970) Transmission Network Estimation Using Linear Programming, *Power Apparatus and Systems, IEEE Transactions on*, vol. PAS-89, pp. 1688–1697, doi: 10.1109/TPAS.1970.292825 URL: http://ieeexplore.ieee.org/stamp/stamp.jsp?tp=&arnumber=4074249&isnumber=4074207
32. Georgilakis PS (2010) Market-based transmission expansion planning by improved differential evolution. Int J Electr Power Energy Syst 32:450–456
33. Haffner S, Monticelli A, Garcia A, Romero R (2001) Specialised branch-and-bound algorithm for transmission network expansion planning. IEE Proc Gener Transm Distrib 148:482–488
34. Hashimoto SHM, Romero R, Mantovani JRS (2003) Efficient linear programming algorithm for the transmission network expansion planning problem. IEE Proc Gener Transm Distrib 150: 536–542
35. Hogan W, Rosellon J, Vogelsang I (2010) Toward a combined merchant-regulatory mechanism for electricity transmission expansion. J Regul Econ 38:113–143

36. Holland JH (1975) Adaptation in natural and artificial systems: an introductory analysis with applications to biology, control, and artificial intelligence. The University of Michigan Press, Ann Arbor, p 228
37. Jalizadeh S, Kazemi A, Shayeghi H, Madavi M (2009) Technical and economic evaluation of voltage level in transmission expansion planning using GA. Energy Convers Manage 49: 1119–1125
38. Jalizadeh S, Shayeghi H, Madavi M, Hadadian H (2009) A GA based transmission expansion planning considering voltage level, network losses, and number of bundle lines. Am J Appl Sci 6: 970–977
39. Kazerooni AK, Mutale J (2010) Transmission network planning under security and environmental constraints. IEEE Trans Power Syst 25:1169–1178
40. Kristiansen T, Rosellon J (2006) A merchant mechanism for electricity transmission expansion. J Regul Econ 29:167–193
41. Kristiansen T, Rosellon J (2010) Merchant electricity transmission expansion: a European case study. Energy 35:4107–4115
42. Latorre G, Dario R, Mauricio J, Villegas A (2003) Classification of publications and models on transmission expansion planning. IEEE Trans Power Syst 18:938–946
43. López JA, Ponnambalam K, Quintana VH (2007) Generation and transmission expansion under risk using stochastic programming. IEEE Trans Power Syst 22:1369–1378
44. Maghouli P, Hosseini SH, Buygi MO, Shahidehpour M (2009) A multi-objective framework for transmission expansion planning in deregulated environments. IEEE Trans Power Syst 24: 1051–1061
45. Mahdavi M, Shayeghi H, Kazemi A (2009) DCGA based evaluating role of bundle lines in TEP considering expansion of substations from voltage level point of view. Energy Convers Manage 50:2067–2073
46. Motto AL, Arroyo JM, Galiana FD (2005) A mixed-integer LP procedure for the analysis of electric grid security under disruptive threat. IEEE Trans Power Syst 20:1357–1365
47. de Oliveira EJ, Da Silva IC Jr, Pereira JLR, Carneiro S Jr (2000) Transmission system expansion planning using a Sigmoid function to handle integer investment variables. IEEE Trans Power Syst 20:1616–1621
48. Paulun T (2006) Strategic expansion planning for electrical networks considering uncertainties. Eur Trans Electr Power 16:661–671
49. Pereira M, Granville S (1985) Analysis of the linearized power flow model in Benders decomposition. SOL Lab, Dept. of Oper. Research, Stanford University, Technical Report SOL 85-04
50. Power Systems Test Case Archive. http://www.ee.washington.edu/research/pstca/April 8, 2011
51. Rider MJ, Garcia AV, Romero R (2008) Transmission system expansion planning by a branch-and-bound algorithm. IET Gener Transm Distrib 2:90–99
52. Ron JH, Shahidehpour M, Wu L (2009) Market-based generation and transmission planning with uncertainties. IEEE Trans Power Syst 24:1587–1598
53. Romero R, Monticelli A (1994) A hierarchical decomposition approach for transmission network expansion planning. IEEE Trans Power Syst 9:373–380
54. Romero R, Gallego RA, Monticelli A (1996) Transmission system expansion planning by simulated annealing. IEEE Trans Power Syst 11:364–369
55. Rosellon J (2003) Different approaches towards electricity transmission expansion. Rev Netw Econ 2:238–269
56. Sauma E, Oren S (2006) Proactive planning and valuation of transmission investments in restructured electricity markets. J Regul Econ 30:261–290
57. da Silva EL, Gil HA, Areiza JM (2000) Transmission network expansion planning under an improved genetic algorithm. IEEE Trans Power Syst 15:1168–1174
58. da Silva EL, Ortiz JMA, de Oliveira GC, Binato S (2001) Transmission network expansion planning under a tabu search approach. IEEE Trans Power Syst 16:62–68

59. de Silva IJ, Rider MJ, Romero R, Garcia AV, Murari CA (2005) Transmission network expansion planning with security constraints. IEE Proc Gener Transm Distrib 152:828–836
60. de Silva IJ, Rider MJ, Romero R, Murari CAF (2006) Transmission network expansion planning considering uncertainty in demand. IEEE Trans Power Syst 21:1565–1573
61. da Silva AML, da Fonseca Manso, LA, Resende LC, Rezende LS (2008) Tabu Search Applied to Transmission Expansion Planning Considering Losses and Interruption Costs, *Probabilistic Methods Applied to Power Systems, 2008. PMAPS '08. Proceedings of the 10th International Conference on*, vol., pp.1–7, 25–29, URL: http://ieeexplore.ieee.org/stamp/stamp.jsp?tp=&arnumber=4912634&isnumber=4912596
62. da Silva AML, Rezende LS, Honorio LM, Manso LAF (2011) Performance comparison of metaheuristics to solve the multi-stage transmission expansion planning problem. IET Gener Transm Distrib 5:360–367
63. Storn R, Price K (1997) Differential evolution – a simple and efficient heuristic for global optimization over continuous spaces. J Global Optimiz 11:341–359
64. Sum-Im T, Taylor GA, Irving MR, Song YH (2006) A Comparative Study of State-of-the-Art Transmission Expansion Planning Tools, *Universities Power Engineering Conference, 2006. UPEC '06. Proceedings of the 41st International*, vol.1, pp. 6–8, 267–271, doi: 10.1109/UPEC.2006.367757 URL: http://ieeexplore.ieee.org/stamp/stamp.jsp?tp=&arnumber=4218686&isnumber=4218631
65. Sum-Im T, Taylor GA, Irving MR, Song YH (2009) Differential evolution algorithm for static and multistage transmission expansion planning. Gener Transm Distrib 3:365–384
66. Torre S, Conejo AJ, Contreras J (2008) Transmission expansion planning in electricity markets. IEEE Trans Power Syst 23:238–248
67. Villanasa R (1984) Transmission network planning using linear and mixed linear integer programming. Ph.D. Dissertation, Ressenlaer Polytechnic Institute
68. Villasana R, Garver LL, Salon SJ (1985) Transmission network planning using linear programming. In: IEEE Transactions on power apparatus and systems, PAS-104, pp 349–356, 1985
69. Wang C, Cheng H (2009) Transmission network optimal planning based on plant growth simulation algorithm. Eur Trans Electr Power 19:291–301
70. Yu H, Chung CY, Wong KP, Zhang JH (2009) A chance constrained transmission network expansion planning method with consideration of load and wind farm uncertainties. IEEE Trans Power Syst 24:1568–1576
71. Zhao JH, Dong ZY, Lindsay P, Wong KP (2009) Flexible transmission expansion planning with uncertainties in an electricity market. IEEE Trans Power Syst 24:479–488
72. Zolezzi JM, Rudnick H (2002) Transmission cost allocation by cooperative games and coalition formation. IEEE Trans Power Syst 17:1008–1015

An Approximate Dynamic Programming Algorithm for the Allocation of High-Voltage Transformer Spares in the Electric Grid

Johannes Enders, Warren B. Powell, and David Egan

Abstract This paper addresses the problem of allocating high-voltage transformer spares (not installed) throughout the electric grid to mitigate the risk of random transformer failures. With this application we investigate the use of approximate dynamic programming (ADP) for solving large scale stochastic facility location problems. The ADP algorithms that we develop consistently obtain near optimal solutions for problems where the optimum is computable and outperform a standard heuristic on more complex problems. Our computational results show that the ADP methodology can be applied to large scale problems that cannot be solved with exact algorithms.

Keywords Approximate dynamic programming • location analysis • spare transformer allocation • transformer replacement • transformer spares • two-stage stochastic optimization

1 Introduction

High-voltage transformers are an integral part of the electric transmission system. A catastrophic transformer failure (where a transformer can literally explode without warning) constitutes the most severe failure event and usually requires the replacement of the transformer. Catastrophic failures can be extremely costly if they require more expensive generation to be brought online in order to relieve system congestion. We refer to the additional costs that are due to a transformer

J. Enders • W.B. Powell (✉)
Department of Operations Research and Financial Engineering, Princeton University, Princeton, New Jersey
e-mail: powell@princeton.edu

D. Egan
PJM Interconnection, Philadelphia, Pennsylvania

A. Sorokin et al. (eds.), *Handbook of Networks in Power Systems I*,
Energy Systems, DOI 10.1007/978-3-642-23193-3_17,

failure are part of the congestion costs. As many of the high-voltage transformers in the U.S. have been installed in the 1960s and 1970s, transmission owners and operators have become increasingly concerned with the growing failure risk of these older transformers.

We study the system of 500 kV to 230 kV transformers that are operated by PJM Interconnection (PJM). PJM operates the electric grid in all or parts of Delaware, Illinois, Indiana, Kentucky, Maryland, Michigan, New Jersey, North Carolina, Ohio, Pennsylvania, Tennessee, Virginia, West Virginia, and the District of Columbia. This service area has a population of about 51 million. The approximately 200 500/230 kV transformers in this area are owned by 11 different transmission owners (TOs).

We address the problem of planning spare transformers to respond to random failures. Spares are crucial because the lead time to obtain a new transformer to replace a failed one can be 12–18 months or longer depending on the order books of transformer manufacturers. For high-voltage transformers the issue of where to locate transformer spares in the transmission network is of particular importance. These transformers weigh up to 200 t and as a result the cost and time involved in moving them are significant. Their transportation needs special permits, and may require the reinforcement of bridges and the restoration of rail access to a transformer substation. Considerable congestion costs may be incurred during the time it takes to transfer a spare to a failure site. Hence, moving spares quickly is an important task. In this paper we consider the problem of placing a given number of spares in the network such that the expected costs associated with random transformer failures are minimized. By running our model repeatedly for different numbers of spares we can also address the issue of spare quantity which is of considerable practical importance.

The problem of planning spare transformers can be formulated as a multistage stochastic, dynamic program of a very large size. A growing body of research in approximate dynamic programming has demonstrated that these techniques scale to very large problems, with a virtually unlimited ability to handle complex operational details (see, for example, Bertsekas and Tsitsiklis [1], Powell et al. [2], Topaloglu and Powell [17]). The basic strategy in ADP is to simulate forward in time, iteratively updating estimates of value functions that approximate the value of being in a state. As with many stochastic optimization algorithms, multistage problems are reduced to sequences of two-stage problems. In our setting, this two-stage problem consists of allocating transformers to different locations, followed by a random realization of failures around the network. If our technique is going to be successful for multistage problems, it has to be able to provide good solutions to the two-stage problem. In this paper, we consider only the two-stage problem, but we use techniques that generalize easily to multistage problems, where we will be interested in 50 year horizons.

The two-stage allocation problem has the behavior of a two-stage stochastic facility location problem where we allocate the spares to locations, then observe random failures after which we have to move the spares to the locations where the failures occurred. The framework of two-stage stochastic programming is presented in Birge and Louveaux [3] and Kall and Wallace [4]. The allocation problem we present here is integer in nature and thus the stochastic integer programming literature is relevant. Sen [5] provides a comprehensive overview of the state of the art in stochastic integer programming. Stochastic facility location problems [6–8] are among the many applications of stochastic integer programming. In that context the problem is to open a number of facilities that can be used to satisfy random client demand. This corresponds to allocating spare transformers that can be used to respond to random transformer failures. Our problem instances are far larger than what appears to be solvable exactly with current technology, where problem size is measured by the number of possible facility locations. For example, the largest instances solved in Ntaimo and Sen [8] have ten and 15 possible facility locations whereas we solve instances with up to 71 candidate locations. Our problem becomes much larger when we add dimensions such as transformer type (e.g. one-phase vs. three-phase), time of arrival (relevant to multistage applications) and other features (e.g. self-monitoring maintenance). It is clear from the CPU times reported in Ntaimo and Sen [8] that the number of possible facility locations has a decisive impact on run times. These findings are consistent with the results reported in Powell et al. [9] (see also Topaloglu [10]) which show that Benders decomposition becomes quite slow as the dimensionality of the second stage resource vector grows.

If we assume that there can be at most one failure and that we always meet that failure with a spare then our problem is equivalent to a generalization of the classic deterministic p-median problem given by Mirchandani [11] and Labbé et al. [12]. This equivalence breaks down as soon as we allow more than one failure or if we allow that a failure is left unmet because the transfer cost is higher than the avoided congestion cost. Nevertheless we will make extensive use of the p-median model as a benchmark for our algorithms.

Several papers in the electric power literature [13–15] present techniques to determine the number of transformer spares to be held. None of these papers considers the issue of spare location which is central to our approach.

Prior research in stochastic resource allocation problems has shown that separable, piecewise linear function approximations work extremely well for two-stage resource allocation problems (see Powell et al. [9], Godfrey and Powell [16], Topaloglu and Powell [17]), providing near-optimal results while scaling easily to very large scale, multistage applications (see Powell and Topaloglu [18]). This work suggests that this strategy might be very practical for the problem of managing spare transformers. However, all of this work was in the context of managing large fleets of vehicles which exhibits very different problem characteristics. The problem of allocating spare transformers, where there may be a half dozen spares spread among 70 locations, is quite different and it was not clear that the same strategies would work (our experiments confirmed this).

Our paper makes three contributions. (1) We present an approximate dynamic programming algorithm that provides very good solutions to realistic instances of the two-stage spare transformer allocation problem. (2) In the context of approximate dynamic programming, we illustrate the shortcomings of standard linear value function approximations when the true value function is not separable.

We introduce new value function approximations that take into account this nonseparability. (3) We contribute insights into spare transformer allocation issues of practical interest. The location and number of spares for the PJM system are important questions as are the value of sharing spares among TOs, the role of ordering lead times, and the influence of transformer transportation costs.

Section 2 contains the basic notation and the model formulation of the spare transformer allocation problem. Section 3 introduces the algorithmic framework that we adopt to solve the problem. The main ingredients of our approach are suitable value function approximations which we present in Sect. 4. We validate our algorithmic approach and apply it to PJM's system performing a series of computational experiments which are described in Sect. 5. The results of the experiments are presented in Sect. 6. We state conclusions and directions for further research in Sect. 7.

2 Model Formulation

Each transformer that is in operation in the electric grid belongs to a transformer bank. A bank is a set of transformers that together handle all three power phases. Each bank consists of either a single three-phase transformer, or three single phase transformers. Failures are modeled at the level of banks since if a single-phase transformer fails, the entire bank has to be shut down. Once the failed transformer is replaced the bank is brought back online.

If a bank fails, then electricity has to be routed through other transformers. However, there is a limit to how much electricity can be routed along specific links in the network. As a result, if a bank fails, then it may be necessary to obtain power from a utility that may be more expensive, but whose location allows electricity to be routed through paths that have available capacity. The additional cost of acquiring power from a more expensive utility due to capacity constraints (resulting from a failed transformer) is part of the congestion costs. The goal of our problem is to minimize the total expected cost incurred as a result of bank failures.

Transformers and banks share the same attribute space $\mathcal{A}$. For the purposes of this paper an element $a \in \mathcal{A}$ is a three-dimensional vector indicating a bank identifier, the location, and the transmission owner. An example for an element $a \in \mathcal{A}$ is

$$a = \begin{pmatrix} a_1 \\ a_2 \\ a_3 \end{pmatrix} = \begin{pmatrix} 1 \\ \text{Branchburg} \\ \mathit{PSEG} \end{pmatrix}.$$

In this example PSEG is the transmission owner. The Branchburg substation belongs to PSEG's service area and one identifies the bank within the Branchburg substation. There are 71.500/230 kV banks in PJM's system and that is the largest attribute space we consider in our numerical work. The notation and our model are completely general and equipped to include other transformer attributes. We are not using any technical transformer attributes as part of our attribute vector because we assume a universal type of transformer spare – one that from a technical standpoint can be used to replace any failed transformer anywhere in the system. However, our methodology can be naturally extended to handle these additional attributes.

The state of the system is defined as

R_{ta} = number of spares with attribute at time $t \in \{0, 1\}$ after the time decisions are made.

$R_t = (R_{ta})_{a \in \mathcal{A}}$ = resource state vector.

We denote a decision by $d \in \mathcal{D}$ where $\mathcal{D}$ is the set of decisions that can be used to act on a transformer. The set $\mathcal{D}$ contains three decision types, i.e. $\mathcal{D} = \mathcal{D}^{buy} \cup \mathcal{D}^{move} \cup d^{\emptyset}$ where the subsets are defined as follows:

$\mathcal{D}^{buy}$ = decision to buy a spare transformer. There is one element in this set for every possible purchasing source, i.e *manufacturer or manufacturing facility*.

$\mathcal{D}^{move}$ = decisions to move a spare to a particular failure site and use it to meet the failure. There is one decision for every possible failure site, i.e. for every element of $\mathcal{A}$.

$d^{\emptyset}$ = decision to hold a transformer (do nothing).

Using the set notation introduced above we define the generic decision variables $xt = (x_{tad})_{a \in \mathcal{A},\, d \in \mathcal{D}}$, where

x_{tad} = the number of transformers with attribute a acted on with decision d at time t and

c_{tad} = the cost parameter associated with x_{tad}.

The parameter B specifies the number of spares that we want to maintain, which we assume is determined by a policy decision made by management. In our model we fix the number of spares and concentrate on the location decisions. However, in our numerical work we run the model for different values of B and can thereby gain insight into the optimal number of spares as well. For example, varying B changes the risk that the system might incur significant failure costs, an issue that is best addressed subjectively by management. We varied B over a range that allowed us to observe congestion costs that ranged from less than \$100 thousand dollars to several million dollars. This analysis can be used to find the number of spares where the marginal value of a spare matches its marginal cost, but the risk of even higher congestion costs is an issue to be addressed by management.

Randomness is introduced in our model via the 0/1 random variables $\hat{W}_1 = (\hat{W}_{1a})_{a \in \mathcal{A}}$ where $\hat{W}_{1a}$ indicates if transformer bank a failed in time period 1. The model underlying $\hat{W}_1$ is the finite probability space $(\Omega, \mathcal{F}, \mathbb{P})$ where Ω is the set of all possible failure scenarios, ω, $\mathcal{F}$ is the discrete σ-algebra on Ω, and P is the probability measure that assigns a given probability to each element $\omega \in \Omega$.

$V_O(R_O)$ is the value function, which is the expectation of the second stage costs as a function of the information of the first stage. We can now present the optimization model. The problem is to find

$$\min_{x_0} \left\{ \sum_{a \in \mathcal{A}} \sum_{d \in \mathcal{D}^{buy}} c_{0ad} x_{0ad} + V_0(R_0) \right\} \tag{1}$$

where

$$V_0(R_0) = \mathbb{E}\left\{ \min_{x_1} \left[\sum_{a \in \mathcal{A}} \left(\sum_{d \in \mathcal{D}^{move}} c_{1ad} x_{1ad} + c_{1ad^{\emptyset}} x_{1ad^{\emptyset}} \right) \right] R_0 \right\} \tag{2}$$

subject to:

$$\sum_{a \in \mathcal{A}} \sum_{d \in \mathcal{D}^{buy}} x_{0ad} = B \tag{3}$$

$$\sum_{d \in \mathcal{D}^{buy}} x_{0ad} - R_{0a} = 0 \ \forall\ a \tag{4}$$

$$x_{0ad} \in \{0, 1\} \ \forall\ a, d \in \mathcal{D}^{buy} \tag{5}$$

$$\sum_{d \in \mathcal{D}^{move} \cup d^{\emptyset}} x_{1ad}(\omega) - R_{0a} = 0 \ \forall\ a, \omega \tag{6}$$

$$x_{1ad}(\omega) \leq \hat{W}_{1a}(\omega) \ \forall\ a, d \in \mathcal{D}^{move}, \omega \tag{7}$$

$$x_{1ad}(\omega) \in \{0, 1\} \ \forall\ a, d \in \mathcal{D}^{move}, \cup\, d^{\emptyset}, \omega \tag{8}$$

In this model formulation a movement decision implies (a) that the spare is moved to a failure site and (b) that the spare is used at the destination location to replace a failed transformer. Therefore, the coefficients c_{lad} for $d \in \mathcal{D}^{move}$ contain two components: the cost associated with the movement, which we call transfer cost, and the avoided congestion costs due to the replacement of the failed transformer. Thus, this model minimizes the sum of transformer purchase costs, transfer costs, avoided congestion costs, and inventory holding costs of spares.

Equation 3 is the budget constraint that fixes the number of spares to be acquired. Equation 4 defines the resource state. The acquisition variables are binary as expressed in Eq. 5. Equation 6 ensures flow conservation in the first stage. Equation 7 states that a spare can only be used to meet a failure if in fact a failure occurred.

3 Basic Algorithm

Solving model (1)–(8) directly is computationally infeasible for problems of realistic size. The experimental evidence in Louveaux and Peeters [7], Laporte et al. [6], and Ntaimo and Sen [8] shows this for integer stochastic programming based algorithms. Interestingly, the results in Ntaimo and Sen [8] suggest that computational difficulties do not necessarily arise with a large sample space but with a high dimensional first-stage decision vector $x_0 = (x_{0a})a \in \mathcal{A}$.

Classic stochastic dynamic programming [19] is also out of the question as a solution approach as it suffers from the well-known "curse of dimensionality" [20] caused by a large action space (x_0), a large state space (R_0), and a large sample space (Ω).

In order to address these computational problems we turn to approximate dynamic programming (ADP). This set of techniques has recently proven useful in finding very good approximate solutions for large-scale multi-period resource allocation problems [17, 21]. Powell and Van Roy [22] give an introduction to ADP in the context of resource allocation problems.

The central idea in ADP is to replace the value function $V(R_0)$ with an approximation $\bar{V}(R_0)$ that depends only on R_0 – the information known at time zero. In order to illustrate the main idea let us for the moment assume a linear value function approximation

$$\bar{V}_0(R_0) = \sum_{a\in\mathcal{A}} \bar{v}_a R_{0a} \tag{9}$$

where the $\bar{v}_a$ are estimates of the unknown parameters v_a which – in this case – can be interpreted as the marginal value of a resource with attribute a.

Using this approximation, the model becomes:

$$\min_{x_0} \left\{ \sum_{a\in\mathcal{A}} \sum_{d\in\mathcal{D}^{buy}} c_{0ad} x_{0ad} + \sum_{a\in\mathcal{A}} \bar{v}_a R_{0a} \right\} \tag{10}$$

subject to:

$$\sum_{a\in\mathcal{A}} \sum_{d\in\mathcal{D}^{buy}} x_{0ad} = B \ \forall \ a \tag{11}$$

$$\sum_{d \in \mathcal{D}^{buy}} x_{0ad} - R_{0a} = 0 \ \forall \ a \tag{12}$$

$$0 \leq x_{0ad} \leq 1 \ \forall \ , a, d \ \in \ \mathcal{D}^{buy} \tag{13}$$

As we can see the model is now radically simplified. The random variable $\hat{W}_1$, the expectation with respect to $\hat{W}_1$, and the random recourse decisions x_I have been eliminated from the model as the entire second stage has been replaced by an approximation. The resulting approximate model can be easily solved using commercially available optimization software such as CPLEX.

Using a value function approximation (VFA) comes at a price. Once a particular functional form of the approximation is chosen the challenge lies in the estimation of the parameters $\bar{v}_a$. In this sense we have replaced an optimization problem with an estimation problem.

In the following we give a high level description of an ADP algorithm that uses stochastic gradients to perform this estimation. For illustrative purposes we continue to use the linear VFA of Eq. 9. Detailed algorithms using different value function approximations are given in Sect. 4.

The core of the algorithm has three steps that are iterated N times. In our notation index n denotes the iteration. The first step consists of solving the approximate problem (10)–(13). In the second step the algorithm takes a failure sample $\hat{W}_1(\omega^n)$ and determines sample gradients of the true value function with respect to the R^n_{0a}. These sample gradients are used to obtain a sample realization $\hat{v}^n_a$ of the parameter value. Sampling failures and solving the second stage problem for that sample realization prepares the ground for the gradient calculation. Let

$m(d)$ = the attribute resulting from modifying a resource with decision d.

The second stage problem is to find

$$\hat{F}\left(R^n_0, \hat{W}_1(\omega^n)\right) = \min_{x_1} \left\{ \sum_{a \in \mathcal{A}} \left(\sum_{d \in \mathcal{D}^{move}} c_{ad} x_{1ad} + c_{ad^\emptyset} x_{1ad^\emptyset} \right) \right\} \tag{14}$$

subject to:

$$\sum_{d \in \mathcal{D}^{move} \cup d^\emptyset} x_{1ad} = R^n_{0a} \ \forall \ a \tag{15}$$

$$\sum_{a' \in \mathcal{A}} x_{1a'd} \leq \hat{W}_{1,m(d)}(\omega^n) \ \forall \ d \in \mathcal{D}^{move} \tag{16}$$

$$0 \leq x_{1ad} \leq 1 \ \forall \ a, d \in \mathcal{D}^{move} \cup d^\emptyset \tag{17}$$

Step 0: Initialize $\bar{V}_0^0$ and set $n = 1$.

Step 1: Solve the approximate problem to get x_0^n and R_0^n. If $n = N + 1$ stop. x_0^{N+1} is the solution.

Step 2: Sample $\hat{W}_1(\omega^n)$ and determine sample realizations of VFA parameters using sample gradients.

Step 3: Update VFA parameter estimates as in equation (19).
Increment n and go to Step 1.

Fig. 1 General ADP algorithm for the spare allocation problem

Note that in this model R_0^n is determined by x_0^n, the solution to the first-stage problem. $\hat{F}$ is also a function of the failure sample $\hat{W}_1(\omega^n)$. For the sake of notational simplicity we will omit $\hat{W}_1(\omega^n)$ as an argument of $\hat{F}$ henceforth. Suppose now that R_{0a}^n is 1 for a particular a. Then the sample gradient of the true value function with respect R_{0a}^n is a left gradient of the form:

$$\hat{\upsilon}_a^n = \hat{F}(R_0^n) - \hat{F}(R_0^n - e_a). \tag{18}$$

where e_a is a vector of zeros with a one at element a.

In the third step of the algorithm we use the newly obtained sample realization $\hat{\upsilon}_a^n$ to update our estimate $\bar{\upsilon}_a^n$. The updating follows the general formula

$$\bar{\upsilon}_a^n = (1 - \alpha_{n-1})\bar{\upsilon}_a^{n-1} + \alpha_{n-1}\hat{\upsilon}_a^n, \tag{19}$$

where α_{n-1} is the step size in iteration n. Figure 1 lists the steps of the ADP algorithm. The more detailed algorithms of Sect. 4 are variations of this general approach. Using this algorithmic approach successfully hinges on finding good value function approximations. This means finding appropriate functional forms that can be estimated with a reasonable amount of effort. This is the task of the next section.

4 Value Function Approximations

In this section we present different value function approximations. We start with the linear VFA as a natural starting point and progress to somewhat more sophisticated approximations that address the limitations of the linear approach.

4.1 Linear Approximation

We have used a linear value function approximation as an example before and repeat the definition here for convenience. The linear approximation has the following form:

$$\bar{V}_0(R_0) = \sum_{a \in A} \bar{v}_a R_{0a}. \tag{20}$$

A linear VFA has been shown to be effective in the context of certain types of resource allocation problems (see Powell et al. [21]). It also has a very intuitive interpretation: $\bar{v}_a$ is the estimate of the value of a transformer with attributes a. It is intuitively clear that spares can have different values. For example, a spare in a central network location might have a higher value than a spare in a remote network location.

The use of this approximation leads to the approximate problem (10)–(13). If R^n_{0a} is 1 then the sample gradient of the true value function with respect R^n_{0a} is a left gradient of the form:

$$\hat{v}_a^{left,n} = \hat{F}(R_0^n) - \hat{F}(R_0^n - e_a). \tag{21}$$

If R^n_{0a} is 0 then the sample gradient is a right gradient. The right gradients require special consideration. If we just added a spare a to calculate $\hat{F}(R_0^n + e_a)$ then we would have $B + 1$ spares in the system. The right gradient $\hat{F}(R_0^n + e_a) - \hat{F}(R_0^n)$ would be the marginal value of spare a if it was the B + 1st spare. Clearly, this gradient is not the best choice if we are restricted to placing B spares. The gradient is influenced by two effects, namely by the attributes of the spare (location effect) and by the fact that we have $B + 1$ spares (quantity effect). But if we are placing B transformers we are only interested in the location effect.

To address this we calculate the right gradients as

$$\hat{v}_a^{right,n} = \hat{F}(R_0^n - e_{a^*} + e_a) - \hat{F}(R_0^n - e_{a^*}) \tag{22}$$

where $a*$ is the marginal spare, i.e. the least valuable of all the B selected spares. Now the right gradients have the same interpretation as do left gradients. They measure the marginal value of a as the Bth spare.

The updating equations for the parameter estimates are

$$\hat{v}_0^n = \begin{cases} (1 - \alpha_{n-1})\bar{v}_a^{n-1} + \alpha_{n-1}\hat{v}_a^{right,n} & \text{if } R^n_{0a} = 0, \\ (1 - \alpha_{n-1})\bar{v}_a^{n-1} + \alpha_{n-1}\hat{v}_a^{left,n} & \text{if } R^n_{0a} = 1. \end{cases} \tag{23}$$

Step 0: Initialize $\bar{V}_0^0$ and set $n = 1$.

Step 1: Solve (10)-(13) to get x_0^n and R_0^n. If $n = N + 1$ stop. x_0^{N+1} is the solution.

Step 2: Sample $\hat{W}_1(\omega^n)$ and determine gradients $\hat{v}_a^n$ $\forall$ $a \in \mathcal{A}$ as in (21) and (22).

Step 3: Calculate $\bar{v}_a^n$ $\forall$ $a \in \mathcal{A}$ as in (23).
Increment n and go to Step 1.

Fig. 2 Spare allocation algorithm with linear VFA

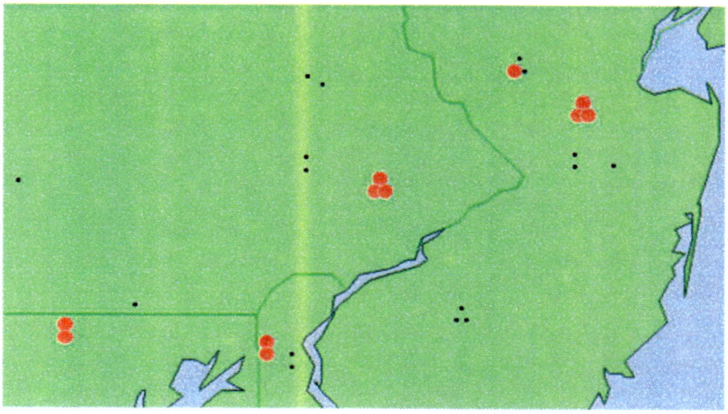

Fig. 3 Partial solution obtained using a linear VFA. The *small spots* are bank locations which are slightly perturbed to show multiple banks in the same location. Transformer spares–indicated by *circles* – are bunched in "good" locations with multiple banks

Figure 2 gives a step by step view of the ADP algorithm using a linear VFA.

Our numerical experiments will provide detailed insight into the solution quality obtained using a linear VFA. We expose its main limitation here to motivate the presentation of our other approximations. Using the linear VFA the ADP algorithm produces solutions as shown in Fig. 3, which depicts part of PJM's service area. Shown is the spare allocation for part of PJM's service area. Overlapping circles indicate substation locations with multiple spares.

The algorithm correctly identifies good spare locations, but when a location has multiple banks, it often allocates multiple spares leading to bunching.

The reason for this behavior is the separability of the linear VFA with respect to the elements of R_0. Separability means that the value of a spare transformer with certain attributes is assumed to be independent of the spare allocation in the rest of the network. This is certainly not true. For example, the value of the first spare in a location is very different from the value of the second or the third. Clearly, the true value function has non-separable effects with respect to the elements of R_0. In the following we present two approximation strategies that incorporate this non-separability.

4.2 Quadratic Approximation

We wish to find VFAs that allow us to make the contribution of a spare dependent on the remaining spare allocation. One way to introduce such dependence is by considering pairs $\{a,a'\}$ of spare attributes. Let us define the relevant set of such pairs as

$$\mathcal{P} = \text{set of unordered attribute pairs } \{\{a,a'\} | a \in \mathcal{A}, a' \in \mathcal{A}, a \neq a'\}.$$

We assume the following model based on attribute pairs:

$$\hat{F}(R_0) = \sum_{\{a,a'\}\in\mathcal{P}} \beta^{00}_{aa'} 1^{00}_{aa'} + \beta^{10}_{aa'} 1^{10}_{aa'} + \beta^{01}_{aa'} + \beta^{11}_{aa'} 1^{11}_{aa'} + \in \tag{24}$$

where we write $1^{00}_{aa'}$ for $1_{\{R_{0a}=0, R_{0a'}=0\}}$ and where $\in$ stands for i.i.d. error terms with expectation zero. Note that in this model every pair of attributes $\{a,a'\}$ contributes to the value of the allocation with one β that depends on the values of R_{0a} and R_{0a}. This is fundamentally different from the linear approximation where the value of an allocation is determined by looking at single attributes rather than attribute pairs. By replacing the indicator functions in Eq. 24 we obtain

$$\begin{aligned}\hat{F}(R_0) = \sum_{\{a,a'\}\in\mathcal{P}} & \beta^{00}_{aa'}(1-R_{0a})(1-R_{oa'}) + \beta^{10}_{aa'} R_{0a}(1-R_{0a'}) \\ & + \beta^{01}_{aa'}(1-R_{0a})R_{0a'} + \beta^{11}_{aa'} R_{0a} R_{0a'} + \in\end{aligned}$$

Multiplying out and collecting terms results in:

$$\hat{F}(R_0) = K + \sum_{a\in\mathcal{A}} \theta_a R_{0a} + \sum_{\{a,a'\}\in\mathcal{P}} \theta_{aa'} R_{0a} R_{0a'} + \in \tag{25}$$

where

$$\begin{aligned} K &= \sum_{\{a,a'\}\in\mathcal{P}} \beta^{00}_{aa'} \\ \theta_a &= \sum_{a'\neq a} \beta^{10}_{aa'} - \beta^{00}_{aa'} \\ \theta_{aa'} &= \beta^{00}_{aa'} - \beta^{10}_{aa'} - \beta^{01}_{aa'} + \beta^{11}_{aa'} \end{aligned}$$

This model gives rise to the value function approximation

$$\bar{V}_0(R_0) = K + \sum_{a\in\mathcal{A}} R_{0a} + \sum_{\{a,a'\}\in\mathcal{P}} \bar{\theta}_{aa'} R_{0a} R_{0a'}. \tag{26}$$

We see that our assumption leads to a quadratic non-separable value function approximation. The $\bar{\theta}_a$ can not be interpreted as the value of spare a because the value of spare a depends on other allocations across the network. However, the $\bar{\theta}_{aa'}$ have an intuitive interpretation. $\bar{\theta}_{aa'}$ is a penalty for allocating spares to a and a' simultaneously. The higher $\bar{\theta}_{aa'}$ the less desirable it is to have a spare in both places.

We now have to show that we can solve the resulting approximate problem and that we can estimate the parameters of Eq. 26. We first show how to solve the approximate problem, which is to find

$$\min_{x_0} \left\{ \sum_{a \in \mathcal{A}} \sum_{d \in \mathcal{D}^{buy}} c_{0ad} x_{0ad} + \sum_{a \in \mathcal{A}} \bar{\theta}_a R_{0a} + \sum_{\{a,a'\} \in \mathcal{P}} \bar{\theta}_{aa'} R_{0a} R_{0a'} \right\} \tag{27}$$

subject to:

$$\sum_{a \in \mathcal{A}} \sum_{d \in \mathcal{D}^{buy}} x_{0ad} = B \ \forall \ a \tag{28}$$

$$\sum_{d \in \mathcal{D}^{buy}} x_{0ad} - R_{0a} = 0 \ \forall \ a \tag{29}$$

$$x_{0ad} \ \in \ \{0, 1\} \ \forall \ a, d \ \in \ D^{buy}. \tag{30}$$

Note that we omit the constant K in the objective function because it does not affect the solution. This model is a quadratic mixed 0–1 program which is much harder to solve than the linear model (10)–(13). In order to facilitate the computational treatment of (27)–(30) we linearize it to obtain an equivalent linear mixed 0–1 program (see Helmberg [23]). We define $y_{aa'} = R_{0a} R_{0a'}$. The linearized approximate problem then is to find

$$\min_{x_0} \left\{ \sum_{a \in \mathcal{A}} \sum_{d \in \mathcal{D}^{buy}} c_{0ad} x_{0ad} + \sum_{a \in \mathcal{A}} \bar{\theta}_a R_{0a} + \sum_{\{a,a'\} \in \mathcal{P}} \bar{\theta}_{aa'} y_{aa'} \right\} \tag{31}$$

subject to:

$$R_{0a} + R_{0a'} - y_{aa'} \leq \ 1 \ \forall \ \{a, a'\} \in \mathcal{P} \tag{32}$$

$$y_{aa'} \leq \ R_{0a} \ \forall \ \{a, a'\} \in \mathcal{P} \tag{33}$$

$$y_{aa'} \leq \ R_{0a'} \ \forall \ \{a, a'\} \ \in \ \mathcal{P} \tag{34}$$

$$y_{aa'} \geq \ 0 \ \forall \ \{a, a'\} \in \mathcal{P} \tag{35}$$

$$x_0 \in \ \mathcal{X}_0 \tag{36}$$

where the last constraint expresses Eqs. 28–30 from before. Note that only the x_{0ab} need a binary constraint; the $y_{aa'}$ do not.

We now turn to the issue of estimating the parameters of the quadratic approximation. Below we show how to obtain sample realizations of the parameters θ_a and$\theta_{aa'}$. In these computations we use second stage objective function values $\hat{F}$, where the resource state vector is perturbed around the initial solution R_0. Let, for example, $R_{0a} = 0$ and $R_{0a'} = 0$ be part of the initial solution. Then

$$\begin{aligned}\hat{F}(R_0) &= \hat{F}(R_{0a^{(1)}}, R_{0a^{(2)}}, \ldots, R_{0a}, \ldots, R_{0a'}, \ldots, R_{0a^{|A|}}) \\ &= \hat{F}(R_{0a^{(1)}}, R_{0a^{(2)}}, 0, \ldots, 0, \ldots, R_{0a^{|A|}}).\end{aligned}$$

We define

$$\hat{F}^{11}_{aa'} = \hat{F}\left(R_{0a^{(1)}}, R_{0a^{(2)}}, \ldots, 1, \ldots, 1, \ldots, R_{0a|A|}\right).$$

For the example case $R_{0a} = 0$ and $R_{0a'} = 0$ this corresponds to the perturbation

$$\hat{F}^{11}_{aa'} = \hat{F}(R_0 + e_a + e_{a'}).$$

$\hat{F}^{10}_{aa'}$, $\hat{F}^{01}_{aa'}$, and $\hat{F}^{00}_{aa'}$ are defined analogously. Let $\mathcal{A}_1 = \{a'' \in \mathcal{A} \setminus a | R_{oa''} = 1\}$ which is the set of all transformer banks that are different from a and have a spare. We define

$$\hat{\theta}_a = \begin{cases} \sum_{a'' \in \mathcal{A}_1} (\hat{F}^{10}_{aa''} - \hat{F}^{00}_{aa''}) - (|\mathcal{A}_1| - 1)\left(\hat{F}(R_0 + e_a) - \hat{F}(R_0)\right), & \text{for } R_{0a} = 0, \\ \sum_{a'' \in \mathcal{A}_1} (\hat{F}^{10}_{aa''} - \hat{F}^{00}_{aa''}) - (|\mathcal{A}_1| - 1)\left(\hat{F}(R_0) - \hat{F}(R_0 - e_a)\right), & \text{for } R_{0a} = 1, \end{cases} \tag{37}$$

$$\hat{\theta}_{aa'} = \hat{F}^{11}_{aa'} - \hat{F}^{10}_{aa'} - \hat{F}^{01}_{aa'} + \hat{F}^{00}_{aa'} \tag{38}$$

In iteration n the $\hat{\theta}^n_a$ and $\hat{\theta}^n_{aa'}$ are calculated and smoothed into the current estimates in the standard way:

$$\bar{\theta}^n_a = (1 - \alpha_{n-1})\bar{\theta}^{n-1}_a + \alpha_{n-1}\hat{\theta}^n_a, \tag{39}$$

$$\bar{\theta}^n_{aa'} = (1 - \alpha_{n-1})\bar{\theta}^{n-1}_{aa'} + \alpha_{n-1}\hat{\theta}^n_{aa'}. \tag{40}$$

The complete ADP algorithm using the quadratic value function approximation is given in Fig. 4.

With the formulas for $\hat{\theta}_a$ and $\hat{\theta}_{aa'}$ we can correctly separate the effect of the variable R_{0a} from the effects of the cross terms $R_{0a} R_{0a'}$. The following proposition makes this statement precise.

Proposition: *Assume the model of Eq. 25 and let θ_a and $\theta_{aa'}$ be estimated with the procedure in Fig. 4. Then $\bar{\theta}^n_a$ and $\bar{\theta}^n_{aa'}$ are unbiased estimators of θ_a and $\theta_{aa'}$ respectively for all $a \in \mathcal{A}, \{a,a'\} \in \mathcal{P}$ and $n = 1, \ldots, N$*

Proof. We start by showing that $\hat{\theta}_a$ is an unbiased estimator of θ_a for all values of R_0. We use Eq. 37 to calculate $\hat{\theta}_a$ and show the proof for $R_{0a} = 0$. The case $R_{0a} = 1$ follows the same arguments. Pick $a'' \in \mathcal{A}_1$.

Step 0: Initialize $\bar{V}_0$ to $\bar{V}_0^0$. Choose a steps size rule α_n such that $\alpha_0 = 1$ and $0 \leq \alpha_n \leq 1$ for $n = 1, ..., N$. Set $n = 1$.

Step 1: Solve (31)-(36) to get x_0^n and R_0^n. If $n = N + 1$ stop. x_0^{N+1} is the solution.

Step 2: Sample $\hat{W}_1(\omega^n)$ and determine sample realizations $\hat{\theta}_a^n \;\forall\; a \in \mathcal{A}$ and $\hat{\theta}_{aa'}^n \;\forall\; \{a, a'\} \in \mathcal{P}$ as in (38) and (37).

Step 3: Calculate $\bar{\theta}_a^n \;\forall\; a \in \mathcal{A}$ and $\bar{\theta}_{aa'}^n \;\forall\; \{a, a'\} \in \mathcal{P}$ as in equations (39) and (40).
Increment n and go to step 1.

Fig. 4 Spare allocation algorithm with quadratic VFA

$$\begin{aligned}\hat{F}^{10}_{aa''} - \hat{F}^{00}_{aa''} &= \hat{F}(R_0 + e_a - e_{a''}) - \hat{F}(R_0 - e_{a''}) \\ &= \theta_a + \sum_{a^* \in A_1 \backslash a'} \theta_{aa^*} + \in^*\end{aligned}$$

Where $\in^*$ is a linear combination of error terms. Summing over all such a$''$ gives:

$$\sum_{a'' \in \mathcal{A}_\infty} \left(\hat{F}(R_0 + e_a - e_{a''}) - \hat{F}(R_0 - e_{a''})\right) = |\mathcal{A}_1|\theta_a + (|\mathcal{A}_1| - 1) \sum_{a^* \in \mathcal{A}_1} \theta_{aa^*} + \in^*. \quad (41)$$

Furthermore,

$$\hat{F}(R_0 + e_a) - \hat{F}(R_0) = \theta_a + \sum_{a* \in A_1} \theta_{aa*} + \in^* \quad (42)$$

Inserting (41) and (42) into Eq. 37 and canceling terms gives

$$\hat{\theta}_a = \theta_a + \in^*. \quad (43)$$

Note that according to (43) $\hat{\theta}_a$ is not a function of R_O. Hence we can take expectations on both sides to get

$$\mathbb{E}\left[\hat{\theta}_a\right] = \theta_a. \quad (44)$$

Having shown unbiasedness of $\hat{\theta}_a$ we proceed by induction. Let $n = 1$. Since $\alpha_O = 1$ we have by (39)

$$\bar{\theta}_a^1 = \hat{\theta}_a^1,$$

and the unbiasedness of $\bar{\theta}_a^1$ follows from the unbiasedness of $\hat{\theta}_a^1$. Assume $\bar{\theta}_a^n$ is unbiased for $n = k-1$ and let $n = k$. We have by (39)

$$\bar{\theta}_a^k = (1 - \alpha_{k-1})\,\bar{\theta}_a^{k-1} + \alpha_{k-1}\,\hat{\theta}_a^k.$$

Since $\bar{\theta}_a^{k-1}$ is unbiased by assumption and $\hat{\theta}_a^k$ is unbiased by (44) we obtain

$$\mathbb{E}\left[\bar{\theta}_a^k\right] = (1 - \alpha_{k-1})\theta_a + \alpha_{k-1}\theta_a = \theta_a.$$

To show unbiasedness of $\bar{\theta}_{aa'}^n$ we start by showing that $\hat{\theta}_{aa'}$ is an unbiased estimator of $\theta_{aa}{}'$ for all values of R_0. We use Eq. 38 to calculate $\hat{\theta}_{aa'}$ and show the proof for the case where $R_{oa} = 1$ and $R_{0a'} = 0$. The other three cases, i.e. $\{R_{0a}, R_{0a'}\} = \{1,1\}$,$\{R_{0a}, R_{0a'}\} = \{0,1\}$and $\{R_{0a}, R_{0a'}\}\{0,0\}$ follow the same arguments. Let $R_{0a} =1$ and $R_{0a} = 0$. By Eq. 38

$$\begin{aligned}\hat{\theta}_{aa'} &= \hat{F}(R_0 + e_{a'}) - \hat{F}(R_0) - \hat{F}(R_0 - e_a + e_{a'}) + \hat{F}(R_0 - e_a)\\ &= \theta_{a'} + \sum_{a'' \in \mathcal{A}_1 \cup a} \theta_{a''a'} - \theta_{a'} - \sum_{a'' \in \mathcal{A}_1} \theta_{a''a'} + \epsilon^*\\ &= \theta_{aa'} + \epsilon^* \end{aligned} \tag{45}$$

where ε^* is a linear combination of error terms. Equation 45 shows that $\hat{\theta}_{aa'}$ is not a function of R_0. Thus, taking expectations on both sides gives

$$\mathbb{E}\left[\hat{\theta}_{aa'}\right] = \theta_{aa'}.$$

Unbiasedness of $\bar{\theta}_{aa'}^n$ follows by induction.

4.3 *Piecewise Linear Approximation and Aggregation*

Another tool to incorporate non-separable effects in the VFA is aggregation. Aggregation means that the set $\mathcal{A}$ is partitioned into groups $\mathcal{A}_1$, $\mathcal{A}_2$, ..., $\mathcal{A}_K$ and that these groups are used in the VFA in a useful way. It is important to make clear that aggregation applies only to the VFA. We do not aggregate data elements such as failure probabilities or congestion costs, and we also do not aggregate purchasing decisions. The simulation of the system does not change and the optimization is affected only in so far as the VFA now has a different form.

Using the partition of $\mathcal{A}$ we define the aggregate resource state variable $R_0^g = (R_{0k}^g)_{k=1,2,...,K}$, where R_{0k}^g denotes the number of spares in group k. Formally, we have

$$R_{0k}^g = \sum_{a \in \mathcal{A}_k} R_{0a}. \tag{46}$$

Using the aggregate resource state variable we introduce the aggregate VFA component $\bar{V}_0^g(R_0^g)$. The key advantage of aggregation is that $\bar{V}_0^g(R_0^g)$ can

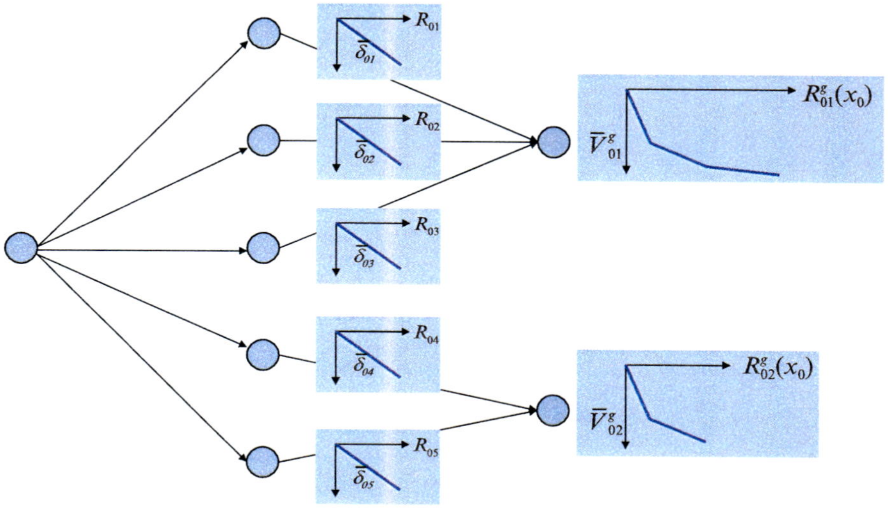

Fig. 5 VFA with nonlinear components. The *arcs on the left* represent the purchasing decisions. The *arcs in the middle* are linear value function terms on the disaggregate level; on the *right* are piecewise linear aggregate VFA terms for every group

consist of nonlinear functions which allow us to capture the declining marginal value of spares in a group. If, for example, a group corresponds to a location then the aggregate VFA component can capture the fact that the first spare in a location is worth more than the second one which is worth more than the third and so forth.

We use piecewise linear, convex functions to approximate value functions of spares at an aggregated level. The usefulness of piecewise linear functions in VFAs was first documented in Godfrey and Powell [24] and Godfrey and Powell [16]. Figure 5 illustrates the VFA with piecewise linear components. Note that the VFA contains both linear terms on the disaggregate level (i.e. the bank level) and aggregate piecewise linear terms. The existence of aggregate and disaggregate VFA components comes from the fact that under the aggregation paradigm we look at the marginal value of a spare transformer, υ_a, as containing two components. We can write for $a \in \mathcal{A}_k$:$\upsilon_a = \upsilon^g{}_k + \delta_{\mathrm{a}}$. One value component, υ^g_k, is derived from the group level and the other, δ_{a}, is a correction term that differentiates spares in a group. Of course, none of these values is known, so we set out to describe how they can be estimated.

We define

y_{kr} = the flow on the rth segment of the piecewise linear function of group k,
$\bar{\upsilon}^g_{kr}$ = estimate of the slope of the rth segment of the piecewise linear function of group k,
$\bar{\delta}_a$ = slope of the disaggregate VFA term a.

The approximate problem is to find

$$\min_{x_0} \left\{ \sum_{a\in\mathcal{A}} \sum_{d\in\mathcal{D}^{buy}} c_{0ad}x_{0ad} + \sum_{a\in\mathcal{A}} \bar{\delta}_a R_{0a} + \sum_{k=1}^{K} \sum_{r=0}^{|\mathcal{A}_k|-1} \bar{v}^g_{kr}\, y_{kr} \right\} \tag{47}$$

subject to:

$$\sum_{a\in\mathcal{A}_k} R_{0a} - \sum_{r=0}^{|\mathcal{A}_k|-1} y_{kr} = 0 \;\forall\, k \tag{48}$$

$$0 \le y_{kr} \le 1 \;\forall\, r, k \tag{49}$$

$$x_0 \in \chi'_0 \tag{50}$$

where χ'_0 is the feasible region defined by (11)–(13). Note that the double sum in (47) is the aggregate VFA component $\bar{V}^g_0\,(R^g_0)$ and that (48) ensures flow conservation in the aggregate nodes.

The raw materials for the estimation of VFA slopes are the left and right sample gradients as defined in Eqs. 21 and 22. The right gradient can be calculated if R^n_{0a} is 0, the left gradient if R^n_{0a} is 1. We define the sample gradient on the group level as:

$$\hat{v}^{g,right,n}_k = \min_{a\in\mathcal{A},R^n_{0a}=0} \hat{v}^{right,n}_a \tag{51}$$

$$\hat{v}^{g,left,n}_k = \min_{a\in\mathcal{A},R^n_{0a}=1} \hat{v}^{left,n}_a\,. \tag{52}$$

The aggregate gradients are the minimum of the left and right gradients in the group. We find that using the minimum function to aggregate sample gradients produces much better solutions than using simple averages. Now we can calculate the sample correction terms $\hat{\delta}^n_a$ as the difference between the disaggregate sample gradient and the aggregate sample gradient:

$$\hat{\delta}^n_a = \begin{cases} \hat{v}^{right,n}_a - \hat{v}^{g,right,n}_k, \text{if } a \in \mathcal{A}_k \text{ and } R^n_{0a} = 0 \\ \hat{v}^{left,n} - \hat{v}^{g,left,n}_k, \text{if } a \in \mathcal{A}_k \text{ and } R^n_{0a} = 1. \end{cases} \tag{53}$$

It remains to be shown how to use $\hat{v}^{g,right,n}_k$, $\hat{v}^{g,left,n}_k$, and $\hat{\delta}^n_a$ to update their respective VFA terms. The update of $\bar{\delta}^{n-1}_a$ follows the usual scheme

$$\bar{\delta}^n_a = (1-\alpha_{n-1})\,\bar{\delta}^{n-1}_a + \alpha_{n-1}\,\hat{\delta}^n_a\,. \tag{54}$$

Updating the piecewise linear VFA components $\bar{V}^g_{0k}(R^g_{0k})$ requires a more involved procedure. Godfrey and Powell [24], Topaloglu and Powell [25], and Powell et al. [9] propose different methods. Common to all methods is a preliminary step where $\hat{v}^{g,left,n}_k$ is used to update the slope to the left of R^{gn}_{0k} and $\hat{v}^{g,right,n}_k$ is used to update the slope to the right of R^{gn}_{0k}. The formula for this preliminary step is

$$u^n_{kr} = \begin{cases} (1-\alpha_{n-1})\,\hat{v}^{g,n-1}_{kr} + \alpha_{n-1}\hat{v}^{g,right,n}_k & \text{if } r = R^{gn}_{0k}, \\ (1-\alpha_{n-1})\,\hat{v}^{g,n-1}_{kr} + \alpha_{n-1}\hat{v}^{g,left,n}_k & \text{if } r = R^{gn}_{0k} - 1, \\ \hat{v}^{g,n-1}_{kr} & \text{otherwise.} \end{cases} \tag{55}$$

Now we would be done except that the updated estimate might not be convex and a procedure to restore convexity is needed. We follow the SPAR algorithm of Powell et al. [9] concave functions and apply it to our convex case. This method restores convexity by projecting the updated estimate of the slopes onto the space of monotone increasing functions. The projection operation is an optimization problem of the form

$$\min_{\bar{v}^{gn}_k} \| \bar{v}^{gn}_k - u^n_k \|^2 \tag{56}$$

subject to:

$$\bar{v}^{gn}_{k,r+1} - \bar{v}^{gn}_{kr} \geq 0 \tag{57}$$

which is easily solved since it involves simply averaging slopes around R^{gn}_{0k}.

Figure 6 shows the steps of the SPAR algorithm for a convexity violation to the left of $R^{gn}_{0k} = 3$. Figure 7 shows the steps of the ADP algorithm with piecewise linear VFA components and aggregation.

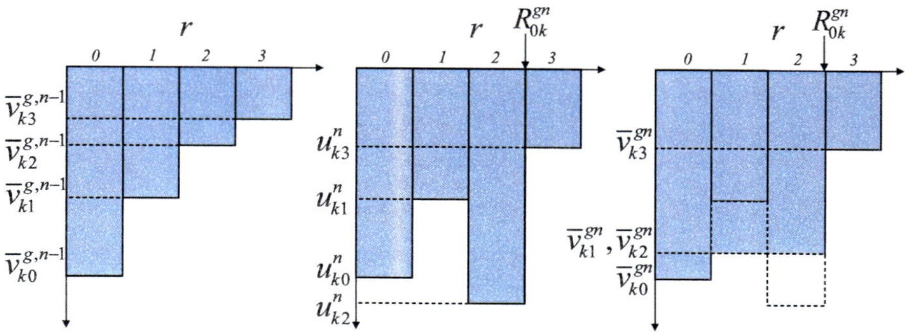

Fig. 6 Illustration of SPAR with a convexity violation to the left of $R^{gn}_{0k} = 3$. The original VFA is on the *left*, the intermediate update in the *middle*, and the result of the projection on the *right*. The functions represent the slopes of the piecewise linear VFA term of group k

Step 0 Initialize $\bar{V}_0^0$ and set $n = 1$.

Step 1 Solve (47)-(50). If $n = N + 1$ stop. x_0^{N+1} is the solution.

Step 2a Sample $\hat{W}_1(\omega^n)$ and determine gradients:

$$\hat{v}^{right,n} = (\hat{v}_a^{right,n})_{a\in\mathcal{A},R_{0a}^n=0}$$
$$\hat{v}^{left,n} = (\hat{v}_a^{left,n})_{a\in\mathcal{A},R_{0a}^n=1}$$

as defined in (21) and (22)

Step 2b Determine:

$$\hat{v}^{g,right,n} = (\hat{v}_k^{g,right,n})_{k=1,2,..,K,R_{0k}^{gn}<|\mathcal{A}_k|}$$
$$\hat{v}^{g,left,n} = (\hat{v}_k^{g,left,n})_{k=1,2,...,K,R_{0k}^{gn}>0}$$
$$\hat{\delta}^n = (\hat{\delta}_a^n)_{a\in\mathcal{A}}$$

as defined in (51), (52), and (53) respectively.

Step 3a For every $k = 1, 2, ..., K$ calculate vector $u_k^n = (u_{kr}^n)_{r<|\mathcal{A}_k|}$ as in (55).

Step 3b For every k project u_k^n onto the space of monotone functions to get $\bar{v}_k^{gn} = (\bar{v}_{kr}^{gn})_{r<|\mathcal{A}_k|}$ using (56)-(57).

Step 3c Calculate $\bar{\delta}^n = (\bar{\delta}_a^n)_{a\in\mathcal{A}}$ as in (54).
Set $n = n + 1$ and go to Step 1.

Fig. 7 ADP algorithm for the spare allocation problem with aggregation

5 Experimental Design

This section describes the setup of the numerical experiments that we perform to evaluate the presented ADP algorithms and to analyze the PJM system. We are interested in the solution quality produced by four different value function approximations. We study the linear approximation of Sect. 4.1 (abbreviated LINEAR), the quadratic approximation of Sect. 4.2 (QUAD) and the piecewise linear approximation of Sect. 4.3 with two different aggregations. We choose to aggregate by transmission owner (PLTO) and by location (PLLO). These are the two most natural aggregations and they differ in a key point. The aggregation by transmission owner requires a mechanism to allocate spares within each group. When aggregating by location the allocation within a group is not important because transfer costs to and from all the banks in a location are identical.

5.1 The p-Median Problem as Reference Solution

How to measure solution quality is a subtle and interesting point in this research. There is a classic deterministic discrete location problem – the p-median problem – that serves very well as a standard to compare against. In fact, with some restrictions our problem reduces to a generalized p-median problem. Since p-median problems of the size encountered in this research can be easily solved using commercial optimization software we are able to find the optimal solution for these simplified problem instances. In these cases the goal for our algorithms is to come close to optimal. In the cases where a problem instance violates p-median assumptions we would expect to outperform the reference solution in some systematic way or at least in most cases.

We refer the reader to Mirchandani [11] and Labbé et al. [12] for a detailed presentation of the p-median problem. It aims to optimally locate p facilities among a discrete set of possible locations. The facilities are used to satisfy demand in a discrete set of demand locations. The objective is to minimize the total sum of transportation costs between facilities and demand locations. At first sight the p-median problem looks very similar to the spare transformer problem. A closer look reveals the differences.

The facilities in the p-median problem are assumed to have infinite capacity and therefore demand is always satisfied and it is always satisfied from the "closest" facility (i.e. the one with the lowest transportation costs).

This is clearly not the case in our problem. If there is a failure and the closest spare is otherwise in use our model attempts to get the next closest spare. If there are more failures than spares then failures are left unmet. If there is a spare available to meet a failure but the transfer is not economical the failure is also left unmet. These are the three most obvious differences to the p-median problem.

We can convert our problem to a p-median problem by assuming there is at most one failure and the congestion costs at a failure location always outweigh the transfer costs of a spare to that location. Making these assumptions puts us in the position to obtain optimal solutions for interesting instances of our problem.

5.2 *Test Data Sets*

Table 1 describes all the data sets used in the experiments. The data sets are meant to comprise an interesting mix of data characteristics. Data sets with the prefix MU include all 71 of PJM's transformer banks in 42 substation locations. That means there are locations with multiple banks. Data sets with the prefix SI include 42 banks in 42 locations. All locations have a single bank.

Data sets MU1, MU2, SI1, SI2, and SI3 are used for testing the solution quality. The others are used in the study of Sect. 6.3. Among the five data sets used for testing MU1, SI1 and SI2 assume a single transformer failure and our model is in this case equivalent to the corresponding p-median problem. Data sets MU2 and SI3 allow for multiple independent failures and we would expect to outperform the reference solutions in these cases. In all five data sets we use hazard rate functions estimated by PJM [26] to determine transformer failure probabilities. The failure probability of a transformer depends on its age and its maintenance status which can be "good", "average", or "watch". We obtain bank failure probabilities from individual transformer failure probabilities by calculating the probability of the event that "at least one transformer in the bank fails" over the period of 1 year.

Table 1 Data sets used for solution quality assessment

Name	No. banks	No. loc.	Shared spares	Failure gener.	Exp. No. failures	Transp. cost (millions)	Transfer time (years)	Cong. cost model	Cong. cost (10_7)
MU1*	71	42	Yes	Single	1	n.a.	min: 0.16 med: 0.26 max: 0.43	n.a.	med: 22.24
MU2	71	42	Yes	Indep.	5.2	n.a.	min: 0.01 med: 0.24 max: 0.63	Uniform Between 12 and 32	med: 24.49
MU3	71	42	Yes	Indep.	2.7	min: 0.23 med: 1.69 max: 3.45	min: 0.16 med: 0.26 max: 0.43	n.a.	med: 22.24
MU4	71	42	No	Indep.	2.7	min: 0.23 med: 1.03 max: 2.15	min: 0.16 med: 0.26 max: 0.43	n.a.	med: 22.24
MU5	71	42	Yes	Indep.	4.1	min: 0.23 med: 1.69 max: 3.45	min: 0.16 med: 0.26 max: 0.43	n.a.	med: 33.36
MU6	71	42	Yes	Indep.	2.7	min: 0.23 med: 1.69 max: 3.45	min: 0.09 med: 0.19 max: 0.36	n.a.	med: 22.24
SI1*	42	42	Yes	Single	1	n.a.	min: 0.01 med: 0.37 max: 0.98	n.a.	med: 22.1
SI2*	42	42	Yes	Single	1	n.a.	min: 0.25 med: 0.25 max: 0.25	n.a.	med: 22.1
SI3	42	42	Yes	Indep.	2.6	n.a.	min: 0.01 med: 0.24 max: 0.63	Uniform Between 12 and 32	med: 23.76

*Equivalent to the corresponding p-median problem.

For the experiments, the inventory holding cost is 0. We also assume a single transformer manufacturer and a fixed transformer purchase price of $5 million. This leaves the parameter C_l for $d\in\mathcal{D}^{move}$ to be considered. The corresponding decision x_{lad} for $d\in\mathcal{D}^{move}$ implies that a spare is moved to a failure site and used to cover a failure. Thus, the cost parameter for that decision is the transfer cost minus the 1-year congestion cost. The former is incurred due to the movement of the spare. The latter represents the benefit of avoided system congestion.

We use two different scenarios for the congestion cost. One scenario consists of congestion costs provided by PJM. In these cases we do not need a congestion cost model (denoted by n.a. in Table 1). Of course we wish to show that our methods work for more than one congestion cost scenario. The second scenario uses randomly generated congestion costs. The median congestion cost of the two scenarios are very close. But otherwise the two scenarios are very different. The real data has high variance and is skewed with a number of outliers that have very high congestion cost. The randomly generated data – coming from a uniform distribution – is much smoother.

The transfer cost contains two components, (a) a transportation cost which is the cost to effect the physical movement of the spare and (b) a congestion cost component that accounts for the system congestion while the spare is being transferred. The transfer time represents the time it takes to organize and execute the spare movement. During the transfer time the failure that initiated the spare movement is left unmet. If the transfer time is, for example, 3 months, then the transfer cost would include one quarter of the 1-year congestion cost at the destination location.

In reality little more than anecdotal evidence on actual transfer times is available. This is why we need a model to generate artificial transfer times. We use different models. Sometimes transfer times are multiples of the distance between locations, sometimes they are fixed, and sometimes they have a fixed component and a linear variable component in the distance. The location of the spares becomes more important as the variability in the transfer times increases. Thus the purely variable case makes for good test cases for our algorithms.

The data sets that are used for the solution quality assessment are set up such that a failure is always met with a spare if a spare is available. In this case the avoided congestion cost is constant for a given sample realization and number of spares. It is therefore legitimate to remove the avoided congestion costs from the objective function when evaluating two competing solutions. By evaluating only the transfer cost we can get a sharper picture of the relative solution quality than by looking at the entire objective function that includes large constant avoided congestion cost terms.

5.3 Algorithm Tuning

ADP algorithms would not work well without parameters that ensure a good learning behavior. The step sizes, α_{n-1}, and the number of training iterations, N, are the most important parameters in terms of controlling the rate of convergence and the quality of the final solution.

Figure 8 shows how the average solution quality changes with the number of training iterations for different problem instances using the PLLO algorithm. To obtain the graph we ran the algorithm for each instance and periodically evaluated the solution of the approximate problem. The evaluation of a solution is performed in slightly different ways. If the sample space is easily enumerable (as in our single failure experiments), then the evaluation uses all the scenarios and their probabilities to calculate the expectation of the relevant second stage costs. If the

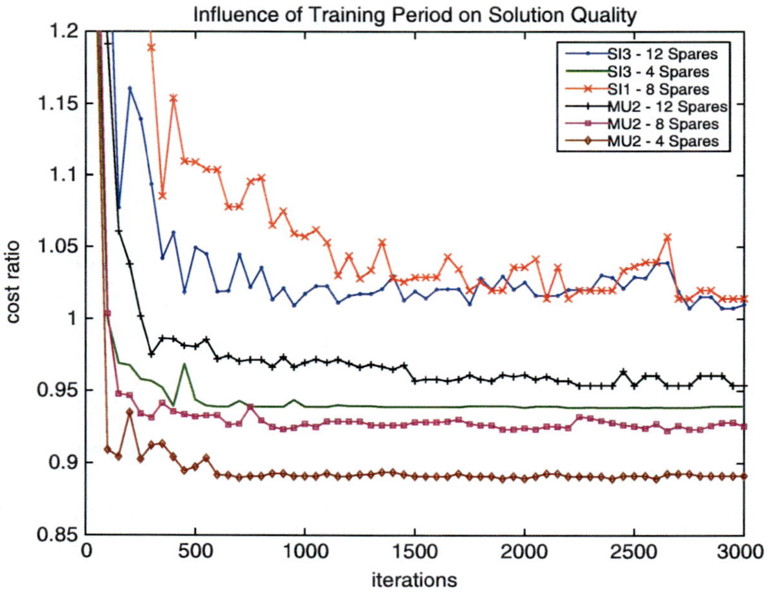

Fig. 8 Average solution quality as a function of the number of training iterations for different problem instances using the PLLO algorithm. Each line is an average of five different sample paths where each sample path (ω^1, ω^2,. . .,ω^{3000}) contains one sample for each iteration. The expected transfer cost is evaluated in increments of 50 iterations and divided by the expected transfer cost of the corresponding p-median solution to obtain the cost ratio

sample space is not enumerable, we use 3,000 randomly drawn, and equally weighted failure scenarios to approximate the expectation. The graphs in Fig. 8 show the cost ratio which is the quality of the PLLO solution divided by the quality of the corresponding p-median solution.

We see that the convergence behavior varies considerably across instances. For data set SI3 with four spares (SI3 – 4 Spares), for example, the solution quality improves and stabilizes within about 500 iterations. SI1 – 8 Spares on the other hand has much slower improvement and the solution quality also varies considerably until around iteration 2,700. Given the variation in convergence behavior it is appropriate to customize the number of training iterations to each problem instance. In order to do so we run PLLO for each problem instance for multiples of 500 iterations and pick N for which the average solution quality is best. Table 3 in Sects. 6.1 and 5 in Sect. 6.2 show the results of this analysis.

Run time considerations prevent us from using more than 600 iterations for QUAD which is adequate according to our empirical tests.

Since convergence is key in ADP algorithms we have chosen the step size rules with great care based on extensive experimental testing. For QUAD we use the quickly decreasing step sizes $\alpha_{n-1} = \frac{1}{n}$ and for LINEAR, PLTO, and PLLO we choose $\alpha_{n-1} = \frac{5}{4+n}$which produces step sizes that decline more slowly.

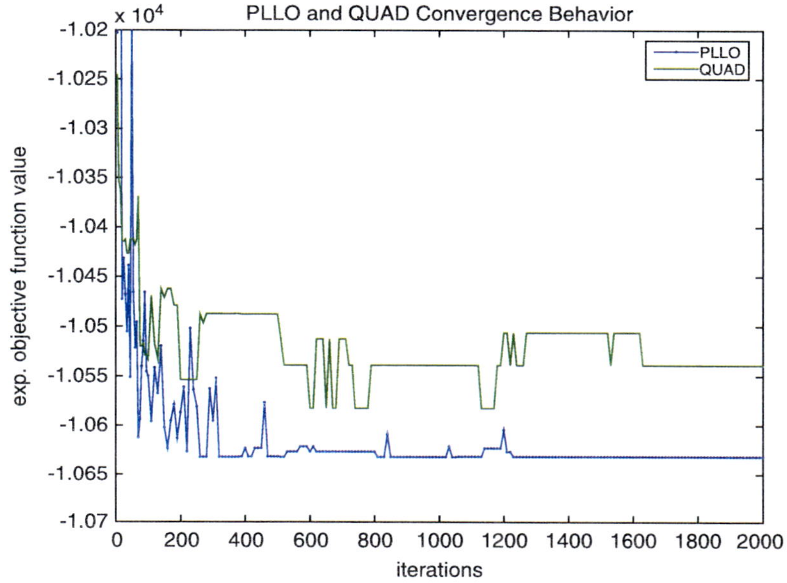

Fig. 9 An example of the convergence behavior of QUAD and PLLO (data set MU3, 10 spares)

As Fig. 8 suggests ADP algorithms do not continuously improve over the training period. For PLLO this characteristic is manageable and obtaining a very good solution is typical. This is not the case for QUAD. Figure 9 shows by example that PLLO converges more uniformly than QUAD. In the example QUAD finds a reasonable solution right around iteration 600 but does not systematically converge to it. This issue needs to be addressed in order to ensure satisfactory solution quality. We employ the simplest possible fix: QUAD periodically evaluates the solution of the approximate problem and remembers the best solution over the course of the algorithm.

QUAD can suffer from excessive run times if the first stage MIP happens to be a difficult problem instance. To address this we cannot rely on MIP warm starts as they are not nearly as effective as LP warm starts. Instead we set a time limit of 17 s for the first stage MIP. Every 50 iterations we increase the limit to 60 s for one iteration to increase the chances of getting an optimal solution. Fortunately, the run time limit on the first stage MIP does not appear to lead to a deterioration of the solution quality.

6 Numerical Results

This section presents the results of three different types of experiments. In Sect. 6.1 we carefully study the solution quality of different VFAs for data sets where the p-median solution is optimal (MU1, SI1, SI2). The goal is to show that our algorithms provide consistently near-optimal solutions. Section 6.2 contains results for more realistic data sets where the p-median solution is not necessarily optimal (MU2, SI3). We wish to show that our algorithms outperforms the reference solution when it is not optimal. Section 6.3 investigates spare transformer allocation issues of practical interest using data sets MU3 – MU6.

6.1 Solution Quality When the Optimal Solution is Known

Table 2 shows the experimental results for problem instances where p-median is optimal. We ran each algorithm for each data set varying the number of transformer spares. The solutions are evaluated as described in Sect. 5.3. r is the cost ratio which is defined as

$$r = \frac{\text{expected transfer cost of ADP solution}}{\text{expected transfer cost of p} - \text{median solution}}.$$

N gives the number of training iterations and s the average run time in seconds. The entries in the r and s columns are averages over five different sample paths $(\omega^1, \omega^2, \ldots, \omega^N)$.

LINEAR works very well in cases with one bank per location. Its bad performance for cases with multiple banks reflects the bunching issue analyzed in Sect. 4.1. PLTO does address the bunching problem to a degree but shows sometimes poor performance when the number of spares is large.

QUAD on the other hand produces generally good solutions even though the run times for the larger instances (MU1) are fairly high. PLLO provides excellent solutions and shows good run-time behavior. The cost ratio is always within 2% of the optimal and the largest instance solves in less than 11 min.

Table 2 Solution quality and computational effort for instances where p-median is optimal. r is the ratio of the transfer cost of the ADP solution to the transfer cost of the p-median solution, N is the number of iterations, and s is the elapsed time in seconds. r and s are averages with sample size 5. Runs are measured on a single processor (Intel P4), 3.06 GHz machine

Data set	No. spares	LINEAR			PLTO			QUAD			PLLO		
		r	N	s	r	N	s	r	N	s	r	N	s
MU1	1	1.00	500	271	1.00	500	168	1.00	100	1,041	1.00	500	168
	4	1.30	2,000	646	1.05	2,000	632	1.02	400	4,195	1.00	1,500	486
	8	1.76	2,000	645	1.15	2,000	634	1.04	400	4,473	1.00	2,000	651
SI1	1	1.00	500	42	1.00	500	41	1.00	100	166	1.00	500	42
	4	1.02	2,000	161	1.07	2,000	158	1.08	600	1,121	1.02	1,000	82
	8	1.04	2,000	159	1.42	2,000	159	1.14	600	2,055	1.01	3,000	242
SI2	1	1.01	500	40	1.01	500	158	1.01	100	161	1.01	500	41
	4	1.01	2,000	161	1.04	2,000	158	1.01	400	654	1.00	1,500	121
	8	1.02	2,000	160	1.16	2,000	158	1.02	400	651	1.00	3,000	241

Table 3 Average solution quality (cost ratio) of PLLO for multiples of 500 iterations and computational effort in seconds per 500 iterations

Data set	No. spares	Iterations						Run time
		500	1,000	1,500	2,000	2,500	3,000	*s*/500 *it*
MU1	1	1.00	1.00	1.00	1.00	1.00	1.00	166
	4	1.03	1.01	1.00	1.00	1.00	1.01	164
	8	1.06	1.04	1.03	1.00	1.01	1.01	161
SI1	1	1.00	1.00	1.00	1.00	1.00	1.00	41
	4	1.04	1.02	1.03	1.02	1.02	1.02	40
	8	1.09	1.06	1.03	1.04	1.04	1.01	40
SI2	1	1.01	1.01	1.01	1.01	1.01	1.01	40
	4	1.02	1.02	1.00	1.01	1.00	1.01	40
	8	1.07	1.04	1.02	1.02	1.02	1.00	40

Table 4 Solution quality and computational effort for instances where p-median is not optimal. r is the cost ratio, N is the number of iterations, and s is the elapsed time in seconds. r and s are averages with sample size 5

Data set	No. spares	LINEAR			PLTO			QUAD			PLLO		
		r	N	s	r	N	s	r	N	s	r	N	s
MU2	1	0.96	500	173	0.96	500	171	0.96	100	1,283	0.96	500	203
	4	0.91	2,000	686	0.96	2,000	669	0.89	400	13,322	0.89	1,000	371
	8	1.11	2,000	661	0.94	2,000	663	0.98	400	12,690	0.93	2,000	688
	12	1.50	2,000	664	1.10	2,000	653	1.04	600	19,571	0.96	2,500	854
SI3	1	1.00	500	43	1.00	500	42	1.00	100	287	0.99	500	60
	4	0.94	2,000	167	0.98	2,000	161	0.99	400	1,872	0.94	500	60
	8	0.92	2,000	168	1.11	2,000	161	1.01	400	1,781	0.92	2,000	179
	12	1.03	2,000	164	1.51	2,000	161	1.10	400	2,699	1.01	3,000	260

Table 3 quantifies the trade off between run time and solution quality for PLLO. We also use these results to pick the best number of training iterations for PLLO as described in Sect. 5.3. This table shows the run times per 500 iterations. Technically, this is an average run time but for each problem instance the run times per 500 iterations is nearly constant. We see that 500 iterations are always enough to bring the solution quality within 10 percentage points of the best achievable value. The rest of the iterations is spent closing this gap.

6.2 *Solution Quality When the Optimal Solution is Unknown*

Table 4 shows the experimental results for problem instances where p-median is not necessarily optimal.

This table documents the problem with QUAD which are the excessive run times for most runs with data set MU2. PLLO is again the best algorithm, outperforming

Table 5 Average solution quality (cost ratio) of PLLO for multiples of 500 iterations and computational effort in seconds per 500 iterations

		Iterations						Run time
Data set	No. spares	500	1,000	1,500	2,000	2,500	3,000	*s*/500 *it*
MU2	1	0.96	0.96	0.96	0.96	0.96	0.96	172
	4	0.90	0.89	0.89	0.89	0.89	0.89	169
	8	0.94	0.93	0.93	0.92	0.93	0.93	165
	12	0.99	0.97	0.96	0.96	0.95	0.95	162
SI3	1	0.99	1.00	1.00	1.00	1.00	1.00	43
	4	0.94	0.94	0.94	0.94	0.94	0.94	42
	8	0.98	0.93	0.93	0.92	0.92	0.92	41
	12	1.05	1.02	1.02	1.03	1.03	1.01	40

the p-median solution almost always and by as much as 11%. PLLO solves instance SI3 – 12 Spares in less than 5 min and MU2 – 12 Spares in less than 15 min which indicates that the algorithm could also handle problems with more locations.

Table 5 shows the results of the convergence analysis. We see that 500 iterations are enough to bring the solution quality within 4 percentage points of the best achievable value.

6.3 Practical Spare Allocation Issues

This section provides insights into four spare transformer allocation issues that are of interest to transmission owners and operators. How many spares to have in order to achieve a balance between failure risk and capital expenditures is a primary concern. We address this question by running our model repeatedly increasing the number of spares and evaluating the solutions. The differences in expected costs are the expected marginal values of spares.

Data sets MU3 – MU6 are used for this analysis. They lay claim to being a realistic – in terms of failure probabilities and cost parameters – for PJM's system. In data set MU3, spares can be used without restrictions throughout the network. This assumption does not always hold in practice. For example, sharing spares among the TOs that own them is not standard practice. Spares can also have technical characteristics that make them unusable in certain network locations. With data set MU4 we analyze the situation where spares may not be shared among TOs. Once a spare is allocated to a TO it can only be used to address failures in this TO's territory. In all other respects MU4 is identical to MU3.

Ordering lead times for new transformers are also a concern in the industry. Production capacity is limited. High demand, labor strikes, copper shortages, and other uncertain events can push ordering lead times beyond 18 or even 24 months. In MU3 the failure probabilities and the avoided congestion costs are calculated for a 12 month horizon which can be interpreted as a 12 month ordering lead time. MU5 uses an 18 month horizon instead.

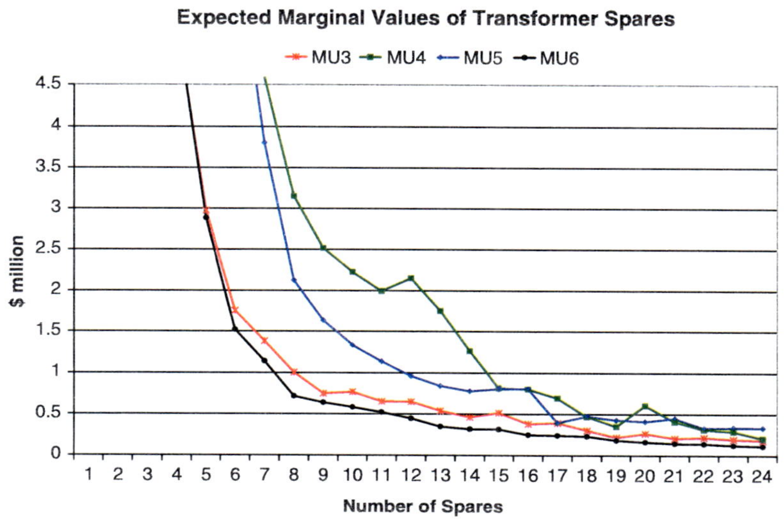

Fig. 10 Expected marginal values of transformer spares as a function of the number of spares. Each data point is an average over five sample paths ($\omega_,^1$ ω1,..., ω^N) where $N = 2{,}000$

Finally, it is of interest to explore the effect of switchable spares. Any spare can be pre-prepared for service at the substation where it is stored. Turning on such a switchable spare is a matter of a few days. If a spare is not switchable, positioning within the substation and putting it in service can take up to 1 month. Data set MU3 assumes switchable spares that take 5 days to put into service. MU6 assumes non-switchable spares which require 30 days of preparation. It is important to stress that this distinction is only relevant for substations with an on-site spare.

Figure 10 shows the results of the marginal value analysis. In order to determine the optimal number of spares one looks for the point where the marginal value hits the marginal cost of a spare. The cost of a spare has to be prorated to match the 1 year time horizon of the model.

As would be expected, the sharing of spares (MU3 vs MU4) has a strong impact on the optimal number. Assuming a marginal cost of $1 million the optimal number is 8 with sharing and between 14 and 15 without. If the marginal cost is $500,000, the numbers are 14 and 18 respectively. The relative effect of transformer sharing decreases as the marginal cost go down.

Comparing MU3 with MU5 we see that the effect of the ordering lead time on the optimal number is relatively small and also shrinks as the marginal cost decreases. If the marginal cost is $1 million/$1.5 million for MU3/MU5, the optimal number is 8 for MU3 and between 9 and 10 for MU5. If the marginal cost is $500,000/$750,000, then the optimal number is approximately 14 for MU3 and 15 for MU5.

The curve for MU3 dominates the curve for MU6. That means that if spares are switchable the optimal number increases. The intuitive interpretation of this behavior is that with switchable spares it is very desirable to have an on-site spare as opposed to bringing in an off-site spare. With additional spares the model can potentially grab more of the on-site/off-site transfer cost difference. Since this difference is big if spares are switchable it can justify more spares.

7 Conclusions and Further Research

In this paper we have introduced a two-stage spare transformer allocation model. Current technology is far from being able to solve large realistic instances of this problem exactly. We use an ADP approach to solve the model approximately and observe that VFAs are needed that can take into account the non-separable behavior of the true value function. We introduce two such VFAs and test the obtainable solution quality. PLLO is the best algorithm. It consistently produces very good solutions and solves very efficiently. We show that the model can be used to answer spare transformer allocation questions of practical interest.

A useful extension of this research is the introduction of multi-period models that reflect the true dynamic nature of high-voltage transformer management. They could be used to model the transformer population and the electric grid as they change over time and answer questions about the timing of transformer purchases and replacements, spare deployment policies, and many others. The performance characteristics of the algorithm used in this paper are so good that it would be a suitable algorithmic starting point for the richer and larger multi-period transformer management models.

References

1. Bertsekas D, Tsitsiklis J (1996) Neuro-dynamic programming. Athena Scientific, Belmont
2. Powell WB, George A, Bouzaiene-Ayari B Simao H (2005) Approximate dynamic programming for high dimensional resource allocation problems. In: Proceedings of the IJCNN, IEEE Press, New York
3. Birge J, Louveaux F (1997) Introduction to stochastic programming. Springer, New York
4. Kall P, Wallace S (1994) Stochastic programming. Wiley, New York
5. Sen S (2005) Algorithms for stochastic mixed-integer programming models. In: Aardal K, Nemhauser GL, Weismantel R (eds) Handbooks in operations research and management science: discrete optimization. North Holland, Amsterdam
6. Laporte G, Louveaux FV, van Hamme L (1994) Excact solution to a location problem with stochastic demands. Transp Sci 28(2):95–103
7. Louveaux FV, Peeters D (1992) A dual-based procedure for stochastic facility location. Oper Res 40(3):564–573
8. Ntaimo L, Sen S (2005) The million-variable "march" for stochastic combinatorial optimization. J Global Optim 32(3):385–400
9. Powell WB, Ruszczyński A, Topaloglu H (2004) Learning algorithms for separable approximations of stochastic optimization problems. Math Oper Res 29(4):814–836

10. Topaloglu H (2001) Dynamic programming approximations for dynamic programming problems. Ph.d. Dissertation, Department of Operations Research and Financial Engineering, Princeton University
11. Mirchandani PB (1990) The p-median problem and generalizations. In: Mirchandani PB, Francis RL (eds) Discrete location theory. Wiley, New York
12. Labbé M, Peeters D, Thisse J-F (1995) Location on networks. In: Ball M, Magnanti TL, Monma CL, Nemhauser GL (eds) Handbooks in operations research and management science: network routing. Elsevier, Amsterdam
13. Chowdhury AA, Koval DO (2005) Development of probabilistic models for computing optimal distribution substation spare transformers. IEEE Trans Ind Appl 41(6):1493–1498
14. Kogan VI, Roeger CJ, Tipton DE (1996) Substation distribution transformers failures and spares. IEEE Trans Power Syst 11(4):1905–1912
15. Li W, Vaahedi E, Mansour Y (1999) Determining number and timing of substation spare transformers using a probabilistic cost analysis approach. IEEE Trans Power Deliver 14 (3):934–939
16. Godfrey G, Powell WB (2002) An adaptive, dynamic programming algorithm for stochastic resource allocation problems I: single period travel times. Transp Sci 36(1):21–39
17. Topaloglu H, Powell WB (2006) Dynamic programming approximations for stochastic, time-staged integer multicommodity flow problems. Informs J Comput 18(1):31–42
18. Powell WB, Topaloglu H (2004) Fleet management. In: Wallace S, Ziemba W (eds) Applications of stochastic programming, SIAM series in optimization. Math Programming Society, Philadelphia
19. Puterman ML (1994) Markov decision processes. Wiley, New York
20. Powell WB (2007) Approximate dynamic programming: solving the curses of dimensionality. Wiley, New York
21. Powell WB, Shapiro JA, Simão HP (2002) An adaptive dynamic programming algorithm for the heterogeneous resource allocation problem. Transp Sci 36(2):231–249
22. Powell WB, Van Roy B (2004) Approximate dynamic programming for high dimensional resource allocation problems. In: Si J, Barto AG, Powell WB, Wunsch D II (eds) Handbook of learning and approximate dynamic programming. IEEE Press, New York
23. Helmberg C (2000) Semidefinite programming for combinatorial optimization. Technical report, Konrad-Zuse-Zentrum fuer Informationstechnik Berlin, Berlin
24. Godfrey GA, Powell WB (2001) An adaptive, distribution-free approximation for the newsvendor problem with censored demands, with applications to inventory and distribution problems. Manage Sci 47(8):1101–1112
25. Topaloglu H, Powell WB (2003) An algorithm for approximating piecewise linear concave functions from sample gradients. Oper Res Lett 31(1):66–76
26. Chen QM, Egan DM (2006) A bayesian method for transformer life estimation using perks' hazard function. IEEE Trans Power Syst 21(4):1954–1965

Decentralized Intelligence in Energy Efficient Power Systems

Anke Weidlich, Harald Vogt, Wolfgang Krauss, Patrik Spiess, Marek Jawurek, Martin Johns, and Stamatis Karnouskos

Abstract Power systems are increasingly built from distributed generation units and *smart* consumers that are able to react to grid conditions. Managing this large number of decentralized electricity sources and flexible loads represent a very huge optimization problem. Both from the regulatory and the computational perspective, no one central coordinator can optimize this overall system. Decentralized control mechanisms can, however, distribute the optimization task through price signals or market-based mechanisms. This chapter presents the concepts that enable a decentralized control of demand and supply while enhancing overall efficiency of the electricity system. It highlights both technological and business challenges that result from the realization of these concepts, and presents the state-of-the-art in the respective domains.

Keywords Decentralized control • demand response • distributed generation • load shifting • smart grid

1 Introduction

In an electricity system with increasing shares of fluctuating generation and of flexible loads, centralized power system optimization has limitations in terms of scalability, actuation speed, security and robustness to failures. Besides, there is no central entity that has information about all generation, load and grid components and who could take over the central coordinator role. Well-designed decentralized coordination mechanisms, in contrast, are less vulnerable to attacks or random

A. Weidlich (✉) • H. Vogt • W. Krauss • P. Spiess • M. Jawurek • M. Johns • S. Karnouskos
University of Applied Sciences Offenburg, Germany
e-mail: anke.weidlich@hs-offenburg.de; harald.vogt@sap.com; wolfgang.krauss@sap.com; patrik.spiess@sap.com; marek.jawurek@sap.com; martin.johns@sap.com; stamatis.karnouskos@sap.com

A. Sorokin et al. (eds.), *Handbook of Networks in Power Systems I*,
Energy Systems, DOI 10.1007/978-3-642-23193-3_18,

failures at a central location and can react more quickly and reliably to local conditions. One of the major challenges facing future electricity transmission and distribution grids is the question of how to integrate the fluctuating renewable electricity generation, both from decentralized and from centralized supplies. This requires greater flexibility in voltage maintenance and efficient load flow control than in present electricity systems. On the other hand, with electric vehicles, a new type of load becomes part of the electricity landscape. By their characteristics as mobile electricity storage devices, they can contribute to stabilizing the grid, if they bring their charging patterns in line with the supply side, or even feed electricity back into the grid in critical peak situations.

Information and communication technologies form the basis for realizing an intelligent electronic network of all components of an electricity system. The higher connectivity enables generating units, network components, usage devices, and electricity system users to exchange information among each other and align and optimize their processes on their own. This forms the basis of a market-oriented, service-based, and decentralized integrated system providing potential for interactive optimization. Bi-directional communication flows can go down to the household level, where home energy management systems and smart meters optimize the energy usage of residential customers and commercial buildings, enabling them to reduce their energy consumption or to avoid using electricity during peak load times, thus preventing critical system situations.

The challenge in a smart grid as envisioned in this context partly lies in establishing the device connectivity within households and buildings, and the bi-directional communication infrastructure between the houses and the grid or the energy service company. Commonly agreed standards form the basis for a convenient integration of new appliances into the local energy management system that can interact with the smart grid. However, at least equally important are the applied control mechanisms that deliver value to all parties involved – i.e. the energy retailer or energy service provider, the grid operator, the metering company and the end-customer – and that respect the privacy and flexibility concerns that the customers have. The decision on which appliance operates at what time should be with the energy consumer as much as possible. It should, however, be guided through incentives to behave in a way that is beneficial for the overall energy system efficiency. Market-based mechanisms and price signals are promising ways to combine the two objectives of maximizing energy efficiency and user comfort at the same time. These are, therefore, reviewed in this chapter, along with other concerns that must be regarded when applying decentralized control mechanisms, e.g. privacy and security.

The remainder of this chapter is structured as follows: Sect. 2 sets the basis for the discussion in the subsequent sections by describing the changes that power systems are undergoing currently and in the near future. In Sect. 3, different possible options of decentralized control in power systems are presented. Sects. 4 and 5 then discuss the technological and economic challenges, respectively, that arise with the development of today's power systems towards a smarter grid. Sect. 6 finally summarizes the findings and concludes.

2 Trends in Power Systems

Besides the developments stimulated by policy and regulation measures in different countries, three major trends can be observed that considerably change the framework of the electricity sector: the still ongoing restructuring or liberalization process that forces integrated utilities to unbundle their activities, thus disintegrating the energy value chain (Sect. 2.1), the new loads that have to be integrated into the electricity system, in which electric vehicles might play the most prominent role (Sect. 2.2), and the increasing share of renewable generation, especially from fluctuating sources such as wind power and photovoltaic plants (Sect. 2.3).

2.1 Restructuring and the Disintegration of the Value Chain

The traditional integrated planning of electricity generation and transmission has become obsolete as a result of liberalization and the unbundling of the electricity value chain.[1] From former vertically integrated regional monopolies, a separation of generation, transmission, distribution and retail supply has been mandated. The generation and retail supply parts of the electricity value chain are subject to competition. On the generation side, generating companies compete for selling power to the wholesale markets. On the retail supply side, energy service companies buy on the wholesale markets the electric power that they need to serve their retail customers. The power grids, i.e. the transmission and distribution systems, are regarded as natural monopolies. They are operated by transmission and distribution system operators, respectively, who are regulated and monitored by a responsible governmental authority. Where possible, competitive elements are introduced into this regulation.

The changed regulatory environment is placing greater demands on the energy system's data networks. Following the disintegration of the electricity value chain, different actors along the value-added chain must now communicate and interact using joint interfaces. Furthermore, new rules on standardization, metering, and consumer transparency generate large amounts of data, which require intelligent, automated processes.

Due to the fundamental changes in constraints, it is essential to maintain the functionality of the power grids. This will require, for example, that the transmission grids have a much higher degree of flexibility in the area of voltage maintenance and efficient load flow control than has been the case to date. The fact that the – partly contradictory – requirements are becoming increasingly and ever more

---

[1] In Europe, electricity sector liberalization was introduced through the Directive 96/02/EG of the European Union [1].

complex means that we must strive for integrated, system-wide innovations to the power supply system.

A well-developed power grid plays a key role in a liberalized energy economy. Through increased power trading and the use of renewable energies, the requirements have already changed considerably today. Whereas these were formerly operated using interconnections spaced far apart, mainly to increase system stability, they now increasingly serve to transport loads across long distances. With a change in the electricity system topology, i.e. due to new generation locations in, e.g., offshore wind parks, this trend will continue to gather force. Major restructuring measures and new operating concepts will thus become indispensable.

New requirements are also emerging on the level of lower-voltage networks. Medium- and low-voltage networks, in which network automation has so far not been installed to a large extent, have more and more difficulties to cope with the requirements posed by the increased integration of decentralized generation facilities. In addition, the continual increase in the demand for power will push the well-developed distribution networks to the limit of their load capacity. Although the networks would be able to absorb additional loads in the medium term, peak loads that depend on the time of day are creating problems not only for the power generation companies. Optimizing the daily load flow can help to avoid expensive investments that are needed only for a few hours each day. Therefore, in the future mechanisms are needed that allow for shifting demand in accordance with economical grid operation criteria. In the future, efficient load management of the power grid will, thus, also have to take into account the requirements of the distribution networks and involve all market players, down to private customers.

2.2 Continued Electrification and Electric Mobility

Global electricity demand has always been rising in the past decades, and is projected to continue to do so in the future [2]. This pattern is observable not only at the global level, but also at all regional and most national levels. As, in many cases, electric appliances allow for a more efficient use of energy to provide the desired final energy service (such as heating by heat pumps, motion by electric motors, light from efficient LEDs etc.), a more sustainable energy system would require even higher shares of electricity in relation to other energy carriers such as liquid fuels or natural gas [3]. Electricity grids are thus facing the connection of new consumption devices in the future, constituting a considerable additional load.

While electric vehicles have been developed since decades, and have been hyped as the future mobility solution several times in the past, there is reason to believe that the current enthusiasm for the use of battery electric vehicles, and also of plug-in hybrid electric vehicles, is sustainable. Announcements of policy-makers and car manufacturers in Asia, Europe, America and elsewhere suggest that relevant stakeholders see a viable future for electric cars, and that these will gain a significant share of the overall car fleet within the next decade. Examples for current electric car promotion initiatives are the planned U.S. "Electric Vehicle Deployment Act of 2010" or the EU "Roadmap on regulations and standards for the electrification of cars", and further national support schemes.

If a significant number of plug-in (hybrid) electric cars are plugged to the grid simultaneously, this imposes a considerable additional stress to the power grid, especially at the distribution level. In order to avoid costly investments into the physical grid infrastructure, intelligent algorithms need to be developed that balance the load caused by car charging events. These algorithms would animate cars to charge preferably when renewable energy is available or when the overall load is low, thus avoiding massive car charging during peak demand. Through such algorithms, car batteries could be used as storage devices in the grid, which are able to serve as buffers for capturing renewable energy supplies and for optimizing the load profiles. They could even re-inject power into the grid in order to level out demand peaks, thus delivering *Vehicle-to-Grid* services [4].

2.3 Decentralization and Renewable Generation

Driven by the need to reduce the dependency on fossil fuels, political and economical pressure are fostering the use of renewable resources and increased efficiency in the generation of usable energy, most importantly electricity and heat. Technology is being pushed by large investments into research and engineering. The market for generation based on renewable resources is the fastest growing branch in the energy sector. The United States have adopted the Energy Independence and Security Act of 2007 [5], which does not demand detailed action about the generation of usable energy, but tries to foster increased efficiency in the consumption of energy. Instead, progress in the generation based on renewable resources shall be honored, and grants for innovation projects are given. The European Commission has issued guidelines for achieving and measuring energy efficiency on the national level in its Directive 2006/32/EC as of April 5, 2006 [6].

Renewable generation is a key element in achieving a truly sustainable energy supply. Wind and solar power are the most important of these technologies, outperforming geothermal energy, tidal power generation and other technologies.

Renewable resources are inherently volatile. Their total power output might be sufficient to provide for demand, but sufficient availability is limited to certain times. Therefore, renewable power generation must be complemented by storage capabilities as well as backup systems and demand side management. Storage of electrical power is still very expensive, although electric car batteries could provide some storage capacity (see Sect. 2.2). Demand side management is depending on shiftable loads, which can be deferred to times of high production without significantly hampering the consumer. Thus, the major part of complementing volatile renewable power falls on backup systems.

For backup functionality, especially systems for decentralized generation are well-suited. These facilitate many desirable characteristics of modern energy systems. One important quality is the resilience against outside threats (e.g. terrorism), which is facilitated by the good isolation properties of decentralized systems, which allow the isolation of (accidental or induced) faults. Another important quality is the high level of efficiency in the use of primary energy. Combined heat and power (CHP) plants can deliver 90% and more of the used primary energy (natural gas, biofuel) to the consumer, in the form of electricity and heat.

In order to establish an energy system with a high degree of decentralization, the political and economical framework must be provided. An increase in decentralized generation diminishes the (economic and political) power of big power providers, shifting profits from traditional utilities to smaller businesses. Such a prospect is likely to spark resistance, which has to be overcome by emphasizing the greater societal benefit of a decentralized infrastructure.

Power generation based on renewable resources complemented by decentralized generation (also by fossil fuels) seems like the perfect pair for achieving energy efficiency on the production side. Technical and societal challenges have to be addressed, however, before they can become the new gold standard for an electricity infrastructure.

3 Decentralized Coordination in Power Systems

With increasing numbers of decentralized generation, it becomes more difficult to ensure the efficiency, reliability and security of power supply. Real-time communication between the grid components, allowing generators and consumers to become an active part of the system can be a way of balancing the grid. However, it can be assumed that a direct control of electrical appliances through a grid operator or an energy service company will not be acceptable for the end user, especially not for private households.

More intelligent alternatives deliver price signals or incentives to customers who can then optimize their energy consumption based on these inputs. We assume that prices and payments or fees are the best way to make generators and loads produce and consume in a way that contributes to overall system efficiency. Price-based coordination can be realized through time-varying tariffs, such as time-of-use rates or real-time pricing (Sect. 3.1), or through dedicated payments to loads and generating units that turn a device on or off at the request of a coordinator (incentive-based mechanism, Sect. 3.2). A more radical implementation of this would be to let all loads and generators participate in a market mechanism to which they submit bids at any time. The decision to switch a device on or off would then be bound to the market result (Sect. 3.3). These concepts are described in the following.

3.1 Time-of-Use and Real-Time Pricing

An electricity tariff for end customers today typically comprises a fixed monthly customer charge and a variable energy charge for the amount of electricity consumed [7]. The energy charge is usually a fixed rate per kWh, independent of the time at which electricity is consumed. While this is comfortable for the customer, who can easily predict energy costs (by multiplying the fixed unit price with the units consumed), it hides the information of how valuable electricity is at different points in time. Consumers, thus, have no incentives to avoid consumption at times of expensive generation, and shift it to less expensive time slots.

Energy retailers have to ensure that the consumption of their customers is constantly matched by equal generation. In most European countries, retailers have to make sure that the energy amounts bought and sold are balanced within every 15 min time interval. The most common strategy to ensure this balance is to apply structured procurement, a three-phase process with narrowing time horizon (see Fig. 1). Each procurement phase can be supported by energy exchanges or can be carried out bilaterally between generators and retailers (over-the-counter – OTC – trade).

In the first phase, a retailer estimates the cumulative base load of his customers (i.e. the lowest load they exhibit in the long term) and usually buys generation of this load for months or even years in advance at a discount price. The second procurement phase would be day-ahead trading. By knowing details of the next day's events, he will create a detailed consumption forecast for his customers for the next day. Events could be e.g. the weather forecast (influencing both the consumption and the amount of renewable energy generated), local and regional holidays, or a deviation from regular demand that has been pre-announced by a larger business customer. The forecasted consumption that was not already covered by the base load will be procured in the second phase on energy markets. If it becomes evident during a day that the energy procured is insufficient or exceeds the load in a given slot (as the result of an incorrect forecast), blocks of energy can still be bought and sold in the intraday market, the third procurement phase.

Eventually, the retailer pays a different price for the energy delivered in each 15 min time slot (resulting from the payments for each of the three phases of procurement). Since this price variability usually is not reflected in end customer's tariffs, the retailer conservatively sets a high price in order to cover his costs, to guard against risks, and to secure his desired margin.

With the introduction of time-of-use pricing or real-time-pricing, i.e. time-dependent prices per energy unit, the electricity retailer can hand over parts of the markets' price fluctuations to the end-customer. With time-of-use (TOU) pricing, fixed time intervals are defined in which different prices are valid. These intervals usually reflect long-time experience of when electricity is more expensive to procure and when it is less expensive, and the prices for each TOU block are fixed for long-term periods. The most common time-of-use pricing is *on-peak* and

Fig. 1 Three phase procurement process

off-peak rates; however, more fine-grained TOU blocks are possible. If the pricing for different consumption time intervals changes frequently and is announced on shorter notice, i.e. day-ahead or even within a day, this is referred to as real-time pricing (RTP). An example for real-time pricing with day-ahead notice is the *Bi-directional Energy Management Interface* as presented in [8].

Hybrids of time-of-use and real-time pricing are also conceivable; they usually referred to as *Critical Peak Pricing* [9]. These concepts have a basic rate structure like in TOU pricing combined with a provision for replacing the normal peak price with a much higher critical peak event price under specified trigger conditions (e.g. when system reliability is compromised or supply prices are very high).

Time-dependent pricing of electricity consumption leads to two benefits. The more obvious one is that this will lead to higher efficiency. The reason for this is market prices accurately reflect the current supply and demand situation. In times of high supply and low demand (e.g. on a windy and sunny Sunday morning) energy is abundant and prices will be low. In this situation it would be good to trigger time-shiftable consumption events (like turning on washing machines or cooling down deep freezers) and turn down conventional consumption from fossil or nuclear sources. In the opposite situation where energy is scarce and market prices are high, the signalization of the high price to the user will motivate him to abstain from avoidable consumption. The higher the price on the market, the more economically inefficient generation equipment will be activated.

The second benefit is more indirect. The distribution of long-term trading vs. day-ahead vs. intraday is strongly inclined towards longer-term trading, i.e. most of the trading is made long in advance.[2] If price risks could be handed over to end customers, more retailers would probably choose to engage in more short-term trade. Given the fact that volatile generation from wind can be predicted very accurately in a time-scale of 3–4 h ahead, this would lead to prices highly correlated with renewable generation, which would ensure the most effective incentive for customers to adjust their load to the current supply of renewable generation. Generally prices are expected to be lower, as indicated by consumer behavior on the Norwegian retail market, where approximately three-quarters of consumers have entered into some form of variable retail-price contract (such as a spot-market contract or a standard variable power-price contract) [12].

Flexible pricing, especially in the form of real-time pricing, also has its disadvantages. From the perspective of the consumers, it exposes them to a higher risk and makes forecasting of energy costs less predictable. Looking at the global system, some experts anticipate avalanche effects: At time of extreme prices, many customers may choose to adapt their consumption (or have automated systems that act accordingly), leading to overcompensation and reversal of the situation. However, this can also be seen as normal market events that are only a problem

[2] To give an example, in Germany day-ahead and intra-day trading volumes at the European Energy Exchange currently only account for roughly one quarter of the total national power consumption [10, 11].

if such short-term changes affect system stability. Generally, the probability of avalanche events might be low, since not all consumers adapt at the very same time and incentives to adapt become more and more unattractive as the extreme price converges to a usual price level as the first consumers adapt.

3.2 *Incentive-Based Load Control*

Incentive-based demand response programs give customers load reduction incentives that are separate from, or additional to, their retail electricity rate, which may be fixed or time-varying. The load reductions can be requested by the grid operator in order support his task of maintaining grid stability. They can also be activated by an energy service provider when prices are very high. Most demand response programs specify a method for establishing customers' baseline energy consumption level, so observers can measure and verify the magnitude of their load response. Examples for such incentive-based curtailment programs are given in the following [9].

- Demand bidding – in which customers offer bids to curtail their loads based on wholesale electricity market prices or an equivalent.
- Capacity market programs – in which customers offer load curtailments as system capacity to replace conventional generation or delivery resources. Customers typically receive day-of notice of events. Incentives usually consist of up-front reservation payments.
- Ancillary services market programs – in which customers bid load curtailments to the grid operator as operating reserves. If their bids are accepted, they are paid the market price for committing to be on standby. If their load curtailments are needed, they are called by the grid operator, and may be paid the spot market energy price.

It must be noted that the regulatory framework of a specific country may hinder the establishment of one or more of these incentive-based mechanisms. In addition, they are usually only practically feasible for large consumers, such as industrial or large commercial sites.

3.3 *Market-Based Coordination*

Centralized wholesale trading at power exchanges has established since many years, because it offers high liquidity and it delivers valuable price information to the energy sector [13]. As there are multiple generators and consumers in the energy market, the dominant market institution for electricity trading is the double-auction format. In a sealed bid double-auction, both buyers and sellers submit bids specifying the prices at which they are willing to buy or sell a certain good. Buying bids are then ranked from the highest to the lowest, selling bids from the lowest to the highest bid price. The intersection of the so formed stepwise supply and demand functions determines the market clearing quantity and gives a range of possible prices from which the market clearing price is chosen according to some arbitrary rule [29]. Double-auctions deliver efficient allocations if the number of sellers and buyers is sufficiently large [14].

While the efficiency of wholesale power markets is generally assumed, market-based mechanisms usually do not play a role at the retail level. However, concepts have been formulated to apply market-based coordination for the intelligent operation of virtual power plants or aggregations of distributed generation units or of flexible loads down to the household level, e.g. [15–17]. These concepts are motivated by the formal proof that the market-based solution is identical to that of a centralized omniscient optimizer, without requiring relevant information such as local state histories, local control characteristics or objectives [18]. The equilibrium price resulting from the market mechanism is, thus, used as the control signal for all units.

In a typical application of market-based coordination for power system scenarios, there are several entities producing and/or consuming electricity; extending the mechanism to allow for combinatorial trade with complementary products, such as natural gas or heat, is not considered here. Each of these entities can communicate with a (centralized) market mechanism. In each market round, the control agents create their market bids, dependent on their current state, and send these to the market. A market is generally defined by three components: a bidding language, which specifies how bids can be formulated, a clearing scheme, which determines who gets which resource, and a payment scheme, which defines the payments the individual users have to make depending on the allocation [19].

The bidding language defines the preferences that an agent can reveal to the market. Bids in a power system market can, e.g., be Walrasian demand functions, stating the amount of the commodity $d(p)$ the agent wishes to consume or generate at a price of p, where a positive and negative amount can be interpreted as consumption and generation, respectively [16]. Bidding languages can also allow for specifying technical constraints, such as minimum levels of generation/consumption, or for expressing how valuation changes depending on time. However, more expressive bidding languages usually lead to higher complexity and may also require the bidder to reveal more (private) information than she wants. Thus, market mechanisms in power system scenarios usually rely on restricted bidding languages, like the example given above.

After collecting all bids, the market agent searches for the equilibrium price, thus defining which agent will buy/sell which amount of electricity.

In practice, one challenging problem when implementing market-based control in real-world applications is to define the agent's policies for defining the bids. These policies differ between different types of appliances. Six different categories of appliances can be defined that can participate in the market [16]:

- *Stochastic operation devices*, where the timing and amount of output cannot be controlled. Examples: fluctuating generation such as wind energy converters or photovoltaic systems
- *Shiftable operation devices* that run for a certain duration, where the starting point can be shifted over time. Examples: washing, drying or ventilation processes
- *External resource buffering devices* that display a storage characteristic without direct electricity storage. Examples: heating or cooling processes
- *Electricity storage devices* such as batteries, capacitors. Examples: electric cars
- *Freely-controllable devices* that can be flexibly deployed within certain limits. Examples: thermal power plants
- *User-action devices* whose operation is defined by the user's needs and desires. Examples: lighting or entertainment devices

The bidding strategies must always take their specific characteristics into account.

4 Technological Challenges

The emerging smart grid is expected to be very much dependent on modern information and communication technologies (ICT). As near real-time communication and information dissemination among all entities is of key importance, a number of technologies that can be integrated into existing processes and enhance them or even provide innovative new services, has been identified. However, in order to realize them, key challenges will need to be adequately tackled. The following subsections provide an overview of the main technological issues related to smart metering (Sect. 4.1), interoperability and standardization (Sect. 4.2), real-time communication (Sect. 4.3), distributed data management and processing (Sect. 4.4) and security and privacy (Sect. 4.5).

4.1 Smart Metering

The true power of smart grids can be realized once fine-grained monitoring i.e. metering of energy consumption or production is in place. Real-time pricing or market-based operation, for example, can only provide incentives for the user to shift loads if her consumption is measured in small time intervals and billed with the according variable tariff. The promise of an advanced metering infrastructure (AMI) is that it will allow provide measurements and analyses of energy usage from advanced electricity meters through various communication media, on request or on a pre-defined schedule. These are usually referred to as *smart meters* and can feature advanced technologies. Today, many utilities have already deployed or are currently deploying smart meters in order to enable the benefits of the AMI. One example is the world's largest smart meter deployment that was undertaken by Enel in Italy, with more than 27 million installed electronic meters. AMI is empowering the next generation of electricity network as envisioned by, e.g., [20, 21] vision.

Smart meters will be able to react almost in real time, provide fine-grained energy production or consumption info and adapt their behavior proactively. These smart meters will be multi-utility ones and will be able to cooperate, and their services will be interacting with various systems not only for billing, but for other value added services as well [22]. Smart meters provide new opportunities and challenges in networked embedded system design and electronics integration. They will be able not only to provide (near) real-time data, but also process them and take decisions based on their capabilities and collaboration with external services. That in turn will have a significant impact on existing and future energy management models. Decision makers will be able to base their actions on real-world, real-time data and not on general less well-grounded predictions. Households and companies will be able to react to market fluctuations by increasing or decreasing consumption or production, thus directly contributing to increased energy efficiency.

4.2 Interoperability and Standardization

The smart grid is a vast ecosystem, composed of a large number of heterogeneous systems that have to interact in order to deliver the envisioned functionality. Up to today, the heterogeneity was hidden in islanded solutions. However, the opening up of the infrastructure as well as the high complexity of the new introduced concepts mean that interoperability will be the key issue that needs to be addressed. Several standards exist today, although many of them still require revisions, especially when it comes to the inter-operation with other standards and systems.

In a recently released report, the U.S. National Institute of Standards and Technology, NIST, provides an overview of standards and problems that will need to be tackled [23]. The priority areas where standards need to be developed and interoperability is required are:

- *Demand response and consumer energy efficiency*, i.e. mechanisms and incentives for electricity generators and consumers to cut energy use during peak times or to shift it to other times (concepts described in Sect. 3).
- *Wide-area situational awareness*, i.e. monitoring and display of power-system components and performance across interconnections and over large geographic areas in near real-time.
- *Energy storage*, which today mainly consists of pumped hydroelectric storage, but can also be millions of *electric car* batteries in the future.
- *Advanced metering infrastructure* as described in Sect. 4.1.
- *Distribution grid management*, which focuses on maximizing performance of feeders, transformers, and other components of networked distribution systems and integrating with transmission systems and customer operations; as smart grid capabilities, such as AMI and demand response, are developed, and as large

numbers of distributed energy resources and plug-in electric vehicles are deployed, the automation of distribution systems becomes increasingly more important to the efficient and reliable operation of the overall power system.
- *Cyber security*, which encompasses measures to ensure the privacy protection, integrity and availability of the electronic information communication systems and the control systems necessary for the management, operation, and protection of the respective energy, information technology, and telecommunications infrastructures.
- *Network communications* – given the variety of networking environments used in a smart grid, the identification of performance metrics and core operational requirements of different applications, actors, and domains in addition to the development, implementation, and maintenance of appropriate security and access controls becomes more and more important.

As in all standardization activities, a great effort is needed in order to develop and actively maintain standards via a collaborative, consensus-driven process that is open to participation by all relevant and materially affected parties, and not dominated by, or under the control of, a single organization or group of organizations [23].

4.3 Real-Time Communication

The systems concerned with the physical parameters of the grid always require real-time communication. If generation and consumption do not match, the quality parameters of the electricity delivered (like voltage and frequency) immediately deteriorate. This is why, already today, real-time systems constantly monitor electricity flows and other parameters, and automatically take action when detecting an unusual situation.

Unlike these critical core systems that ensure the physical stability of the grid, the more or less virtual trading layer on top of these systems only takes an *ex ante* and *ex post* view. The *ex ante* view, i.e. the trade phase until a certain deadline before physical execution of generation and consumption, ensures that the expected generation will match the expected consumption. During the execution phase, the trading systems are not involved in the effort maintaining grid stability. Only *ex ante*, i.e. after execution, the actual generation and consumption is assessed and the trading systems do the accounting. Unfortunate retailers that deviated from their announced schedule in the direction that harmed grid stability are punished. As an example, if there was not enough supply and a retailer's contracted generators generated less than announced or his customers consumed more, the retailer would have to pay penalties. The more lucky ones that deviated in a way that stabilized the grid would not pay penalties.

This strict distinction between trading and technical systems is challenged by new systems like market-based control, as explained in Sect. 3.3. This leads to challenges for the current vendors of today's trade systems, who are usually not familiar with real-time software engineering.

4.4 Distributed Data Management and Processing

As already motivated in the introduction, the operation of the electricity system is such a huge optimization problem that it cannot be solved centrally. The complexity can be handled by simplification and aggregation, and by pushing processing to the edge of the network. All the paradigms shown in Sect. 3 manifest in systems that do little coordination centrally and let most of it be done at the edge. The central controlling unit merely sets a price or an incentive and lets the end user (or an automated system on his behalf) decide how to react to the external stimuli. Core to all these systems are home, office, or factory gateways that receive the external signals. They have built-in, customizable logic that triggers if-then rules, e.g. if the price is below a certain threshold, then shiftable devices start operating.

There might be one scenario, however, where it makes sense to propagate and apply a central decision right down to a single device: the charging of plug-in (hybrid) electric vehicles. This scenario delivers the highest benefit if controlled centrally and executed strictly according to the central decision (of cause taking user preferences like his or her desired time of full charge into account). We base our judgment on the following assumptions: (1) a given topological area of the low-voltage distribution grid cannot support simultaneous charge of a car fleet that is highly electrified, (2) the possible time window for charging is longer than the time window needed to charge a single car battery, (3) users have set a desired mileage and the time it should be made available in the car's battery, (4) user preferences are communicated to a system that is responsible for the topological area, (5) the charging schedule for each car (i.e. the load curve needed to charge the battery) is known and communicated. If all these assumptions are valid, coordination of loading start points by the central system is an alternative to deploying more conductive material (i.e. new power lines) that is expected to come with a far lower price tag.

An alternative to direct control could be market splitting, leading to different prices in different topological areas of the grid, as transport line capacity is considered when matching demand with generation. This is also a general approach to consider physical capacity limitations in the economics of grid operation.

4.5 Security and Privacy

The trends (see Sect. 2), the paradigm changes in end-user market participation (see Sect. 3) and the enabling or implicated technological changes (see Sect. 4) necessitate development of new security and privacy measures and a review of the existing ones. Security and privacy of the used IT systems and protocols ensure trust in the market itself and therefore form an important factor for its successful operation.

In this section, we derive the security and privacy challenges that are implied by the emerging organizational and technical changes described in this chapter.

4.5.1 New Paradigms in Energy Trade and Their Security Implications

New paradigms in the implementation of energy trade will cause significant implications for the involved systems' security:

The most apparent change is that communication between customers and suppliers of energy will become bidirectional. Before, consumers of energy reported the amount of used energy to their supplier once a year. With the emergence of time- or load-dependentor incentive-based tariffs prices, incentives or other coordination activities (see Sect. 3) have to be communicated back to the customer. In turn, the customer negotiates parameters or reacts to the received signals by adapting his electricity usage or generation. The utilized IT-systems must account for the resulting security and privacy implications: Data transmitted to the customer is highly sensitive as it influences the customer's behavior and is relevant for billing. Therefore, its integrity, authenticity and non-reputability must be ensured, allowing the customer to verify the received data's soundness.

The decentralized coordination approaches mentioned in Sect. 3 might involve another significant change in how communication works in the smart grid: Automated communication relationships with quickly changing heterogeneous partners will emerge. As customers will also become providers of services (control of appliances, reduction/increase in energy consumption), they can potentially have energy related communication relationships with multiple parties. Market-based coordination might even implicate that those parties are not fixed over time but change rapidly. Ensuring authenticity of a communication partner and ensuring integrity and timeliness of communication in such systems is not trivial to accomplish neither by organizational nor by technical means.

The mobility of energy consumers (see Sect. 2.2) represents another paradigm change and opens up a whole new field for IT. The mobility requires an authorization and billing infrastructure that features high-availability and confidentiality and potentially spans several countries or whole continents. The actual charging procedures and systems must ensure that neither involved parties can commit fraud (charging point operator by simulating a charging procedure, the customer by repudiation, supplier by claiming false charging records) nor that outsiders threaten the acceptance of electric vehicles by attacking the availability/credibility of the system.

The previously mentioned more frequent and bi-directional communication (see Sect. 4.5.1) implies that a huge amount of privacy-related data will be accumulated. This is also new for a field where, at least for consumers, only few data was gathered throughout a year. Smart meters will accumulate and transfer data that can be used to create personal profiles [24] of residents and can be subject to national data privacy laws. It can even be used to deduce the individual use of appliances [25].

Electric mobility creates information about the position of past and future charges that could be used for extortion (husband at unambiguous location) or industrial espionage (employee of company Y at headquarters of company X). It is crucial that architectures (or organizational measures) that are developed for the handling of this data account for its importance and prevent leakage of data to unauthorized parties and ensure retention times longer than necessary.

4.5.2 Requirements for Secure and Privacy-Preserving Energy Systems

All areas of the energy sector, from generation over transmission and distribution to consumption, will eventually be connected technologically in order to foster efficiency by communication and more cooperation. The necessary overarching architectures will probably face security challenges that are very hard to predict. It is safe to say that it will face the same challenges that all distributed systems face with regard to security.

Large-scale identity management measures can pose one building block to enable grid-wide trust relationships and to tackle the security problems associated with bi-directional communication, frequently changing heterogeneous communication partners and with the mobility of communication partners. The solutions to cope with the huge amount of privacy related data will certainly be twofold: Technologically, data gathering and sharing must be mitigated as far as possible while the remaining risk must be minimized organizationally.

However, the solutions to the aforementioned problems look like the move from previously confined devices with limited external interfaces to networking systems increases the resulting attack surface of the whole system significantly. In turn, this leads to the requirement that all software which is created must be designed and implemented using state-of-the-art secure software processes, to avoid potential implementation level vulnerabilities [26] as otherwise, software insecurities could expose the systems. For instance, [27] documented a buffer overflow vulnerability in a smart meter firmware. Based on this finding, it was demonstrated that this vulnerability could enable an adversary to create a bot-net like structure on these devices via self-replicating malware. A scenario in which an attacker fully controls a large number of smart meters could lead to potentially serious consequences.

In addition to secure development practices, properly defined processes for secure and timely software updates of all rolled-out devices are needed for risk mitigation. The sheer number of potentially affected devices will probably rule out on-premise updating of defective firmwares. Consequently, reliable mechanisms for updates over the network have to be investigated. This, in turn, requires sound proof of the authenticity and integrity of the transmitted firmware which has to be done via code-signing.

In addition to the security challenge, it is also very hard to create such a system to be safe and reliable in the first place. Reliability and safety are two attributes that, at least in history, have always been very high priorities for electrical grid operators but are also very heavily dependent on security.

One point that should be stressed here is the following: When the smart grid is fully realized, it will probably be the largest logical network of embedded devices (charging cars and smart meters), control systems (ICS) and traditional IT systems with a real impact on our everyday-life [28]. This means that a failure of such a system, however it was produced, would lead to a complete standstill of our society, unlike with similar networks (mobile phones, the Internet). Containment strategies in terms of organizational and technical means have to be devised in order to limit the impact range of attacks (security) or failures (safety).

5 Economic and Business Challenges

Distributed energy generation systems are usually located in close proximity to the actual consumption of the energy required and can be supported by storage or demand side management measures. Due to the close proximity additional energy i.e. thermal energy can be utilized, further losses through distribution and transmission are reduced making these systems overall often more efficient than central energy systems. In contrast, distributed energy systems operate at a much smaller scale, possibly making the marginal price for a single kilowatt-hour (kWh) more expensive than as this would apply for a central energy system. As production unit numbers of small scale systems increase these systems are becoming more and more economic to operate. Looking alone at the marginal price for a single kWh is in many cases not sufficient. Extended related energy applications for distributed energy systems take a further approach by providing more than an analysis of the marginal price perspective. As the systems are installed locally, many challenges of the past can now be tackled with greater precision on this local level. This includes the following: energy security concerns (power availability), power quality issues, tighter emissions standards and possible transmission and distribution bottlenecks.

The cost effectiveness of any power system can generally be characterized by comparison of revenue generated and costs involved. The aim of any business venture can be defined as to maximize profits. Especially for the energy sector investments in capital assets are involved with a high degree of initial investments. Furthermore, these assets have a long depreciation which makes it specifically important to best understand the capital streams involved over the lifecycle of such a system. The approach should be to look at economic viability of power systems by means of the cash flow involved, allowing to use valuation methods based on discounted cash flows.

For the involved stakeholder investing and operating, this party is interested in a capital return on his investments, whereas in a broader sense the opportunistic costs shall also be taken into comparison. Typical questions that need to be answered upon investing in such a power system as described above can be rendered as follows:

- Which kind of costs arise (prime costs, maintenance, commodities, site, emissions, etc.) ?
- When do they arise and how can they be valued?
- How long can the assets be utilized?

- What is the finance structure of the investment?
- For methods of how the investment can be financed this involves the following possible sources of initial or continues capital streams:
- Sales of energy feed into the electric grid
- Avoided costs for grid access
- Opportunistic costs of energy that would have been utilized otherwise.

The challenge of realizing decentralized control mechanisms as described in Sect. 3 is that potential benefits are distributed across the whole value chain of electricity delivery, whereas under the impact of unbundling, single companies may only be active in one or two of these activities. If overall benefits can be gained by a technology, but the party that has to invest into it is not the same as the one who is profiting most from it, regulation must step into the game and set the framework in a way as to give incentives for the former party.

6 Summary and Conclusion

Power systems are currently undergoing considerable changes. In order to be able to accommodate the growing number of fluctuating renewable generation, the current consumption-driven generation pattern must make room for the opposite paradigm, i.e. a generation-driven consumption. Through flexible demand that can react to the scarcity situation (in real-time) generation, it can be possible to avoid large investments into stand-by power plants that balance out the fluctuations caused by renewable sources and into expensive grid-reinforcements, all along with safeguarding a high level of security and reliability of supply.

This flexible reaction to different grid situation requires several prerequisites. First, information about the grid status needs to be made available consumers and prosumers in order to make them aware of what the best times for their consumption and generation is. Second, customers also need incentives to behave in the desirable way, as given by the grid status. This involves detailed measurements of consumption and generation feed-in in for small time intervals, and a variable pricing scheme that pushes low and high prices down to the customer. Some possible ways of designing such pricing schemes or mechanisms that involve the customers more actively are described in this chapter.

Besides, a safe and interoperable communication infrastructure that allows for bi-directional communication between different actors within the electricity system is needed. It must also be ensured that the customers' privacy is preserved. At the economic level, it must be ensured that companies in different parts of the value chain can profit from an investment into the enabling technologies required for the necessary transition of the energy system. This may require a change in regulation or policies, and it also requires creativity for discovering new business models in the changed framework.

All these aspects have been discussed in this chapter, and were put into relation to the trends that can be perceived in today's energy systems.

References

1. European Union (1996) Directive 96/92/EC of the European Parliament and of the Council of 19 December 1996 concerning common rules for the internal market in electricity. Official Journal of the European Union, L 027
2. International Energy Agency (2008) World energy outlook 2008. Technical report, OECD/IEA
3. MacKay DJ (2009) Sustainable Energy – without the hot air. UIT Cambridge Ltd, Cambridge
4. Kempton W, Tomić J (2005) Vehicle-to-grid power implementation: from stabilizing the grid to supporting large scale renewable energy. J Power Sources 144(1):280–294
5. United States (2007) Energy independence and security act of 2007. U.S. G.P.O., Washington DC, pp 110–140
6. European Union (2006) Directive 2006/92/EC of the European Parliament and of the Council of 5 April 2006 on energy end-use efficiency and energy services and repealing Council Directive 93/76/EEC. Official Journal of the European Union, L 114/64
7. Doty S, Turner WC (eds) (2009) Energy management handbook, 7th edn. Fairmont Press, Lilburn
8. Nestle D, Ringelstein J (2009) Application of bidirectional energy management interfaces for distribution grid services. In: 20th international conference on electricity distribution CIRED, Prague
9. U.S. Department of Energy (2006) Benefits of demand response in electricity markets and recommendations for achieving them. Technical report, U.S. DOE. http://eetd.lbl.gov/ea/EMP/reports/congress-1252d.pdf
10. European Energy Exchange (2010) Market data
11. German Federal Ministry of Economics and Technology (2010) Energiedaten
12. Bye T, Hope E (2005) Deregulation of electricity markets – The Norwegian experience. Discussion papers 433, Research Department of Statistics Norway. http://ideas.repec.org/p/ssb/dispap/433.html. Accessed Sept 2005
13. Weidlich A (2008) Engineering interrelated electricity markets – an agent-based computational approach, Contributions to management science. Springer Physica, Heidelberg
14. Wilson R (1985) Incentive efficiency of double auctions. Econometrica 53(5):1101–1115
15. Franke M, Rolli D, Kamper A, Dietrich A, Geyer-Schulz A, Lockemann P, Schmeck H, Weinhardt C (2005) Impacts of distributed generation from virtual power plants. In; Proceedings of the 11th annual international sustainable development research conference, Helsinki, pp 1–12
16. Kok K, Scheepers M, Kamphuis R (2009) Intelligent infrastructures, chapter intelligence in electricity networks for embedding renewables and distributed generation, Intelligent systems, control and automation: science and engineering series. Springer, Dordrecht, pp 179–209
17. Lamparter S, Becher S, Fischer J-G (2010) An agent-based market platform for smart grids. In: Proceedings of the 9th international conference on autonomous agents and multiagent system AAMAS, Toronto, pp 1689–1696
18. Akkermans H, Schreinemakers J, Kok K (2004) Microeconomic distributed control: theory and application of multi-agent electronic markets. In: CRIS 2004 – 2nd international conference on critical infrastructures, Grenoble, pp 163–176
19. Schnizler B, Neumann D, Veit D, Weinhardt C (2008) Trading grid services – a multi-attribute combinatorial approach. Eur J Oper Res (EJOR) 187(3):943–961
20. Block C, Fraunhofer FB, Fraunhofer PB, Briegel F, Burger N, Drzisga T, Fey B, Frey H, Hartmann J, Kern C, Muhs M, Plail B, Schetters GPL, Schöpf F, Schumann D, Schwammberger F, Terzidis O, Thiemann R, van Dinther C, von Sengbusch K, Weidlich A, Weinhardt C (2010) Internet of energy: ICT for energy markets of the future. BDI publication No. 439, Federation of German Industries (BDI e.V.), Berlin. www.bdi.eu. http://www.bdi.eu/BDI_english/download_content/ForschungTechnikUndInnovation/BDI_initiative_IoE_ us-IdE-Broschure.pdf. Accessed Feb 2010

21. SmartGrids European Technology Platform (2008) Smartgrids: strategic deployment document for Europe's electricity networks of the future. http://www.smartgrids.eu/documents/sra/sra_finalversion.pdf
22. Karnouskos S, Terzidis O (2007) Towards an information infrastructure for the future internet of energy. In: Kommunikation in Verteilten Systemen (KiVS 2007) Conference, VDE Verlag, 26 Feb 2007–02 Mar 2007
23. NIST (2010) NIST framework and roadmap for smart grid interoperability standards. Technical Report NIST Special Publication 1108, National Institute of Standards and Technology (NIST). http://www.nist.gov/public_affairs/releases/smartgrid_interoperability_final.pdf. Accessed Jan 2010
24. Sultanem F (1991) Using appliance signatures for monitoring residential loads atmeter panel level. IEEE Trans Power Deliv 6(4):1380–1385. doi:10.1109/61.97667, ISSN 0885–8977
25. Bauer G, Stockinger K, Lukowicz P (2009) Recognizing the use-mode of kitchen appliances from their current consumption. In: EuroSSC, Heidelberg, pp 163–176
26. McGraw G (2010) Software [In]security: The smart (electric) grid and dumb cybersecurity. [online], http://www.informit.com/articles/article.aspx?p=1577441. Accessed Mar 2010
27. Davis M (2009) Smart grid device security – adventures in a new medium. Talk at the black hat USA 2009 conference, Las Vegas. http: //www.blackhat.com/presentations/bh-usa-09/MDAVIS/BHUSA09-Davis-AMI-SLIDES.pdf. Accessed July 2009
28. CISCO (2009) Securing the smart grid. Whitepaper, CISCO. http://www.cisco.com/web/strategy/docs/energy/SmartGridSecurity_wp.pdf
29. McAfee RP, McMillan J (1987) Auctions and bidding. J Econ Lit 25(2):699–738

Realizing an Interoperable and Secure Smart Grid on a National Scale

George W. Arnold

Abstract The structure of the electrical system has not changed much since it was first developed: it is characterized by the one-way flow of electricity from centralized power generation plants to users. The smart grid will enable the dynamic, two-way flow of electricity and information needed to support growing use of distributed green generation sources (such as wind and solar), widespread use of electric vehicles, and ubiquitous intelligent appliances and buildings that can dynamically adjust power consumption in response to real-time electricity pricing. The realization of the smart grid is a huge undertaking requiring an unprecedented level of cooperation and coordination across the private and public sectors. A robust, interoperable framework of technical standards is critical to making it happen.

Keywords Cyber security • Electric transportation • Energy management • Interoperability • Renewable energy • Smart grid • Standards

1 Introduction: Why Is the Smart Grid Needed?

Modernization of the electric power grid is central to national efforts to reduce greenhouse gas emissions, achieve greater security in energy supply, and increase the reliability and security of the electric system. Around the world, billions of dollars are being spent to build elements of what ultimately will be "smart" electric power grids.

Fossil fuels that are burned to produce electricity represent a significant source of greenhouse gas emissions that contribute to global warming. Most electricity is

---

G.W. Arnold (✉)
U.S. Department of Commerce, National Institute of Standards and Technology, Gaithersburg, MD, USA
e-mail: george.arnold@nist.gov

A. Sorokin et al. (eds.), *Handbook of Networks in Power Systems I*,
Energy Systems, DOI 10.1007/978-3-642-23193-3_19,

generated from coal, oil and natural gas. On a global basis in 2007, 68% of electricity was generated from these sources [1]. For the United States, the proportion was somewhat higher: 72% [2]. In the United States, electric-power generation accounts for about 40% of human-caused emissions of carbon dioxide, the primary greenhouse gas [3]. If the current power grid were just 5% more efficient, the resultant energy savings would be equivalent to permanently eliminating the fuel consumption and greenhouse gas emissions from 53 million cars [4]. The need to reduce carbon emissions has become an urgent global priority to mitigate climate change.

Many nations that rely heavily on imported oil are concerned about the security of their energy supply. While oil represents less than 2% of the fuel used to generate electricity in the U.S. [2], transportation is heavily dependent on oil. Substituting "green" electricity for oil to provide heating and to power transportation has a double benefit by reducing carbon emissions while also increasing the security of energy supply.

Modern society has become highly dependent on a reliable electrical system. Interruptions to power supply are estimated to cost the U.S. economy $80 billion annually [5]. With the pervasive application of electronics and microprocessors, reliable and high quality electric power is becoming increasingly important. However, the basic architecture of the aging electrical system has changed little over the last century. Improvements to the reliability and quality of electricity supply are needed to meet the demands of twenty-first century society.

In summary, the development of the smart grid is intended to support the following goals:

- Help reduce energy use overall and increase grid efficiency;
- Enable increased use of renewable "green" sources of energy such as wind and solar;
- Provide the electrical infrastructure needed to support widespread use of electric vehicles; and
- Enhance the reliability and security of the electric system.

2 Characteristics of the Present Electric Grid

The electric grid in the U.S. is owned and operated by over 3,100 electric utilities which are interconnected nationally through ten Independent System Operators (ISO)/Regional Transmission Organizations (RTO) that coordinate the bulk power system and wholesale electricity market.

The structure of the present electric grid was designed to support a one-way flow of electricity from centralized bulk generation facilities through a transmission and distribution network to customers (See Fig. 1). Most electricity is generated by coal, natural gas, nuclear and hydroelectric plants whose output under normal conditions is predictable and controllable. Demand for electricity varies considerably

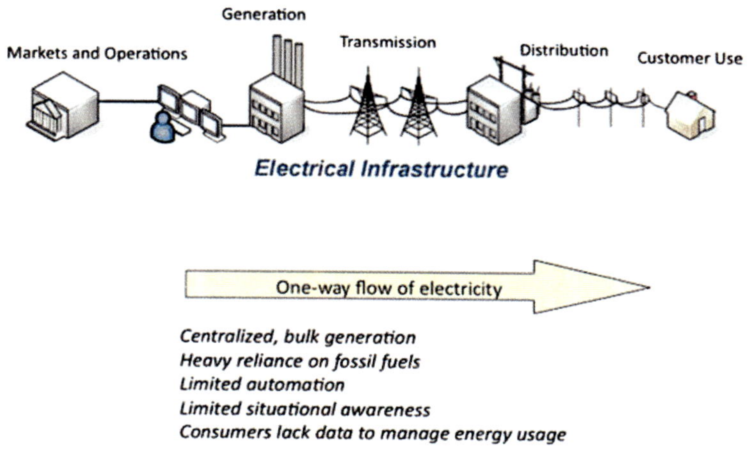

Fig. 1 Characteristics of the present electric grid

according to time of day and season. Generating capacity must be provided to handle peak periods. During periods of low demand, that generating capacity is idle. Some generation facilities, which cannot be dispatched on demand, serve as "spinning reserves", operating continuously even if their output is not needed to satisfy demand.

The transmission and distribution networks that carry electricity from generating plants to the customer have limited capability to monitor and report on their condition in real time. Advanced sensors called phasor measurement units (PMU) that can measure the condition of transmission facilities are not yet widely deployed. At the distribution level, in many areas, the only indication that an electric utility receives of an outage is the customer trouble report. There is limited ability to remotely re-route power around a failed line.

Customers receive limited information about their own energy use that is helpful in monitoring and reducing their energy consumption. In most cases that information is limited to monthly usage readings. "Smart meters" that capture electricity usage data in near-real time and can transmit the data electronically to the utility and the customer are just beginning to be deployed.

3 Smart Grid Benefits

Following are a few examples of the benefits that will be enabled by a modernized, "smart" electric grid.

Reduce Peak Usage. Electric grids must be in balance at all times, matching generated electricity with load. Hourly and seasonal variability in load requires that sufficient generating capacity be available to handle peak periods. A significant fraction of this capacity is idle most of the time, resulting in inherent inefficiency. If usage during the peak hours could be shifted to non-peak periods or otherwise curtailed (a capability referred to as Demand Response), a significant fraction of generation capacity could be saved. The U.S. Federal Energy Regulatory Commission estimates the potential for peak electricity demand reductions to be equivalent to up to 20% of national peak demand – enough to eliminate the need to operate hundreds of back-up power plants [6]. Demand response can be implemented either through direct load control by the utility, or indirectly through market forces with dynamic or time-of-use pricing of electricity. A smart grid would enable dynamic adjustment of load as well as generation capacity, by providing near real time information on usage and price to a customer's energy management system or smart appliances.

Enable Large-Scale Use of Renewable Sources of Energy. U.S. Administration energy policies are intended to double renewable energy generating capacity, to 10%, by 2012 [7] – an increase in capacity that is enough to power 6 million American homes. Legislative proposals being considered envision a further increase in renewable energy capacity to 15–20% by 2021. Some states have set higher targets – California, for example, has a goal to achieve 33% by 2020 [8].

Solar and wind represent an abundant source of clean, renewable energy. However, unlike traditional energy sources, solar and wind are variable and intermittent. Integrating solar and wind into the grid presents a new challenge as the penetration reaches the target levels because of the need to continually balance load with a varying and less predictable supply. Energy storage technologies that can buffer varying supply will become an important element of the smart grid.

While some solar and wind generation will be deployed in centralized, large-scale "farms", a growing proportion of renewable generation will be distributed locally, for example solar rooftop panels. Some communities will function as "micro-grids" capable of generating, at times, enough power to satisfy their own needs, or selling excess power back into the grid at other times, or buying power from the grid at other times. The smart grid will have to support much more distributed power generation and two-way flow of electricity.

Provide Infrastructure for Electric Transportation. Over the long term, the integration of the power grid with the nation's transportation system has the potential to yield huge energy savings and other important benefits. Estimates of associated potential benefits include:

- Displacement of about half of the nation's net oil imports;
- Reduction in U.S. carbon dioxide emissions by about 25%; and
- Reductions in emissions of urban air pollutants of 40–90%.

A DoE study found that the idle capacity of today's electric power grid could supply 70% of the energy needs of today's cars and light trucks without adding to generation or transmission capacity – if the vehicles charged during off-peak times [9]. The smart grid will provide capabilities to monitor and manage the charging of electric vehicles to avoid overloading the grid and minimize cost for consumers.

Provide Tools for Customers to Manage and Reduce Energy Use. Presently customers have little information available to understand or manage their energy use. An advanced metering infrastructure (AMI), a key element of the smart grid, will allow near real-time measurement of customer energy use. The smart grid will also provide customers with information management capabilities that permit smart appliances and energy management systems to minimize energy use and shift demand to less costly non-peak periods, saving money.

4 Vision of the Smart Grid

While definitions and terminology vary somewhat, all notions of an advanced power grid for the twenty-first century hinge on adding and integrating many varieties of digital computing and communication technologies and services with the power-delivery infrastructure. Bi-directional flows of energy and two-way communication and control capabilities will enable an array of new functionalities and applications that go well beyond "smart" meters for homes and businesses (see Fig. 2). Following are some additional descriptive characteristics of the future smart grid:

- High penetration of renewable energy sources: 20–35% by 2020;
- Distributed generation and microgrids;
- "Net" metering – selling local power into the grid;
- Distributed storage;
- Smart meters that provide near-real time usage data;
- Time of use and dynamic pricing;
- Ubiquitous smart appliances communicating with the grid;

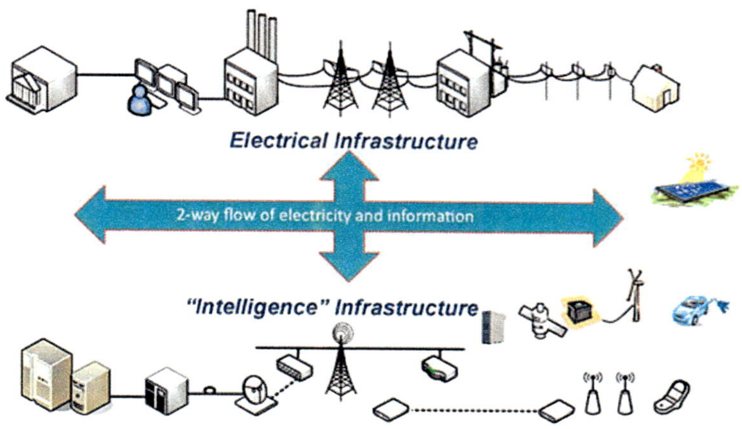

Fig. 2 Characteristics of the smart grid

- Energy management systems in homes as well as commercial and industrial facilities linked to the grid;
- Growing use of plug-in electric vehicles; and
- Networked sensors and automated controls throughout the grid.

Developing and deploying the smart grid is also expected to have a positive effect on the economy by creating significant numbers of new jobs and opportunities for new businesses. A consultant study performed for the GridWise Alliance estimates that 280,000 new jobs will be created during the early deployment of the smart grid in the U.S. (2009–2012) and 140,000 new jobs in the steady state (2013–2018) [10]. The numbers represent new jobs in electric utilities, their contractors and supply chain, as well as new businesses enabled by the smart grid.

5 Smart Grid National Policy in the United States

The electric grid is often described as the largest and most complex system ever developed. The effort required to transform this critical national infrastructure to the envisioned smart grid is unprecedented in its scope and breadth. It will demand unprecedented levels of cooperation to achieve the ultimate vision. In the United States, the Energy Independence and Security Act (EISA) of 2007 [11], states that support for creation of a smart grid is the national policy. Distinguishing characteristics of the smart grid cited in the act include:

- Increased use of digital information and controls technology to improve reliability, security, and efficiency of the electric grid;
- Dynamic optimization of grid operations and resources, with full cyber security;
- Deployment and integration of distributed resources and generation, including renewable resources;
- Development and incorporation of demand response, demand-side resources, and energy-efficiency resources;
- Deployment of "smart" technologies for metering, communications concerning grid operations and status, and distribution automation;
- Integration of "smart" appliances and consumer devices;
- Deployment and integration of advanced electricity storage and peak-shaving technologies, including plug-in electric and hybrid electric vehicles, and thermal-storage air conditioning;
- Provision to consumers of timely information and control options; and
- Development of standards for communication and interoperability of appliances and equipment connected to the electric grid, including the infrastructure serving the grid.

In the United States, the transition to the smart grid already is under way, and it is gaining momentum as a result of both public and private sector investments. The American Recovery and Reinvestment Act of 2009 (ARRA) included a Smart Grid

Table 1 Department of energy smart grid investment grants [12]

Category	US $ Millions	Examples of equipment (across all categories)
Integrated/crosscutting	$ 2,150	18 million smart meters
AMI	$ 818	1.2 million in-home displays
Distribution	$ 254	206,000 smart transformers
Transmission	$ 148	177,000 load control devices
Customer systems	$ 32	170,000 smart thermostats
Manufacturing	$ 26	877 networked phasor measurement units
Total	$ 3,429	671 automated substations
		1.2 million in-home displays
		100 PEV charging stations

Investment Grant Program (SGIG) which provides $3.4 billion for cost-shared grants to support manufacturing, purchasing and installation of existing smart grid technologies that can be deployed on a commercial scale (Table 1).

6 Standards Framework for the Smart Grid

A significant aspect of the EISA legislation is the recognition of the critical role of technical standards in the realization of the smart grid.

Nearly 80% of the U.S. electrical grid is owned and operated by about 3,100 private sector utilities and the equipment and systems comprising the grid are supplied by hundreds of vendors. Transitioning the existing infrastructure to the smart grid requires an underlying foundation of standards and protocols that will allow this complex "system of systems" to interoperate seamlessly and securely. Establishing standards for this critical national infrastructure is a large and complex challenge.

Recognizing this, Congress assigned the responsibility for coordinating the development of interoperability standards for the U.S. smart grid to the National Institute of Standards and Technology (NIST) in the Energy Independence and Security Act of 2007. NIST, a non-regulatory science agency within the U.S. Department of Commerce, has a long history of working collaboratively with industry, other government agencies, and national and international standards bodies in creating technical standards underpinning industry and commerce.

The DOE announcement instructs grant applicants that their project plans should describe their technical approach to "addressing interoperability," including a "summary of how the project will support compatibility with NIST's emerging smart grid framework for standards and protocols."

There is an urgent need to establish standards. Some smart grid devices, such as smart meters, are moving beyond the pilot stage into large-scale deployment. The DoE Smart Grid Investment Grants will accelerate deployment. In the absence of standards, there is a risk that these investments will become prematurely obsolete or, worse, be implemented without adequate security measures. Lack of standards may also impede the realization of promising applications, such as smart appliances that are responsive to price and demand response signals. In early 2009, recognizing the urgency, NIST intensified and expedited efforts to accelerate progress in identifying and actively coordinating the development of the underpinning interoperability standards.

NIST developed a three-phase plan [13] to accelerate the identification of standards while establishing a robust framework for the longer-term evolution of the standards and establishment of testing and certification procedures. In May 2009, U.S. Secretary of Commerce Gary Locke and U.S. Secretary of Energy Steven Chu chaired a meeting of nearly 70 executives from the power, information technology, and other industries at which they expressed their commitment to support NIST's plan.

Phase 1 of the NIST plan engaged over 1,500 stakeholders representing hundreds of organizations in a series of public workshops over a six month period to create a high-level architectural model for the smart grid, analyze use cases, identify applicable standards, gaps in currently available standards, and priorities for new standardization activities. The result of this phase, "NIST Special Publication 1108 – NIST Framework and Roadmap for Smart Grid Interoperability Release 1.0" was published in January 2010 [14].

Phase 2 established a more permanent public-private partnership, the Smart Grid Interoperability Panel, to guide the development and evolution of the standards. This body is also guiding the establishment of a testing and certification framework for the smart grid, which is Phase 3 of the NIST plan.

7 NIST Smart Grid Interoperability Framework Release 1.0

Reference Model The smart grid is a very complex system of systems. There needs to be a shared understanding of its major building blocks and how they inter-relate (an architectural reference model) in order to analyze use cases, identify interfaces for which interoperability standards are needed, and to develop a cyber security strategy. The reference model partitions the smart grid into seven domains (bulk generation, transmission, distribution, markets, operations, service provider, and customer) as illustrated in Fig. 3.

Underlying the conceptual model is a legal and regulatory framework that includes policies and requirements that apply to various actors and applications and to their interactions. Regulations, adopted by the Federal Energy Regulatory Commission at the federal level and by public utility commissions at the state and local levels, govern many aspects of the Smart Grid. Such regulations are intended to ensure that electric rates are fair and reasonable and that security, reliability, safety, privacy, and other public policy requirements are met. The transition to the Smart Grid introduces new regulatory considerations, which may transcend jurisdictional boundaries and require increased coordination among federal, state, and

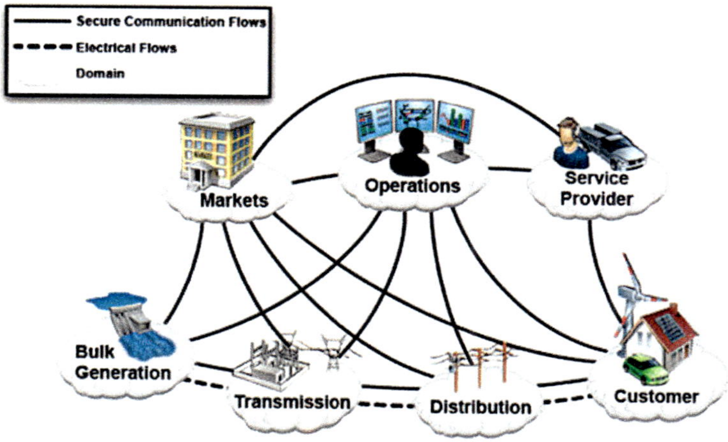

Fig. 3 Smart grid domains

local lawmakers and regulators. The conceptual reference model must be consistent with the legal and regulatory framework and support its evolution over time.

The reference model also identifies major actors and applications within each domain and interfaces among them over which information must be exchanged and for which interoperability standards are needed (see Fig. 4). The reference model is being further developed and maintained by a Smart Grid Architecture Committee within the Smart Grid Interoperability Panel.

One aspect of the reference model related to metering is the distinction made between the "meter" and the "energy services interface." At a minimum, meters need to perform the traditional metrology functions (measuring electricity usage), connect or disconnect service, and communicate over a field area network to a remote meter data management system. These basic functions are unlikely to change during the meter service life of 10 years or more. More advanced functions such as communication of pricing information, demand response signaling, and providing energy usage information to a home display or energy management system are likely to undergo significant change as innovations enabled by the smart grid occur and new applications appear in the market. The reference model associates these functions with the energy services interface to allow for the possibility of rapid innovation in such services without requiring that they be embedded in the meter.

Initial Standards The Release 1 framework identifies 75 standards or families of standards that are applicable or likely to be applicable to support smart grid development. The standards address a range of functions, such as basic communication protocols (e.g. IPv6), meter standards (ANSI C12), interconnection of distributed energy sources (IEEE 1547), information models (IEC 61850), cyber security (e.g. the NERC CIP standards) and others. The standards identified are produced by 27 different standards development organizations at the national and

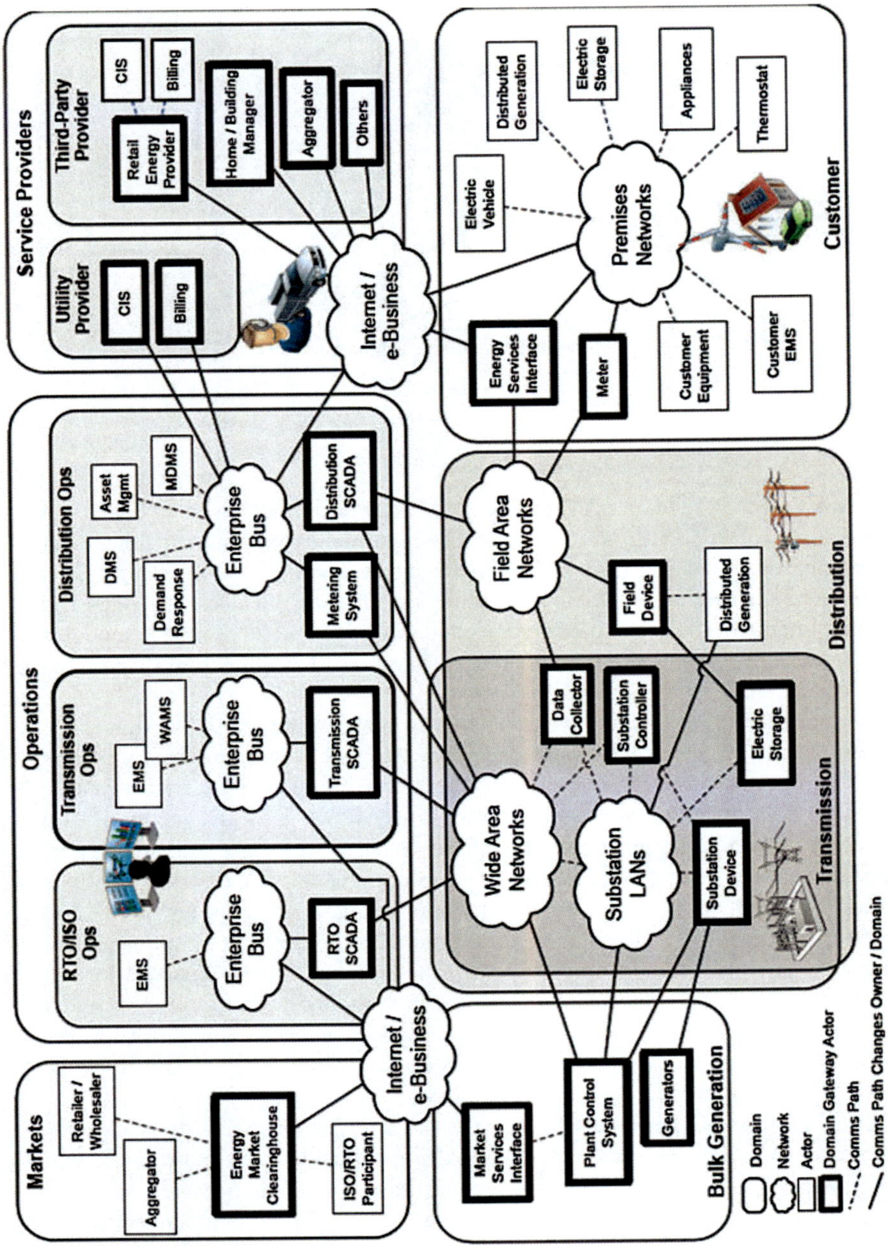

Fig. 4 Smart grid conceptual reference diagram

international level, such as IEC, ISO, IEEE, SAE, IETF, NEMA, NAESB, and many others.

Roadmap In the course of reviewing the standards during the NIST workshops, 70 gaps and issues were identified pointing to existing standards that need to be revised or new standards that need to be created. NIST has worked with the standards development community to initiate 16 priority action plans to address the most urgent of the 70 gaps.

An example of one of these issues pertains to smart meters. The ANSI C12.19 standard, which defines smart meter data tables, is one of the most fundamental standards needed to realize the smart grid. Unless the data captured by smart meters is defined unambiguously, it will be impossible to create smart grid applications that depend on smart meter data. The existing ANSI C12.19 standard defines over 200 data tables but does not indicate which are mandatory. Different manufacturers have implemented various subsets of the standard, presenting a barrier to interoperability. In addition, the standard permits manufacturer-defined data tables with proprietary functionality that is not interoperable with other systems. To address this problem, one of the 16 priority action plans defined in the NIST roadmap was established to update the ANSI C12.19 standard to define common data tables that all manufacturers must support to ensure interoperability.

Manufacturers require lead-time to implement the revised standard. In the meantime, smart meters are in the process of being deployed and public utility commissions are concerned that they may become obsolete. To address the issue, NIST requested the National Electrical Manufacturers Association to lead a fast-track effort to develop a meter upgradeability standard. Developed and approved in just 90 days, the NEMA Smart Grid Standards Publication SG-AMI 1-2009, "Requirements for Smart Meter Upgradeability," is intended to provide reasonable assurance that meters conforming to the standard will be securely field-upgradeable to comply with anticipated revisions to ANSI C12.19.

Other priority action plans that are underway to accelerate and coordinate the work of standards bodies include:

- Standard protocols for communicating pricing information, demand response signals, and scheduling information across the smart grid;
- Standard for access to customer energy usage information;
- Guidelines for electric storage interconnection;
- Common object models for electric transportation;
- Guidelines for application of internet protocols to the smart grid;
- Guidelines for application of wireless communication protocols to the smart grid;
- Standards for time synchronization;
- Common information model for distribution grid management;
- Transmission and distribution systems model mapping;
- IEC 61850 objects/DNP3 mapping; and
- Harmonize power line carrier standards for appliance communications in the home.

Cyber security Ensuring cyber security of the smart grid is a critical priority. Security must be designed in at the architectural level, not added on later. Information technology (IT) and telecommunications infrastructures play a critical role in the smart grid. Therefore, the security of systems and information in the IT and telecommunications infrastructures must be addressed by an increasingly diverse electric sector. Security must be included at the design phase to ensure adequate protection.

Cyber security must address not only deliberate attacks, such as from disgruntled employees, industrial espionage, and terrorists, but also inadvertent compromises of the information infrastructure due to user errors, equipment failures, and natural disasters. Vulnerabilities might allow an attacker to penetrate a network, gain access to control software, and alter load conditions to destabilize the grid in unpredictable ways. Additional risks to the grid include:

- Increasing the complexity of the grid could introduce vulnerabilities and increase exposure to potential attackers and unintentional errors;
- Interconnected networks can introduce common vulnerabilities;
- Increasing vulnerabilities to communication disruptions and introduction of malicious software could result in denial of service or compromise the integrity of software and systems;
- Increased number of entry points and paths for potential adversaries to exploit; and
- Potential for compromise of data confidentiality, including the breach of customer privacy.

The need to address potential vulnerabilities has been acknowledged across the federal government.

A NIST-led Cyber Security Coordination Task Group consisting of more than 400 participants from the private and public sectors was formed to develop a cyber security strategy and requirements for the smart grid. Activities of the task group included identifying use cases with cyber security considerations; performing a risk assessment including assessing vulnerabilities, threats and impacts; developing a security architecture linked to the smart grid conceptual reference model; and documenting and tailoring security requirements to provide adequate protection (see Fig. 5). Results of the task group's work are described in a publication NIST IR 7628 [15].

Additional Considerations There are many additional issues and considerations that must be addressed in developing the smart grid. Several of these are discussed below.

Standards for the Smart Grid should consider electromagnetic disturbances, including severe solar (geomagnetic) storm risks and Intentional Electromagnetic Interference (IEMI) threats such as High-Altitude Electromagnetic Pulse (HEMP). Our modern high-tech society is built upon a foundation vulnerable to electromagnetic disturbances. The existence and potential impacts of such threats provide impetus to evaluate, prioritize, and protect/harden the new Smart Grid.

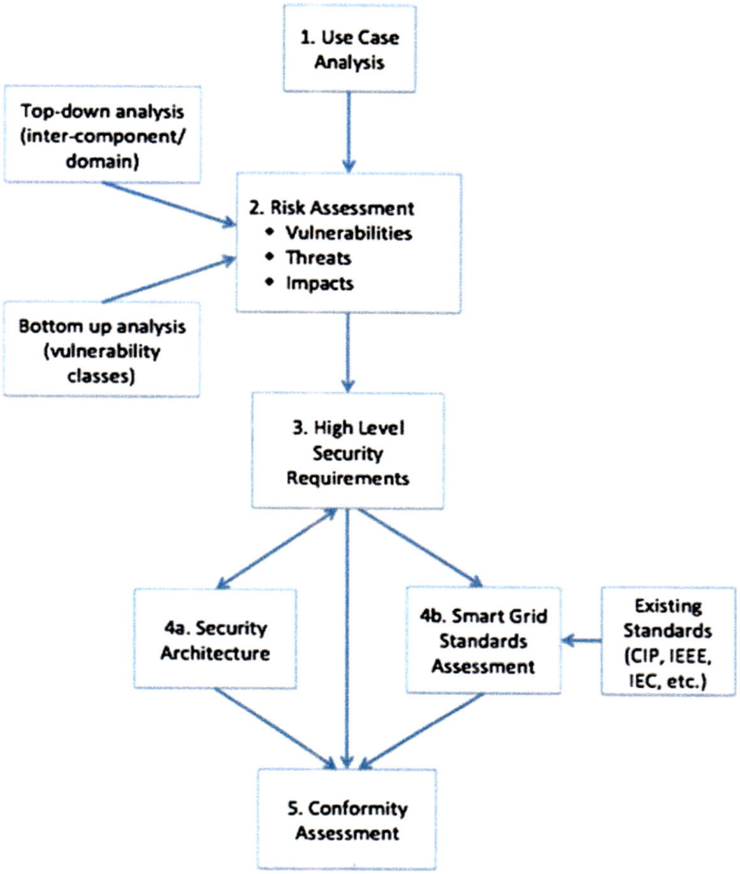

Fig. 5 Activities in developing the smart grid cyber security strategy

The burgeoning of communications technologies, both wired and wireless, used by Smart Grid equipment can lead to EMC interference, which represents another standards issue requiring study. Additionally, new options may be considered, such as the allocation of dedicated spectra for utility communications. Support of multiple standards is appropriate to meet different real-world requirements and coincides with Congress's requirement that the NIST Interoperability Framework be technology-neutral to encourage innovation. However, some communications technologies perform better in some environments than others. Additional research is needed to identify and evaluate potential interference issues, to offer technical guidance to mitigate interference, and to inform utilities' communications technology choices.

In addition to the wireless transmitters discussed above, electromagnetic interference sources include electrostatic discharge, fast transients, and surges, which can lead to interruptions of service. The ability to withstand this interference with sufficient immunity without causing interference to other devices or systems is generally termed electromagnetic compatibility (EMC). There are significant benefits, including minimizing overall costs, to incorporating EMC up front in system development through modeling, simulation, and testing to appropriate standards. EMC standards and testing issues relating to the Smart Grid need to be addressed.

The benefits anticipated by Smart Grid systems also come with privacy risks that must be addressed. The ability to access, analyze, and respond to a much wider range of data from all levels of the electric grid poses a significant concern from a privacy viewpoint, particularly when the data, resulting analysis and assumptions, are associated with individual consumers or dwellings. The privacy implications of frequent meter readings being fed into the Smart Grid networks could provide a detailed time line of activities occurring inside the home. This data may point to a specific individual or give away privacy sensitive data.

The constant collection and use of smart meter data has also raised potential surveillance possibilities posing physical, financial, and reputational risks that must be addressed. Many more types of data are being collected, generated and aggregated within the Smart Grid than when the only data collected was through monthly meter readings by the homeowner or utility employee. Numerous additional entities outside of the energy industry may also be collecting, accessing, and using the data, such as entities that are creating applications and services specifically for smart appliances, smart meters and other yet-to-be-identified purposes. Additionally, privacy issues arise from the question of the legal ownership of the data being collected. With ownership comes both control and rights with regard to usage. If the consumer is not considered the owner of the data obtained from metering and home automation systems, the consumer may not receive the privacy protections provided to data owners under existing laws.

Adaptation of well-established methods for protecting consumer privacy is necessary to keep up with the multitude of use cases of the various technologies and business processes created for the Smart Grid. Legal and regulatory frameworks can be further harmonized and updated as the Smart Grid becomes more pervasive. A potential additional measure of protection for consumers' privacy would be in the design of Smart Grid applications and devices that allows consumers to have control of their personal information to the greatest extent possible.

The safe operation of the smart grid is of primary importance to all stakeholders; thus it is critical to incorporate appropriate safety procedures, criteria, and considerations into the relevant Smart Grid standards. For example, without proper attention to safety in standards, utility crews or first responders could find themselves in situations where they are potentially exposed to live wires connected to such sources as energy storage units or photovoltaic solar panels. These and other related issues need to be addressed in a comprehensive manner across the smart grid. Safety considerations must not only be addressed in transmission and distribution systems, but also other devices and systems (such as the operation of Smart

Grid consumer products in the home). The smart grid must be reflected in relevant standards and codes such as the National Fire Protection Association's National Electric Code and IEEE's National Electric Safety Code. Only through a coordinated effort that includes a demonstrated compliance to safety criteria will it be possible to ensure that the Smart Grid operates in a manner that does not threaten life or property.

8 Conformance Testing and Certification

Standards are critical to enabling interoperable systems and components. Mature, robust standards are the foundation of mass markets for the millions of components that will have a role in the future Smart Grid. Standards enable innovation where components may be constructed by thousands of companies. They also enable consistency in systems management and maintenance over the life cycles of components.

While standards are necessary for achieving interoperability, they are not sufficient. A conformance testing and certification regime is essential. In order to support interoperability of Smart Grid systems and products, smart grid products developed to conform to the interoperability framework should undergo a rigorous standards conformity and interoperability testing process. NIST has initiated a program to develop a Smart Grid Conformity Testing Framework within the Smart Grid Interoperability Panel which is described below.

9 Evolution of the Standards Framework

The reference model, standards, gaps and action plans described in the NIST Release 1.0 Smart Grid Framework and Roadmap provided an initial foundation for a secure, interoperable smart grid. However this initial document represents only the beginning of an ongoing process that is needed to create the full set of standards that will be needed and to manage their evolution in response to new requirements and technologies.

In Phase 2 of the NIST smart grid program, a public-private partnership, the Smart Grid Interoperability Panel (SGIP) was formed to provide a more permanent organizational structure to support the ongoing evolution of the framework. The SGIP provides an open process for stakeholders to participate in the ongoing coordination, acceleration and harmonization of standards development for the smart grid. The SGIP does not write standards, but serves as a forum to coordinate the development of standards and specifications by many standards development organizations. The SGIP reviews use cases, identifies requirements, coordinates and accelerates smart grid testing and certification, and proposes action plans for achieving these goals.

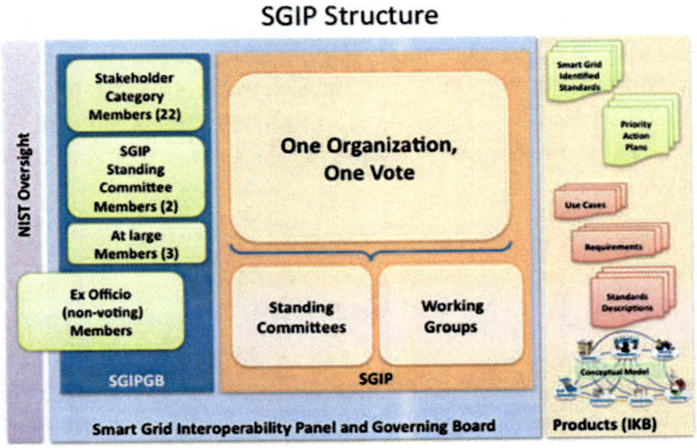

Fig. 6 Smart grid interoperability panel structure

The structure of the SGIP is illustrated in Fig. 6. The SGIP has two permanent committees. One committee is responsible for maintaining and refining the architectural reference model, including lists of the standards and profiles necessary to implement the vision of the smart grid. The other permanent committee is responsible for creating and maintaining the necessary documentation and organizational framework for testing interoperability and conformance with these smart grid standards and specifications.

The SGIP is managed and guided by a Governing Board that approves and prioritizes work and arranges for the resources necessary to carry out action plans. The Governing Board's responsibilities include facilitating a dialogue with standards development organizations to ensure that the action plans can be implemented.

The SGIP and its governing board are an open organization dedicated to balancing the needs of a variety of smart grid related organizations. Any organization may become a member of the SGIP. Members are required to declare an affiliation with an identified Stakeholder Category (22 have thus far been identified by NIST and are listed in Table 2). Members may contribute multiple Member Representatives, but only one voting Member Representative. Members must participate regularly in order to vote on the work products of the panel.

10 International Collaboration

Many countries have begun or are planning to modernize their electric grids, and significant smart grid programs are underway all over the world. Within Europe, for example, the dominant utility in Italy has fully deployed smart meters to its subscriber base, and experience with integration of variable renewable resources is being

Table 2 Smart grid stakeholder categories

Appliance and consumer electronics providers	Power equipment manufacturers and vendors
Commercial and industrial equipment manufacturers and automation vendors	Professional societies, users groups, trade associations and industry consortia
Consumers – residential, commercial, and industrial	R&D organizations and academia
Electric transportation industry stakeholders	Relevant federal government agencies
Electric utility companies – investor owned utilities and publicly owned utilities	Renewable power producers
Electric utility companies – municipal	Retail service providers
Electric utility companies – rural electric association	Standard and specification development organizations
Electricity and financial market traders (includes aggregators)	State and local regulators
Independent power producers	Testing and certification vendors
Information and communication technologies (ICT) infrastructure and service providers	Transmission operators and independent system operators
Information technology (IT) application developers and integrators	Venture capital

gained through large-scale deployments of wind and solar generation in countries such as Denmark, Portugal and Spain. In Japan a "Smart Community Alliance" has been formed to broaden the concept of the smart grid to encompass energy efficiency and management of other resources such as water, gas, and transportation. China is making significant investments to realize a "strong and smart grid" including ultra-high voltage transmission lines. The Australia federal government has invested AU $100 million for a National Energy Efficiency Initiative to develop an innovative Smart-Grid energy network. The South Korean government has announced a plan to establish a national smart grid. Brazil's energy regulator ANEEL has announced plans for a nationwide deployment of smart meters.

The United States' electric grid interconnects with Canada and Mexico, and the equipment and systems used in the grid is supplied by companies that address a global market. In addition, utilities and their customers benefit from lower prices that result when there is global supplier competition. Therefore a major goal of the NIST program is to utilize international standards wherever possible, and to ensure U.S. participation in the development of smart grid standards by international organizations. The NIST program works closely with IEC Strategic Group 3 on smart grid, and looks to various IEC TCs, such as TC57, which is working on a Common Information Model for the smart grid, to provide key parts of the NIST Smart Grid Framework. Other international organizations whose standards play an important role in the NIST framework include IEEE, IETF, ISO, ITU-T, SAE and others.

The process of international harmonization is also facilitated through bilateral communication and information exchange. To encourage international harmonization, participation in the NIST Smart Grid Interoperability Panel is open to organizations outside the U.S.

11 Conclusion

Realization of the smart grid represents one of the greatest engineering challenges of the twenty-first century. Its development and deployment in the U.S. is being accomplished within a national policy framework enacted in federal legislation. A robust foundation of standards is critical to achieving an interoperable and secure smart grid. This foundation is being developed through an innovative public/private partnership model.

References

1. International Energy Agency (2009) Key world energy statistics. http://www.iea.org/textbase/nppdf/free/2009/key_stats_2009.pdf. Accessed 26 Nov 2009
2. Energy Information Administration (2009) Official energy statistics from the U.S. Government. http://www.eia.doe.gov/fuelelectric.html. Accessed 26 Nov 2009
3. Energy Information Administration (2009) U.S. carbon dioxide emissions from energy sources, 2008 flash estimate. http://www.eia.doe.gov/oiaf/1605/flash/pdf/flash.pdf. Accessed 1 Dec 2009
4. U.S. Department of Energy (2008) The smart grid: an introduction. http://www.oe.energy.gov/SmartGridIntroduction.htm. Accessed 1 Dec 2009
5. LaCommare K, Eto J (2004) Understanding the cost of power interruptions to U.S. Electricity customers. Lawrence Berkeley National Laboratory LBNL-55718.
6. The Brattle Group (2009) A national assessment of demand response potential. http://www.brattle.com/_documents/UploadLibrary/Upload775.pdf. Accessed 1 Dec 2009
7. White House (2009) Progress report: the transformation to a clean energy economy. http://www.whitehouse.gov/administration/vice-president-biden/reports/progress-report-transformation-clean-energy-economy. Accessed 14 May 2010
8. Governor of the State of California (2009) Executive order S-21-09. http://gov.ca.gov/executive-order/13269/. Accessed 1 Dec 2009
9. Kintner-Meyer M, Schneider K, Pratt R (2006) Impacts assessment of plug-in hybrid vehicles on electric utilities and regional U.S. power grids part 1: technical analysis. Pacific Northwest National Laboratory, U.S. Department of Energy. http://www.ferc.gov/about/com-mem/wellinghoff/5-24-07-technical-analy-wellinghoff.pdf. Accessed 1 Dec 2009
10. KEMA, Inc (2009) The U.S. smart grid revolution KEMA's perspectives for job creation. http://www.kema.com/services/consulting/utility-future/job-report.aspx. Accessed 1 Dec 2009
11. Energy Independence and Security Act of 2007 (2007) [Public Law No: 110-140] Title XIII, Sec. 1301. http://frwebgate.access.gpo.gov/cgi-bin/getdoc.cgi?dbname=110_cong_public_laws&docid=f:publ140.110.pdf. Accessed 1 Dec 2009
12. U.S. Department of Energy (2009) Press release October 27, 2009. http://www.energy.gov/news2009/8216.htm. Accessed 31 Jan 2010
13. National Institute of Standards and Technology (2009) NIST announces three-phase plan for smart grid standards, paving way for more efficient, reliable electricity. Press release. http://www.nist.gov/public_affairs/smartgrid_041309.html. Accessed 1 Dec 2009
14. National Institute of Standards and Technology (2010) NIST special publication 1108 – NIST framework and roadmap for smart grid interoperability standards release 1.0. http://www.nist.gov/public_affairs/releases/smartgrid_interoperability.pdf. Accessed 31 Jan 2010
15. National Institute of Standards and Technology (2009) Smart grid cyber security strategy and requirements. NISTIR 7628 (Draft). http://csrc.nist.gov/publications/drafts/nistir-7628/draft-nistir-7628.pdf. Accessed 1 Dec 2009

Power System Reliability Considerations in Energy Planning

Panida Jirutitijaroen and Chanan Singh

Abstract We discuss how to incorporate reliability considerations into power system expansion planning problem. Power system reliability indexes can be broadly categorized as probabilistic and deterministic. Increasingly, the probabilistic criteria have received more attention from the utilities since these can more effectively deal with the uncertainty in system parameters. We propose a stochastic programming framework to effectively incorporate random uncertainties in generation, transmission line capacity and system load for the expansion problem. Favourable system reliability and cost trade off is achieved by the optimal solution. The problem is formulated as a two-stage recourse model where random uncertainties in area generation, transmission lines, and area loads are considered. Power system network is modelled using DC flow analysis. Reliability index used in this problem is the expected cost of load loss as it incorporates duration and magnitude of load loss. Due to exponentially large number of system states (scenarios) in large power systems, we apply sample-average approximation (SAA) concept to make the problem computationally tractable. The method is implemented on the 24-bus IEEE reliability test system.

Keywords Energy planning • power system reliability • sample average approximation • stochastic programming

P. Jirutitijaroen (✉)
National University of Singapore, Singapore
e-mail: elejp@nus.edu.sg

C. Singh
Texas A & M University, College Station, TX, USA
e-mail: singh@ece.tamu.edu

A. Sorokin et al. (eds.), *Handbook of Networks in Power Systems I*,
Energy Systems, DOI 10.1007/978-3-642-23193-3_20,

1 Introduction

Energy planning is the development of policy to ensure medium to long term energy supply and delivery to end consumers. Generation expansion problem addresses critical issues of optimal location for new generation resources. This information can be used by entities like Independent System Operators (ISO) to generate price signals or other incentives for materializing such resources. The decision policy should take into account reliability consideration that may influence possible solutions.

Reliability evaluation of power systems can be considered and characterised in two aspects; deterministic and probabilistic. Deterministic indexes are rules of thumb such as reserve as a percentage of the peak load or equal to the capacity of the largest unit. While the deterministic indexes are easy to assess and implement, they tend to provide too much margin of safety and are not suitable to trade off reliability and cost. Probabilistic indexes such as loss of load probability or expected energy not supplied, on the other hand, are much more complicated to incorporate in the problem but they can represent the quality of potential solutions in a more comprehensive and mathematically based manner. The integration of probabilistic reliability assessment is thus the focus of this chapter.

Several optimization techniques have been proposed for the generation expansion problem [1]. We mainly consider those with explicit uncertainties formulation. Among them, references [2, 3] propose stochastic programming framework to the planning problem and formulate it as a two-stage recourse model. The first stage decision on expansion policy is completed before the random uncertainties in generation, transmission line capacity and load are realized. After the random realization i.e. the generating capacity, load demand, and transmission capacity are known, the second stage decision is to determine the generating capacity from each bus to minimize operation cost with the assumption that the load is satisfied at all times. The problem is solved with large-scale deterministic equivalent problem.

In this chapter, we modify the formulation and incorporate reliability consideration in the planning problem [4, 5]. Reliability index used is expected cost of load loss which is computed in the second stage problem. The term load loss is used to indicate the load that could not be served because of generation or transmission deficiency. The overall objective is then to minimize expansion cost in the first stage and operation cost and expected cost of load loss in the second stage. Generating unit availability of additional units in terms of their forced outage rates and derated forced outage rates can also be included in this formulation.

The deterministic equivalent problem under stochastic programming framework becomes a large-scale linear programming problem with special structure. L-shaped algorithm is a standard algorithm for solving large-scale stochastic programming problem. Interested readers may refer to [6, 7] for details of the two-stage recourse model formulation and available solution algorithms. The algorithm considers entire probability space which, for a very large system, may be impractical and even impossible to enumerate and evaluate. Direct application of L-shaped algorithm cannot thus be achieved in a computationally effective manner.

A technique called sample-average approximation is used to overcome this problem of dimensionality. The sampling technique is employed to reduce the number of system states. The objective function of the second stage, called sample-average approximation (SAA) of the actual expected value, is defined by these samples. This approximation makes it possible to solve the problem with the deterministic equivalent model. The objective function values from the SAA problems are in fact estimates of the actual optimal objective values. These estimates yield upper bounds and lower bounds of the optimal objective value (actual expected value). Upper bounds and lower bounds of the optimal objective values can then be used to analyze the quality of the approximate solutions.

We also analyze in Sect. 4 the performance of optimal solution to reliability distribution of each node to examine if the system-wide reliability maximization can lead to fair reliability enhancement to all customers [8]. When a customer is charged equally the expansion cost, each may not receive the same reliability level. Since the formulation considers reliability in terms of the overall system, the optimal solution yields system reliability maximization, which may not guarantee fair improvement of reliability to customers at each bus.

The chapter is organized as follows. Detailed problem formulation is first introduced. Sample-average approximation technique is described next. Lower bound, Upper bound estimation of the actual objective value, and the approximate solution are presented and discussed. The method is implemented on a 24-bus IEEE-Reliability test system [9]. Comparative study of system-wide reliability maximization is examined. Concluding remarks are given in the last section.

2 Problem Formulation

The objective of energy expansion problem is to minimize the expansion cost while maximizing system reliability under uncertainty in generation, load, and transmission lines. Typically uncertainties in generation and transmission lines capacity are represented by a Two-Stage Markov model where uncertainty in system load is modeled according to its fluctuations throughout a year. When reliability is one of the constituents, a reliability model needs to be incorporated into the problem formulation. This underlying reliability evaluation model requires a flow model for evaluating system states regarding their loss of load status. A commonly accepted approach in composite reliability evaluation is to use a DC load flow model. The capacity of every element in the network is represented by a random variable with its discrete probability distribution. Using system expected cost of load loss as a reliability index, the problem is formulated as a two-stage recourse model.

2.1 A Generation Expansion Planning Problem with Two-Stage Recourse Model

The first stage decision variables are the number of generators, x_i, to be installed at node i while the second stage decision variables are the power at nodes and flows in the network i.e. power generation, power flow in transmission lines and load curtailment in each system state. The first stage decision variables in the expansion policy are determined before the realization of randomness in the problem while the second stage decision variables are evaluated after the random uncertainties are realized. The objective of the problem is given by Eq. 1 to minimize both the expansion and operation cost.

$$\min \sum_{i \in I} c_i x_i + E_{\tilde{\omega}}\{f(x, \tilde{\omega})\} \tag{1}$$

$$s.t. \sum_{i \in I} c_i x_i \leq B \tag{2}$$

$$x_i \geq 0, \text{integer} \tag{3}$$

where c_i is cost of additional generators at bus i and Eq. 2 represents a budget constraint of B. Constraint (3) is requirement for the number of additional generators to be an integer.

The function, $E_{\tilde{\omega}}\{f(x, \tilde{\omega})\}$, in Eq.1 is the expected value of second stage objective function to minimize operation cost and loss of load cost under a realization ω of Ω. The second stage problem is to schedule the generating capacity in order to minimize operation cost as well as the reliability cost incurred from the load curtailment in each state. Power system network constraints are formulated using DC flow model. Second stage decision variables are generation at bus i in state ω, $y_{gi}(\omega)$, load curtailment at bus i in state ω, $y_{li}(\omega)$, and voltage angle at bus i in state ω, $\theta_i(\omega)$.

$$f(x, \omega) = \min \sum_{i \in I} \{c_{li}(\omega) y_{li}(\omega) + c_{oi}(\omega) y_{gi}(\omega)\} \tag{4}$$

$$s.t. y_{gi}(\omega) \leq g_i(\omega) + A_i x_i, \forall i \in I \tag{5}$$

$$b_{ij}(\theta_i(\omega) - \theta_j(\omega)) \leq t_{ij}(\omega), \forall i, j \in I, i \neq j \tag{6}$$

$$y_{li}(\omega) \leq l_i(\omega), \forall i \in I \tag{7}$$

$$\sum_{j \in I} B_{ij} \theta_j(\omega) + y_{gi}(\omega) + y_{li}(\omega) = l_i(\omega), \forall i \in I \tag{8}$$

$$y_{gi}(\omega), y_{li}(\omega) \geq 0, \forall i \in I, \theta_i(\omega) \text{unrestrict} \tag{9}$$

where, $c_{li}(\omega)$ and $c_{oi}(\omega)$ are cost of load loss and cost of operation at bus i in state ω in dollar per MW, A_i is an additional generation capacity at bus i in MW. Parameters $g_i(\omega)$, $t_{ij}(\omega)$ and $l_i(\omega)$ are generation capacity at bus i in MW, tie line capacity between bus i and j in MW, and load at bus i in MW in state ω. $\mathbf{B} = [B_{ij}]$ is an augmented node susceptance matrix and b_{ij} is tie-line susceptances between bus i and bus j. It should be noted that the cost of load loss coefficient depends on system states. The calculation of this coefficient is performed separately and is shown next.

Constraints (5), (6), and (7) are maximum capacity flows in the network under uncertainty in generation, tie line, and load respectively. Constraint (8) constitutes conservation of flow in the network and (9) presents variable restrictions in the model. Note that the decision on the expansion policy is done before the realization ω of Ω.

The failure probability of additional generators can be taken into account by using their effective capacities or by explicitly incorporating the unit availability of additional units in terms of their forced outage rates and derated forced outage rates in the formulation. The first stage problem is slightly modified in as follows.

$$\min \sum_{i \in I} \sum_{q \in Q_i} c_{iq} x_{iq} + E_{\tilde{\omega}}\{f(x, \tilde{\omega})\} \tag{10}$$

$$s.t. \sum_{i \in I} \sum_{q \in Q_i} c_{iq} x_{iq} \leq B \tag{11}$$

$$x_{iq}, \text{binary} \tag{12}$$

where c_{iq} is the cost of additional generators q at bus i, x_{iq} is equal to 1 if generating unit q is installed in area i and 0 otherwise, and Q_i is the total number of additional generators at bus i. Constraint (5) in the second stage problem only needs to be modified as follow.

$$y_{gi}(\omega) \leq g_i(\omega) + \sum_{q \in Q_i} A_{iq}(\omega) x_{iq}, \forall i \in I \tag{13}$$

where A_{iq} is additional generating capacity of unit q at bus i in state ω in MW. It should be noted that the number of system states without unit availability of additional generators is much less than that with unit availability consideration as we shall see from Sect. 4.2.

2.2 *Loss of Load Cost (LOLC) Coefficient Calculation*

Loss of load cost depends on interruption duration as well as the type of interrupted load. The most common approach to represent power interruption cost is through customer damage function (CDF) [5]. This function relates different types of load and interruption duration to cost per MW. In order to accurately calculate system expected LOLC, LOLC coefficient needs to be evaluated according to the duration of each state (ω). We use mean duration of a state to calculate the cost coefficient. The cost coefficient can be improved further by assuming that the duration has exponential or some other distribution. The cost coefficient is then calculated as the expected value for different possibilities.

Mean duration of each stage can be calculated by the reciprocal of equivalent transition rate from that state to others. State mean duration is presented in (14). Equivalent transition rate of all components can be calculated using the recursive formula in (14) when constructing probability distribution function.

$$D_\omega = \frac{1}{\sum_{i\in I} \lambda_{gi}^{\omega+} + \sum_{i\in I} \lambda_{gi}^{\omega-} + \sum_{i,j\in I,i\neq j} \lambda_{lij}^{\omega+} + \sum_{i,j\in I,i\neq j} \lambda_{lij}^{\omega-} + \sum_{l\in L} \lambda_{l}^{\omega}} \tag{14}$$

D_ω mean duration of state ω in hours

$\lambda_{gi}^{\omega+}$ equivalent transition rate of generation in area i from a capacity of state ω to higher capacity in per hour

$\lambda_{gi}^{\omega-}$ equivalent transition rate of generation in area i from a capacity of state ω to lower capacity in per hour

$\lambda_{lij}^{\omega+}$ equivalent transition rate of transmission line from area i to area j from a capacity of state ω to higher capacity in per hour

$\lambda_{lij}^{\omega-}$ equivalent transition rate of transmission line from area i to area j from a capacity of state ω to lower capacity in per hour

λ_{l}^{ω} equivalent transition rate of area load from state ω to other load states in per hour

Customer damage function used in this chapter is taken from [10], however, other damage functions if known could be used. The function was estimated from electric utility cost survey in the US. For residential loads, interruption cost in dollars per kWh can be described, as a function of outage duration, by (15).

$$c_{li}(\omega) = e^{0.2503+0.2211D_\omega-0.0098D_\omega^2} \tag{15}$$

2.3 Reliability-Constrained Consideration

We include reliability consideration in the expansion problem by limiting the expected loss of load upto a pre-specified value as shown in Eq. 16.

$$E_{\tilde{\omega}}\{y_{li}(\omega)\} \leq \alpha \tag{16}$$

where α is the upper limit of expected load loss. This reliability constraint together with a budget constraint may cause the problem to be infeasible. Instead of directly imposing the reliability constraint, the objective function is modified using Lagrangian relaxation.

$$\min \sum_{i \in I} c_i x_i + E_{\tilde{\omega}}\{f(x, \tilde{\omega})\} + P \times (E_{\tilde{\omega}}\{y_{li}(\omega)\} - \alpha) \tag{17}$$

where P is a penalty factor if the expected load loss violates the limit α. Due to the budget constraint, it is possible that the resulting expected load loss is higher than the upper limit.

3 Sample Average Approximation (SAA)

The expected cost of load loss can be approximated by means of sampling. Let $\omega_1, \omega_2, \ldots, \omega_N$ be N realizations of random vector for all uncertainties in the model, the expected cost of load loss can be replaced by expression (18).

$$\tilde{f}_N(x) = \frac{1}{N} \sum_{k=1}^{N} f(x, \omega_k) \tag{18}$$

This function is a SAA of the expected cost of load loss. The problem can then be transformed into deterministic equivalent model as follows.

$$\min \sum_{i \in I} c_i x_i + \frac{1}{N} \sum_{k=1}^{N} \left\{ \sum_{i \in I} \{c_{li}(\omega_k) y_{lik}(\omega_k) + c_{oi}(\omega_k) y_{gik}(\omega_k)\} \right\} \tag{19}$$

$$s.t. \sum_{i \in I} c_i x_i \leq B \tag{20}$$

$$\frac{1}{N} \sum_{k=1}^{N} y_{lik}(\omega_k) \leq \alpha \tag{21}$$

For all $k \in \{1, 2, \ldots, N\}$,

$$y_{gik}(\omega_k) \leq g_i(\omega_k) + A_i x_i, \forall i \in I \tag{22}$$

$$b_{ij}(\theta_{ik}(\omega_k) - \theta_{jk}(\omega_k)) \leq t_{ij}(\omega_k), \forall i, j \in I, i \neq j \tag{23}$$

$$y_{lik}(\omega_k) \leq l_{ik}(\omega_k), \forall i \in I \tag{24}$$

$$\sum_{j \in I} B_{ij}\theta_{jk}(\omega_k) + y_{gik}(\omega_k) + y_{lik}(\omega_k) = l_{ik}(\omega_k), \forall i \in I \tag{25}$$

$$x_i \geq 0, \text{integer} \tag{26}$$

$$y_{gik}(\omega_k), y_{lik}(\omega_k) \geq 0, \forall i \in I \tag{27}$$

$$\theta_i(\omega_k) \quad \text{unrestrict} \tag{28}$$

Note that, by virtue of the nature of sampling, a solution obtained from this sample-based approach does not necessarily guarantee optimality in the original problem. The optimal sample-based solutions, when obtained with different sample sets, rather provide statistical inference of a confidence interval of the actual optimal solution.

The reliability-constrainted problem can be formulated by modifying the objective function with constraint (21) according to Eq. 17. Equation 13 can also be modified to incorporate unit availability as follows. For each scenario k,

$$y_{gik}(\omega_k) \leq g_i(\omega_k) + \sum_{q \in Q_i} A_{iq}(\omega_k)x_{iq}, \forall i \in I. \tag{29}$$

Let x_N^* be the optimal solution and z_N^* be the optimal objective value of an approximated problem. Generally, x_N^* and z_N^* varies by the sample size N. If x^* is the optimal solution and z^* is the optimal objective value of the original problem, then obviously,

$$z^* \leq z_N^*. \tag{30}$$

Therefore, z_N^* constitutes an upper bound of the optimal objective value. Since z_N^* is the optimal solution of the approximated problem, then the following is true.

$$z_N^* = z_N^*(x_N^*) \leq z_N^*(x^*) \tag{31}$$

Taking expectation on both sides, Eq. 31 becomes

$$E[z_N^*(x_N^*)] \leq E[z_N^*(x^*)]. \tag{32}$$

Since the SAA is an unbiased estimator of the population mean,

$$E[z_N^*(x_N^*)] \leq E[z_N^*(x^*)] = z^*. \tag{33}$$

which constitutes a lower bound of the optimal objective value. In the following, details on obtaining lower bound and upper bound estimates are discussed. The derivation of lower and upper bound confidence interval was presented in [11] and has been applied in [12, 13].

3.1 *Lower Bound Estimates*

The expected value of z_N^*, $E[z_N^*]$, can be estimated by generating M_L independent batches, each of N_L samples. For each sample set s, solve the SAA problem which gives $z_{N_L^s}^*$ and the lower bound can be found from Eq. 34.

$$L_{N_L,M_L} = \frac{1}{M_L} \sum_{i=1}^{M_L} z_{N_L,i}^* \tag{34}$$

By the central limit theorem, the distribution of the lower bound estimate converges to a normal distribution $\mathcal{N}(\mu_L, \sigma_L^2)$ where $\mu_L = E[z_{N_L}^*]$, which can be approximated by a sample mean L_{N_L,M_L}, and σ_L^2 can be approximated by a sample variance.

$$s_L^2 = \frac{1}{M_L - 1} \sum_{i=1}^{M_L} (z_{N_L,i}^* - L_{N_L,M_L})^2 \tag{35}$$

The two-sided $100(1-\beta)\%$ confidence interval of the lower bound is found from Eq. 36.

$$\left[L_{N_L,M_L} - \frac{z_{\beta/2} s_L}{\sqrt{M_L}}, L_{N_L,M_L} + \frac{z_{\beta/2} s_L}{\sqrt{M_L}} \right] \tag{36}$$

where $z_{\beta/2}$ satisfies $Pr\{z_{-\beta/2} \leq \mathcal{N}(0,1) \leq z_{\beta/2}\} = 1 - \beta$. It should be noted that the lower bound confidence interval is computed by solving M_L independent SAA problems of sample size N_L.

3.2 *Upper Bound Estimates*

Given a sample-based solution x_N^*, the upper bound of the actual optimal objective can be estimated by generating M_U independent batches, each of N_U samples. Since the solution is set to x_N^*, Eq. 19 can be decomposed based on system state ω to N_U independent linear programming (LP) problems. For each sample batch s, solving the LP problems gives $z_{N_U,i}^*(x_N^*)$. Then, the upper bound is approximated using Eq. 37.

$$U_{N_U,M_U}(x_N^*) = \frac{1}{M_U} \sum_{i=1}^{M_U} z_{N_U,i}^*(x_N^*) \tag{37}$$

By central limit theorem, the distribution of the upper bound estimate converges to a normal distribution $\mathcal{N}(\mu_U, \sigma_U^2)$ where $\mu_U = E[z_{N_U,i}^*(x_N^*)]$, which can be approximated by a sample mean U_{N_U,M_U}, and σ_U^2 can be approximated by a sample variance.

$$s_U^2(x_N^*) = \frac{1}{M_U - 1} \sum_{i=1}^{M_U} \left(z_{N_U,i}^*(x_N^*) - U_{N_U,M_U}(x_N^*)\right)^2. \tag{38}$$

The two-sided $100(1-\beta)\%$ confidence interval of the lower bound is found from Eq. 39.

$$\left[U_{N_U,M_U}(x_N^*) - \frac{z_{\beta/2}s_U}{\sqrt{M_U}}, U_{N_U,M_U}(x_N^*) + \frac{z_{\beta/2}s_U}{\sqrt{M_U}}\right] \tag{39}$$

where $z_{\beta/2}$ satisfies $Pr\{z_{-\beta/2} \le N(0,1) \le z_{\beta/2}\} = 1 - \beta$.

A solution x_N^* is found from each batch s of M_L batches in lower bound SAA problems and used to compute the upper bound estimates. It should be noted that the upper bound confidence interval depends on the chosen approximate solution x_N^* from SAA problems. Thus, M_L upper bound intervals are computed.

3.3 Optimal Solution Approximation

An optimal solution can be extracted when a unique solution is obtained from solving several SAA problems with different samples of a given size, N. In theory, optimality should be attained with sufficiently large N. This means that it may be possible that each sample yields different solutions for small sample size. If an identical solution is found from solving SAA problems with these samples, it is highly likely that the solution is optimal or close to optimal. In addition to obtaining identical solutions, confidence intervals of the lower bound and upper bound estimates are also used to validate the approximate solution. If the intervals of both lower and upper bound estimates are close enough then the approximate solution tends to be close to the optimal solution. The sample average approximation offers a solution to the large-scale problems that are otherwise computationally intractable. The optimal solution validation is still a fairly open research topic. The number of samples as well as number of batches also plays an important part. Large sample sizes increase the computation burden while small sample sizes do not seem to represent the entire state space well. Interested readers are referred to reference [11–13] for more information about the optimal solution verification.

4 Computational Results

We report the results of three studies in this section. The first study is to determine optimal generation planning solution where the lower and upper bounds of the objective functions are estimated. The second study is to incorporate unit availability in the generation expansion problem. The third study is to analyze the

Table 1 Additional generation parameters

Bus	Unit capacity (MW)	Cost ($m)
101	20	20
102	20	20
107	100	100
115	12	12
122	50	50

performance of system-wide reliability maximization. The test system of all three studies is the 24-bus IEEE-RTS. The generator and transmission line parameters can be found in [9]. In order to reduce the number of system states, system load is grouped into 20 clusters using clustering algorithm. Cost of additional units is assumed and shown in Table 1. Five buses are chosen as possible locations for adding units. With original load, the expected load loss (α) is 0.07 MW and this is used as the specified limit in the optimization.

The penalty factor (P) of 10^6 is used. The load is increased by 10% to represent projected demand growth. The budget is 100 million dollars. The total number of system states ($|\Omega|$) of this problem is 9×10^{18}. To simplify the problem, operation cost in the second stage objective function is neglected. This is also due to the lack of data for operation cost of existing units in IEEE-RTS. There is however no inherent limitation in the methodology to include the operation cost in the problem.

4.1 *Optimal Generation Planning Problem*

To compare the effectiveness of the sample average approximation using Monte Carlo sampling, four different sample sizes are chosen, namely, 500, 1,000, 2,000, and 5,000. Lower bound estimate of each sample size is calculated by solving SAA problems with data generated by five different batches of the sample. Therefore, M_L is five and N_L are 500, 1,000, 2,000, and 5,000. Note that, at this point, each sample size will produce five solutions, which may or may not be identical, from five batches of sample. These solutions are then used to calculate upper bound estimate. The upper bound estimate of each sample size is obtained by substituting the solution obtained from that particular SAA problem. This will transform SAA problem into independent linear programming problem which makes it faster to solve than SAA problem. In this study, five batches of sample of size 10,000 are used to estimate upper bound. Thus, M_U is five and N_U is 10,000.

The 95% confidence intervals of the lower bound from different sample sizes and the 95% confidence intervals of upper bound from different batches of sample size (N_L) are shown in Fig. 1. For each sample size, the best upper bound estimate is chosen from the tightest confidence interval. If the interval is the same, the best upper bound is found from minimum average value.

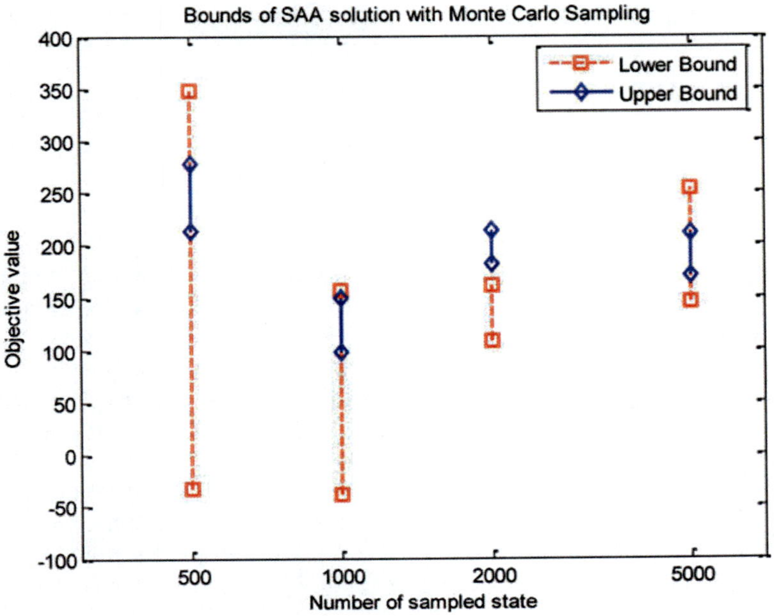

Fig.1 Bounds of SAA solution

The lower bound intervals overall tend to be higher and tighter when sample size increases except for sample size of 5,000. This is due to the nature of sampling. When sample sizes are smaller such as 500 and 1,000, the lower bound intervals may be higher than the upper bound. This may be due to the fact that the upper bound is found from sample size of 10,000. When a sample size is small, the duration of sampled states may be overestimated, which results in higher cost coefficient and expected loss of load cost.

The solution obtained from different batches of sample size can be found from Table 2. An optimal solution is approximated by solving SAA problems with increased sample sizes. In this study, the optimal solution is assumed to be reached when identical solutions are found within five consecutive batches of sample of the same size. It can be seen from Table 2 that the solutions are identical when sample sizes are 5,000. Therefore, the solution to this problem is to install five units at bus 102.

Even though the original problem has a large number of system states (9×10^{18}), sample average approximation requires only a small manageable sample size of 5,000 to solve the optimization problem. With the budget of $100 million dollars, the optimal solution yields expected load loss of 0.08 MW. Note that this expected load loss is greater than the initial limit of 0.07 MW. This is due to the specified budget constraint.

Table 2 Approximate solutions

Sample size	Batch	Number of additional units at bus				
		101	102	107	115	122
500	1	0	0	0	0	0
	2	0	0	0	0	0
	3	0	5	0	0	0
	4	0	0	1	0	0
	5	0	1	0	0	0
1,000	1	0	5	0	0	0
	2	0	2	0	0	0
	3	0	0	0	0	0
	4	0	0	0	0	0
	5	0	0	0	0	0
2,000	1	0	5	0	0	0
	2	4	0	0	1	0
	3	0	5	0	0	0
	4	0	5	0	0	0
	5	0	5	0	0	0
5,000	1	0	5	0	0	0
	2	0	5	0	0	0
	3	0	5	0	0	0
	4	0	5	0	0	0
	5	0	5	0	0	0

4.2 Availability Considerations of Additional Units

In this analysis, a two-state Markov model is assumed to represent unit availability of additional generators using forced outage rate data of the IEEE-RTS. The formulation is rather general in that any three or more state Markov model can be accommodated. With the same budget of 100 million dollars, maximum number of additional generating units in each bus is shown in Table 3. Total number of system states ($|\Omega|$) of this problem is 1.9×10^{25}, compared to 9×10^{18} without availability consideration of additional units. The optimal solution approximation is found from four different sample sizes, which are 1,000, 2,000, 8,000, and 12,000 using Monte Carlo sampling.

Lower bound estimate of the objective function from each sample size is calculated by solving SAA problems with data generated by four different batches of sample. Therefore, M_L is four and N_L are 1,000, 2,000, 8,000, and 12,000. Note that, at this point, each sample size will produce four solutions, which may or may not be identical, from four batches of sample. The 95% confidence intervals of lower bound of the objective function from different sample sizes are shown in Table 4.

The objective function is expansion cost and expected power loss, which are shown separately. The solution obtained from different batches of sample size can be also seen from Table 5.

Table 3 Unit availability data of additional generation parameters

Bus	Unit capacity (MW)	Forced outage rate	Cost ($m)	Number of units
101	20	0.1	20	5
102	20	0.1	20	5
107	100	0.04	100	1
115	12	0.02	12	8
122	50	0.01	50	2

Table 4 Lower bound estimates

Sample size	Objective function value	
	Expansion cost ($m)	Expected power loss (MW)
1,000	70 ± 37.53	0.0249 ± 0.0455
2,000	100	0.1083 ± 0.0301
8,000	100	0.0812 ± 0.0373
12,000	100	0.0720 ± 0.0169

Table 5 Approximate solutions with availability considerations of additional units

Sample size	Batch	Number of additional units at bus				
		101	102	107	115	122
1,000	1	1	4	0	0	0
	2	0	3	0	0	0
	3	0	5	0	0	0
	4	0	1	0	0	0
2,000	1	0	5	0	0	0
	2	1	4	0	0	0
	3	2	3	0	0	0
	4	3	2	0	0	0
8,000	1	1	4	0	0	0
	2	2	3	0	1	0
	3	2	3	0	0	0
	4	1	4	0	0	0
12,000	1	2	3	0	0	0
	2	2	3	0	0	0
	3	2	3	0	0	0
	4	2	3	0	0	0

It can be seen from Table 5 that the solutions are identical when sample sizes are 12,000. Therefore, the solution of the problem with availability considerations of additional units is to install two units at bus 101 and three units at bus 102, which yields expected power loss of 0.099 MW. This study shows that unit availability consideration can affect the solution of the generation expansion problem.

4.3 Comparative Study of System-Wide Reliability-Constrained Generation Expansion Problem

Generation and peak load data at each bus are shown in Table 6. All reliability indexes are found from Monte Carlo simulation with coefficient of variation of 0.05. Loss sharing policy is implemented in this study. We ignore unit availability consideration in this case study.

An optimal solution, found in the first case study, is to install five units at bus 102 which yields system expected power loss of 0.08. Expected unserved energy (EUE) of each area before and after optimal planning are shown in Table 7.

It can be seen from Table 7 that the system-wide reliability maximization planning does not necessarily yield equally distributed reliability level in each bus. However, it does tend to improve the reliability at almost all buses, especially those with high level of unserved energy. In some cases, bus 113, 119, and 120 sacrifices their reliability for the system since the reliability level of these buses are worse after the optimal planning. The EUE of these buses is, however, very small to start with. EUE percentage reduction is shown in Table 8.

The percentage EUE reductions of each bus vary from 2% to 87%. Most of the buses with large EUE reductions, for example, bus 103–110 are in the neighboring area with the additional units at bus 102. Other distance buses from bus 102 seem to have smaller reliability improvement. The results indicate that though the reliability improvement may not be proportionately distributed across buses, but most of the

Table 6 Generation and peak load

Bus	Generation (MW)	Peak load (MW)
101	192	118.8
102	192	106.7
103	–	198
104	–	81.4
105	–	78.1
106	–	149.6
107	300	137.5
108	–	188.1
109	–	192.5
110	–	214.5
113	591	291.5
114	–	213.4
115	215	348.7
116	155	110
118	400	366.3
119	–	199.1
120	–	140.8
121	400	–
122	300	–
123	660	–

Table 7 Expected unserved energy before and after optimal planning in MWh/year

Bus	Before optimal planning	After optimal planning
103	41.50	17.91
104	47.23	27.76
107	13.59	5.86
108	2.50	0.31
109	1022.06	591.24
110	9.23	5.08
113	0.00	0.88
114	6.39	6.25
115	33.27	21.51
116	3.95	2.27
118	20.48	17.55
119	0.00	0.61
120	0.00	1.04

Table 8 Expected unserved energy reduction

Bus	EUE reduction in MWh/year	Percentage reduction (%)
103	23.58	56.82
104	19.47	41.22
107	7.73	56.88
108	2.19	87.60
109	430.81	42.15
110	4.15	44.96
113	−0.88	–
114	0.14	2.19
115	11.75	35.32
116	1.69	42.78
118	2.93	14.31
119	−0.61	–
120	−1.04	–

buses do experience improvement, especially the most unreliable ones experience a considerable gain.

5 Conclusion and Discussion

A stochastic programming approach with sample average approximation is presented for the composite-system generation adequacy planning problem. The problem is formulated as a two-stage recourse model with the objective to minimize expansion cost in the first stage and operation and reliability cost in the second stage. Reliability is included in terms of expected cost of load loss in the objective function and expected load loss in the constraint. Availability of additional units can also be incorporated in the formulation, which may results in exponentially large number of system states.

Due to numerous system states, straightforward implementation of L-shaped method is impractical, if not impossible. To overcome this, exterior sampling method is proposed in this study. Reliability function of the problem is approximated by the sample-average using Monte Carlo sampling. Generation expansion planning is implemented on the 24-bus IEEE-RTS. Results show that even though the problem itself has a huge number of system states the proposed method can effectively estimate the optimal solution with a relatively small number of samples. The planning problem includes system reliability consideration which may be of interest to Independent System Operators when designing price incentive program for the generation companies.

A comparative study on customer reliability level at each bus before and after optimal planning is conducted. Results show that system-wide reliability optimization may not equally improve reliability level at each bus but most buses experience improvement in reliability, especially those suffering the most. It is likely that buses within close distance to the additional generators benefit more from the optimal planning. Perhaps that is the reason for the placement of the additional generators in those locations. This information is useful for the Independent System Operators in discharging their responsibility to overlook the electric market and design appropriate mechanism in order to promote fair pricing for reliability.

References

1. Zhu J, Chow M (1997) A review of emerging techniques on generation expansion planning. IEEE Trans Power Syst 12(4):1722–1728
2. Infanger G (1993) Planning under uncertainty: solving large-scale stochastic linear programs. Boyd & Fraser, Danvers
3. Jirutitijaroen P, Singh C (2008) Reliability constrained multi-area adequacy planning using stochastic programming with sample-average approximations. IEEE Trans Power Syst 23(2): 504–513
4. Jirutitijaroen P, Singh C (2008) Composite-system generation adequacy planning using stochastic programming with sample-average approximation. In: Proceedings of the 16th power systems computation conference, Glasgow, 2008
5. Jirutitijaroen P, Singh C (2008) Unit availability considerations in composite-system generation planning. In: Proceedings of the 10th international conference on probabilistic methods applied to power systems, Rincon, 2008
6. Birge JR, Louveaux F (1997) Introduction to stochastic programming. Duxbury, Belmont
7. Higle JL, Sen S (1996) Stochastic decomposition: a statistical method for large scale stochastic linear programming. Kluwer Academic, The Netherlands
8. Jirutitijaroen P, Singh C (2008) Comparative study of system-wide reliability-constrained generation expansion problem. In: Proceedings of the 3th international conference on electric utility deregulation and restructuring and power technologies, Nanjing, 2008
9. IEEE APM Subcommittee (1999) The IEEE Reliability Test System-1996. IEEE Trans Power Syst 14(3):1010–1020
10. Lawton L, Sullivan M, Liere KV, Katz A, Eto J (2003) A framework and review of customer outage costs: integration and analysis of electric utility outage cost surveys. Lawrence Berkeley National Laboratory. Paper LBNL-54365. http://repositories.cdlib.org/lbnl/LBNL-54365. Accessed 1 Nov 2003

11. Mak WK, Morton DP, Wood RK (1999) Monte Carlo bounding techniques for determining solution quality in stochastic programs. Oper Res Lett 24:47–56
12. Linderoth JT, Shapiro A, Wright SJ (2006) The empirical behavior of sampling methods for stochastic programming. Ann Oper Res 142(1):215–241
13. Verweij B, Ahmed S, Kleywegt AJ, Nemhauser G, Shapiro A (2003) The sample average approximation method applied to stochastic routing problems: a computational study. Comput Optim Appl 24:289–333

Flexible Transmission in the Smart Grid: Optimal Transmission Switching

Kory W. Hedman, Shmuel S. Oren, and Richard P. O'Neill

Abstract There is currently a national push to create a smarter, more flexible electrical grid. Traditionally, network branches (transmission lines and transformers) in the electrical grid have been modeled as fixed assets in the short run, except during times of forced outages or maintenance. This traditional view does not permit reconfiguration of the network by system operators to improve system performance and economic efficiency. However, it is well known that the redundancy built into the transmission network in order to handle a multitude of contingencies (meet required reliability standards, i.e., prevent blackouts) over a long planning horizon can, in the short run, increase operating costs. Furthermore, past research has demonstrated that short-term network topology reconfiguration can be used to relieve line overloading and voltage violations, improve system reliability, and reduce system losses. This chapter discusses the ways that the modeling of flexible transmission assets can benefit the multi-trillion dollar electric energy industry. Optimal transmission switching is a straightforward way to leverage grid controllability; it treats the state of the transmission assets, i.e., in service or out of service, as a decision variable in the optimal power flow problem instead of treating the assets as static assets, which is the current practice today. Instead of merely dispatching generators (suppliers) to meet the fixed demand throughout the network, the new problem co-optimizes the network topology along with generation.

K.W. Hedman (✉)
School of Electrical, Computer, and Energy Engineering at Arizona State University, Tempe, AZ, USA
e-mail: kory.hedman@asu.edu

S.S. Oren
Industrial Engineering and Operations Research Department, University of California at Berkeley, Berkeley, CA, USA
e-mail: oren@ieor.berkeley.edu

R.P. O'Neill
Federal Energy Regulatory Commission, Washington, DC, USA
e-mail: richard.oneill@ferc.gov

A. Sorokin et al. (eds.), *Handbook of Networks in Power Systems I*,
Energy Systems, DOI 10.1007/978-3-642-23193-3_21,

By harnessing the choice to temporarily take transmission assets out of service, this creates a superset of feasible solutions for this network flow problem; as a result, there is the potential for substantial benefits for society even while maintaining stringent reliability standards. On the contrary, the benefits to individual market participants are uncertain; some will benefit and other will not. Consequently, this research also analyzes the impacts that optimal transmission switching may have on market participants.

Keywords Mixed integer programming • Optimal power flow • Power generation dispatch • Power system economics • Power system reliability • Power transmission control

1 Introduction

The physics that govern the flow of electric energy across the electric transmission network create a complex and unique network flow problem. The flow of electricity across the network follows Kirchhoff's laws. These unique physical laws imply that changing a transmission asset's impedance changes how the power flows throughout the network. Moreover, electric energy is instantaneously consumed and it is currently too expensive to store. These factors, along with the many stability constraints, reliability constraints, generator dispatch constraints, etc., make this a very difficult network flow problem. However, the mathematical modeling of the network is not as complex as it could be and various control mechanisms have yet to be acknowledged as well as harnessed within the optimization formulation.

Traditionally, the system operator treats transmission assets (lines or transformers) as static assets within Optimal Power Flow (OPF) problems, which are the network flow problem for the electrical grid. The OPF dispatches generators to minimize cost subject to satisfying the fixed demand throughout the network, ensuring that reliability standards are met, and satisfying all of the network flow constraints for the transmission network problem. This traditional view does not describe transmission assets as assets that operators have the ability to control. However, it is acknowledged, both formally and informally, that system operators can and do change the grid topology to improve voltage profiles, increase transfer capacity, and even improve system reliability. These ad-hoc procedures are determined by the system operators, rather than in an automated or systematic way. Furthermore, such flexibility is not incorporated into dispatch optimization problems today. This is a shortcoming regarding today's electric grid operations; due to the physics that govern the flow of electric energy and due to the complexities within this network flow problem, it is extremely unlikely that there is a single optimal network topology for all periods and possible market realizations over a long time horizon.

The electric grid is built to be a redundant network in order to ensure mandatory reliability standards and these standards require protection against worst-case scenarios. However, it is well known that these network redundancies can cause dispatch inefficiency and, furthermore, a network branch that is required to be built in order to meet reliability standards during specific operational periods may not be required to be in service during other periods. Consequently, due to the interdependency between network branches (transmission lines and transformers), it is possible to temporarily take a branch out of service during certain operating conditions and improve the efficiency of the network while maintaining reliability standards.

Past research has identified how the control of transmission assets can be used to benefit the network. These papers have generally focused on the switching of transmission lines when a line is overloaded, when there are voltage violations, as well as other factors related to using the control of transmission assets to alleviate an active network constraint. These approaches, however, do not attempt to use the control of transmission assets optimally by co-optimizing the network topology with generation to improve the dispatch efficiency during steady-state operations. Optimal transmission switching formally introduces the control of transmission assets into the classical formulation of dispatch optimization problems common in system and market operation procedures employed by vertically integrated utilities and Independent System Operators (ISOs).

There is currently a national push to model the grid in a more sophisticated, smarter way as well as to introduce advanced technologies and control mechanisms into grid operations. In particular, there are national directives that call on researchers to examine topics in this general area of research. The US Energy Policy Act of 2005 includes a directive for federal agencies to "encourage...deployment of advanced transmission technologies," including "optimized transmission line configuration."[1] This research is also in line with FERC Order 890: to improve the economic operations of the electric transmission grid. It also addresses the items listed in Title 13 "Smart Grid" of the Energy Independence and Security Act of 2007: (1) "increased use of... controls technology to improve reliability, stability, and efficiency of the grid" and (2) "dynamic optimization of grid operations and resources." This research examines the smart grid application of harnessing the control of transmission assets by incorporating their discrete state into the network optimization problem and it analyzes the benefits and market implications of this concept.

The rest of this chapter is broken down to include six main sections. The following section provides a thorough overview of the literature that is relevant to this research as well as a discussion on current industry practices that demonstrate the benefit of transmission control. Section 3 discusses the impact that transmission switching has on the feasible set of dispatch solutions, its affect on reliability, and how it differs from transmission expansion planning. Section 4 then presents a mathematical overview of OPF problems and optimal transmission switching. Section 5 presents results on the potential economic savings as a result of optimal transmission switching. Section 5 also focuses on the market implications

---

[1] See Sec.1223.a.5 of the US Energy Policy Act of 2005.

when these optimization models are modified to include the control of transmission assets. While optimal transmission switching can improve economic efficiency of grid operations, the implementation of this new technology may have unpredictable distributional effects on market participants and undermine some prevalent market design principles that rely on the premise of a fixed network topology. Section 6 provides an overview of future research topics and Sect. 7 concludes this chapter.

2 Literature Review

2.1 Transmission Switching as a Corrective Mechanism

Past research has explored transmission switching as a control method for a variety of problems. The primary focus of past research has been on proposing transmission switching as a corrective mechanism when there is line overloading, voltage violations, etc. While this past research acknowledges certain benefits of harnessing the control of transmission, they do not use the flexibility of the transmission grid to co-optimize the generation along with the network topology during steady-state operations. Such co-optimization, as will be shown by this research, can provide substantial economic savings even while maintaining N-1 reliability standards. Furthermore, the use of transmission switching as a corrective mechanism to respond to a contingency has been acknowledged in some of the past research to have an impact on the cost of generation rescheduling due to the contingency. However, it has not been acknowledged that such flexibility should be accounted for when solving for the steady-state optimal dispatch.

Glavitsch [1] gives an overview of the use of transmission switching as a corrective mechanism in response to a contingency. He discusses the formulation of such a problem and provides an overview on search techniques to solve the problem. Mazi et al. [2] propose a method to alleviate line overloading due to a contingency by the use of transmission switching as a corrective mechanism and use a heuristic technique to solve the problem. Gorenstin et al. [3] study a similar problem concerning transmission switching as a corrective mechanism; they use a linear approximate Optimal Power Flow (OPF) formulation and solve the problem based on branch and bound. Bacher et al. [4] further examine transmission switching in the AC setting to relieve line overloads; however, they assume that the generation dispatch is already determined and fixed thereby not capturing the benefit of co-optimizing the network topology with generation. Bakitzis et al. [5] examine transmission switching as a corrective mechanism both with a continuous variable formulation for the switching decision as well as with discrete control variables.

Schnyder et al. [6, 7] proposed a fast corrective switching algorithm to be used in response to a contingency. The benefit of this algorithm over past research is that they simultaneously consider the control over the network topology and the ability to redispatch generation whereas other methods would assume that the generation is fixed when trying to determine the appropriate switching action. Due to the complexity of this problem for its time, this method does not search for the actual optimal topology but rather considers limited switching actions. Rolim et al. [8] provide a review of past transmission switching methods, the solution techniques used, the objective at hand, etc. Shao et al. [9] continued previous research on the use of transmission switching as a corrective mechanism to relieve line overloads and voltage violations. They propose a new solution technique to find the best switching actions. Their technique employs a sparse inverse technique and involves a fast decoupled power flow in order to reduce the number of required iterations. In Shao et al. [10], a binary integer programming technique is used for the same motivation: to use switching actions as a corrective mechanism to relieve line overloads and voltage violations.

Granelli et al. [11] propose transmission switching as a tool to manage congestion in the electrical grid. They discuss ways to solve this problem by genetic algorithms as well as deterministic approaches. However, their approach does not consider the impact the topology has on the choice of steady-state dispatch solutions.

2.2 Transmission Switching To Minimize Losses

In Bacher et al. [12], they propose switching to minimize system losses. This paper demonstrates that, contrary to general belief, it is possible to reduce electrical losses in the network by temporarily opening a transmission line. Fliscounakis et al. [13] proposed a mixed integer linear program to determine the optimal transmission topology with the objective to minimize losses. Unlike past research, this model does search for the optimal topology but it does not co-optimize the generation with the network topology in order to maximize the market surplus. It is even possible that the solution that maximizes the market surplus has an increase in losses but by accounting for the influence between generation and transmission, the overall costs may still be lower. In contrast to these approaches, the optimal transmission switching concept maximizes the market surplus by co-optimizing the transmission topology along with generation.

2.3 Ad-Hoc Transmission Switching Protocols

One of the most common industry practices of transmission line switching involves the common protocol to switch specific lines offline during lightly loaded hours. The capacitive component of a transmission line is the predominant component during low load levels whereas the reactive component is predominant at higher load levels. Consequently, during low load levels there can be situations where a transmission line causes voltage violations in the network, i.e., the voltage levels are too high. Therefore, one simple protocol that operators are aware of is to select key transmission lines that are not currently needed for reliability considerations and they take these lines out of service. This reduces the capacitance and can help alleviate voltage violations. Such a protocol is acknowledged as a procedure within the PJM network and by Excelon. Likewise, the Northeast Power Coordinating Council includes "switch out internal transmission lines" in the list of possible actions to avoid abnormal voltage conditions, [14, 15]. Another ad-hoc transmission switching protocol that is at times used by grid operators is to identify key transmission lines that can be taken out of service in order to improve the transfer capability on other high voltage transmission lines. This is a protocol implemented in the PJM network.

2.4 Implementation of Transmission Switching in Special Protection Schemes

Special Protection Schemes (SPSs), also known as special protection systems or remedial action schemes, are becoming a mainstream protocol in electric grid operations. Grid operators identify specific grid conditions where it can be advantageous to implement an automatic, predetermined corrective action in response to specific abnormal grid operations. SPSs can be used to solve a variety of issues from maintaining voltage stability to a corrective action that is taken once a specific contingency occurs; the main motivation is to maintain proper reliable operations of the grid. These actions may involve changes in generation, reduction in load if necessary, as well as grid topology modifications. The PJM system uses SPSs to implement transmission switching protocols; this includes both pre-contingency transmission switching as well as post-contingency transmission switching. There can be situations where the operator will take a line out of service temporarily during steady-state operations but may switch the line back into service once a specific contingency occurs. Likewise, there are situations where opening a transmission line once there is a contingency can help the system recover from the contingency without causing a blackout. Further information on SPSs that implement transmission switching can be found in [16].

2.5 Transmission Line Maintenance Scheduling

The focus of past transmission line maintenance scheduling was on the effect on reliability. However, just as transmission lines affect reliability they also affect the operational costs of the electrical grid. Operators are now acknowledging the importance of transmission line maintenance scheduling not only regarding its affect on reliability but on operational costs. For instance, the Independent System

Operator of New England (ISONE) recently released a report stating that they saved $72 million in 2008 by considering the impact of transmission line maintenance scheduling on the overall operational costs, [17]. This study, however, was done by estimating prices instead of determining an optimal maintenance schedule, which further emphasizes the need for research on network topology optimization.

2.6 *Optimal Transmission Switching*

The initial concept of a dispatchable network was first proposed by O'Neill et al. [18]. Fisher et al. [19], further developed and examined the concept of incorporating the control of transmission assets into dispatch optimization formulations. This chapter is based on past and current research by the authors on this concept; further information can be found in [18–25].

3 Discussion of Optimal Transmission Switching, Feasible Sets, Reliability, and Transmission Planning

3.1 *Transmission Switching's Impact on the Feasible Set*

Even though this research is based on examining optimal transmission switching with the DCOPF and not the ACOPF, optimal transmission switching still has the ability to provide substantial benefits by providing more control to the operator. One simple way to demonstrate this fact is by understanding how optimal transmission switching affects the feasible set of dispatch solutions for any OPF. For simplicity, the following example is based on the DCOPF. However, the set of feasible solutions for any optimal power flow problem, be it the DCOPF, the ACOPF, or a security constrained OPF, depends on the characteristics of the network branches. This example demonstrates what happens in the DCOPF by changing the characteristics of a line (opening a line is equivalent to changing the impedance to infinity) and based on Kirchhoff's laws it is known that a similar result can be demonstrated for any OPF.

If a transmission asset's impedance is changed, this changes the feasible set of dispatch solutions. Since optimal transmission switching allows for the selection of any network topology, this gives the operator the choice to choose any dispatch that is feasible for any of these individual topologies instead of being restricted to choosing a dispatch that is feasible for only the static topology. As a result, optimal transmission switching creates a superset of feasible dispatch solutions and, therefore, it improves the operational flexibility of the grid in order to potentially improve reliability, stability, and/or economic operations. Obviously, optimal transmission switching will not provide additional flexibility to the operator in situations where the electric grid is not congested, i.e., there are no active network constraints (except the node balance constraints), since all dispatch solutions that satisfy the generator constraints are feasible.

Figure 1 provides a simple three-bus example; each transmission line has the same impedance but their thermal capacity limits differ. The feasible sets in Fig. 2 are defined by the thermal transmission constraints. For the original topology, there are three equations that represent the network constraints, (1)- to (3). Opening any line will change these constraints and, thus, change the feasible set. With all lines closed, i.e., in service, the feasible set is defined by the set of vertices {0, 1, 2, 3} in Fig. 2. If line A-B is opened, i.e., out of service, the feasible set is {0, 4, 5, 6}.

$$-80 \leq \frac{1}{3}G_A - \frac{1}{3}G_B \leq 80 \quad (1)$$

$$-80 \leq \frac{1}{3}G_A + \frac{2}{3}G_B \leq 80 \quad (2)$$

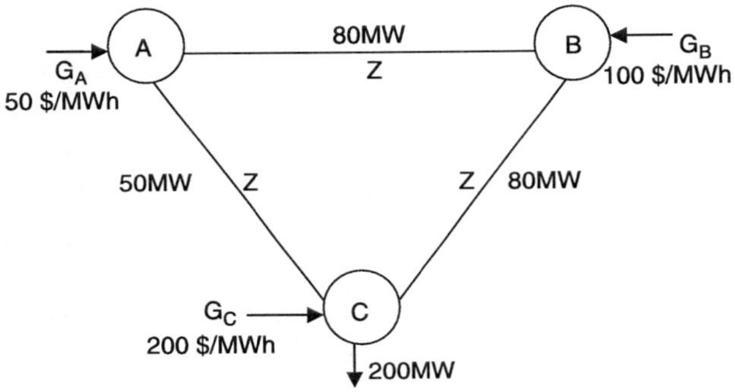

Fig. 1 3-Node example

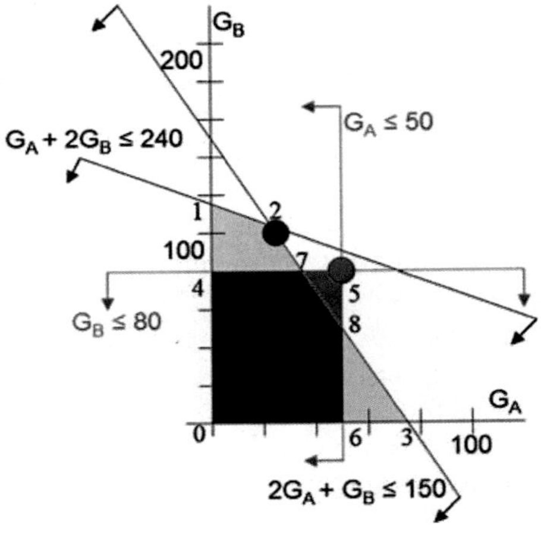

Fig. 2 Feasible sets for Gen A and Gen B

$$-50 \leq \frac{2}{3}G_A + \frac{1}{3}G_B \leq 50 \tag{3}$$

The advantage of optimal switching is that it gives the operator the choice to choose any dispatch solution defined by the set {0, 1, 2, 7, 5, 8, 3}, which is the non-convex union of the nomograms corresponding to the two operating states for the line between buses A and B. Though this example does not enforce reliability standards and though it is not based on the ACOPF model, the example demonstrates the flexibility that optimal transmission switching give to the operator. No matter what type or form of constraint that is a part of an OPF problem, harnessing the flexibility of whether to keep a transmission asset in service or not will create a superset of feasible dispatch solutions. It is possible that even with this control the operator would choose the original, static topology. However, due to the nature of the electrical grid and its complexity over a wide range of operating conditions over a long time horizon, it is highly unlikely that there is one single topology that is always preferred no matter the current operating state of the grid. An electrical transmission network can have over 10,000 transmission assets, thereby creating $2^{10,000}$ possible network configurations. Obviously, many of those configurations would result in an infeasible dispatch solution; however, many configurations would still be possible and it is near impossible for there to be one perfect topology for every operating condition. This is further confirmed by the well-known practice to open high-voltage lines during the night to improve the voltage profile.

With optimal transmission switching, there is a guarantee that the solution will not be worse off than before since the original network configuration can always be chosen. For this example, if the objective were to minimize the total dispatch cost, the original solution would be located at {2} where the corresponding dispatch is 20 MW from GEN A, 110 MW from GEN B, and 70 MW from GEN C for a total cost of $26,000 per hour. By opening line A-B, there is a new feasible solution, {5}, where the corresponding dispatch is 50 MW from GEN A, 80 MW from Gen B, and 70 MW from GEN C. This places the total cost at $24,500, which translates into a $1,500 savings.

3.2 Transmission Switching and Reliability

Transmission switching would not be implemented if it were to violate established reliability standards. The previous example in Sect. 3.1 shows graphically how optimal transmission switching adds flexibility to the dispatch choice for congested networks. Even though that stylized example does not enforce N-1, the conclusions would not change if reliability constraints were enforced in the OPF. Optimal transmission switching adds another layer of control to the OPF and, thus, it creates a superset of possible dispatch solutions.

Even still, it is often thought that reliability must diminish if you take a line out of service, that reliability is something that is judged purely on the topology of the network. This is, in fact, not so. The research papers and practical examples on the use of transmission switching in the literature review section demonstrate that, during certain operational states, the system reliability can improve by the removal of a line. For instance, transmission switching is used today as a post-contingency corrective action in some SPSs, [16]. Reliability cannot be judged purely on the network topology alone. Reliability depends on the network topology but it also depends on the generation's commitment schedule, ramping capabilities, and their available capacity. With optimal transmission switching, it is possible to switch to a topology that has fewer available paths to transfer energy but that the different generation schedule, which can only be obtained if the topology is altered, provides more available capacity and these generators are, overall, faster than the generation dispatch solution that would have been chosen if the topology was not altered. In Sect. 2.3 of [24], there is an example demonstrating this possibility. It is then possible that the combination of changing the topology with generation improves system reliability, which again emphasizes that the grid topology itself cannot be used as the only indicator to examine system reliability.

Furthermore, the true issue is not whether the system reliability in general diminishes; rather, the concern is whether the required reliability standards are met. The objective of the grid operator is to serve the load at least cost subject to maintaining set reliability standards as well as satisfying the operational constraints. Thus, no preference is given to solutions that improve system reliability beyond required levels; rather, with multiple solutions that satisfy these requirements, the operator chooses the least cost solution. Optimal transmission switching is consistent with this conventional policy to serve load at least cost while maintaining established reliability requirements as it can improve the market surplus while meeting the required reliability standards. If the network is initially N-3[2] reliable but with the optimal transmission switching solution it is only N-2 reliable, then no required reliability standard has been violated. The correct decision is, therefore, to implement the optimal transmission switching concept to be able to obtain superior economic dispatch solutions that also meet set reliability standards, e.g., N-1. Furthermore, with a truly smarter, more advanced electrical grid, transmission lines that are temporarily taken out of service during no-contingency operating states can be switched back into service if there is a contingency. This would bring the grid back to its redundant state during a contingency state; this concept is further discussed in Sect. 6.7. As was discussed in Sect. 2.4, similar corrective actions are used today by ISOs as there are SPSs that implement pre-contingency and post-contingency switching actions, [16].

---

[2] N-k reliability means that the system can survive the simultaneous failure of any k elements without violating any constraints on the surviving network and without the need for load shedding.

3.3 *Optimal Transmission Switching and Transmission Planning*

The previous sections demonstrate how optimal transmission switching creates a superset of feasible dispatch solutions and how this added flexibility in dispatch choice can be used to improve dispatch efficiency and/or improve system reliability. However, there is an underlying question as to why a line that is taken out of service would have been built in the first place. Applying transmission switching to reduce costs may seem counter-intuitive as it seems to contradict the purpose of transmission planning. Transmission lines are built because they are needed to maintain reliability and they are built to facilitate additional trading of energy, i.e., increase the transfer capability between areas in order to access cheaper energy. As a result, there is the common misconception that optimal transmission switching can only benefit system operations when there was previously poor transmission planning. If transmission switching were never beneficial and never feasible, such a result would mean that a single topology is the optimal topology for every possible network condition over a long planning horizon; to have one perfect topology out of a vast number of possible network configurations over such varying operational situations is unlikely.

Optimal transmission switching and transmission planning are two distinct problems with different goals. Even if the network is optimally planned, optimal transmission switching can still be beneficial. First, the basis for optimality of transmission expansion planning is the aggregate of benefits due to building a transmission element over a long time horizon. This is distinctly different than optimal transmission switching, which is a short-term optimization problem that determines the optimal topology for very specific operating conditions over a short time horizon. There is no guarantee that the optimal transmission expansion project is necessary to meet reliability standards during every period throughout its lifecycle, e.g., the line may only be required during peak-hours. Furthermore, there is also no guarantee that the line provides an economic benefit to the system for each period over its lifecycle. Based on optimization theory, it is known that the optimal investment over a long planning horizon, i.e., choosing one investment for all periods, need not be the same as the optimal investment for each individual period. In fact, the optimal transmission expansion project could propose the building of a line that is a detriment, regarding system reliability and economic efficiency, to the system during a few specific hours but is overall the best investment choice over a long planning horizon. Consequently, the fact that transmission switching may be beneficial and feasible does not guarantee inefficient transmission planning since they are two distinct problems. Moreover, it is well known that the redundancies built into the grid in order to handle a multitude of contingencies over a long planning horizon causes dispatch inefficiency. The purpose of optimal transmission switching is to remedy this issue by solving for the best topology to have for specific short-term operating conditions.

The concept of short-term network reconfiguration is further supported by the fact that transmission expansion planning is a very granular process. Due to the high level of uncertainty regarding future network conditions, it is next to impossible to determine the optimal topology over such a long planning horizon. As system conditions change, it should be expected that the optimal topology may change from one period to the next and the choice regarding which topology is best for a specific period is better answered just prior to the period since there is less operational uncertainty. Finally, transmission expansion planning is a very difficult optimization problem, which limits the modeling complexity and further decreases the validity of the solution. These factors further argue in support of short-term network reconfiguration.

4 Optimal Transmission Switching

4.1 *Optimal Power Flow*

The flow of electric energy follows Kirchhoff's laws. The Alternating Current Optimal Power Flow (ACOPF) problem is the network flow problem for the AC electric transmission grid and it is used to dispatch generation optimally subject to the network flow constraints and reliability constraints. The ACOPF optimization problem is a non-convex optimization problem involving trigonometric functions and, thus, it is a difficult problem to solve. Equations 4 and 5 represent the equations for the flow of electric power into bus n from transmission line k (line k is connected from bus m to bus n), see [26].

$$P_k = {V_m}^2 G_k - V_m V_n (G_k \cos(\theta_m - \theta_n) + B_k \sin(\theta_m - \theta_n)), \forall k \tag{4}$$

$$Q_k = -V_m V_n (G_k \sin(\theta_m - \theta_n) - B_k \cos(\theta_m - \theta_n)) - B_k {V_m}^2, \forall k \tag{5}$$

Due to the difficulty with solving the ACOPF problem, a linear approximation is commonly used in its place, both by academia and by the industry. This problem is referred to as the Direct Current Optimal Power Flow (DCOPF) problem, which contains all linear constraints. Many assumptions are made to go from the equations listed by (4) and (5) to produce the DCOPF line flow constraint, (6). The voltage variables are assumed to take on a per unit value of one, the angle difference between buses m and n is assumed to be small so that the cosine terms are assumed to be one and the sine terms are assumed to be the angle difference itself, the reactive power flow constraint, (5), is ignored, and the resistance is assumed to be negligible. Note that the basic DCOPF model does not account for reactive power flow or losses; however, there are ways to account for reactive power and losses in the DCOPF. These assumptions produce the crude approximation that is listed by (6) below.

$$P_k = B_k(\theta_n - \theta_m), \forall k \tag{6}$$

4.2 Direct Current Optimal Power Flow Problem

For the purpose of this research, it is assumed that generators' cost functions are linear, i.e. constant marginal cost[3], and, hence, the DCOPF problem is a Linear Programming (LP) problem since all of the constraints are linear. The objective is to minimize the total generation cost, (7); note that since the load throughout the network is assumed to be fixed, i.e., perfectly inelastic, minimizing the total cost is the same as maximizing the total market surplus. Constraint (8) restricts the difference of the bus voltage angles for any two buses that are connected by a transmission element. Constraint (9) specifies the operational constraints for generator g; for the basic DCOPF formulation, it is assumed that the minimum operating level for the generator is zero even though most generators do not have a zero minimum operating level. In order to enforce the true minimum operating levels of generators, i.e., if their minimum operating level is not zero, requires the use of a binary unit commitment variable thereby changing the linear program into a mixed integer linear program. Constraint (10) is the node balance constraint that specifies that the power flow into a bus must equal the power flow out of a bus. Generator supplies at a bus are injections while the load is a withdrawal. Constraint (11) represents the thermal capacity constraint on transmission line k; it is generally the case that $P_k^{max} = -P_k^{min}$. Finally, constraint (6) represents the linear approximation of the real power flow on transmission asset k.

OPF formulations generally include lower and upper bound constraints on the voltage angle difference, $\theta_n - \theta_m$, for any two buses that are connected by a transmission asset, see (8). In the DCOPF formulation, P_k is equal to the susceptance times the angle difference thereby allowing (8) to be subsumed by (11), i.e., placing lower and upper bounds on the angle difference for a line places bounds on the power flow for that line. Therefore, the thermal capacity lower and upper bounds, P_k^{min} and P_k^{max}, can be replaced by $B_k\theta^{max}$ and $B_k\theta^{min}$ if those bounds are tighter than the thermal capacity constraints for the lines. Therefore, constraint (8) is not included in the optimal transmission switching DCOPF formulations that are presented in the following section; instead, we update P_k^{min} and P_k^{max} accordingly.

Minimize:

$$\sum_g c_g P_g \tag{7}$$

s.t.

$$\theta^{\min} \leq \theta_n - \theta_m \leq \theta^{\max}, \forall k \tag{8}$$

[3] In reality, generator cost functions are quadratic in output (aside from startup and no load costs); however, in practice, such cost functions are approximated by piecewise linear functions represented as block offers at different marginal prices. The DCOPF formulation with piecewise linear cost functions is also a linear programming problem.

$$0 \leq P_g \leq P_g^{\max}, \forall g \tag{9}$$

$$\sum_{\forall k(n,.)} P_k - \sum_{\forall k(.,n)} P_k + \sum_{\forall g(n)} P_g = d_n, \forall n \tag{10}$$

$$P_k^{\min} \leq P_k \leq P_k^{\max}, \forall k \tag{11}$$

$$P_k = B_k(\theta_n - \theta_m), \forall k \tag{6}$$

4.3 Mathematical Modeling of Optimal Transmission Switching

In order to introduce optimal transmission switching into the DCOPF, a binary variable is first needed to reflect the state of the transmission line. z_k is used as a binary variable for transmission asset k; it takes on a value of one when the asset is in service (circuit breakers are closed) and it takes on a value of zero when the asset is out of service (circuit breakers are open). The first and easiest modification to make is to multiply the lower and upper bounds in (11) by the binary variable. Thus, when z_k equals zero, the line flow variable will be forced to be zero; otherwise, if z_k is one, the constraint (11) will appear in the OPF formulation in its original form. This modification is shown in the later formulations as (11.1).

The modification to (11), however, is not sufficient. If zk equals zero and Pk equals zero, then (6) will force the bus angles to equal each other. This is not the desired outcome; if the line is taken out of service by the opening of an electrical switch, there should be no constraint on the angle difference for two buses that are not directly connected (unless there will be a breaker reclosing procedure to bring the line back into service). This creates a situation where (6) is modified into what is known as an indicator constraint. There are different ways that this relationship can be modeled. A simple way to implement this relationship is to break (6) into two inequality constraints, (6.1) and (6.2), and use what is known as a big M value. As a result, when zk equals one, the Mk in each inequality is multiplied by zero and these inequalities will then enforce (6) as desired. When zk equals zero, the value of Mk is large enough such that it allows θn and θm to take on different values as desired. The value of Mk does place an indirect bound on the difference between these two angles; however, this is the desired outcome. When there is a breaker reclosing procedure to bring a line back into service, the operator must limit the angle difference between the two buses that are about to be directly reconnected. The use of Mk provides a way to model this relationship. Furthermore, it is important to have Mk be as small as possible as it is well known that the use of a big M value to create such relationships substantially impacts the computational performance of the MIP. This reclosing rule provides that needed minimum value on Mk and for this formulation, Mk = |Bkθrec|.

Formulation 1:
Minimize:

$$\sum_{g} c_g P_g \tag{7}$$

s.t.

$$0 \le P_g \le P_g^{\max}, \forall g \tag{9}$$

$$\sum_{\forall k(n,.)} P_k - \sum_{\forall k(.,n)} P_k + \sum_{\forall g(n)} P_g = d_n, \forall n \tag{10}$$

$$P_k^{\min} z_k \le P_k \le P_k^{\max} z_k, \forall k \tag{11.1}$$

$$B_k(\theta_n - \theta_m) - P_k + (1 - z_k)M_k \ge 0, \forall k \tag{6.1}$$

$$B_k(\theta_n - \theta_m) - P_k - (1 - z_k)M_k \le 0, \forall k \tag{6.2}$$

$$z_k \in \{0, 1\}, \forall k \tag{12}$$

There are additional ways to introduce the state of the transmission asset into the DCOPF formulation. The second formulation introduces a new variable to the formulation to allow P_k to be replaced by $B_k(\gamma_k - \theta_m)$. Note that in (10.1), the θ_m is the voltage angle corresponding to the *from bus* for line *k*. Then, (6) is replaced with constraints (6.3) and (6.4) to force γ_k to equal θ_n if the line is in service; if the line is out of service, γ_k is not forced to equal θ_n by (6.3) and (6.4). Due to (11.2), γ_k will equal θ_m when the line is out of service. Thus, the new variable, γ_k, equals the *to bus* voltage angle value, θ_n, when the line is in service and it equals the *from bus* voltage angle value, θ_m, when the line is out of service. Once again, there is a reclosing rule that places a restriction on θ_n minus θ_m through (6.3) and (6.4) since γ_k equals θ_m when the line is out of service. The big M value in this formulation is represented by θ^{rec} to enforce this reclosing rule.

Formulation 2:
Minimize:

$$\sum_{g} c_g P_g \tag{7}$$

s.t.

$$0 \le P_g \le P_g^{\max}, \forall g \tag{9}$$

$$\sum_{\forall k(n,.)} B_k(\gamma_k - \theta_m) - \sum_{\forall k(.,n)} B_k(\gamma_k - \theta_m) + \sum_{\forall g(n)} P_g = d_n, \forall n \tag{10.1}$$

$$P_k^{\min} z_k \le B_k(\gamma_k - \theta_m) \le P_k^{\max} z_k, \forall k \tag{11.2}$$

$$\theta_n - \gamma_k + (1 - z_k)\theta^{rec} \ge 0, \forall k \tag{6.3}$$

$$\theta_n - \gamma_k - (1 - z_k)\theta^{rec} \le 0, \forall k \tag{6.4}$$

$$z_k \in \{0, 1\}, \forall k \tag{12}$$

5 Economic and Market Implications of Optimal Transmission Switching

5.1 Economic Savings Resulting from Optimal Transmission Switching

Optimal transmission switching has been researched for various test cases and formulations, [19–25]; it has been studied with a DCOPF, an N-1 DCOPF, and a unit commitment N-1 DCOPF formulation with the IEEE 73-bus test case (RTS96 system), the IEEE 118-bus test case, and two large scale, 5000-bus test cases provided to the authors by the Independent System Operator of New England (ISO-NE). Table 1 presents the best found economic savings for these various formulations and test cases. Additional sensitivity studies were done with these test cases as well, see [19–25], with all studies showing that optimal transmission switching can provide substantial economic benefits. Since these test cases vary in size and generator costs, the best indicator of the potential of optimal transmission switching is the percent savings instead of the dollar savings. Of the solutions in Table 1, the only solution proven to be the optimal solution is the IEEE 118-bus DCOPF result. Consequently, the true optimal solutions for the rest of the test cases may provide even more economic savings. If optimal transmission switching can be practically implemented and save even a fraction of the savings that are shown here, such would be a remarkable result for the three-hundred billion dollar electric industry in the USA.

Table 1 Economic savings from transmission switching for various test cases and formulations

Formulation		IEEE 73-Bus (RTS 96)	IEEE 118-Bus	ISONE 5000-Bus
1HR DCOPF	% Savings	–	25% [19, 20]	13% [23]
	$ Savings	–	$512 [19, 20]	$62,000 [23]
1HR	% Savings	8% [21]	16% [21]	–
N-1 DCOPF	$ Savings	$8,480 [21]	$530 [21]	–
24HR UC	% Savings	3.7% [22]	–	–
N-1 DCOPF	$ Savings	$120,000 [22]	–	–

For the N-1 results, it was assumed that if a line is temporarily taken out of service for a given steady-state (no contingency) period and if a contingency occurs during that period, then the operator would not have the choice to reclose the line, i.e., place the line back in service immediately after the contingency. Thus, all of the N-1 solutions in Table 1 enforce N-1 without acknowledging the possibility to implement post-contingency transmission switching actions. This is a conservative approach since it is well known that there are SPSs that exist today involving post-contingency switching actions, [16].

The DCOPF results in Table 1 do not enforce N-1. However, these results still provide very useful information. First, these results demonstrate that the redundancy built into the grid, in order to handle a multitude of contingencies over a long planning horizon, does cause dispatch inefficiency. Second, the results estimate the potential savings for the concept of just-in-time transmission, [23]. The concept basically states that we should be able to co-optimize the topology with generation while accounting for the ability to implement actions similar to the SPSs that involve post-contingency switching actions. Transmission assets that are a detriment to dispatch efficiency can be kept offline during steady-state operating periods but they can be switched back into service, if needed, *just-in-time* once there is a contingency in order to bring the grid back to its redundant, reliable state. This concept is discussed in more detail in Sect. 6.7.

In [24], the authors examined the potential yearly savings for the IEEE 118-bus test case with the DCOPF optimal transmission switching formulation. The unconstrained economic dispatch solution[4], which is a lower bound to the optimal transmission switching problem, for the IEEE 118-bus test case was 3.07% below the DCOPF solution. This is likely a unique result corresponding to this particular IEEE test case as most systems have a much larger gap between the unconstrained economic dispatch solution and the DCOPF solution. This gap defines the maximum potential savings for the optimal transmission switching DCOPF problem; for this yearly case study in [24], optimal transmission switching saved 3.05% out of this 3.07% gap.

5.2 Generation Cost, Generation Rent, Congestion Rent, and Load Payment

Harnessing the control over transmission can be used for a variety of operational benefits; optimal transmission switching suggests that operators should co-optimize the generation with the network topology, while meeting reliability requirements, in order to reduce the overall system operating costs. This approach is not

---

[4] The unconstrained economic dispatch problem is a dispatch problem without transmission network constraints.

controversial for a vertically integrated utility that takes the role of serving its region at least cost as the savings would be passed on to the consumers. In standard Independent System Operator (ISO) markets that are based on a nodal pricing system, i.e., they use Locational Marginal Prices (LMPs), the goal of the operator is to maximize the market surplus while ensuring a reliable system (note that when load is perfectly inelastic minimizing the total cost achieves the same objective as maximizing the market surplus). LMPs are the dual variables (shadow prices) on the node-balance equations in the OPF formulation, the dual variable on Eq. 10 in Sect. 4.2; it reflects the marginal cost to deliver another unit of energy to that location in the network. With an LMP pricing system, generators are paid their LMPs and the load pays their LMP to consume. Even though optimal transmission switching increases the surplus in the market, there is no guarantee that, with the implementation of this new technology, all market participants will be better off than before.

Figure 3 demonstrates the unpredictable impact optimal transmission switching can have on groups of market participants, [20]. The generation rent is the short term generation profit for all generators and, thus, the generation revenue is equal to the generation rent plus the generation cost, which can also be determined by summing each generator's production times its LMP (note that we are not including other payments made to generators, i.e., uplift payments). The load payment is defined as the sum of each load times its LMP. The congestion rent is defined as the sum of each line's flow times the dual variable on its capacity constraint, (11); this dual variable is often referred to as the flowgate marginal price. Since the DCOPF is an LP, it has a well defined dual. Based on duality theory, complementary

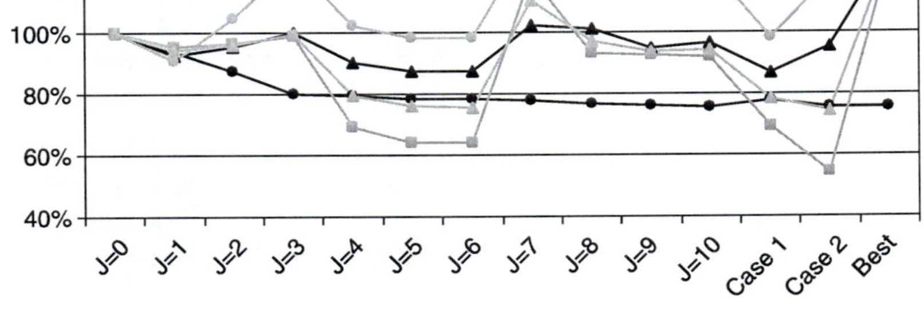

Fig. 3 Generation cost, generation rent, congestion rent, and load payment for various transmission switching solutions – IEEE 118-bus test case, [20]

slackness, strong duality, and by identifying the parts of the dual problem that reflect these four terms, it can be shown that the following identity holds: load payment – generation rent – congestion rent (objective of dual) = generation cost (objective of primal). The congestion rent is often also labeled as the cost to send energy from a source to a sink location, which translates into the difference in LMP between these two locations times the quantity.

The J index on the x-axis is a reflection of how many lines were opened; $J = 0$ reflects the base case where all lines are in service; all values for $J = 0$ are normalized to 100% in the graph to reflect each term's value from the optimal DCOPF solution for the original topology. The plots reflect how each term varies from one transmission switching solution to the next as compared to that term's value in the base case when all lines are in service. For instance, the load payment for the $J = 0$ DCOPF solution is $7,757/h and the load payment is 79% of that value for the $J = 4$ optimal transmission switching solution (with the restriction that only four lines can be opened). Case 1 and case 2 reflect solutions found by a heuristic technique and the "best" solution represents the optimal transmission switching solution for this IEEE 118-bus test case when there is no restriction on how many transmission assets can be temporarily taken out of service.

This figure identifies the following interesting results: first, the majority of the savings are first obtained by only opening a few lines; this is an important result in term of computational complexity because it states that good solutions can be found by only searching for a few lines to open instead of considering all possible topology configurations (the other studies we have conducted agree with this statement). Next, there is a plateau affect in the sense that many transmission switching solutions are extremely close to the optimal solution. There are many solutions that are very close in objective and yet the results show that there can be drastically different outcomes for the market participants with these solutions. This is an interesting result as it is highly unlikely that this concept will be implemented and that optimality will be proven. It is also interesting to note that each term, except for the objective of the primal, the generation cost, is at some point below 100% as well as above 100%. Finally, the optimal solution ends up providing the generators with the highest generation rent out of all solutions and every category outside of the generation cost is at least 20% higher than the corresponding DCOPF solution, $J = 0$, for the original topology.

5.3 LMPs

MIPs do not have well defined duals; the LMPs from the optimal transmission switching MIP problem come from the node, which is an LP, in the branch and bound tree where the optimal solution was found or they can be reproduced by fixing the integer variables to their optimal values and then by solving the

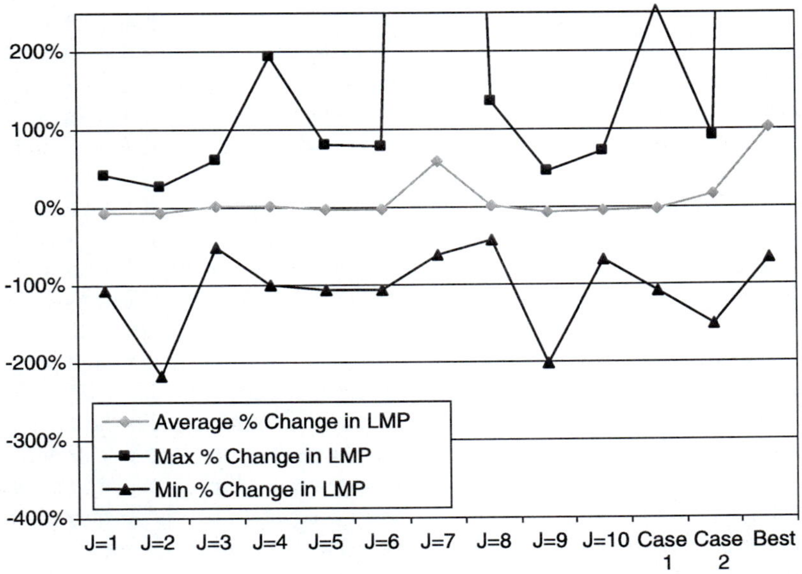

Fig. 4 Maximum, average, and minimum change in LMP – IEEE 118-bus test case, [20]

corresponding LP. LMPs can vary substantially between what is obtained from the optimal transmission switching problem as compared to the LMPs from the traditional DCOPF problem. Figure 4 below shows the percent change between the LMPs that result from the traditional DCOPF with the original topology as compared to these various optimal transmission switching solutions' LMPs. These solutions correspond to the same solutions presented in Fig. 3.

Although the total generation cost of the system decreases with optimal transmission switching, there is no guarantee as to the impact on the LMPs. By reconfiguring the network as well as by changing the generation dispatch solution, LMPs can decrease substantially as well as increase substantially. As a result, there are unpredictable wealth effects for the various market participants due to implementing optimal transmission switching. Finally, it is interesting to note that the average LMP for the optimal solution, i.e., the "best" solution, (when there is no restriction on the number of lines that can be opened) has a substantial increase in LMP. This test case from [20] did not enforce reliability so the main cause of this result is that the network is stripped down to a much less redundant network that is more economically efficient. By doing so, the marginal cost to deliver an additional unit of energy to various locations in the network would cause a substantial redispatch cost to the system, which is the reason the LMPs have increased. This is not a general result of optimal transmission switching but a specific result for this particular test case, which emphasizes the possibility of such a dramatic result to occur.

5.4 Unit Commitment Schedule

Network topology reconfigurations can be beneficial for many different situations. During real time operations, if there is an unexpected outage in the system, which causes a network constraint violation, transmission switching can be used as an effective corrective mechanism to alleviate this constraint violation. The concept of optimal transmission switching, a co-optimization of generation resources and the network topology configuration, is geared towards the day-ahead setting when the operator is forecasting market conditions for each period tomorrow and will solve for the optimal generation and unit commitment schedule.

Reference [22] formulated the problem of co-optimizing unit commitment with an N-1 reliable optimal transmission switching DCOPF formulation[5] and this was tested on the IEEE 73-bus test case, [22]. By co-optimizing the topology with unit commitment, the overall costs decreased by 3.7% (3.2% optimality gap), which saved $120,000 for this 24-h, base load day. If similar savings were obtained for every day, this would produce $40 million in a year while this is only a small IEEE test case, which does not compare in size to the actual electrical grid.

This research also demonstrated that the grid topology changes for the various operating periods. Each period included a switching action, even the low load periods. This further confirms the statement that there is no single topology configuration that is optimal for any and all possible market conditions (load levels, commitment schedules, etc.). Furthermore, no line was opened for the entire 24-h period. It is also interesting to note that there were only a few adjacent periods that had the same topology configuration; this is a result of having the same load levels as well as the same unit commitment schedule for those adjacent hours as opposed to a result that would suggest that there is one topology that is optimal for multiple market conditions. The optimal unit commitment schedule for the original, fixed network topology dispatched three peaker units (short-term, flexible, expensive generators) for only 1 h in the day. With optimal transmission switching, these units were always offline. This result demonstrates that, by harnessing the control of the network topology, the operator can avoid the need to turn on peaker units for a short time interval. This should not be seen as a general result; rather, it is a result for this particular test case.

[5] Ideally, unit commitment models would endogenously represent all N-1 contingencies. Instead, reserve constraints are generally used as surrogates since solving a unit commitment problem while modeling every single contingency is computationally very challenging. The research in [16] modeled this much more robust and difficult problem since there is the underlying question as to whether reserve requirements created for the original topology will work for the reconfigured topology, which is a topic for future research as indicated in Sect. 6.2.

5.5 Financial Transmission Rights Market

Financial Transmission Rights (FTRs) are financial instruments used in electricity markets as a mechanism to manage congestion risk by market participants. An FTR has a defined source node and sink node along with a specific quantity (MW); the holder of the FTR is then entitled to the LMP difference between the sink and source locations times the FTR quantity. For electricity markets that are based on an LMP pricing system, the congestion charges for sending 1 MW of power from an injection node to a withdrawal node is equal to the LMP difference. Hence, FTRs can be used to create a perfect hedge against congestion charges.

The FTR holders are compensated by the ISO. The ISO covers the FTR obligations by the additional revenue they collect, the congestion rent. Revenue adequacy occurs when the ISO has enough congestion rent to fully compensate the FTR holders. Revenue inadequacy occurs when the ISO does not have adequate congestion rents to fully compensate the FTR holders. In order to ensure that the FTR market is revenue adequate, the ISO runs a Simultaneous Feasibility Test (SFT) when the ISO allocates or auctions off the FTRs. This SFT ensures that if these financial rights were physically exercised by actual transfers then the network would be able to satisfy these transfers without any network constraint violations. Reference [27] showed that given a set of assumptions, the SFT can guarantee revenue adequacy for the FTR market. This proof works for the DC load flow model and requires that the set of feasible solutions is convex. Since topology reconfigurations create a superset of feasible solutions, which can be non-convex, the SFT relies on the assumption that the network topology will not be altered.

Unfortunately, even though optimal transmission switching improves the social welfare, it may be incompatible with prevailing market design practices. In this case, optimal transmission switching may cause revenue inadequacy in the FTR market since it modifies the network topology. The most common practice employed by ISOs to handle revenue inadequacy is to de-rate the payments to the FTR holders, which undermines one of the purposes of FTRs, to create a hedge against congestion risk. The ideal solution would be to redesign the SFT mechanism to account for optimal transmission switching; however, it would be difficult to predict the chosen topologies in future operating periods at the time that the SFT is conducted. Consequently, we are currently presented with a choice to implement a new technology that can benefit the common good and then deal with a potential revenue inadequacy problem or to leave operations as is and pass on potential societal improvements. Obviously, such a situation raises an interesting policy debate; further discussion on revenue inadequacy and optimal transmission switching can be found in [24].

6 Future Research

6.1 ACOPF Optimal Transmission Switching

The majority of the high-voltage electrical grid operates under an AC setting, except for a few DC lines. This research is not based on the ACOPF formulation but rather it is based on an AC approximate formulation, the DCOPF. The ACOPF problem is a non-convex optimization problem while the DCOPF is a linear program. For the concept of optimal transmission switching to be practically implemented, it is imperative to examine its affect on the ACOPF as well as to analyze the effects on reactive power, voltage stability, etc.

The main difficulty in solving an ACOPF problem with transmission switching is because the ACOPF is an extremely difficult non-convex optimization problem itself. There are a number of nonlinear programming solvers that can handle the ACOPF problem, e.g., PowerWorld [28] and Knitro [29]; however, adding binary variables to the ACOPF problem in order to incorporate transmission switching would increase the complexity of this problem immensely. With today's optimization software, solving a large-scale mixed integer non-linear program containing trigonometric functions is very difficult. Mainstream MIP software today are primarily limited to mixed integer linear programming, thereby indicating the significant practical challenge to incorporate non-linear functions into MIP. For that reason, operators today use the DCOPF formulation with unit commitment, which is an integer program, as opposed to an ACOPF. The MIP optimal solution, with the integer variables fixed to their optimal MIP solution values, is used as an initial solution that is fed into an ACOPF solver to obtain the best feasible AC solution possible. It is likely that a similar approach would be taken if transmission switching is to be implemented. By using such an approach, there is no guarantee that an AC feasible solution will be obtained when using an AC approximate formulation to produce an initial solution and, thus, future research is needed to investigate this concern.

6.2 Proxy Constraints and Reserve Requirements

Due to the difficulty in solving the ACOPF problem, proxy constraints are commonly used within the DCOPF. Voltage stability is not something that can be directly modeled within the DCOPF since the DCOPF does not contain voltage variables; however, it is possible to approximate voltage issues through proxy constraints and then apply these constraints within the DCOPF problem. Similarly, reserve requirements are a surrogate way to ensure the system is reliable as opposed to explicitly mathematically modeling each single contingency within the OPF formulation. With the implementation of new grid operations and technology, like optimal transmission switching, current market mechanisms and operational protocols may be undermined. If these proxies are based on the original topology, they may not be able to achieve the desired goal if the topology is altered. Future research is needed to determine if such proxy constraints depend on the topology of the network and, if so, how they are affected by optimal transmission switching.

6.3 Circuit Breaker Control, Maintenance, and Cost of Switching

Since optimal transmission switching increases the frequency of transmission switching actions, further research is needed as to the effect on breakers, whether there will be additional required maintenance, and whether more advanced, new breakers would be required to be installed. Any additional capital and maintenance costs should be considered in future research. Furthermore, the optimal transmission switching formulation may need to be adjusted to reflect the marginal cost of switching a circuit breaker. However, any such additional costs are likely to be minor in comparison to the substantial potential economic savings that have been demonstrated thus far and, hence, such costs are unlikely to change the findings of this research.

6.4 Protective Relay Settings

It is imperative to have protective relays set and function correctly if the grid is to be capable of surviving a contingency. Relay settings are used to trip, i.e., switch, specific circuit breakers when a fault is detected so that the fault can be cleared. It is essential that the relay settings properly identify the fault in order to trip the appropriate breakers. If either the wrong circuit breaker is tripped or no breaker is tripped when needed, then this can significantly complicate the situation and may even cause a blackout. If optimal transmission switching is implemented, then grid operators will have to examine whether the relay settings need to be adjusted based on the chosen topology. Similar operational procedures already exist today since relay settings are reset after a contingency occurs or when lines are down due to maintenance. However, future research is needed to address this practical barrier, to determine if operators will be able to conduct such studies on a more frequent basis and within a limited timeframe.

6.5 Computational Performance

As more is learned about the network and transmission switching, operators will know which transmission elements are candidates for switching. It may not be necessary to represent every transmission element within the network with a binary decision variable reflecting whether the element will be in service or not. Rather, the operator may be able to focus on a subset of transmission elements that are key candidates for switching, which will greatly reduce the number of binary variables in the optimization problem and, thus, reduce the computational complexity of the problem. Future research should focus on methods to identify transmission elements that are candidates for switching as well as focus on new solution techniques for this problem. One particular technique that should be investigated is Benders' decomposition. Due to the structure of the optimal transmission switching MIP problem, the use of a big M value, Benders' decomposition is a viable decomposition technique to be considered in order to improve the computational performance of this problem.

6.6 *Transmission Expansion Planning*

Just as network topology optimization has a substantial impact on operations, it may also have an affect on investment planning; in particular, future research should consider incorporating transmission switching with transmission expansion planning. By considering the possibility of transmission switching in the expansion problem, this may change which line is optimal to build or it can delay the need to build a new line. However, incorporating transmission switching into transmission expansion planning will not be easy. First, there is the problem with solving a transmission expansion problem. Since it is such a hard problem to solve, many simplifications are made to make the problem more tractable. Adding transmission switching, an operations based model that should be solved on an hourly basis, to such a complex MIP problem would increase the difficulty immensely. Second, many approximations are made with transmission expansion planning since it is very difficult to predict future network conditions. These imprecise predictions may be more problematic for a transmission expansion plus transmission switching model, as the transmission switching solutions may never be realized while the transmission expansion solution is likely to depend heavily on exactly which topology is used for actual operations.

6.7 *Just in Time Transmission*

The optimal transmission switching concept presented in this chapter is not incompatible with reliable grid operations as research has shown that it is possible to maintain N-1 standards while improving the dispatch efficiency. For the N-1 optimal transmission switching studies, it was assumed that if a line is chosen to be taken out of service temporarily for a specific hour during steady-state operations, then it will remain out of service if a contingency occurs during that period and the system must still be able to maintain N-1. More precisely, it was assumed that the operator would treat the transmission assets as flexible for steady-state operations but that the operator would not change the network topology after a contingency occurs.

Just as it was pointed out that it is better to treat transmission as a controllable asset for steady-state operations, the exact same is true for emergency operations. Furthermore, it has already been established that there are SPS protocols in place in PJM where there are pre-contingency and post-contingency switching actions in order to protect against a contingency as well as to respond to a contingency, [20]. These facts motivate this new concept called just-in-time transmission; operators should co-optimize the electrical grid for any state of the network, be it for steady-state operations or for contingency-states, and transmission that is offline can be switched back into service *just-in-time* in order to respond to a contingency and return the network to its redundant structure.

The electrical grid is built to be redundant in order to satisfy the reliability requirements that are established to make sure that blackouts are rare; however, these redundancies are known to cause dispatch inefficiency. With just-in-time transmission, transmission elements that are a detriment to dispatch efficiency can be kept offline during steady-state operations but they can be switched back into service if needed in case of a contingency, which is similar to some of PJM's SPSs. Such an operation would require adequate ancillary services and generator ramping capabilities to be able to reach a feasible dispatch solution once the topology is changed after the contingency.

The just-in-time transmission concept incorporates additional flexibility as compared to the optimal transmission switching with contingency analysis concept presented by [21]. It acknowledges, in the day-ahead optimization problem, the operator's ability to implement a corrective switching action after a contingency occurs. In the day-ahead setting, the operator would co-optimize the generation and network topology for each steady-state period but would also simultaneously determine the required corrective switching actions to take if a specific contingency were to occur along with the required generator dispatches. By capturing this operational flexibility, the use of transmission switching as a corrective mechanism when there is a contingency, the operator can improve the economic efficiency of the system for steady-state operations as well as improve system reliability.

Additional research is needed to further develop and test this concept with the main difficulty being the computational complexity of this problem. Optimal transmission switching is already a complex problem; just-in-time transmission switching would require many more binary variables as a binary variable would be needed for each transmission line as well as for each contingency state that is enforced. Though this problem may be hard to solve, the potential economic savings from this concept may be significant, as this concept would ideally allow the operator to optimize the topology for steady-state operations while ignoring N-1 reliability constraints knowing that the transmission can be put back into service if there is a contingency.

7 Conclusion

Previous research has demonstrated that harnessing the control of transmission assets can provide substantial benefits. The use of transmission switching as a corrective mechanism has been the most frequently proposed use of transmission control; such research has demonstrated its ability to help alleviate line overloading, voltage violations, as well as other constraint violations. Furthermore, past research has demonstrated the ability for transmission switching to help reduce system losses, increase transfer capability, and manage congestion. Even though past research has emphasized the substantial benefits that can be obtained from transmission switching, for the most part transmission assets are still viewed as static assets as transmission switching is primarily limited today to ad-hoc procedures and special protection schemes.

There is currently a national push to create a smarter, more flexible electric grid. The true grid of the future should include advanced technologies, be more flexible, and include new operational protocols. This research has proposed that the way in which transmission assets are viewed in economic dispatch optimization models should change, that the state of a transmission asset should be seen as a discrete decision variable in optimal power flow formulations. The concept of optimal transmission switching has been tested on three different dispatch optimization problems: the DCOPF, the N-1 DCOPF, and the multi-period unit commitment N-1 DCOPF problem. The results demonstrated that, indeed, the network redundancies that are built into the electrical grid, in order to survive a multitude of contingencies, cause economic dispatch inefficiency. By incorporating the control of transmission assets into OPF formulations, a superior optimization problem is created, as the feasible set of dispatch solution from the optimal transmission switching problem is a superset of the feasible set of dispatch solutions for the traditional OPF formulation. Moreover, by harnessing the control of transmission assets, redundancies that are not currently required to maintain reliability standards can be kept offline in order to improve dispatch efficiency.

Furthermore, this research has demonstrated that the perception that taking a transmission line out of service must be a detriment to reliable operations is, indeed, a common misconception; optimal transmission switching has been shown to provide substantial economic savings even while satisfying strict N-1 reliability standards. Overall, this concept has been tested on a variety of standard IEEE test cases as well as on two real-world test cases provided by ISONE. All results demonstrate the potential of this concept to reduce operational costs significantly for the multi-trillion dollar electric industry.

While the optimal transmission switching concept maximizes the benefits to society as a whole by maximizing the market surplus, it unfortunately can cause unpredictable distributional effects for market participants and it can create substantial wealth transfers between market participants. Implementation of this concept will create winners and losers. Results demonstrated that LMPs can vary significantly between two network topology solutions even if their objectives vary by trivial amounts. As a result, implementation of this concept can have unpredictable effects on the generation rent, congestion rent, and load payment. It is even possible that the load ends up paying more even though the system as a whole is more economically efficient. Unfortunately, optimal transmission switching also undermines a current market design protocol that relies on the assumption that the transmission grid is treated as a static asset; the implementation of optimal transmission switching has the potential to cause revenue inadequacy in the FTR market. As a result, it may be necessary to rethink past market design principles and market mechanisms that are made obsolete by emerging technologies that are implemented to create a smarter electrical grid.

In summary, this research has demonstrated how to modify economic dispatch optimization problems to incorporate the flexibility of transmission assets' states, i.e., in service or out of service. By doing so, there can be substantial economic savings since it creates a superset of feasible generation dispatch solutions as compared to traditional economic dispatch optimization problems. Furthermore, such savings can be obtained even while enforcing reliability standards. If the savings are even a fraction of the current findings, this would still be huge for the multi-trillion dollar electric industry. The main drawbacks of this concept include the substantial wealth transfers among market participants that can occur as well as dealing with the computational complexity of this problem. The concept of optimal transmission switching is consistent with the national push to create a smarter electrical grid; it challenges traditional views and misconceptions held concerning transmission assets and how they should be treated in operational procedures. The results from this research confirm that if we wish to develop a truly smarter, more flexible grid, the treatment of transmission assets should change.

Appendix

Notation	
Indices and Sets	
g	Generator
$g(n)$	Set of generators at bus n
k	Transmission element (line or transformer)
$k(n,.)$, $k(.,n)$	Set of transmission elements with bus n as the to bus and the set with bus n as the from bus respectively
m, n	Nodes
Parameters	
B_k	Susceptance of transmission element k
c_g	Production cost for generator g
d_n	Real power load at node n
G_k	Conductance of transmission element k
M_k	Big M value for transmission element k

(continued)

P_g^{max}, P_g^{min}	Max and min capacity of generator g
P_k^{max}, P_k^{min}	Max and min rating of transmission element k; typically $P_k^{max} = -P_k^{min}$
$\theta^{max}, \theta^{min}$	Max and min bus voltage angle difference; typically $\theta^{max} = -\theta^{min}$
θ^{rec}	Max voltage angle difference when reclosing breakers to bring a line back into service
Variables	
P_g	Real power supply from generator g at node n
P_k	Real power flow from node m to node n for transmission element k
Q_k	Reactive power flow from node m to node n for transmission element k
V_n	Bus voltage at node n
z_k	Binary switching variable for transmission element k (0 open/not in service, 1 closed/in service)
γ_k	Bus voltage angle variable that is equal to transmission element k's *to bus* angle value when the line is in service but equal to transmission element k's *from bus* angle value when the line is out of service
θ_n	Bus voltage angle at node n

Biographies

Kory W. Hedman received the B.S. degree in electrical engineering and the B.S. degree in economics from the University of Washington, Seattle, in 2004 and the M.S. degree in economics and the M.S. degree in electrical engineering from Iowa State University, Ames, in 2006 and 2007, respectively. He received the M.S. and Ph.D. degrees in industrial engineering and operations research from the University of California, Berkeley in 2007 and 2010 respectively.

Currently, he is an assistant professor in the school of electrical, computer, and energy engineering at Arizona State University. He previously worked for the California ISO (CAISO), Folsom, CA, on transmission planning and he has worked with the Federal Energy Regulatory Commission (FERC), Washington, DC, on transmission switching.

Shmuel S. Oren received the B.Sc. and M.Sc. degrees in mechanical engineering and in materials engineering from the Technion Haifa, Israel, and the MS. and Ph.D. degrees in engineering economic systems from Stanford University, Stanford, CA, in 1972.

He is a Professor of IEOR at the University of California at Berkeley and the Berkeley site director of the Power System Engineering Research Center (PSERC). He has published numerous articles on aspects of electricity market design and has been a consultant to various private and government organizations including the Public Utilities Commission of Texas, The Energy Division of the California Public Utilities Commission, The California ISO and The Bonneville Power Authority. Dr. Oren is a Fellow of INFORMS and of the IEEE.

Richard P. O'Neill has a Ph.D. in operations research and a BS in chemical engineering from the University of Maryland at College Park.

Currently, he is the Chief Economic Advisor in the Federal Energy Regulatory Commission (FERC), Washington, D.C. He was previously on the faculty of the Department of Computer Science, Louisiana State University and the Business School at the University of Maryland at College Park.

References

1. Glavitsch H (1985) State of the art review: switching as means of control in the power system. INTL JNL Elect Power Energy Syst 7(2):92–100
2. Mazi AA, Wollenberg BF, Hesse MH (1986) Corrective control of power system flows by line and bus-bar switching. IEEE Trans Power Syst 1(3):258–264
3. Gorenstin BG, Terry LA, Pereira MVF et al (1986) Integrated network topology optimization and generation rescheduling for power system security applications. IASTED INTL SYMP: High Tech Power Ind 1:110–114
4. Bacher R, Glavitsch H (1986) Network topology optimization with security constraints. IEEE Trans Power Syst 1(4):103–111
5. Bakirtzis AG, Meliopoulos AP (1987) Incorporation of switching operations in power system corrective control computations. IEEE Trans Power Syst 2(3):669–675
6. Schnyder G, Glavitsch H (1988) Integrated security control using an optimal power flow and switching concepts. IEEE Trans Power Syst 3(2):782–790
7. Schnyder G, Glavitsch H (1990) Security enhancement using an optimal switching power flow. IEEE Trans Power Syst 5(2):674–681
8. Rolim JG, Machado LJB (1999) A study of the use of corrective switching in transmission systems. IEEE Trans Power Syst 14:336–341
9. Shao W, Vittal V (2005) Corrective switching algorithm for relieving overloads and voltage violations. IEEE Trans Power Syst 20(4):1877–1885
10. Shao W, Vittal V (2006) BIP-based OPF for line and bus-bar switching to relieve overloads and voltage violations. PSCE 2006 1:2090–2095
11. Granelli G, Montagna M, Zanellini F et al (2006) Optimal network reconfiguration for congestion management by deterministic and genetic algorithms. Electr Power Syst Res 76(6–7):549–556
12. Bacher R, Glavitsch H (1988) Loss reduction by network switching. IEEE Trans Power Syst 3 (2):447–454
13. Fliscounakis S, Zaoui F, Simeant G et al (2007) Topology influence on loss reduction as a mixed integer linear programming problem. IEEE Power Tech 1:1987–1990
14. ISONE (2007) ISO New England Operating Procedure no. 19. Transm Oper 1:7–8
15. Northeast Power Coordinating Council (1997) Guidelines for inter-area voltage control. NPCC Operating Procedure Coordinating Committee and NPCC System Design Coordinating Committee, New York
16. PJM (2010) Manual 3: transmission operations, revision: 35, October 5, 2009. Section 5: index and operating procedures for PJM RTO Operation. PJM. http://www.pjm.com/markets-and-operations/compliance/nerc-standards/~/media/documents/manuals/m03.ashx. Accessed 1 Sep 2010
17. ISONE (2010) ISO New England Outlook: Smart Grid is About Consumers. ISONE http://www.iso-ne.com/nwsiss/nwltrs/outlook/2009/outlook_may_2009_final.pdf. Accessed 1 Sep 2010

18. O'Neill RP, Baldick R, Helman U et al (2005) Dispatchable transmission in RTO markets. IEEE Trans Power Syst 20(1):171–179
19. Fisher EB, O'Neill RP, Ferris MC (2008) Optimal transmission switching. IEEE Trans Power Syst 23(3):1346–1355
20. Hedman KW, O'Neill RP, Fisher EB et al (2008) Optimal transmission switching – sensitivity analysis and extensions. IEEE Trans Power Syst 23(3):1469–1479
21. Hedman KW, O'Neill RP, Fisher EB et al (2009) Optimal transmission switching with contingency analysis. IEEE Trans Power Syst 24(3):1577–1586
22. Hedman KW, Ferris MC, O'Neill RP et al (2010) Co-optimization of generation unit commitment and transmission switching with N-1 reliability. IEEE Trans Power Syst 25(2): 1052–1063
23. Hedman KW, O'Neill RP, Fisher EB et al (2010) Smart flexible just-in-time transmission and flowgate bidding. IEEE Trans Power Syst 26(1):93–102
24. Hedman KW, Oren SS, O'Neill RP (2010) Optimal transmission switching: economic efficiency and market implications. JNL Reg Econ. 40(2):111–140
25. O'Neill RP, Hedman KW, Krall EA et al (2010) Economic analysis of the N-1 reliable unit commitment and transmission switching problem using duality concepts. Eneregy Syst 1 (2):165–195
26. Bergen A, Vittal V (2000) Power systems analysis, vol 2. Prentice Hall, Upper Saddle River
27. Hogan WW (1992) Contract networks for electric power transmission. J Reg Econ 4:211–242
28. PowerWorld Corporation(2010) PowerWorld simulator – optimal power flow analysis tool. http://www.powerworld.com/products/opf.asp. Accessed 3 Nov 2010
29. Ziena Optimization Inc (2010) Knitro optimization software. http://www.ziena.com/knitro.htm. Accessed 3 Nov 2010

Power System Ancillary Services

Juan Carlos Galvis and Antonio Padilha Feltrin

Abstract Ancillary services are essential for the reliably high-quality operation of a power system. These services are provided by network users and procured by the independent system operator – ISO. Due to system requirements and market structures, ancillary services are managed in different ways around the world. In this chapter, we briefly describe the definition, classification, technical requirements and economic issues of ancillary services. Particularly, we compare active power reserves and reactive support ancillary services in different systems. Finally we show two illustrative examples: A co-optimization model with AC network constraints for the energy and reserve dispatch and a modified version of this model that considers the reactive power dispatch.

Keywords Active power reserves • ancillary services • co-optimization • dispatch • power system • reactive support

1 Introduction

Deregulation in the electricity industry brought significant changes in power systems management. Generation and distribution activities began to be performed by different companies trading energy in a competitive market. Transmission activity, however, remained under a monopoly structure. These reforms were realized to improve efficiency and create economic incentives for the expansion of the system [1]. On the other hand, new problems arose with deregulation: definition of responsibilities for network users, energy pricing and cost allocation. In this context, ancillary services became an important issue because they are necessary to support the transmission of energy through the network.

J.C. Galvis (✉) • A.P. Feltrin
UNESP, Avenida Brasil, Ilha Solteira, SP, Brazil
e-mail: juancgalvis@gmail.com; padilha@dee.feis.unesp.br

A. Sorokin et al. (eds.), *Handbook of Networks in Power Systems I*, Energy Systems, DOI 10.1007/978-3-642-23193-3_22,

The deregulation process was different in electric systems due to technical, economic and political aspects. Therefore, markets with different decentralization levels and competitive rules emerged [2]. Since ancillary services are linked with the energy market, they also have different management structures.

Some of the main issues with ancillary services include how to determine ancillary services requirements, how much it should cost, how to allocate the costs and who should pay for these services. Therefore ancillary services involve important topics such as system reliability, operation, regulation and market structure.

In this chapter, we examine some aspects of ancillary service management, focusing on active power reserves and reactive support ancillary services. To this purpose, the content is organized as follows. Section 2 starts with the definition and classification of ancillary services. Next, in Sects. 3 and 4, technical and economic issues related to active power reserves and reactive support are described. Section 5 shows an example of active power reserve dispatch, and Sect. 6 an example of reactive support dispatch. Finally, in Sect. 7, conclusions are outlined.

2 Ancillary Services

2.1 Definition

In 1995, the Federal Energy Regulatory Commission (FERC) established Order 888 with the following definition of ancillary services [3]:

> Ancillary Services: Those services that are necessary to support the transmission of capacity and energy from resources to loads while maintaining reliable operation of the Transmission Provider's Transmission System in accordance with Good Utility Practice.

Other definitions are illustrated in [4]. In general, ancillary services are provided by network users and are mainly constituted of active and reactive power resources. These services in most cases are procured by the ISO in a centralized system. They support the function of the energy market and enable the system to operate even under contingency or emergency conditions.

2.2 Classification

The classification of ancillary services may vary from one system to another depending on market structure and technical requirements. Hirst and Kirby discuss and explain the classification and unbundling of these services [5]. Here we present a short description of the six ancillary services proposed by FERC in its Order 888.

2.2.1 Scheduling and Dispatch

These services are coordinated by the ISO. Scheduling refers to the assignment of generation and transmission resources a week, a day or a few minutes before real-time operation. Dispatch refers to the management and control of these resources in real-time operation.

2.2.2 Regulation and Frequency Response

Regulation and frequency response refers to the continual balancing of generation and load. It is also used to maintain the system's frequency within an acceptable range. The regulation action is performed through generators or by means of interruptible loads with a time response inside the standard requirements.

2.2.3 Reactive Supply and Voltage Control

Reactive power supply is used for voltage control, reducing active power losses or releasing congestion in some points of the system. Due to the tight link between reactive power and bus voltages, reactive supply and voltage control are essentially the same service. Therefore reactive supply is used to maintain system voltages within acceptable margins. Voltage control can be realized through different devices like transformer taps, voltage regulators, capacitors, reactors, static-var compensators, generators and synchronous condensers.

2.2.4 Energy Imbalance

As defined by FERC, an energy imbalance service refers to financial arrangements for compensating differences between the scheduling and actual delivery of energy to a load located within a control area in a single hour.

2.2.5 Operating Reserves – Spinning Reserve Service

Spinning reserve service involves on-line active power resources (generators or interruptible loads already synchronized to the system) that are able to attend to active power imbalances produced by a contingency in the system in real time.

2.2.6 Operating Reserves – Supplemental Reserve Service

Supplemental reserve service is an additional capacity that is not available immediately but is normally available within a short period of time. It may be provided by on-line but unloaded units, quick start generation or interruptible loads.

3 Active Power Reserves and Reactive Support Ancillary Services: Technical Issues

Active power reserves and reactive support are considered common ancillary services in most systems. These services are directly involved in normal operations, to maintain frequency and voltage requirements, and in emergency conditions, to prevent frequency and voltage stability problems.

Active power reserves are classified differently as a function of their time response [6]. For example, in Europe, active power reserves are classified from fastest to slowest as primary, secondary and tertiary control reserves [7]. Primary and secondary control reserves correspond to regulation and frequency response services in North America. On the other hand, in the United Kingdom, operating reserves are differentiated from contingency reserves [8], in direct contrast to the classification by FERC in the United States [3]. In terms of reactive supply, however, definitions are more uniform and do not require any discussion.

Next we will describe some technical issues concerning active power reserves and reactive support ancillary services for the four systems illustrated in Table 1. This table also shows the corresponding regulator agencies and ISOs. The systems are from different regions of the world (South America, North America and Europe) and were chosen to illustrate that technical and economic ancillary service requirements are different. Differences between electrical systems are more common than similarities, because of specific requirements and market structures. More extensive studies that compare the characteristics of these services in different systems can be found in [6, 9–11]. There is also less extensive but more detailed research on specific regions like Spain [12] and the Nordic Countries [13].

3.1 Active Power Reserves

As mentioned before, in Europe, the Union for the Co-ordination of Transmission of Electricity (UCTE) distinguishes active power reserves in primary, secondary and tertiary control reserves [7], as a function of the time response. Primary control reserves are managed locally by each generator/load while secondary and tertiary control reserves are managed centrally by the ISO. Table 2 shows the full availability time of active power reserves for the four systems illustrated in Table 1. Full availability refers to the time it takes to provide 100% of the reserve capacity.

Table 1 Regulator agencies and ISOs in different power systems

System	Abbreviation	Regulator	ISO
PJM	PJM	FERC	PJM ISO
Spain	ES	CNE	REE
Argentina	AR	ENRE	CAMMESA
Brazil	BR	ANEEL	ONS

ISO independent system operator

Table 2 Full availability of active power reserves in several systems

Types	ES [14]	PJM [15]	BR [16]	AR [17]
Primary	<30 s	No rec	<60 s	<30 s (thermal) <60 s (hydraulic)
Secondary	<300–500 s	<10 min	No rec	several min
Tertiary	<15 min	<10 min. $10 < t < 30$ min. 30 min. $< t <$ 8 h	No rec	<5 min. <10 min. <20 min

No rec no recommendation

From Table 2, it can be noted that primary control reserves must be available within a few seconds, secondary control reserve within minutes and tertiary control reserve within a longer time frame. In Argentina, for example, primary control reserve is provided before 30 s for thermal units and 60 s for hydraulic units. On the other hand, tertiary control reserve availability times depend on the kind of reserves. In Argentina, there are 5, 10 and 20 min reserves. In PJM, spinning (regulating and non-regulating) and quick start reserves must be provided within 10 min. Secondary reserve (unlike the UCTE secondary control reserve) is required between 10 and 30 min. Complementary reserve is available between 30 min and 8 h; this is called the beyond secondary reserve.

3.2 *Reactive Supply*

The transmission of reactive power over long distances is highly impractical because the reactive and active power losses may be unacceptable. Therefore reactive supply is considered a local problem. However, the ISO must coordinate these resources to provide reactive compensations at any point and to maintain a specific voltage profile.

Usually, the ISO asks generators to set aside a portion of their reactive power capacity as a compulsory connecting condition. The nearest connecting point to an electrical network from a generator is the point of delivery (POD). From the generator terminals up to the point of delivery (POD), there is a significant loss of reactive power. Typical reactive power losses in auxiliary equipments, the step-up transformer and the transmission line before the POD add up to approximately 15% of apparent power [6]; thus:

$$Q_{POD} \approx Q_{STATOR} - 0.15 \times S_n. \tag{1}$$

Table 3 shows the compulsory reactive requirements for the four systems in Table 1. In this table, reactive power requirements are indicated at the *POD* for Spain and at the terminals of the stator for Brazil. Argentina indicates its reactive requirements as a function of the maximum and minimum limits of the generator capability curve. The literature [12, 21] shows a detailed description of the voltage control in the Spanish case.

Table 3 Reactive power requirements in several systems

	ES [18]	PJM [19]	BR [16]	AR [20]
Absorption capability requirement (Q_g^{lag})	$pf = 0.989$ at the POD for P_n and U_n	No rec	$pf = 0.989$ at the terminals for P_n	$90\%Q^{min}$ at the terminals for every P
Production capability requirement (Q_g^{lead})	$pf = 0.989$ at the terminals for P_n and U_n	No rec	$pf = 0.925$ at the terminals for P_n	$90\%Q^{max}$ at the terminals for every P

No rec no recommendation, P_n rated power, U_n rated voltage

4 Active Power Reserves and Reactive Supply Ancillary Services: Economic Issues

With the unbundling of ancillary services comes the necessity of determining how to pay the providers and charge the users. In this context, it is important to identify some aspects that influence the economics of these services:

- Ancillary service pricing: determining the price of a specific service. Pricing could be direct, in terms of the incurred costs; indirect, based on the benefit provided by the service to the system; or a combination thereof.
- Procuring methods: Compulsory, bilateral contracts or long- or short-term auctions.
- Settlement rule: Regulated pricing, marginal pricing or pay-as-bid pricing.
- Offer structure: simple (availability) or double (availability and use).
- Optimization methods: sequential or co-optimized (joint dispatch).
- Remuneration structure: The cost components that should be recognized: fixed, availability, use and opportunity costs.
- Cost allocation structure: How cost will be charged to users–pro-rata, based on marginal costs or based on incremental costs, with or without consideration of the user's location.

These and other aspects such as the needs of users (expectations about quality and the willingness to pay for this quality), price caps and incentives are related to the market design structure. Some discussion about these topics can be found in [9, 22–27]. The following sections describe some of the above aspects.

4.1 Ancillary Service Cost Components

Finding an accurate methodology for assessing ancillary service costs is a challenging task since they normally make up part of the energy cost structure and the unbundling process may not be easy. In fact, the provision of an ancillary service is frequently affected by energy production or other ancillary services. This aspect

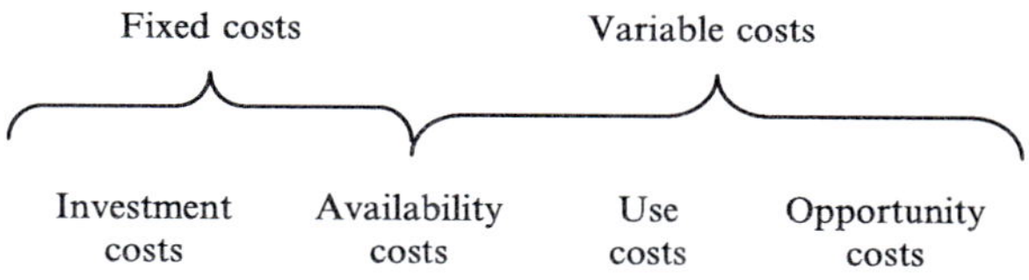

Fig. 1 Ancillary service cost components

highlights the existence of opportunity costs which will be illustrated in examples 1 and 2 (Sects. 4 and 5). Additionally, ancillary services are significant in the system's budget. For example, in the United States the cost of these services was roughly estimated to be between 5% and 25% of total generation and transmission costs [28]. Here, we separate the ancillary service costs as illustrated in Fig. 1. It should be noted that the opportunity cost has been incorporated as a variable cost because it varies depending on the production level of the ancillary service.

The cost structure depends on the kind of ancillary service and its provider. For example, active power reserves may be provided by generators or interruptible loads. In the case of active power reserves provided by generators, investment costs refer to the cost of adaptations in reserve equipment and expenditures in administration, measuring, control, communication and data processing. The costs also include investments in regulation equipment and reserve capacity. Fixed availability costs include the cost of preventive maintenance and other expenditures, such as the cost of tests for regulation equipment. Variable costs include a variable component of the availability cost, use and opportunity costs. Variable availability costs depend on the production level and include predictive and corrective maintenance costs as well as the cost of efficiency losses in production. Use costs represent the cost of the reserve energy provided in real-time operation. Opportunity costs refer to losses in the provider's profit due to non-generated energy available in the form of reserve capacity.

On the other hand, reactive support can be provided by synchronous generators, synchronous condensers, capacitors and static VAR compensators. Some characteristics of these devices are described in [29]. Here we will describe the cost components of reactive support provided from generators. In order to clarify that description, the generator capability curve is illustrated in Fig. 2. This figure shows the armature and field limits of the synchronous generator. The armature limit is a circle with its center in zero and radius $R = V_t \times I_a$, with V_t the voltage at the terminal bus and I_a the steady state armature current. The field limit has center in $\left(0, -V_t^2/X_s\right)$ and radius $R = E_{af} \times V_t/X_s$, with E_{af} the excitation voltage and X_s the synchronous reactance. As mentioned in Sect. 2 there is a compulsory reactive requirement Q_g^{lag} and Q_g^{lead} that is represented through the shaded area on that figure. The intersection point between the armature and field limits indicates the rated MVA and the power factor (pf) of the generator. It should be noted that if $P_g > P_{gR}$, reactive power is limited by armature winding; if $P_g < P_{gR}$, reactive power is limited by field winding. Under-excitation and stability constraints can also be included in that figure [30].

The fixed costs of reactive supply provided by generators are related to expenditures on the excitation system. Unlike active power, the variable costs of

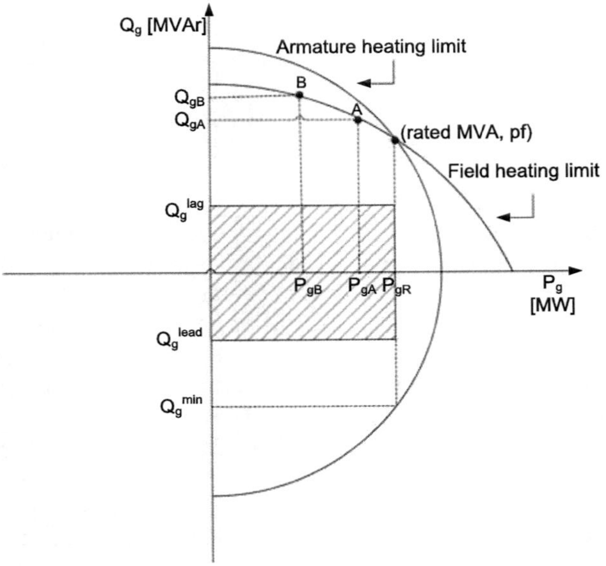

Fig. 2 Synchronous generator capability curve

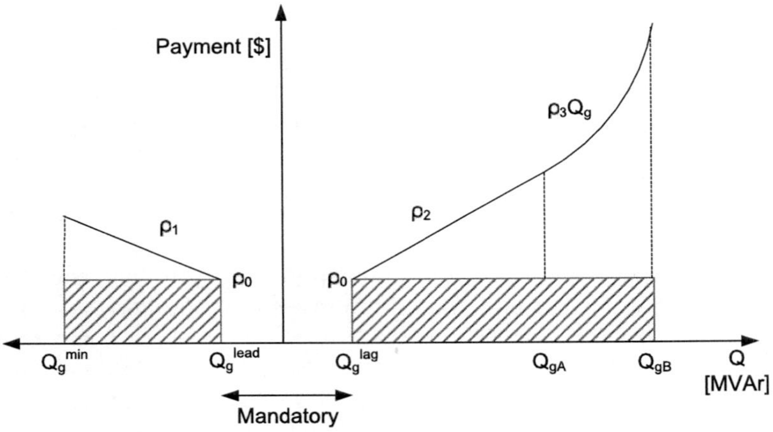

Fig. 3 Reactive power payment components

reactive power are small because it does not involve any fuel consumption [31]. However, reactive support may include an opportunity cost component. We see in Fig. 2 that if a generator is operating at A, supplying Q_{gA}, and needs to supply more reactive power, it will have to reduce its active power production to P_{gB} in order to provide Q_{gB}. Opportunity costs are related to the financial cost through the reduction in generated energy due to reactive support requirements. Figure 3 shows a typical payment structure for reactive supply [73]. This figure illustrates the compulsory reactive requirement from Q_g^{lead} to Q_g^{lag}. It also shows a payment for fixed costs ρ_0 in \$, which is independent of production levels and a use payment (ρ_1, ρ_2 in \$/MVAr) for operation in the under-excitation and over-excitation regions. Use costs are modeled as linear functions and account for the increased winding losses as the reactive power output increases. Finally, payment for the opportunity cost ρ_3 in \$/MVAr/MVAr, is modeled through a quadratic function.

4.2 Procuring Methods

The ISO procures an ancillary service through some of the following mechanisms [9]:

- Compulsory.
- Bilateral contracts.
- Tendering process or long-term auctions (weekly, monthly or annually).
- Spot markets (day-ahead and real-time markets).

Compulsory procurement is the simplest mechanism but it has some disadvantages: the ISO may not make an efficient procurement because the service does not represent any cost for it, and efficient providers do not have enough incentives because they receive the same treatment as do inefficient agents. Bilateral contracts solve the problem of compulsory provision because the system operator procures the appropriate amount of service and negotiates with the cheapest providers. As a disadvantage, the procurement can be inaccurate because this is a long-term mechanism that does not take into consideration short-term variations in price and operating conditions. Tendering processes and spot markets are more transparent and competitive mechanisms. Tendering process is referred here as a mid- or long-term mechanism (weekly, monthly or yearly), while spot market is a day ahead or real-time mechanism. In these cases, the ISO looks for the service from the cheapest provider. These mechanisms are more accurate but demand higher administrative costs, and there is a market power risk. A more complete description of the advantages and disadvantages of these methods can be found in [9, 22].

Table 4 shows the implemented methods for active power reserves and reactive supply in the four systems considered in Table 1. A comparison covering more systems is illustrated in [9].

From this table, we can see that none of the ancillary services are procured by means of bilateral contracts in the considered systems. However, bilateral contracts are used in Australia, New Zealand and France for some of these services [9]. Primary control reserve is compulsory in Spain, Brazil and Argentina [14, 17, 32]. Secondary control reserve is procured on the day-ahead market in Spain and Argentina [14, 17]. In PJM, primary and secondary control reserves are one service called regulation and frequency response and procured through the regulation

Table 4 Procuring methods in several systems

	Compulsory	Bilateral Contracts	Tendering Process	Spot
Primary control reserve	ES, BR AR			PJM
Secondary control reserve	BR			ES, PJM AR
Tertiary control reserve	BR		AR	ES, PJM
Reactive supply	PJM, AR, BR		ES	

market [19]. Tertiary control reserves are procured on a spot market in Spain and PJM [14, 19], while they are accomplished through a tendering process in Argentina [72]. Reactive supply is compulsory in Argentina [20] and PJM [19] and has a tendering process in Spain [18]. In the Brazilian system, even though there are bilateral contract arrangements, there is no price negotiation since all services are compulsory. However, some of the services have cost compensation. For example, the reactive supply provided from generators operating as synchronous condensers is paid through a regulated tariff [32].

5 Active Power Reserve Markets

As illustrated before, ancillary services are procured in the mid or long term through tendering processes or bilateral contracts. Some studies focus on pricing and procuring reserves by the ISO over a mid or long term (weeks, months or years) [33, 34]. These studies are also performed by reserve providers in order to know the financial impact due to reserve supply [35, 36]. In the short term, reserves can be procured through the day-ahead or real-time market.

5.1 Some Issues About Reserve Markets

Two common optimization strategies for reserve markets are the sequential and the co-optimization approach. In the sequential approach, the clearing of energy and reserve markets is separated and sequential. In the co-optimization approach, the clearing of energy and reserve markets is simultaneously. Since these markets are performed a day ahead or close to real time, and involve energy and reserves, we also call them dispatch models. On the other hand, the optimization process can be carried out from the ISO's or the provider's point of view as illustrated in Table 5 [37]. The optimization performed by the ISO aims at clearing the market, while the optimization performed by the provider aims at creating a decisions support tool.

Some examples of sequential models are the current Spanish energy market [12] and the California market in its first years [28]. In the California market, the sequential approach presented problems of market power [38]. The sequential

Table 5 Energy and reserve optimization approach

Point of view	Dispatch model
ISO	Joint (co-optimization) sequential hybrid
Provider	Joint

approach was later improved through the rational buyer approach [25, 27, 28], but it was still separated from the energy market. Currently, co-optimization is more widely accepted, as can be seen in the recent literature [39–45]. Co-optimization has been implemented in many systems such as PJM, Australia, the current California market and New England. In Table 5, an hybrid approach combining the sequential and the joint dispatch can be seen as in [46]. In that work, the hybrid approach was implemented by solving a sequence of joint dispatches. These dispatches provide intermediate solutions in order to calculate reserve opportunity costs. Some works on co-optimization from the provider's point of view, with the aim of finding optimal bidding strategies, can be found in [47, 48]. A provider always uses a joint dispatch model because the main interest is to analyze the effect of its decisions in both, energy and reserve markets.

In short-term markets, pricing schemes for power reserves are usually the same as in energy pricing schemes: the Market Clearing Price (MCP) and the Pay-as-Bid Price (PBP). Under some theoretical assumptions, both rules produce the same revenues for providers and payments for users [49, 50]. However, for real markets, it is not easy to define which rule is best [51].

Reserve bids can be single or multipart bids, as energy offers. In this case, however, a single bid refers to a capacity bid, while a multipart bid refers to a capacity and an energy bid. Capacity bids reflect the availability costs, while energy bids reflect the cost of using that reserve [51]. The opportunity costs can be embedded in the capacity bid because a bidder can decide to offer an available capacity on the reserve market, if the capacity price covers the respective opportunity costs teswider [59].

Reserve markets can consider both up and down reserve [37, 41, 46]. The former is used to balance a shortage of energy supply and the latter to balance a surplus of energy supply. Since down reserves do not require additional capacity, some studies do not consider them to be part of the market [39, 45, 47].

In most of the studies related to reserve markets, reserve requirements are predetermined by the ISO. The reserve requirement is thus distributed among market participants according to the results of the dispatch model. However, reserve requirements are related to the security of the system, and some studies consider the level of security to be variable in the dispatch model [41–44, 52]. In [41], for example, security is considered through the inclusion of contingency and network constraints. Additionally, other security constraints (overload index, voltage drop index and voltage stability margin) and the modeling of system uncertainties (load uncertainties and contingencies) have been recently considered [45].

Controllable loads may also be able to participate in the energy and reserve markets. In the energy market, controllable loads bid their willingness to pay a specified price for a specific amount of energy. In the reserve market, controllable loads bid their willingness to partially or totally interrupt their load. These loads enhance the efficiency and competitiveness of the reserve market while reducing the price and benefit of the reserve provision. A proposal incorporating controllable loads in the reserve market is explored in [53], and examples of load management in different systems are demonstrated in [54].

Finally, the use of primary and secondary control reserves in real time is made automatically. In this sense, reserve supply is determined by the market process but also by the control logic for frequency deviations. Primary control reserve is performed locally, and the reserve provided from each generator is usually defined through the adjustment of the speed governor. Secondary control reserve or automatic generation control AGC [6, 7] is performed centrally and the reserve provided from each generator is usually defined through a participation factor. The modification of AGC logic control has also been studied after deregulation in order to consider power reserve provision and other market issues [55–57].

5.2 A Co-optimized Dispatch

Most ISOs use a linearized optimal power flow DC-OPF for real power market clearing and dispatch [58]. Because of this approximation, other security constraints must be incorporated later in the dispatch process. AC models are more accurate but are not used in real applications because of the computational burden and non-linearity problems.

Here, for illustrative purposes, we use an AC-OPF to show the energy and spinning reserve dispatch. This formulation assumes that there is a centralized dispatch and suppliers can bid simple offers of energy and spinning reserve services for the next 24 h. The AC network constraints are included and the system demand and reserve requirements are supposed to be inelastic. Equations 2–11 show the proposed dispatch model for n_T periods in a system with n buses, n_L lines and n_G generators.

$$Min\, C = \sum_{t=1}^{n_T}\sum_{i=1}^{n_G} \Psi_i(P^t_{gi}) + \sum_{t=1}^{n_T}\sum_{i=1}^{n_G} \Omega_i\left(R^t_{gi}\right) \tag{2}$$

$$\mathbf{P}^t_g - \mathbf{P}^t_d - \mathbf{P}^t(\mathbf{V}^t, \theta^t) = 0 \; t = 1, \ldots, n_T \tag{3}$$

$$\mathbf{Q}^t_g - \mathbf{Q}^t_d - \mathbf{Q}^t(\mathbf{V}^t, \theta^t) = 0 \; t = 1, \ldots, n_T \tag{4}$$

$$\underline{\mathbf{V}} \leq \mathbf{V}^t \leq \bar{\mathbf{V}} t = 1, \ldots, n_t \tag{5}$$

$$\mathbf{S}^t_{from} \leq \overline{\mathbf{S}_f} \; t = 1, \ldots, n_T \tag{6}$$

$$\mathbf{S}^t_{to} \leq \overline{\mathbf{S}_f} \; t = 1, \ldots, n_T \tag{7}$$

$$\underline{\mathbf{P}_g} \le \mathbf{P}_g^t + \mathbf{R}_g^t \le \overline{\mathbf{P}_g}\ t = 1, \ldots, n_T \tag{8}$$

$$\Delta t \times rr_i^{down} \le P_{gi}^t - P_{gi}^{t-1} \le \Delta t \times rr_i^{up} t = 1, \ldots, n_T \tag{9}$$

$$0 \le R_{gi}^t \le min\left(\tau_{SR} \times rr_i^{up}, \overline{R_{gi}}\right)\ t = 1, \ldots, n_T \tag{10}$$

$$R_{req}^t = \sum_{i=1}^{n_G} R_{gi}^t\ t = 1, \ldots, n_T \tag{11}$$

where:

C: Energy and spinning reserve procurement costs.
Ψ_i, Ω_i: Generator energy and spinning reserve offers of the n_G generators.
$\mathbf{P}_g^t$: The $n \times 1$ generated active power vector at each bus n and time interval t.
$\mathbf{P}_d^t$: The $n \times 1$ consumed active power vector at each bus n and time interval t.
$\mathbf{Q}_g^t$: The $n \times 1$ generated reactive power vector at each bus n and time interval t.
$\mathbf{Q}_d^t$: The $n \times 1$ consumed reactive power vector at each bus n and time interval t.
$\mathbf{P}^t\ (V^t, \theta^t)$: The $n \times 1$ active power injection at time interval t.
$\mathbf{Q}^t\ (V^t, \theta^t)$: The $n \times 1$ reactive power injection at time interval t.
$\mathbf{V}^t$: The $n \times 1$ bus voltage magnitude vector at time interval t.
$\boldsymbol{\theta}^t$: The $n \times 1$ bus voltage angle vector at time interval t.
$\underline{\mathbf{V}}, \overline{\mathbf{V}}$: The $n \times 1$ vectors of minimum and maximum bus voltage magnitude.
$\mathbf{S}_{from}^t$, $\mathbf{S}_{to}^t$, $\overline{\mathbf{S}_f}$: The $n_L \times 1$ apparent power flow vectors (MVA) in the lines at time interval t in both terminals and their limits.
$\mathbf{R}_g^t$: The $n \times 1$ spinning reserve vector at time interval t.
$\underline{\mathbf{P}_g}, \overline{\mathbf{P}_g}$: The $n_G \times 1$ vectors of minimum and maximum generator power outputs.
Δt: Duration of each time interval.
rr_i^{down}, rr_i^{up}: Down- and up-ramp rates of generator i.
τ_{SR}: The specified response time for spinning reserve.
$\overline{R_{gi}}$: The maximum amount of reserve capacity offered by generator i.
R_{req}^t: Reserve requirement at each time interval t.

In this formulation $\mathbf{S}_{from}$ and $\mathbf{S}_{to}$ from Eqs. 6 and 7 are given by:

$$S_{from-km} = \sqrt{P_{from-km}^2 + Q_{from-km}^2} \tag{12}$$

$$S_{to-km} = \sqrt{P_{to-km}^2 + Q_{to-km}^2} \tag{13}$$

$$P_{from-km} = V_k^2 g_{km} - V_k V_m [g_{km} cos(\theta_{km}) + b_{km} sin(\theta_{km})] \tag{14}$$

$$P_{to-km} = V_m^2 g_{km} - V_k V_m [g_{km} cos(\theta_{km}) - b_{km} sin(\theta_{km})] \tag{15}$$

$$Q_{from-km} = -V_k^2 \left(b_{km}^{sh} + b_{km}\right) - V_k V_m [g_{km} sin(\theta_{km}) - b_{km} cos(\theta_{km})] \tag{16}$$

$$Q_{to-km} = -V_m^2 \left(b_{km}^{sh} + b_{km}\right) + V_k V_m [g_{km} sin(\theta_{km}) + b_{km} cos(\theta_{km})]. \tag{17}$$

This formulation considers the energy balance constraints (Eqs. 3 and 4), voltage security limits (Eq. 5), power flow limits (Eqs. 6 and 7), the generator capacity limit (Eq. 8), the ramping constraint (Eq. 9), the reserve allocation constraint (Eq. 10) and system reserve requirements (Eq. 11). Zonal reserve constraints can also be incorporated [39].

The Lagrange multipliers of constraints 3 and 11 are the bus marginal prices for energy and spinning reserves respectively. It should be noted that this model only includes a capacity bid that could be based on opportunity costs [51, 59].

5.3 Example 1

According to the co-optimization model illustrated above, this example shows the scheduling of two generators in a day-ahead electricity market considering network constraints. The test system with three buses and two lines was taken from [60]. Figure 4 illustrates this system for the loading conditions at hour 18. Bus 1 is the slack with voltage $1\angle 0^0$ for all hours. Bus 2 is a voltage-controlled bus with $V = 1$ p.u. for all hours.

Data for energy and spinning reserve offers, capacity limits and ramp rates for generators at bus 1 and 2 are illustrated in Table 6. For simplicity, energy and reserve offers are assumed to be the same for all periods. Spinning reserve is supposed to be available in 10 min ($\tau_{SR} = 10$ min.).

System reserve requirement and forecasted demand are illustrated in Fig. 5. Reactive power demand was calculated assuming a power factor of 0.89 for all periods.

The bus marginal prices (BMPs) and the ancillary service clearing price (ASCP) obtained from the co-optimization model (Eqs. 2–11) are illustrated in Figs. 6 and 7.

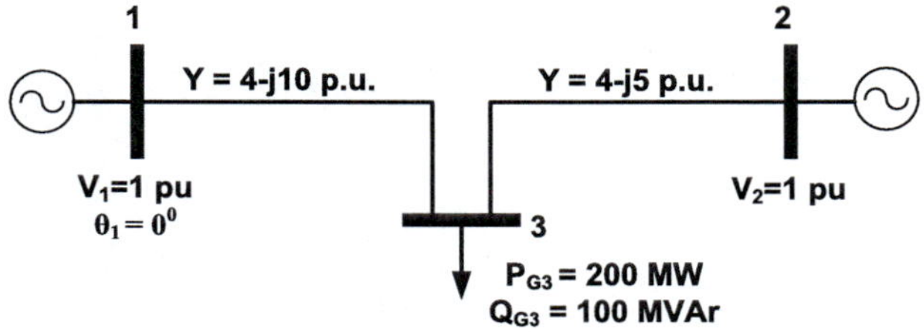

Fig. 4 Three bus system

Table 6 Energy and reserve offers

Generator	Energy offer		Reserve offer		Capacity (MW)	Ramp rate (MW/min)
	Quantity (MW)	Price ($/MWh)	Quantity (MW)	Price ($/MWh)		
G_1	250	14	50	2.66	250	10
G_2	70	13	20	2.55	70	2.2

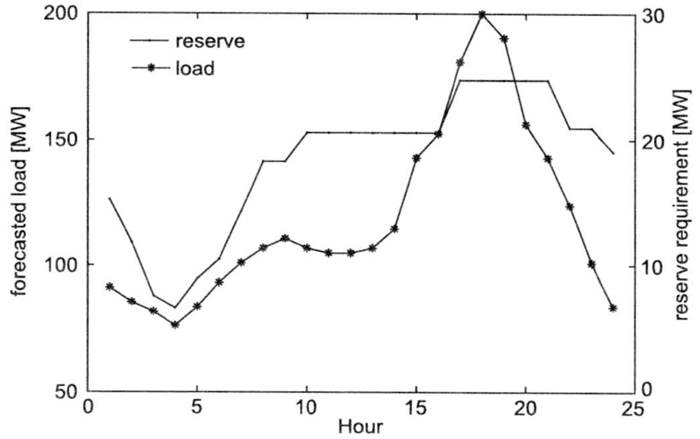

Fig. 5 Forecasted load and spinning reserve requirements

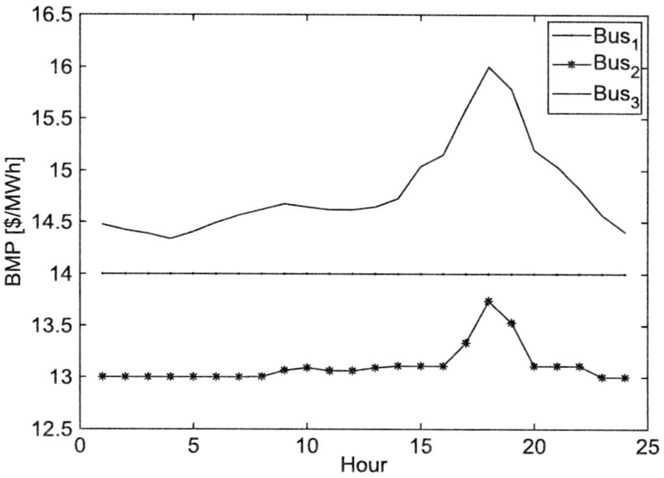

Fig. 6 Bus marginal prices

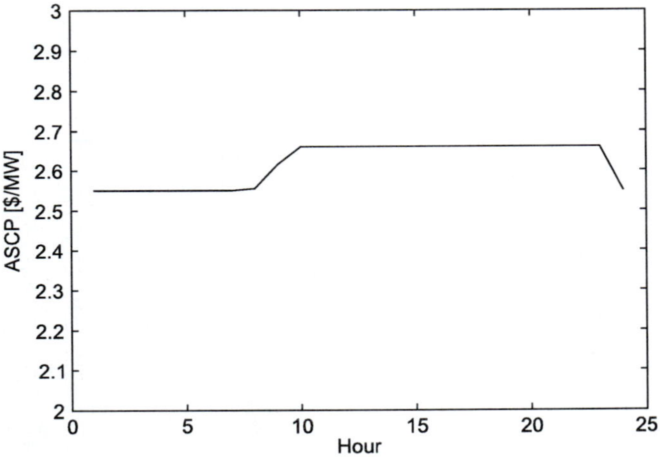

Fig. 7 Ancillary service clearing prices

In this example, the line power flow and voltage constraints are not binding. Therefore BMPs are influenced by generation constraints, energy offers and active transmission losses. For example, the nodal price at bus 2 reflects the offer of generator 2 from hours 1 to 8; however, during hours 8 to 22, this price is also influenced by the offer of generator 1, the capacity constraint of generator 2 and transmission losses.

On the other hand, the ASCP reflects the offers of generators. It could be noted that the reserve price increases as there are more reserve requirements. At hours 14, 15 and 16, the BMP at bus 2 is 13.11$/MWh. Since the offer of generator 2 is 13$/MWh, a profit of 0.11 can be made for each *MW* that it sells in the energy market instead of the reserve market. However, the ASCP in those hours is 2.66 and the reserve offer of generator 2 is 2.55. In this case, the opportunity cost of generator 2 is covered exactly in the reserve market. For generator 2, the reserve market is more attractive from hours 1 to 13, while the energy market is more attractive for hours 17, 18 and 19.

The energy and spinning reserve schedules are illustrated in Figs. 8 and 9. Generator 2, which is cheaper, supplies more energy from hours 1 to 5. For the next hours, supplying energy from generator 1 becomes more attractive because transmission losses are reduced. During the peak hours (17, 18 and 19), the capacity constraint of generator 2 forces the purchase of still more energy from generator 1.

For the spinning reserve market, the bid of generator 2 is more competitive. However, from hour 14 to 21, the reserve from generator 1 starts to become competitive because generator 2 is then more useful in the energy market.

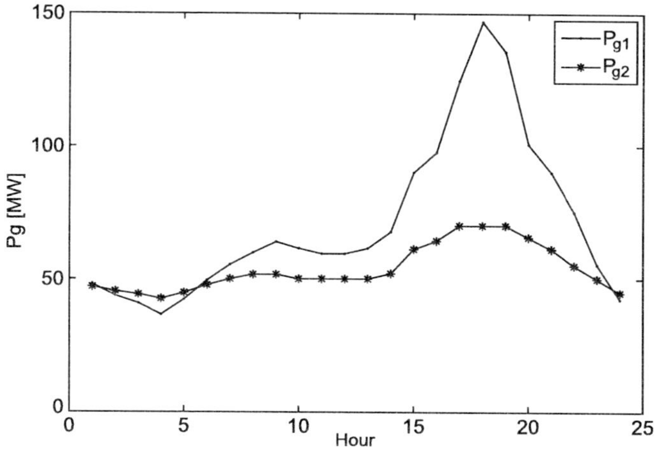

Fig. 8 Generator's schedule

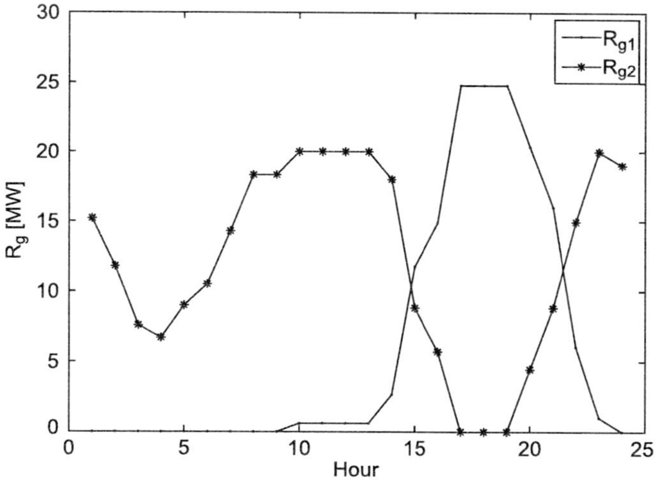

Fig. 9 Spinning reserve schedule

6 Reactive Power Dispatch

In a deregulated environment, an accurate pricing mechanism for reactive power is necessary. If reactive power has a low price, it could create a sudden deficit in the system because production is not encouraged while consumption is. This may cause operation to become unfeasible. On the other hand, if the service has a high price, consumers tend to reduce demand because of the high price, and a surplus of reactive power can result [31].

Reactive power pricing and management have been discussed extensively in the past [61–63] highlighting some important issues. For example, reactive supply is basically a local service and its variable costs are small compared with energy and active power reserve costs. However, the existence of opportunity costs associated with the capability curve has also been recognized. The recent literature indicates that this service is essential for maintaining a security voltage margin in both normal and emergency conditions. Under the current deregulated structure of electricity markets, researchers have proposed an analysis of the reactive power supply problem on two levels: reactive power procurement (in the long term) and reactive power dispatch (in the short term) [64]. Procurement methods for reactive supply are rarely based on short-term markets [9] but on contracts or tendering schemes over the long term [58, 64–67]. Implementing reactive power markets has been difficult due to the local nature of reactive support, which can cause the exercise of market power [68, 69].

In the short term, from the operation point of view, the reactive power dispatch can be optimized by the ISO through an OPF. Some objective functions usually considered in the OPF are

- Minimize active power losses: $0.5 \times \sum_{k,m} g_{km} \times \left[V_k^2 + V_m^2 - 2V_kV_m cos(\theta_k - \theta_m)\right]$
- Minimize production costs: $\sum_{i=1}^{n_G} a_i \times P_i^2 + b_i \times P_i + c_i$
- Minimize energy market costs: $\sum_i \Psi_i(P_i)$
- Minimize reactive injection costs: $\sum_i Q_{Ci} \times C_{Ci} + \sum_i Q_{Li} \times C_{Li}$

6.1 A Joint Active and Reactive Power Dispatch Model

Here, we illustrate a reactive power dispatch through the AC-OPF presented in Sect. 5.2 in which the minimization of energy and reserve market costs was considered. Normally a reactive power dispatch is made under the assumption that the active power dispatch has already been established. Here, the idea is to show the impact of the reactive power dispatch on the energy and reserve dispatch and the existence of the opportunity costs associated with reactive supply. For this purpose we introduced two modifications to the previous energy and reserve dispatch model. The first modification involves bus voltages at the slack and PV nodes, which have been released to allow voltage control through reactive power injection of generators. Voltage control is also done through other devices like taps and capacitor banks and can be incorporated in the OPF [70]. The second modification is related to the incorporation of the capability curve constraints (Eqs. 28 and 29) of Fig. 2 in order to consider the opportunity costs of reactive support [71]. The proposed formulation is illustrated below (Eqs. 18–29).

$$Min\, C = \sum_{t=1}^{n_T}\sum_{i=1}^{n_G} \Psi_i\left(P_{gi}^t\right) + \sum_{t=1}^{n_T}\sum_{i=1}^{n_G} \Omega_i\left(R_{gi}^t\right) \tag{18}$$

$$\mathbf{P}_g^t - \mathbf{P}_d^t - \mathbf{P}^t(\mathbf{V}^t, \theta^t) = 0 \ t = 1, \ldots, n_T \tag{19}$$

$$\mathbf{Q}_g^t - \mathbf{Q}_d^t - \mathbf{Q}^t(\mathbf{V}^t, \theta^t) = 0 \ t =, \ldots, n_T \tag{20}$$

$$\underline{\mathbf{V}} \leq \mathbf{V}^t \leq \overline{\mathbf{V}} \ t = 1, \ldots, n_T \tag{21}$$

$$\mathbf{S}_{from}^t \leq \overline{\mathbf{S}_f} \ t = 1, \ldots, n_T \tag{22}$$

$$\mathbf{S}_{to}^t \leq \overline{\mathbf{S}_f} \ t = 1, \ldots, n_T \tag{23}$$

$$\underline{\mathbf{P}_g} \leq \mathbf{P}_g^t + \mathbf{R}_g^t \leq \overline{\mathbf{P}_g} \ t = 1, \ldots, n_T \tag{24}$$

$$\Delta t \times rr_i^{down} \leq P_{gi}^t - P_{gi}^{t-1} \leq \Delta t \times rr_i^{up} \ t = 1, \ldots, n_T \tag{25}$$

$$0 \leq R_{gi}^t \leq min\left(\tau_{SR} \times rr_i^{up}, \overline{R_{gi}}\right) \ t = 1, \ldots, n_T \tag{26}$$

$$R_{req}^t = \sum_{i=1}^{n_G} R_{gi}^t \ t = 1, \ldots n_T \tag{27}$$

$$Q_{gi}^t \leq \sqrt{\left(\frac{V_{ti} \times E_{afi}}{X_{si}}\right)^2 - \left(P_{gi}^t + R_{gi}^t\right)^2} - \frac{V_{ti}^2}{X_{si}} \tag{28}$$

$$Q_{gi}^t \leq \sqrt{V_{ti} \times I_{ai} - \left(P_{gi}^t + R_{gi}^t\right)^2} \tag{29}$$

6.2 *Example 2*

The three-bus test system presented in example 1 is used here to illustrate the reactive power dispatch. Additional characteristics for generators 1 and 2 are shown in Table 7, including the compulsory reactive requirements. These data are associated with Fig. 2.

For comparison purposes, some results obtained from example 1 are illustrated again. The results from examples 1 and 2 are marked as (1) and (2) on the next figures.

Nodal prices are illustrated in Fig. 10. It can be noted that the nodal price at bus 3 is lower in this case because the reactive demand is supplied at better voltage levels, so the total power losses were reduced. However, the nodal price at bus 2 is higher

Table 7 Characteristics of generators

	Capacity [MVA]	fp	X_s (p.u.)	V_t (p.u.)	E_{af} (p.u.)	$Q_{compulsory}$ (MVAr)
G_1	270	0.85	1.62	1	4.9739	75.4
G_2	75	0.80	0.98	1	1.5563	19.7

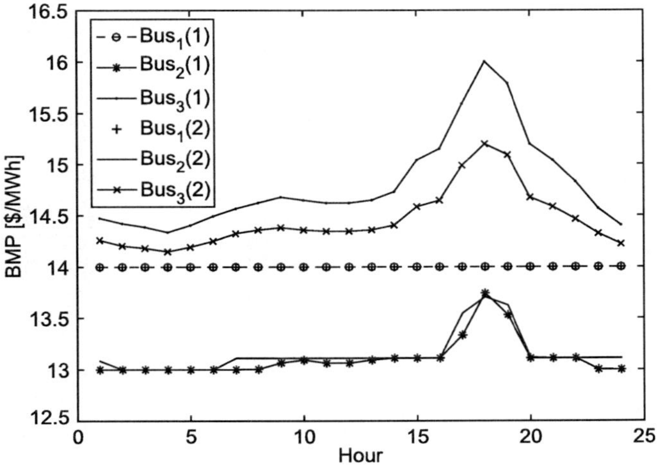

Fig. 10 Nodal prices for example 2

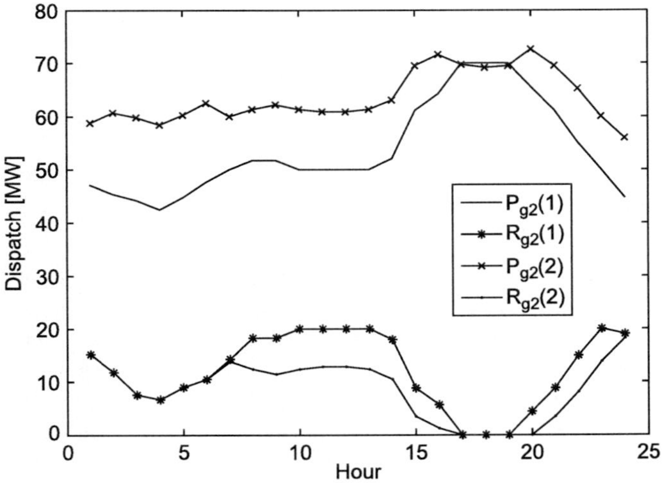

Fig. 11 Energy and reserve dispatch for generator 2

in example 2 than in 1 due to the armature limit of generator 2. The nodal price at bus 1 does not change because the limits of generator 1 are not binding.

Figure 11 shows the energy and reserve dispatch for generator 2. In this case, the energy dispatch was increased while the reserve dispatch was decreased with respect to example 1. That happens because constraint eight was replaced by the capability curve constraints (Eqs. 28 and 29), so generator 2 is enabled to produce more active power. Although not shown, generator 1 decreased its active power

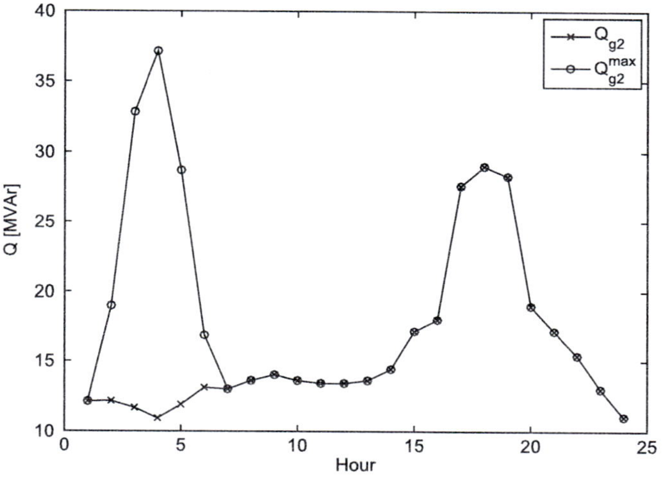

Fig. 12 Reactive power dispatch of generator 2

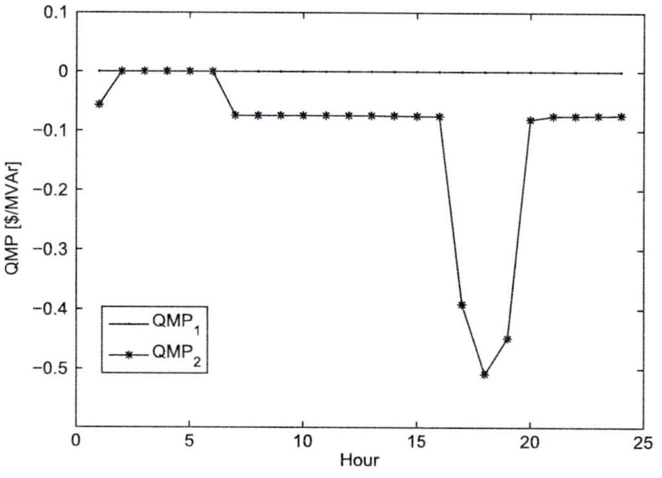

Fig. 13 Reactive marginal costs

production and increased its spinning reserve as a consequence of the dispatch of generator 2.

Figure 12 shows the reactive power dispatch for generator 2. The maximum reactive power limit is also illustrated. In this case, the limit for reactive power was imposed by the armature limit at all hours. It can be observed that this limit varies over time as the operation point of the generator changes. Generator 2 operates at its armature limit in hour 1 and from hours 7 to 24. Consequently, the marginal cost associated with constraint 28 appears as illustrated in Fig. 13. Here we call this reactive marginal cost QMC, because it represents the variation in energy and reserve dispatch costs due to the variation of 1 *MVAr* in the armature limit constraint. These costs are negative in Fig. 13 because one additional *MVAr* from generator 2 reduces the market procurement costs. From the generator point of view, the QMC can be associated with opportunity costs and incorporated into the payment function of Fig. 3.

7 Conclusions

Active power reserves and reactive supply are the most common ancillary services in power systems. In this chapter we described some relevant issues highlighting important differences in the classification, technical requirements and economics of these services. It was shown that primary control reserves are in many cases a compulsory service and that currently there are no spot markets for reactive supply because of the local nature of this service. The principal cost components of ancillary services were also illustrated. However, the unbundling and assessing of these components is not easy because they are both physically and economically coupled with energy production. Two dispatch models for ancillary services were also illustrated. The first one considered a joint energy and reserve dispatch. In this case, it was shown how energy and reserve prices are affected by the generator and network constraints and how the opportunity costs are embedded into these prices. The second model is based on the first one but incorporates voltage control and generator capability curves. It was shown that reactive supply can also affect energy and reserve prices. Finally, the chapter illustrated the existence of an opportunity cost component due to reactive supply when generators are working at their physical limits.

References

1. Rudnick H, Barroso LA, Bezerra B (2008) A delicate balance in South America. IEEE Power Energy Mag 6(4):22–35
2. Barroso LA, Cavalcanti TH, Giesbertz P, Purchala K (2005) Classification of electricity market models worldwide. CIGRE Task Force C5.2.1. http://www.cigre-c5.org/Site/Publications/pa_ex.asp. Accessed 30 Oct 2009

3. FERC – Federal Energy Regulatory Commission (1996) Promoting wholesale competition through open-access nondiscriminatory transmission service by public utilities. Order 888.
4. Rashidinejad M, Song YH, Javidi MH (2003) Ancillary services I: pricing and procurement of reserves. In: Song Y-H, Wang X-F (eds) Operation of market-oriented power systems. Springer, London
5. Hirst E, Kirby B (1996) Electric-power ancillary services. OAK Ridge National Laboratory, Energy Division, Oak Ridge
6. Rebours YG, Kirschen DS, Trotignon M, Rossingnol S (2007) A survey of frequency and voltage control ancillary services – part I: technical features. IEEE Trans Power Syst 22 (1):350–357
7. UCTE – Union for the Coordination of Transmission of Electricity (2004) Load-frequency control and performance: policy 1. http://www.ucte.org/publications/ophandbook/. Accessed 20 Jan 2008
8. NGET, Great Britain (2007) The grid code. http://www.nationalgrid.com/uk/Electricity/Codes/gridcode/gridcodedocs/. Accessed 6 Mar 2009
9. Rebours YG, Kirschen DS, Trotignon M, Rossingnol S (2007) A survey of frequency and voltage control ancillary services – part II: economic features. IEEE Trans Power Syst 22(1):358–366
10. Raineri R, Ríos S, Schiele D (2005) Technical and economic aspects of ancillary services markets in the electric power industry: an international comparison. Electr Power Syst Res 34 (13):1540–1555
11. Granville S, Vieira FX, Pereira MVF, Marzano G, Soto J, Melo ACG, Gorenstin BG, Mello JC, Adapa R, Mansour Y, Messing L, Barry M, Bertoldi O, Doorman G, Jegouzo Y, Pruvot P, Stewart B (2001) Methods and tools for costing ancillary services. Task Force 38.05.07
12. Miguélez EL, Cortés IE, Rodríguez LR, Camino GL (2008) An overview of ancillary services in Spain. Electr Power Syst Res 78(3):3515–3523
13. Kristiansen T (2007) The Nordic approach to market-based provision of ancillary services. Electr Power Syst Res 35(7):3681–3700
14. Ministerio de Industria y Comercio (1998) Resolution of 30 Jul. 1998 (In Spanish). BOE No. 197. http://www.ree.es/operacion/procedimientos_operacion.asp. Accessed 5 Jan 2008
15. PJM (2007) Balancing operations. Manual 12. http://www.pjm.com/contributions/pjm-manuals/pdf/m12.pdf. Accessed 30 Jan 2008
16. ONS Operador Nacional do Sistema (2002) Network procedures: minimal requirements for the connection to basic network (In Portuguese). http://www.ons.org.br/procedimentos/modulo_03.aspx. Accessed 4 Feb 2007
17. CAMMESA (2007) Annex 23: frequency regulation (In Spanish). http://www.cammesa.com/inicio.nsf/marconormativa. Accessed 31 Jan 2008
18. Ministerio de Industria y Comercio (2000) Resolution of 7 Mar. 2000 (In Spanish). BOE No. 67. http://www.ree.es/operacion/procedimientos_operacion.asp. Accessed 5 Jan 2008
19. PJM (2010) Generator operational requirements. Manual 14. http://www.pjm.com/documents/manuals.aspx. Accessed 06 Aug 2010
20. CAMMESA (2007) Annex 4: voltage control and reactive power dispatch (In Spanish). http://www.cammesa.com/inicio.nsf/marconormativa. Accessed 31 Jan 2008
21. Migueléz EL, Cerezo FME, Rodriguez LR (2007) On the assignment of voltage control ancillary services of generators in Spain. IEEE Trans Power Syst 22(1):367–375
22. Rebours YG, Kirschen DS, Trotignon M (2007) Fundamental design issues in markets for ancillary services. The Electricity J 20(6):26–34
23. Strbac G, Kirschen DS (2000) Who should pay for reserve? Electricity J 13(8):32–37
24. Zhong J, Bhattacharya K (2002) Design of competitive markets for spinning reserve services. Power Eng Soc Summer Meet 3:1627–1632
25. Oren SS (2001) Design of ancillary services markets. In: Proceedings of the 34th Hawaii International Conference on System Sciences, Berkeley
26. Zheng D, Zhou W (2003) A design for regional ancillary services auctions markets in China. IEEE Bologna PowerTech Conf 4:5

27. Papalexopoulos A, Singh H (2001) On the various design options for ancillary services markets. In: Proceedings of the 34th Hawaii International Conference on System Sciences, Berkeley
28. Shahidehpour M, Yamin H, Li Z (2002) Market operations in electric power systems. Wiley, New York
29. Zhong J (2003) On some aspects of design of electric power ancillary service markets. PhD Thesis, Department of Electric Power Engineering Chalmers University of Technology, Göteborg
30. Kundur P (1993) Power system stability and control. McGraw-Hill, New York
31. Rabiee A, Shayanfar HA, Amjady N (2009) Reactive power pricing: problems & proposal for a competitive market. IEEE Power Energy Mag, Vol. 7, 18–32
32. ONS Operador Nacional do Sistema (2003) Commercial arrangement for ancillary services provided from generators, transmission and distribution agents (In Portuguese). http://www.ons.org.br/procedimentos/modulo_14.aspx. Accessed 24 Jan 2008
33. Havel P, Horáček P, Černý V, Fantík J (2008) Optimal planning of ancillary services for reliable power balance control. IEEE Trans Power Syst 23(3):1375–1382
34. Verbić G, Gubina F (2004) Cost-based models for the power-reserve pricing of frequency control. IEEE Trans Power Syst 19(4):1853–1858
35. Sousa T, Jardini JA, Masuda M, de Lima RA (2004) Spinning reserve service pricing in hydroelectric power plants. In: IEEE/PES Transmission and Distribution Conference and Exposition, Latin America
36. Sousa T, Tahan CMV, Jardini JA, Rosa JAO, Feltrin AP, Manso JCG (2010) A new approach to remuneration of ancillary services provided by generation agents in Brazil. IEEE Latin America Trans 8(1):38–44
37. García-González J, Roque AMS, Campos FA, Villar J (2007) Connecting the intraday energy and reserve markets by an optimal redispatch. IEEE Trans Power Syst 22(4):2220–2231
38. Brien L (1999) Why the ancillary services markets don't work and what to do about it. Electricity J 12(5):38–48
39. Costa AL, Costa AS (2007) Energy and ancillary service dispatch through dynamic optimal power flow. Electr Power Syst Res 77(8):1047–1055
40. Ongsakul W, Chayakulkheeree K (2006) Coordinated fuzzy constrained optimal power dispatch for bilateral contract, balancing electricity, and ancillary services markets. IEEE Trans Power Syst 21(2):593–604
41. Arroyo JM, Galiana FD (2005) Energy and reserve pricing in security and network-constrained electricity markets. IEEE Trans Power Syst 20(2):634–643
42. Jaefari-Nokandi M, Monsef H (2009) Scheduling of spinning reserve considering customer choice on reliability. IEEE Trans Power Syst 24(4):1780–1789
43. Wong S, Fuller JD (2007) Pricing energy and reserves using stochastic optimization in an alternative electricity market. IEEE Trans Power Syst 22(2):631–638
44. Wang J, Wang X, Wu Y (2005) Operating reserve model in the power market. IEEE Trans Power Syst 20(1):223–229
45. Amjady N, Aghaei J, Shayanfar HA (2009) Stochastic multiobjective market clearing of joint energy and reserve auctions ensuring power system security. IEEE Trans Power Syst 24 (4):1841–1854
46. Cheung KW, Shamsollahi P, Sun D, Milligan J, Potishnak M (2000) Energy and ancillary service dispatch for the interim ISO New England electricity market. IEEE Trans Power Syst 15(3):968–974
47. Arroyo JM, Conejo AJ (2002) Optimal response of a power generator to energy, AGC, and reserve pool-based markets. IEEE Trans Power Syst 17(2):404–410
48. Wen FS, David AK (2002) Optimally co-ordinated bidding strategies in energy and ancillary service markets. IEE Proc Gener Transm Distrib 149(3):331–338
49. Ren Y, Galiana FD (2004) Pay-as-bid versus marginal pricing – part I: strategic generator offers. IEEE Trans Power Syst 19(4):1771–1776

50. Ren Y, Galiana FD (2004) Pay-as-bid versus marginal pricing – part II: market behavior under strategic generator offers. IEEE Trans Power Syst 19(4):1777–1783
51. Stoft S (2002) Power system economics: designing markets for electricity. Wiley, New York
52. Ruiz PA, Sauer PW (2008) Spinning contingency reserve: economic value and demand functions. IEEE Trans Power Syst 23(3):1071–1078
53. Bai J, Gooi HB, Xia LM, Strbac G, Venkatesh B (2006) A probabilistic reserve market incorporating interruptible load. IEEE Trans Power Syst 21(3):1079–1087
54. Jazayeri P, Schellenberg A, Rosehart WD, Doudna J, Widergren S, Lawrence D, Mickey J, Jones S (2005) A survey of load control programs for price and system stability. IEEE Trans Power Syst 20(3):1504–1509
55. Tyagi B, Srisvastava SC (2006) A decentralized automatic generation control scheme for competitive electricity markets. IEEE Trans Power Syst 21(1):312–320
56. Zhong Jin, Bhattacharya Kankar (2003) Frequency linked pricing as an instrument for frequency regulation in deregulated electricity markets. Power Engineering Society General Meeting, Göteborg Sweden
57. Christie RD, Bose A (1996) Load frequency control issues in power system operations after deregulation. IEEE Trans Power Syst 11(3):1191–1200
58. Cañizares C, Bhattacharya K, El-Samahy I, Haghighat H, Pan J, Tang C (2010) Re-defining the reactive power dispatch problem in the context of competitive electricity markets. IET Gener Transm Distrib 4(2):162–177
59. Swider DJ (2007) Efficient scoring-rule in multipart procurement auctions for power systems reserve. IEEE Trans Power Syst 22(4):1717–1725
60. Dommel HW, Tinney WF (1968) Optimal power flow solutions. IEEE Trans PAS 87 (10):1866–1876
61. Lin XJ, Yu CW, Chung CY (2005) Pricing of reactive support ancillary services. IEEE Proc-Gener Transm Distrib 152(5):616–622
62. Hao S, Papalexopoulos A (1997) Reactive power pricing and management. IEEE Trans Power Syst 12(1):95–104
63. PSERC (2001) Reactive power support services in electricity markets. Power System Engineering Research Center. http://www.pserc.wisc.edu/research/public_reports/markets.aspx. Accessed 4 Mar 2009
64. El-Samahy I, Bhattacharya K, Cañizares C, Anjos MF, Pan J (2008) A procurement market for reactive power services considering system security. IEEE Trans Power Syst 23(1):137–149
65. Chicco G, Gross G (2008) Current issues in reactive power management: a critical overview. In: IEEE power and energy society general meeting, Pittsburgh, pp 1–6
66. Frías P, Gómez T, Soler D (2008) A reactive power capacity market using annual auctions. IEEE Trans Power Syst 23(3):1458–1459
67. Zhong J, Nobile E, Bose A, Bhattacharya K (2004) Localized reactive power markets using the concept of voltage control areas. IEEE Trans Power Syst 19(3):1555–1561
68. Chitkara P, Zhong J, Bhattacharya K (2008) Oligopolistic competition of gencos in reactive power ancillary service provisions. IEEE Trans Power Syst 24(3):1256–1265
69. Feng D, Zhong J, Gan D (2008) Reactive market power analysis using must-run indices. IEEE Trans Power Syst 23(2):755–765
70. Taylor GA, Phichaisawat S, Irving MR, Song Y-H (2003) Ancillary services II: voltage security and reactive power management. In: Yong-Hua S, Xi-Fan W (eds) Operation of market-oriented power systems. Springer, London
71. Bhattacharya K, Zhong J (2001) Reactive power as an ancillary service. IEEE Trans Power Syst 16(2):294–300
72. CAMMESA (2007) Annex 36: Sort and medium term reserve services (In Spanish). http://www.cammesa.com/inicio.nsf/marconormativa. Accessed 31 Jan 2008
73. Zhong J, Bhattacharya K (2002) Toward a competitive market for reactive power. IEEE Trans Power Syst 17(4):1206–1215

Index

Page references in Roman denote Vol. I and *Italic* page references denote Vol. II.

A. Sorokin et al. (eds.), *Handbook of Networks in Power Systems I*,
Energy Systems, DOI 10.1007/978-3-642-23193-3,